Springer Proceedings in Mathematics & Statistics

Volume 234

Springer Proceedings in Mathematics & Statistics

This book series features volumes composed of selected contributions from workshops and conferences in all areas of current research in mathematics and statistics, including operation research and optimization. In addition to an overall evaluation of the interest, scientific quality, and timeliness of each proposal at the hands of the publisher, individual contributions are all refereed to the high quality standards of leading journals in the field. Thus, this series provides the research community with well-edited, authoritative reports on developments in the most exciting areas of mathematical and statistical research today.

More information about this series at http://www.springer.com/series/10533

Marston D. E. Conder · Antoine Deza
Asia Ivić Weiss
Editors

Discrete Geometry and Symmetry

Dedicated to Károly Bezdek and Egon Schulte on the Occasion of Their 60th Birthdays

Springer

Editors
Marston D. E. Conder
Department of Mathematics
University of Auckland
Auckland
New Zealand

Asia Ivić Weiss
Department of Mathematics
and Statistics
York University
Toronto, ON
Canada

Antoine Deza
Department of Computing
and Software
McMaster University
Hamilton, ON
Canada

ISSN 2194-1009 ISSN 2194-1017 (electronic)
Springer Proceedings in Mathematics & Statistics
ISBN 978-3-030-08697-8 ISBN 978-3-319-78434-2 (eBook)
https://doi.org/10.1007/978-3-319-78434-2

Mathematics Subject Classification (2010): 05B40, 05B45, 52A10, 52A21, 52A35, 52B10, 52B11, 52B15, 52C15, 52C17, 52C20, 52C35, 52C45, 90C05, 90C27

Printed on acid-free paper

This Springer imprint is published by the registered company Springer International Publishing AG part of Springer Nature
The registered company address is: Gewerbestrasse 11, 6330 Cham, Switzerland

Preface

This volume contains a number of articles on the topics of symmetry and discrete geometry. Most of them were papers presented during the conference 'Geometry and Symmetry', held at the University of Pannonia in Veszprém, Hungary, the week 29 June to 3 July 2015. This conference was arranged in honour of Károly Bezdek and Egon Schulte, on the occasion of the year in which they both turned 60. Many of the papers reflect the remarkable contributions they made to geometry.

The revival of interest in discrete geometry over the past few decades has been influenced by Bezdek and Schulte to a large degree. Although their research interests are somewhat different, one could say that they have complemented each other, and this has resulted in a lively interaction across a wide variety of different fields. Accordingly, the volume includes a range of topics and provides a snapshot of a rapidly evolving area of research. The contributions demonstrate profound interplays between different approaches to discrete geometry.

Kepler was the first to raise the discrete geometry problem of sphere packing. Associated tiling problems were considered at the turn of the century by many researchers, including Minkowski, Voronoi, and Delone. The Hungarian school pioneered by Fejes Tóth in the 1940s initiated the systematic study of packing and covering problems, while numerous other mathematicians contributed to the field, including Coxeter, Rogers, Penrose, and Conway. While the classical problems of discrete geometry have a strong connection to geometric analysis, coding theory, symmetry groups, and number theory, their connection to combinatorics and optimisation has become of particular importance. These areas of research, at the heart of Bezdek's work, play a central role in many of the contributions to this volume.

Kepler, with his discovery of regular non-convex polyhedra, could also be credited with founding of modern polytope theory. The subject went into decline before it was taken up again by Coxeter almost a century ago and later by Grünbaum. Based on their impressive and seminal contributions, the search for deeper understanding of symmetric structures has over the past few decades produced a revival of interest in discrete geometric objects and their symmetries. The rapid development of abstract polytope theory, popularised by McMullen's and

Schulte's research monograph with the same name, has resulted in a rich theory, featuring an attractive interplay of methods and tools from discrete geometry (such as classical polytope theory), combinatorial group theory, and incidence geometry (generators and relations, and Coxeter groups), graph theory, hyperbolic geometry, and topology.

We note with sadness that during the work on this volume, our good friend and colleague Norman W. Johnson (a contributor to this volume) passed away. Since receiving his Ph.D. with Coxeter in 1966, Norman held a position at the Wheaton College in Massachusetts, where he taught until his retirement in 1998.

It is our hope that this volume not only exhibits the recent advances in various areas of discrete geometry, but also fosters new interactions between several different research groups whose contributions are contained within this collection of papers.

Auckland, New Zealand — Marston D. E. Conder
Hamilton, Canada — Antoine Deza
Toronto, Canada — Asia Ivić Weiss

Contents

Contributors

Károly Bezdek Department of Mathematics and Statistics, University of Calgary, Calgary, Canada; Department of Mathematics, University of Pannonia, Veszprém, Hungary

Károly Böröczky Roland Eötvös University, Budapest, Hungary

Károly J. Böröczky Alfréd Rényi Institute of Mathematics, Hungarian Academy of Sciences, Budapest, Hungary; Central European University, Budapest, Hungary

Marston D. E. Conder Mathematics Department, University of Auckland, Auckland, New Zealand

Balázs Csikós Institute of Mathematics, Eötvös Loránd University, Budapest, Hungary; Budapest, Hungary

Antoine Deza Advanced Optimization Laboratory, Faculty of Engineering, McMaster University, Hamilton, ON, Canada; Laboratoire de Recherche en Informatique, CNRS – Université de Paris Sud, Orsay, France

Dimitri Leemans Department of Mathematics, University of Auckland, Auckland, New Zealand; Département de Mathématique, Algèbre et Combinatoire, Université Libre de Bruxelles, Brussels, Belgium

Nikolay Dolbilin Steklov Mathematical Institute of the Russian Academy of Sciences, Moscow, Russia

Ian Douglas Tucson, AZ, USA

István Estélyi FMF, University of Ljubljana, Ljubljana, Slovenia; IAM, University of Primorska, Koper, Slovenia; NTIS, University of West Bohemia, Plzeň 3, Czech Republic

Maria Elisa Fernandes Center for Research and Development in Mathematics and Applications, Department of Mathematics, University of Aveiro, Aveiro, Portugal

D. Frettlöh Bielefeld University, Bielefeld, Germany

Alexey Glazyrin School of Mathematical and Statistical Sciences, The University of Texas Rio Grande Valley, One West University Blvd, Brownsville, USA

Gábor Gévay Bolyai Institute, University of Szeged, Szeged, Hungary

Ákos G. Horváth Department of Geometry, Budapest University of Technology, Budapest, Hungary

Isabel Hubard Instituto de Matemáticas, Universidad Nacional Autónoma de México Circuito Exterior, Mexico D.F., Mexico

Norman W. Johnson Norton, USA

Muhammad A. Khan Department of Mathematics and Statistics, University of Calgary, Calgary, Canada

Ágnes Kovács Roland Eötvös University, Budapest, Hungary

Undine Leopold Fakultät für Mathematik, Technische Universität Chemnitz, Chemnitz, Germany

Zoran Lučić Department of Mathematics, University of Belgrade, Belgrade, Serbia

George Manoussakis Laboratoire de Recherche en Informatique, CNRS – Université de Paris Sud, Orsay, France

Horst Martini Fakultät für Mathematik, Technische Universität Chemnitz, Chemnitz, Germany

Peter McMullen University College London, London, UK

Emil Molnár Institute of Mathematics, Department of Geometry, Budapest University of Technology and Economics, Budapest, Hungary

Márton Naszódi Department of Geometry, Lorand Eötvös University, Budapest, Hungary

Shmuel Onn Operations Research, Davidson faculty of IE & M, Technion - Israel Institute of Technology, Haifa, Israel

János Pach EPFL Lausanne and Rényi Institute, Budapest, Hungary

Daniel Pellicer Centro de Ciencias Matemáticas, Universidad Nacional Autónoma de México, Morelia, Michoacán, Mexico

Tomaž Pisanski IAM, University of Primorska, Koper, Slovenia; FAMNIT, University of Primorska, Koper, Slovenia

Dmitry Ryabogin Department of Mathematics, Kent State University, Kent, OH, USA

Egon Schulte Department of Mathematics, Northeastern University, Boston, MA, USA

Konrad Swanepoel Department of Mathematics, London School of Economics and Political Science, London, UK

Nebojša Vasiljević Department of Mathematics, University of Belgrade, Belgrade, Serbia

Asia Ivić Weiss Department of Mathematics and Statistics, York University, Toronto, ON, Canada

Steve Wilson Department of Mathematics and Statistics, Northern Arizona University, Flagstaff, AZ, USA

Károly Bezdek—Biosketch

Károly Bezdek was born on 28 May 1955 in Budapest, a son of Károly Bezdek Sr. (who was chief engineer of Hungary's largest steel factory for over 20 years) and Magdolna Cserey (who had a strong interest in the literature and languages). His childhood years were spent in Dunaújváros. This period was challenging for his parents, who had grown up in a totally different Hungary, but despite some of the hardships faced by his family during and after WWII, his parents made all possible efforts to ensure a very educational and enjoyable childhood for Károly and his younger brother András. They encouraged both of their sons to develop interests in learning (across a wide range of subjects) and sports such as fencing and tennis.

Károly and András (who is also a mathematician) scored at the top level in several mathematics and physics competitions for high school and university students in Hungary. The awards won by Károly include first prize in the national KöMal contest (run by the Hungarian mathematics journal for high school students) in 1972/73 and first prize at the National Science Conference for Hungarian Undergraduate Students (TDK) in 1977/78, for his work on optimal circle coverings. As a result of these successes, Károly was admitted to Eötvös Loránd University in Budapest in 1973, without any entrance examination.

His first three years as an undergraduate involved rigorous basic courses, tested in oral exams, but also participation in special seminars on topics representing much of the frontline mathematical research in Hungary. Then in his last two years, he chose to specialise in discrete geometry and completed a Diploma in Mathematics (the equivalent of a master's degree) with a thesis on optimal circle coverings, under the supervision of Professor Károly Böröczky (who held the Chair of Geometry) in 1978.

He was awarded a Ph.D. in 1980 and Candidate of Mathematical Sciences degree in 1985, again with Prof. Károly Böröczky as his advisor in both cases, and later he was awarded a Doctor of Mathematical Sciences degree from the Hungarian Academy of Sciences in 1995 and Habilitation in Mathematics from Eötvös Loránd University in 1997.

Károly became a Faculty Member in the Department of Geometry at Eötvös Loránd University in 1978, served as chair of that department from 1999 to 2006, and earned the position of full professor in 1998. From 1998 to 2001, he served as Széchenyi Professor of Mathematics at Eötvös Loránd University, in a named position awarded to him by the Hungarian government. Although the university never really had a sabbatical system, he was fortunate to be able to travel regularly. During the period 1978 to 2003, he held numerous visiting positions at research institutions in Canada, Germany, the Netherlands, and the USA, including seven years at Cornell University, in Ithaca, NY.

He was invited to take up a Canada Research Chair at the University of Calgary, and he accepted this position in 2003. He is also Director of the Center for Computational and Discrete Geometry in Calgary; for the last few years, he has been an Associate Member of the Alfréd Rényi Institute of Mathematics in Budapest, and he also holds the title of Full Professor at the University of Pannonia in Veszprém.

Károly's research interests are in combinatorial, computational, convex, and discrete geometry, including some aspects of geometric analysis, geometric rigidity, and optimisation. He is the author of more than 110 research papers many of which are highly cited. He also wrote *Classical Topics in Discrete Geometry* (Springer, 2010) and *Lectures on Sphere Arrangements—the Discrete Geometric Side* (Springer, 2013), the monographs that take the reader to the frontiers of the most recent research developments in the relevant parts of discrete geometry.

He has been always interested in teaching, which he finds very rewarding as well. In particular, he has very much enjoyed working with graduate students, who are all very different from each other, but all gifted in many ways, each bringing a new perspective to geometric research. He has supervised five master's students, who he says have become great instructors with the potential to improve mathematics education, and a number of talented undergraduate research students. He has successfully supervised eleven Ph.D. students to date: Tibor Ódor (1991), László Szabó (1995), István Talata (1997), Endre Kiss (2004), Balázs Visy (2002), Márton Naszódi (2007), Zsolt Lángi (2008), Peter Papez (2009), Mate Salat (2009), Ryan Trelford (2014), and Muhammad A. Khan (2017).

Károly says that his work was influenced by a number of great mathematicians, colleagues, and friends, including 1978–1988 by Károly Böröczky, Aladar Heppes, Gábor Fejes Tóth, László Fejes Tóth, Kurt Leichtweiss, Keith Ball, Ted Bisztriczky, Robert Connelly, Oded Schramm, Joerg Wills, Thomas Hales, Alexander Litvak, Oleg Musin, Rolf Schneider, Marjorie Senechal, Egon Schulte, and Elisabeth Werner. He has also enjoyed travelling, often together with his wife Éva and their family, as well as inviting visitors for dinner in their home.

Károly is grateful to Éva for being 'such a fantastic partner and supporter'. Currently, Éva is Director and Teacher at the Gabor Bethlen Hungarian Language School in Calgary, and they have three sons: Dániel, Máté, and Márk. Márk is a third-year undergraduate student majoring in Public Relations at Mount Royal University in Calgary; Dániel has a degree in finance and is now completing a second undergraduate major in Computer Science at the University of Calgary; Máté is a third-year doctoral student in Chemistry at Princeton University.

We are very happy to pay tribute to Károly to his successful career and many contributions to mathematics, especially in geometry.

Egon Schulte—Biosketch

Egon Schulte was born on 7 January 1955 in Heggen (Finnentrop), North Rhine-Westfalia, Germany, to parents Egon and Gisela Schulte. He attended the Volksschule Lenhausen (Finnentrop) and the Katholische Volksschule Herdecke from 1961 to 1965, and the Städtisches Gymnasium Wetter (in the Ruhr region) from 1965 to 1973, completing the Abitur qualification in 1973. It was not until the last year or two in high school that Egon decided to study mathematics. In school, he was always good in mathematics, but was also very much interested in sports. He played very actively in a (European-style) handball team in Herdecke until about 1976 or so. Sports have always been an important part of his life; he has even run marathons.

From 1973 to 1978, he studied at the University of Dortmund, graduating with a 'Diploma' in Mathematics in 1978. Egon's Diplom thesis was on *Konstruktion regulärer Hüllen konstanter Breite* (regular hulls of constant width), a topic in convex geometry, and was published as his first paper in *Monatshefte der Mathematik.* His advisor was Ludwig Danzer, who also was advisor for his doctoral dissertation on *Regular Inzidenzkomplexe* (regular incidence complexes), which began Egon's lifelong interest in regular abstract polytopes. Egon graduated as a Doctor of Natural Sciences (in Mathematics) at the University of Dortmund in 1980. All three of Egon's main qualifications (Abitur, Diplom, and Doctorate) were awarded 'Auszeichnung' (distinction).

Prospects for academic positions in Germany were not good in the late 1970s and 1980s, especially in pure mathematics. Egon took a position as Wissenschaftlicher Assistent at the University of Dortmund from 1978 to 1983, and again from 1984 to 1987, but the period in between was very important for him, in that he found a very clear direction for himself, thanks largely to a visit by Branko Grünbaum to Dortmund in 1982. This had a profound influence on Egon, both mathematically and career-wise. He spent the 1983/84 academic year at the University of Washington, Seattle, and he describes the year as 'fantastic'. It introduced him to life in the USA and ultimately set him on a path towards a career there.

After Seattle, he returned to Germany for three years, gained Habilitation in Mathematics at the University of Dortmund in 1985, with a thesis on *Monotypische Pflasterungen und Komplexe* (monotypic tilings and complexes), and gained the title of 'Privatdozent'. Then, 1987 marked a new beginning for Egon, by moving to Boston, where he has been ever since. He worked as Visiting Assistant Professor at the Massachusetts Institute of Technology from 1987 to 1989 and then as an Associate Professor at Northeastern University from 1989 to 1992. Since 1992, he has been a Professor of Mathematics at Northeastern University, with tenure since 1993.

A few years after moving to Boston, Egon married Ursula Waser. They had two children: Sarah Marlen Schulte (born in 1992) and Isabelle Sophie Schulte (born in 1994), and both have studied at Northeastern. Sarah studied International Affairs and is now in her third year of Law School, and Isabelle graduated in 2017 with a major in Chemistry. Egon and Ursula separated in 2013 but remain good friends.

Mathematics has been Egon's passion ever since he began university. Looking back, he would say that over the years there were four people who strongly influenced his mathematical work and development: Ludwig Danzer, Branko Grünbaum, Harold Scott MacDonald (Donald) Coxeter, and Peter McMullen. Of course, he was positively influenced by many others as well. He is co-author with Peter McMullen of the outstanding book *Abstract Regular Polytopes*, has published well over 100 research articles (on a range of topics spanning discrete geometry, combinatorics and group theory), and edited six special issues of journals.

He is a popular invited lecture at conferences, has also organised or co-organised several conferences and workshops (or special sessions), and served on the editorial boards of many journals. He has won several grants, including many from the NSA and NSF in the USA, and a recent one from the Simons Foundation. And to date, he has successfully supervised 12 Ph.D. students: Barbara Nostrand (1993), Sergey Bratus (1999), Daniel Pellicer (2007), Anthony Cutler (2009), Mark Mixer (2010), Gabriel Cunningham (2012), Ilanit Helfand (2013), Andrew Duke (2014), Undine Leopold (2014), Ilya Scheidwasser (2015), Abigail Dalton-Williams (2015), and Nicholas Matteo (2015).

On top of all this, Egon is well-liked and highly respected by his friends and colleagues around the world for his positive attitude, his enthusiasm for mathematics, his engaging personality, and his encouragement of the next generation.

As far as choice of research topics is concerned, he says he usually followed his own interests and instincts and did not pay too much attention to trends and fashions. This had its rewards, but he says at times it came at a high price: 'It might have been smarter to follow more trendy mathematics', but we have the impression he does not regret his choices.

The Geometry of Homothetic Covering and Illumination

Károly Bezdek and Muhammad A. Khan

Abstract At a first glance, the problem of illuminating the boundary of a convex body by external light sources and the problem of covering a convex body by its smaller positive homothetic copies appear to be quite different. They are in fact two sides of the same coin and give rise to one of the important longstanding open problems in discrete geometry, namely, the Illumination Conjecture. In this paper, we survey the activity in the areas of discrete geometry, computational geometry and geometric analysis motivated by this conjecture. Special care is taken to include the recent advances that are not covered by the existing surveys. We also include some of our recent results related to these problems and describe two new approaches – one conventional and the other computer-assisted – to make progress on the illumination problem. Some open problems and conjectures are also presented.

Keywords Illumination number · Illumination conjecture · Covering conjecture · Separation conjecture · X-ray number · X-ray conjecture · Illumination parameter · Covering parameter · Covering index · Cylindrical covering parameters · ϵ-net of convex bodies

MSC (2010): 52A37 · 52A40 · 52C15 · 52C17

1 Shedding Some 'Light'

> ... N_k bezeichne die kleinste natürliche Zahl, für welche die nachfolgende Aussage richtig ist: Ist A ein eigentlicher konvexer Körper des k-dimensionalen euklidischen Raumes, so gibt es n mit A translations-gleiche Körper A_i mit $n \leq N_k$ derart, dass jeder Punkt von A ein innerer Punkt der Vereinigungsmenge $\bigcup_i A_i$ ist,... Welchen Wert hat N_k für $k \geq 3$? [51]

K. Bezdek (✉) · M. A. Khan
Department of Mathematics and Statistics, University of Calgary, Calgary, Canada
e-mail: bezdek@math.ucalgary.ca

M. A. Khan
e-mail: muhammkh@ucalgary.ca

K. Bezdek
Department of Mathematics, University of Pannonia, Veszprém, Hungary

M. D. E. Conder et al. (eds.), *Discrete Geometry and Symmetry*,
Springer Proceedings in Mathematics & Statistics 234,
https://doi.org/10.1007/978-3-319-78434-2_1

The above statement roughly translates to "Let N_k denote the smallest natural number such that any k-dimensional convex body can be covered by the interior of a union of at the most N_k of its translates. What is N_k for $k \geq 3$?" When Hadwiger raised this question in 1957 he probably did not imagine that it would remain unresolved half a century later and become a central problem in discrete geometry. Apparently, Hadwiger had a knack of coming up with such questions[1]. However, he was not the first one to study this particular problem. In fact, its earliest occurrence can be traced back to Levi's 1955 paper [62], who formulated and settled the 2-dimensional case of the problem. Later in 1960, the question was restated by Gohberg and Markus[2] [49] in terms of covering by homothetic copies. The equivalence of both formulations is relatively easy to check and details appear in Sect. 34 of [35].

Conjecture 1.1 (Covering Conjecture) *We can cover any d-dimensional convex body by 2^d or fewer of its smaller positive homothetic copies in Euclidean d-space, $d \geq 3$. Furthermore, 2^d homothetic copies are required only if the body is an affine d-cube.*

The same conjecture has also been referred to in the literature as the Levi–Hadwiger Conjecture, Gohberg–Markus Covering Conjecture and Hadwiger Covering Conjecture. The condition $d \geq 3$ has been added as the statement is known to be true in the plane [52, 62].

Let us make things formal. A d-dimensional *convex body* $\mathbf{K}$ is a compact convex subset of the Euclidean d-space, $\mathbb{E}^d$ with nonempty interior. Let $\mathbf{o}$ denote the origin of $\mathbb{E}^d$. Then $\mathbf{K}$ is said to be $\mathbf{o}$-*symmetric* if $\mathbf{K} = -\mathbf{K}$ and *centrally symmetric* if some translate of $\mathbf{K}$ is $\mathbf{o}$-symmetric. Since the quantities studied in this paper are invariant under affine transformations, we use the terms $\mathbf{o}$-symmetric and centrally symmetric interchangeably. A *homothety* is an affine transformation of $\mathbb{E}^d$ of the form $\mathbf{x} \mapsto \mathbf{t} + \lambda \mathbf{x}$, where $\mathbf{t} \in \mathbb{E}^d$ and λ is a non-zero real number. The image $\mathbf{t} + \lambda \mathbf{K}$ of a convex body $\mathbf{K}$ under a homothety is said to be its *homothetic copy* (or simply a *homothet*). A homothetic copy is *positive* if $\lambda > 0$ and *negative* otherwise. Furthermore, a homothetic copy with $0 < \lambda < 1$ is called a smaller positive homothet. In terms of the notations just introduced the Covering Conjecture states that for any $\mathbf{K} \subseteq \mathbb{E}^d$, there exist $\mathbf{t}_i \in \mathbb{E}^d$ and $0 < \lambda_i < 1$, for $i = 1, \ldots, 2^d$, such that

$$\mathbf{K} \subseteq \bigcup_{i=1}^{2^d} (\mathbf{t}_i + \lambda_i \mathbf{K}). \tag{1}$$

[1]The Hadwiger conjecture in graph theory is, in the words of Bollobás et al. [28], "*one of the deepest unsolved problems in graph theory*". Hadwiger even edited a column on unsolved problems in the journal *Elemente der Mathematik*. On the occasion of Hadwiger's 60th birthday, Victor Klee dedicated the first article in the Research Problems section of the *American Mathematical Monthly* to Hadwiger's work on promoting research problems [47, pp. 389–390].

[2]Apparently, Gohberg and Markus worked on the problem independently without knowing about the work of Levi and Hadwiger [32].

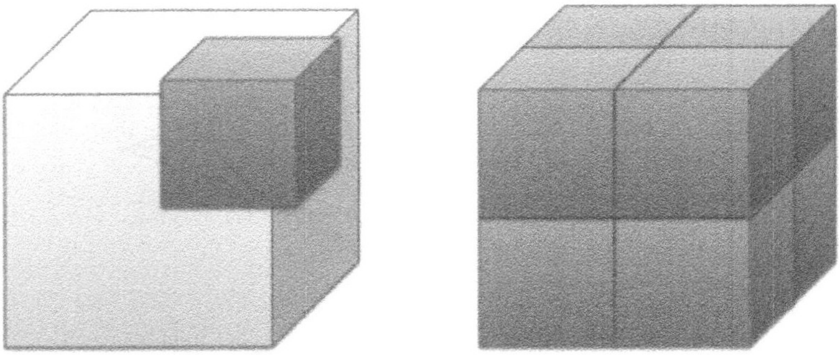

Fig. 1 A cube can be covered by 8 smaller positive homothets and no fewer

A light source at a point $\mathbf{p}$ outside a convex body $\mathbf{K} \subset \mathbb{E}^d$, *illuminates* a point $\mathbf{x}$ on the boundary of $\mathbf{K}$ if the halfline originating from $\mathbf{p}$ and passing through $\mathbf{x}$ intersects the interior of $\mathbf{K}$ at a point not lying between $\mathbf{p}$ and $\mathbf{x}$. The set of points $\{\mathbf{p}_i : i = 1, \ldots, n\}$ in the exterior of $\mathbf{K}$ is said to *illuminate* $\mathbf{K}$ if every boundary point of $\mathbf{K}$ is illuminated by some $\mathbf{p}_i$. The *illumination number* $I(\mathbf{K})$ of $\mathbf{K}$ is the smallest n for which $\mathbf{K}$ can be illuminated by n point light sources (Fig. 1).

One can also consider illumination of $\mathbf{K} \subset \mathbb{E}^d$ by parallel beams of light. Let $\mathbb{S}^{d-1}$ be the unit sphere centered at the origin $\mathbf{o}$ of $\mathbb{E}^d$. We say that a point $\mathbf{x}$ on the boundary of $\mathbf{K}$ is illuminated in the direction $\mathbf{v} \in \mathbb{S}^{d-1}$ if the halfline originating from $\mathbf{x}$ and with direction vector $\mathbf{v}$ intersects the interior of $\mathbf{K}$.

The former notion of illumination was introduced by Hadwiger [52], while the latter notion is due to Boltyanski[3] [29]. It may not come as a surprise that the two concepts are equivalent in the sense that a convex body $\mathbf{K}$ can be illuminated by n point sources if and only if it can be illuminated by n directions. However, it is less obvious that any covering of $\mathbf{K}$ by n smaller positive homothetic copies corresponds to illuminating $\mathbf{K}$ by n points (or directions) and vice versa (see [35] for details). Therefore, the following Illumination Conjecture [29, 35, 52] of Hadwiger and Boltyanski is equivalent to the Covering Conjecture (Figs. 2, 3 and 4).

Conjecture 1.2 (Illumination Conjecture) *The illumination number $I(\mathbf{K})$ of any d-dimensional convex body $\mathbf{K}$, $d \geq 3$, is at most 2^d and $I(K) = 2^d$ only if $\mathbf{K}$ is an affine d-cube.*

The conjecture also asserts that affine images of d-cubes are the only extremal bodies. The conjectured bound of 2^d results from the 2^d vertices of an affine cube, each requiring a different light source to be illuminated. In the sequel, we use the titles Covering Conjecture and Illumination Conjecture interchangeably, shifting between the covering and illumination paradigms as convenient.

[3]Vladimir Boltyanski (also written Boltyansky, Boltyanskii and Boltjansky) is a prolific mathematician and recepient of Lenin Prize in science. He has authored more than 220 mathematical works including, remarkably, more than 50 books!

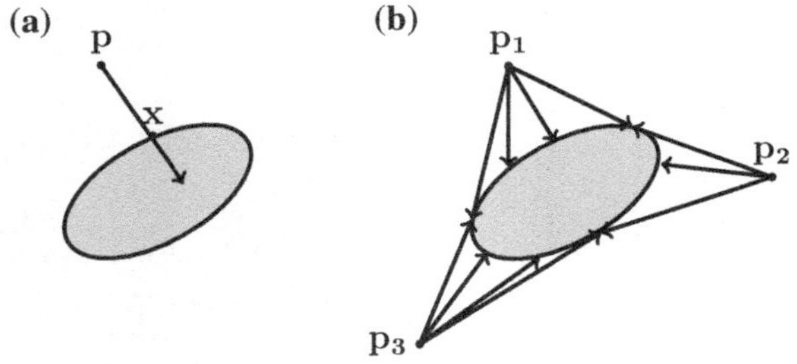

Fig. 2 **a** Illuminating a boundary point $\mathbf{x}$ of $\mathbf{K} \subset \mathbb{E}^d$ by the point light source $\mathbf{p} \in \mathbb{E}^d \setminus \mathbf{K}$, **b** $I(K) = 3$

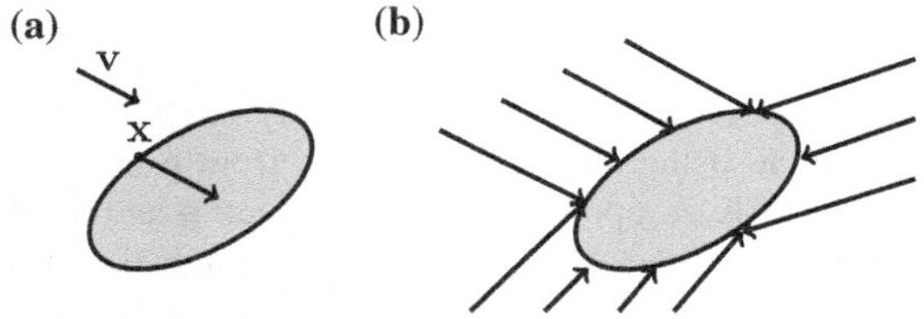

Fig. 3 **a** Illuminating a boundary point $\mathbf{x}$ of $\mathbf{K} \subset \mathbb{E}^d$ by a direction $\mathbf{v} \in \mathbb{S}^{d-1}$, **b** $I(K) = 3$

Fig. 4 Vladimir Boltyanski (left, courtesy Annals of the Moscow University) and Hugo Hadwiger (right, courtesy Oberwolfach Photo Collection), two of the main proponents of the illumination problem

We have so far seen three equivalent formulations of the Illumination Conjecture. But there are more. In fact, it is perhaps an indication of the richness of this problem that renders it to be studied from many angles, each with its own intuitive significance. We state one more equivalent form found independently by P. Soltan and V. Soltan [75], who formulated it for the **o**-symmetric case only and the first author [9, 10].

Conjecture 1.3 (Separation Conjecture) *Let* **K** *be an arbitrary convex body in* $\mathbb{E}^d$, $d \geq 3$, *and* **o** *be an arbitrary interior point of* **K**. *Then there exist* 2^d *hyperplanes of* $\mathbb{E}^d$ *such that each intersection of* **K** *with a supporting hyperplane, called a face of* **K**, *can be strictly separated from* **o** *by at least one of the* 2^d *hyperplanes. Furthermore,* 2^d *hyperplanes are needed only if* **K** *is the convex hull of* d *linearly independent line segments which intersect at the common relative interior point* **o**.

Over the years, the illumination conjecture has inspired a vast body of research in convex and discrete geometry, computational geometry and geometric analysis. There exist some nice surveys on the topic such as the papers [16, 64] and the corresponding chapters of the books [22, 35]. However, most of these are a bit dated. Moreover, we feel that the last few years have seen some interesting new ideas, such as the possibility of a computer-assisted proof, that are not covered by any of the above-mentioned surveys. The aim of this paper is to provide an accessible introduction to the geometry surrounding the Illumination Conjecture and a snapshot of the research motivated by it, with special emphasis on some of the recent developments. At the same time we describe some of our new results in this area.

We organize the material as follows. Section 2 gives a brief overview on the progress of the Illumination Conjecture. In Sect. 3, we mention some important relatives of the illumination problem, while Sect. 4 explores the known important quantitative versions of the problem including a new approach to make progress on the Illumination Conjecture based on the covering index of convex bodies (see Problem 3 and the discussion following it in Sect. 4.2). Finally, in Sect. 5 we present Zong's computer-assisted approach [85] for possibly resolving the Illumination Conjecture in low dimensions.

2 Progress on the Illumination Conjecture

2.1 *Results in* $\mathbb{E}^3$ *and* $\mathbb{E}^4$

Despite its intuitive richness, the illumination conjecture has been notoriously difficult to crack even in the first nontrivial case of $d = 3$. The closest anything has come is the proof announced by Boltyanski [36] for the 3-dimensional case. Unfortunately, the proof turned out to have gaps that remain to date. Later, Boltyanski [37] modified his claim to the following.

Theorem 2.1 *Let* **K** *be a convex body of* $\mathbb{E}^3$ *with* md **K** $= 2$. *Then* $I(\mathbf{K}) \leq 6$.

Here md is a functional introduced by Boltyanski in [30] and defined as follows for any d-dimensional convex body: Let $\mathbf{K} \subseteq \mathbb{E}^d$ be a convex body. Then md($\mathbf{K}$) is the greatest integer m for which there exist $m + 1$ regular boundary points of $\mathbf{K}$ such that the outward unit normals $\mathbf{v}_0, \ldots, \mathbf{v}_m$ of $\mathbf{K}$ at these points are minimally dependent, i.e., they are the vertices of an m-dimensional simplex that contains the origin in its relative interior.[4]

So far the best upper bound on illumination number in three dimensions is due to Papadoperakis [68].

Theorem 2.2 *The illumination number of any convex body in* $\mathbb{E}^3$ *is at most 16.*

However, there are partial results that establish the validity of the conjecture for some large classes of convex bodies. Often these classes of convex bodies have some underlying symmetry. Here we list some such results. A convex polyhedron $\mathbf{P}$ is said to have affine symmetry if the affine symmetry group of $\mathbf{P}$ consists of the identity and at least one other affinity of $\mathbb{E}^3$. The first author obtained the following result [9].

Theorem 2.3 *If* $\mathbf{P}$ *is a convex polyhedron of* $\mathbb{E}^3$ *with affine symmetry, then the illumination number of* $\mathbf{P}$ *is at most 8.*

Recall that a convex body $\mathbf{K}$ is said to be centrally symmetric if it has a point of symmetry. Furthermore, a body $\mathbf{K}$ is symmetric about a plane p if a reflection across that plane leaves $\mathbf{K}$ unchanged. Lassak [57] proved that under the assumption of central symmetry, the illumination conjecture holds in three dimensions.

Theorem 2.4 *If* $\mathbf{K}$ *is a centrally symmetric convex body in* $\mathbb{E}^3$, *then* $I(\mathbf{K}) \leq 8$.

Dekster [43] extended Theorem 2.3 from polyhedra to convex bodies with plane symmetry.

Theorem 2.5 *If* $\mathbf{K}$ *is a convex body symmetric about a plane in* $\mathbb{E}^3$, *then* $I(\mathbf{K}) \leq 8$.

It turns out that for 3-dimensional bodies of constant width – that is bodies whose width, measured by the distance between two opposite parallel hyperplanes touching its boundary, is the same regardless of the direction of those two parallel planes – we get an even better bound.

Theorem 2.6 *The illumination number of any convex body of constant width in* $\mathbb{E}^3$ *is at most 6.*

Proofs of the above theorem have appeared in several papers [18, 60, 79]. It is, in fact, reasonable to conjecture the following even stronger result.

Conjecture 2.7 *The illumination number of any convex body of constant width in* $\mathbb{E}^3$ *is exactly 4.*

---

[4]In fact, it is proved in [30] that md($\mathbf{K}$) = him($\mathbf{K}$) holds for any convex body $\mathbf{K}$ of $\mathbb{E}^d$ and therefore one can regard Theorem 2.1 as an immediate corollary of Theorem 2.16 for $d = 3$ in Sect. 2.2.

The above conjecture, if true, would provide a new proof of Borsuk's conjecture [41] in dimension three, which states that any set of unit diameter in $\mathbb{E}^3$ can be partitioned into at most four subsets of diameter less than one. We remark that although it is false in general [54], Borsuk's conjecture has a long and interesting history of its own and the reader can look up [31, 35, 50] for detailed discussions.

Now let us consider the state of the Illumination Conjecture in $\mathbb{E}^4$. It is well known that neighbourly d-polytopes have the maximum number of facets among d-polytopes with a fixed number of vertices (for more details on this see for example, [27]). Thus, it is natural to investigate the Separation Conjecture for neighbourly d-polytopes (see also Theorem 2.18). Since interesting neighbourly d-polytopes exist only in $\mathbb{E}^d$ for $d \geq 4$, it is particularly natural to first restrict our attention to neighbourly 4-polytopes. Starting from a cyclic 4-polytope, the sewing procedure of Shemer (for details see [27]) produces an infinite family of neighbourly 4-polytopes each of which is obtained from the previous one by adding one new vertex in a suitable way. Neighbourly 4-polytopes obtained from a cyclic 4-polytope by a sequence of sewings are called *totally-sewn*. The main result of the very recent paper [27] of Bisztriczky and Fodor is a proof of the Separation Conjecture for totally-sewn neighbourly 4-polytopes.

Theorem 2.8 *Let* **P** *be an arbitrary totally-sewn neighbourly* 4*-polytope in* $\mathbb{E}^4$*, and* **o** *be an arbitrary interior point of* **P**. *Then there exist* 16 *hyperplanes of* $\mathbb{E}^4$ *such that each face of* **P**, *can be strictly separated from* **o** *by at least one of the* 16 *hyperplanes.*

However, Bisztriczky [26] conjectures the following stronger result.

Conjecture 2.9 *Let* **P** *be an arbitrary totally-sewn neighbourly* 4*-polytope in* $\mathbb{E}^4$*, and* **o** *be an arbitrary interior point of* **P**. *Then there exist* 9 *hyperplanes of* $\mathbb{E}^4$ *such that each face of* **P**, *can be strictly separated from* **o** *by at least one of the* 9 *hyperplanes.*

2.2 General Results

Before we state results on the illumination number of convex bodies in $\mathbb{E}^d$, we take a little detour. We need Rogers' estimate [69] of the infimum $\theta(\mathbf{K})$ of the covering density of $\mathbb{E}^d$ by translates of the convex body **K**, namely, for $d \geq 2$,[5]

$$\theta(\mathbf{K}) \leq d(\ln d + \ln \ln d + 5)$$

and the Rogers–Shephard inequality [70]

$$\mathrm{vol}_d(\mathbf{K} - \mathbf{K}) \leq \binom{2d}{d} \mathrm{vol}_d(\mathbf{K})$$

[5]The bound on $\theta(\mathbf{K})$ has been improved to $\theta(\mathbf{K}) \leq d \ln d + d \ln \ln d + d + o(d)$ by Fejes Tóth [46].

on the d-dimensional volume $\mathrm{vol}_d(\cdot)$ of the difference body $\mathbf{K} - \mathbf{K}$ of $\mathbf{K}$.

It was rather a coincidence, at least from the point of view of the Illumination Conjecture, when in 1964 Erdős and Rogers [45] proved the following theorem. In order to state their theorem in a proper form we need to introduce the following notion. If we are given a covering of a space by a system of sets, the *star number* of the covering is the supremum, over sets of the system, of the cardinals of the numbers of sets of the system meeting a set of the system (see [45]). On the one hand, the standard Lebesgue brick-laying construction provides an example, for each positive integer d, of a lattice covering of $\mathbb{E}^d$ by closed cubes with star number $2^{d+1} - 1$. On the other hand, Theorem 1 of [45] states that the star number of a lattice covering of $\mathbb{E}^d$ by translates of a centrally symmetric convex body is always at least $2^{d+1} - 1$. However, from our point of view, the main result of [45] is the one under Theorem 2 which (combined with some observations from [44] and the Rogers–Shephard inequality [69]) reads as follows.

Theorem 2.10 *Let* $\mathbf{K}$ *be a convex body in the d-dimensional Euclidean space* $\mathbb{E}^d$, $d \geq 2$. *Then there exists a covering of* $\mathbb{E}^d$ *by translates of* $\mathbf{K}$ *with star number at most*

$$\frac{\mathrm{vol}_d(\mathbf{K} - \mathbf{K})}{\mathrm{vol}_d(\mathbf{K})}(d \ln d + d \ln \ln d + 5d + 1) \leq \binom{2d}{d}(d \ln d + d \ln \ln d + 5d + 1).$$

Moreover, for sufficiently large d, $5d$ can be replaced by $4d$.

The periodic and probabilistic construction on which Theorem 2.10 is based gives also the following.

Corollary 2.11 *If* $\mathbf{K}$ *is an arbitrary convex body in* $\mathbb{E}^d$, $d \geq 2$, *then*

$$\begin{aligned} I(\mathbf{K}) &\leq \frac{\mathrm{vol}_d(\mathbf{K} - \mathbf{K})}{\mathrm{vol}_d(\mathbf{K})} d(\ln d + \ln \ln d + 5) \leq \binom{2d}{d} d(\ln d + \ln \ln d + 5) \\ &= O(4^d \sqrt{d} \ln d). \end{aligned} \tag{2}$$

Moreover, for sufficiently large d, $5d$ can be replaced by $4d$.

Note that the bound given in Corollary 2.11 can also be obtained from the more general result of Rogers and Zong [71], which states that for d-dimensional convex bodies $\mathbf{K}$ and $\mathbf{L}$, $d \geq 2$, one can cover $\mathbf{K}$ by $N(\mathbf{K}, \mathbf{L})$ translates[6] of $\mathbf{L}$ such that

$$N(\mathbf{K}, \mathbf{L}) \leq \frac{\mathrm{vol}_d(\mathbf{K} - \mathbf{L})}{\mathrm{vol}_d(\mathbf{L})} \theta(\mathbf{L}).$$

For the sake of completeness we also mention the inequality

$$I(\mathbf{K}) \leq (d + 1)d^{d-1} - (d - 1)(d - 2)^{d-1}$$

---

[6] $N(\mathbf{K}, \mathbf{L})$ is called the *covering number of* $\mathbf{K}$ *by* $\mathbf{L}$.

due to Lassak [59], which is valid for an arbitrary convex body $\mathbf{K}$ in $\mathbb{E}^d$, $d \geq 2$, and is (somewhat) better than the estimate of Corollary 2.11 for some small values of d.

Since, for a centrally symmetric convex body $\mathbf{K}$, $\frac{\mathrm{vol}(\mathbf{K}-\mathbf{K})}{\mathrm{vol}_d(\mathbf{K})} = 2^d$, we have the following improved upper bound on the illumination number of such convex bodies.

Corollary 2.12 *If* $\mathbf{K}$ *is a centrally symmetric convex body in* $\mathbb{E}^d$, $d \geq 2$, *then*

$$I(\mathbf{K}) \leq \frac{\mathrm{vol}_d(\mathbf{K}-\mathbf{K})}{\mathrm{vol}_d(\mathbf{K})} d(\ln d + \ln\ln d + 5) = 2^d d(\ln d + \ln\ln d + 5) = O(2^d d \ln d). \tag{3}$$

The above upper bound is fairly close to the conjectured value of 2^d. However, most convex bodies are far from being symmetric and so, in general, one may wonder whether the Illumination Conjecture is true at all, especially for large d. Thus, it was important progress when Schramm [73] managed to prove the Illumination Conjecture for all convex bodies of constant width in all dimensions at least 16. In fact, he proved the following inequality.

Theorem 2.13 *If* $\mathbf{W}$ *is an arbitrary convex body of constant width in* $\mathbb{E}^d$, $d \geq 3$, *then*

$$I(\mathbf{W}) \leq 5d\sqrt{d}(4 + \ln d)\left(\frac{3}{2}\right)^{\frac{d}{2}}.$$

By taking a closer look of Schramm's elegant paper [73] and making the necessary modifications, the first author [23] somewhat improved the upper bound of Theorem 2.13, but more importantly he succeeded in extending that estimate to the following family of convex bodies (called the family of *fat spindle convex bodies*) that is much larger than the family of convex bodies of constant width. Thus, we have the following generalization of Theorem 2.13 proved in [23].

Theorem 2.14 *Let* $X \subset \mathbb{E}^d$, $d \geq 3$ *be an arbitrary compact set with* $\mathrm{diam}(X) \leq 1$ *and let* $\mathbf{B}[X]$ *be the intersection of the closed d-dimensional unit balls centered at the points of* X. *Then*

$$I(\mathbf{B}[X]) < 4\left(\frac{\pi}{3}\right)^{\frac{1}{2}} d^{\frac{3}{2}}(3 + \ln d)\left(\frac{3}{2}\right)^{\frac{d}{2}} < 5d^{\frac{3}{2}}(4 + \ln d)\left(\frac{3}{2}\right)^{\frac{d}{2}}.$$

On the one hand, $4\left(\frac{\pi}{3}\right)^{\frac{1}{2}} d^{\frac{3}{2}}(3 + \ln d)\left(\frac{3}{2}\right)^{\frac{d}{2}} < 2^d$ for all $d \geq 15$. (Moreover, for every $\epsilon > 0$ if d is sufficiently large, then $I(\mathbf{B}[X]) < \left(\sqrt{1.5} + \epsilon\right)^d = (1.224\cdots + \epsilon)^d$.) On the other hand, based on the elegant construction of Kahn and Kalai [54], it is known (see [1]), that if d is sufficiently large, then there exists a finite subset X'' of $\{0, 1\}^d$ in $\mathbb{E}^d$ such that any partition of X'' into parts of smaller diameter requires more than $(1.2)^{\sqrt{d}}$ parts. Let X' be the (positive) homothetic copy of X'' having unit diameter and let X be the (not necessarily unique) convex body of constant width one containing X'. Then it follows via standard arguments that $I(\mathbf{B}[X]) > (1.2)^{\sqrt{d}}$ with $X = \mathbf{B}[X]$.

Recall that a convex polytope is called a *belt polytope* if to each side of any of its 2-faces there exists a parallel (opposite) side on the same 2-face. This class of polytopes is wider than the class of zonotopes. Moreover, it is easy to see that any convex body of $\mathbb{E}^d$ can be represented as a limit of a covergent sequence of belt polytopes with respect to the Hausdorff metric in $\mathbb{E}^d$. The following theorem on belt polytopes was proved by Martini in [63]. The result that it extends to the class of convex bodies, called belt bodies (including zonoids), is due to Boltyanski [33–35]. (See also [38] for a somewhat sharper result on the illumination numbers of belt bodies.)

Theorem 2.15 *Let* **P** *be an arbitrary d-dimensional belt polytope (resp., belt body) different from a parallelotope in* $\mathbb{E}^d$, $d \geq 2$. *Then*

$$I(\mathbf{P}) \leq 3 \cdot 2^{d-2}.$$

Now, let $\mathbf{K}$ be an arbitrary convex body in $\mathbb{E}^d$ and let $\mathcal{T}(\mathbf{K})$ be the family of all translates of $\mathbf{K}$ in $\mathbb{E}^d$. The *Helly dimension* $\text{him}(\mathbf{K})$ of $\mathbf{K}$ [74] is the smallest integer h such that for any finite family $\mathcal{F} \subseteq \mathcal{T}(\mathbf{K})$ with cardinality greater than $h+1$ the following assertion holds: if every $h+1$ members of $\mathcal{F}$ have a point in common, then all the members of $\mathcal{F}$ have a point in common. As is well known $1 \leq \text{him}(\mathbf{K}) \leq d$. Using this notion Boltyanski [37] gave a proof of the following theorem.

Theorem 2.16 *Let* **K** *be a convex body with* $\text{him}(\mathbf{K}) = 2$ *in* $\mathbb{E}^d$, $d \geq 3$. *Then*

$$I(\mathbf{K}) \leq 2^d - 2^{d-2}.$$

In fact, in [37] Boltyanski conjectures the following more general inequality.

Conjecture 2.17 *Let* **K** *be a convex body with* $\text{him}(\mathbf{K}) = h > 2$ *in* $\mathbb{E}^d$, $d \geq 3$. *Then*

$$I(\mathbf{K}) \leq 2^d - 2^{d-h}.$$

The first author and Bisztriczky gave a proof of the Illumination Conjecture for the class of dual cyclic polytopes in [14]. Their upper bound for the illumination numbers of dual cyclic polytopes has been improved by Talata in [77]. So, we have the following statement.

Theorem 2.18 *The illumination number of any d-dimensional dual cyclic polytope is at most* $\frac{(d+1)^2}{2}$, *for all* $d \geq 2$.

3 On Some Relatives of the Illumination Number

3.1 Illumination by Affine Subspaces

Let $\mathbf{K}$ be a convex body in $\mathbb{E}^d$, $d \geq 2$. The following definitions were introduced by the first named author in [13] (see also [9] that introduced the concept of the first definition below).

Let $L \subset \mathbb{E}^d \setminus \mathbf{K}$ be an affine subspace of dimension l, $0 \leq l \leq d-1$. Then L illuminates the boundary point $\mathbf{q}$ of $\mathbf{K}$ if there exists a point $\mathbf{p}$ of L that illuminates $\mathbf{q}$ on the boundary of $\mathbf{K}$. Moreover, we say that the affine subspaces $L_1, L_2, \ldots, L_n$ of dimension l with $L_i \subset \mathbb{E}^d \setminus \mathbf{K}$, $1 \leq i \leq n$ illuminate $\mathbf{K}$ if every boundary point of $\mathbf{K}$ is illuminated by at least one of the affine subspaces $L_1, L_2, \ldots, L_n$. Finally, let $I_l(\mathbf{K})$ be the smallest positive integer n for which there exist n affine subspaces of dimension l say, $L_1, L_2, \ldots, L_n$ such that $L_i \subset \mathbb{E}^d \setminus \mathbf{K}$ for all $1 \leq i \leq n$ and $L_1, L_2, \ldots, L_n$ illuminate $\mathbf{K}$. Then $I_l(\mathbf{K})$ is called the *l-dimensional illumination number* of $\mathbf{K}$ and the sequence $I_0(\mathbf{K}), I_1(\mathbf{K}), \ldots, I_{d-2}(\mathbf{K}), I_{d-1}(\mathbf{K})$ is called the *successive illumination numbers* of $\mathbf{K}$. Obviously, $I(\mathbf{K}) = I_0(\mathbf{K}) \geq I_1(\mathbf{K}) \geq \cdots \geq I_{d-2}(\mathbf{K}) \geq I_{d-1}(\mathbf{K}) = 2$.

Recall that $\mathbb{S}^{d-1}$ denotes the unit sphere centered at the origin of $\mathbb{E}^d$. Let $HS^l \subset \mathbb{S}^{d-1}$ be an l-dimensional open great-hemisphere of $\mathbb{S}^{d-1}$, where $0 \leq l \leq d-1$. Then HS^l illuminates the boundary point $\mathbf{q}$ of $\mathbf{K}$ if there exists a unit vector $\mathbf{v} \in HS^l$ that illuminates $\mathbf{q}$, in other words, for which it is true that the halfline emanating from $\mathbf{q}$ and having direction vector $\mathbf{v}$ intersects the interior of $\mathbf{K}$. Moreover, we say that the l-dimensional open great-hemispheres $HS_1^l, HS_2^l, \ldots, HS_n^l$ of $\mathbb{S}^{d-1}$ illuminate $\mathbf{K}$ if each boundary point of $\mathbf{K}$ is illuminated by at least one of the open great-hemispheres $HS_1^l, HS_2^l, \ldots, HS_n^l$. Finally, let $I'_l(\mathbf{K})$ be the smallest number of l-dimensional open great-hemispheres of $\mathbb{S}^{d-1}$ that illuminate $\mathbf{K}$. Obviously, $I'_0(\mathbf{K}) \geq I'_1(\mathbf{K}) \geq \cdots \geq I'_{d-2}(\mathbf{K}) \geq I'_{d-1}(\mathbf{K}) = 2$.

Let $L \subset \mathbb{E}^d$ be a linear subspace of dimension l, $0 \leq l \leq d-1$ in $\mathbb{E}^d$. The *l-codimensional circumscribed cylinder* of $\mathbf{K}$ generated by L is the union of translates of L that have a nonempty intersection with $\mathbf{K}$. Then let $C_l(\mathbf{K})$ be the smallest number of translates of the interiors of some l-codimensional circumscribed cylinders of $\mathbf{K}$ the union of which contains $\mathbf{K}$. Obviously, $C_0(\mathbf{K}) \geq C_1(\mathbf{K}) \geq \cdots \geq C_{d-2}(\mathbf{K}) \geq C_{d-1}(\mathbf{K}) = 2$.

The following theorem, which was proved in [13], collects the basic information known about the quantities just introduced.

Theorem 3.1 *Let $\mathbf{K}$ be an arbitrary convex body of $\mathbb{E}^d$. Then*

(i) $I_l(\mathbf{K}) = I'_l(\mathbf{K}) = C_l(\mathbf{K})$, *for all* $0 \leq l \leq d-1$.
(ii) $\lceil \frac{d+1}{l+1} \rceil \leq I_l(\mathbf{K})$, *for all* $0 \leq l \leq d-1$, *with equality for any smooth* $\mathbf{K}$.
(iii) $I_{d-2}(\mathbf{K}) = 2$, *for all* $d \geq 3$.

The Generalized Illumination Conjecture was phrased by the first named author in [13] as follows.

Conjecture 3.2 (Generalized Illumination Conjecture) *Let* $\mathbf{K}$ *be an arbitrary convex body and* $\mathbf{C}^d$ *be a d-dimensional affine cube in* $\mathbb{E}^d$. *Then*

$$I_l(\mathbf{K}) \le I_l(\mathbf{C}^d)$$

holds for all $0 \le l \le d-1$.

The above conjecture was proved for zonotopes and zonoids in [13]. The results of parts (i) and (ii) of the next theorem are taken from [13], where they were proved for zonotopes (resp., zonoids). However, in the light of the more recent works in [34] and [38] these results extend to the class of belt polytopes (resp., belt bodies) in a rather straightforward way so we present them in that form. The lower bound of part (iii) was proved in [13] and the upper bound of part (iii) is the major result of [55]. Finally, part (iv) was proved in [12].

Theorem 3.3 *Let* $\mathbf{M}$ *be a belt polytope (resp., belt body) and* $\mathbf{C}^d$ *be a d-dimensional affine cube in* $\mathbb{E}^d$. *Then*

(i) $I_l(\mathbf{M}) \le I_l(\mathbf{C}^d)$ *holds for all* $0 \le l \le d-1$.
(ii) $I_{\lfloor \frac{d}{2} \rfloor}(\mathbf{M}) = \cdots = I_{d-1}(\mathbf{M}) = 2$.
(iii) $\frac{2^d}{\sum_{i=0}^{l}\binom{d}{i}} \le I_l(\mathbf{C}^d) \le K(d,l)$, *where* $K(d,l)$ *denotes the minimum cardinality of binary codes of length d with covering radius l,* $0 \le l \le d-1$.
(iv) $I_1(\mathbf{C}^d) = \frac{2^d}{d+1}$, *provided that* $d+1 = 2^m$.

3.2 'X-raying' the Problem

In 1972, the *X-ray number* of convex bodies was introduced by P. Soltan as follows (see [64]). Let $\mathbf{K}$ be a convex body of $\mathbb{E}^d$, $d \ge 2$, and $L \subset \mathbb{E}^d$ be a line through the origin of $\mathbb{E}^d$. We say that the boundary point $\mathbf{x} \in \mathbf{K}$ is X-rayed along L if the line parallel to L passing through $\mathbf{x}$ intersects the interior of $\mathbf{K}$. The X-ray number $X(\mathbf{K})$ of $\mathbf{K}$ is the smallest number of lines such that every boundary point of $\mathbf{K}$ is X-rayed along at least one of these lines. Clearly, $X(\mathbf{K}) \ge d$. Moreover, it is easy to see that this bound is attained by any smooth convex body. On the other hand, if $\mathbf{C}^d$ is a d-dimensional (affine) cube and F is one of its $(d-2)$-dimensional faces, then the X-ray number of $\mathbf{C}^d \setminus F$, the convex hull of the set of vertices of $\mathbf{C}^d$ that do not belong to F, is $3 \cdot 2^{d-2}$.

In 1994, the first author and Zamfirescu [17] published the following conjecture.

Conjecture 3.4 (X-ray Conjecture) *The X-ray number of any convex body in* $\mathbb{E}^d$ *is at most* $3 \cdot 2^{d-2}$.

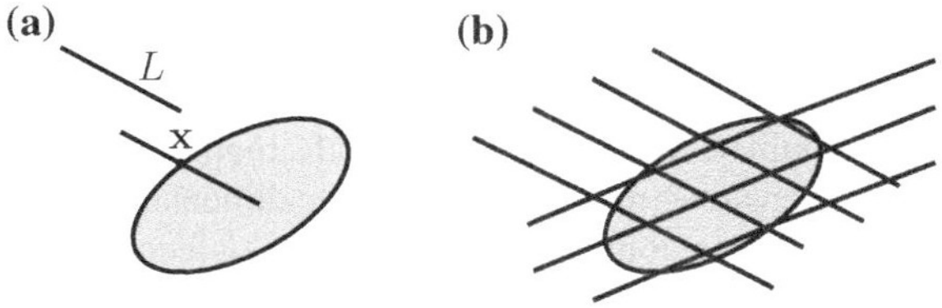

Fig. 5 **a** X-raying a boundary point $\mathbf{x}$ of $\mathbf{K}$ along a line L, **b** $X(\mathbf{K}) = 2$

The X-ray Conjecture is proved only in the plane and it is open in higher dimensions. Here we note that the inequalities

$$X(\mathbf{K}) \leq I(\mathbf{K}) \leq 2X(\mathbf{K})$$

hold for any convex body $\mathbf{K} \subset \mathbb{E}^d$. In other words, any proper progress on the X-ray Conjecture would imply progress on the Illumination Conjecture and vice versa. We also note that a natural way to prove the X-ray Conjecture would be to show that any convex body $\mathbf{K} \subset \mathbb{E}^d$ can be illuminated by $3 \cdot 2^{d-2}$ pairs of pairwise opposite directions (Fig. 5).

The main results of [21] on the X-ray number can be summarized as follows. In order to state them properly we need to recall two basic notions. Let $\mathbf{K}$ be a convex body in $\mathbb{E}^d$ and let F be a face of $\mathbf{K}$. The *Gauss image* $\nu(F)$ of the face F is the set of all points (i.e., unit vectors) $\mathbf{u}$ of the $(d-1)$-dimensional unit sphere $\mathbb{S}^{d-1} \subset \mathbb{E}^d$ centered at the origin $\mathbf{o}$ of $\mathbb{E}^d$ for which the supporting hyperplane of K with outer normal vector $\mathbf{u}$ contains F. It is easy to see that the Gauss images of distinct faces of $\mathbf{K}$ have disjoint relative interiors in $\mathbb{S}^{d-1}$ and $\nu(F)$ is compact and spherically convex for any face F. Let $C \subset \mathbb{S}^{d-1}$ be a set of finitely many points. Then the *covering radius* of C is the smallest positive real number r with the property that the family of spherical balls of radii r centered at the points of C covers $\mathbb{S}^{d-1}$.

Theorem 3.5 *Let $\mathbf{K} \subset \mathbb{E}^d$, $d \geq 3$, be a convex body and let r be a positive real number with the property that the Gauss image $\nu(F)$ of any face F of $\mathbf{K}$ can be covered by a spherical ball of radius r in $\mathbb{S}^{d-1}$. Moreover, assume that there exist $2m$ pairwise antipodal points of $\mathbb{S}^{d-1}$ with covering radius R satisfying the inequality $r + R \leq \frac{\pi}{2}$. Then $X(\mathbf{K}) \leq m$. In particular, if there are $2m$ pairwise antipodal points on $\mathbb{S}^{d-1}$ with covering radius R satisfying the inequality $R \leq \pi/2 - r_{d-1}$, where $r_{d-1} = \arccos\sqrt{\frac{d+1}{2d}}$ is the circumradius of a regular $(d-1)$-dimensional spherical simplex of edge length $\pi/3$, then $X(\mathbf{W}) \leq m$ holds for any convex body $\mathbf{W}$ of constant width in $\mathbb{E}^d$.*

Theorem 3.6 *If $\mathbf{W}$ is an arbitrary convex body of constant width in $\mathbb{E}^3$, then $X(\mathbf{W}) = 3$. If $\mathbf{W}$ is any convex body of constant width in $\mathbb{E}^4$, then $4 \leq X(\mathbf{W}) \leq 6$. Moreover, if $\mathbf{W}$ is a convex body of constant width in $\mathbb{E}^d$ with $d = 5, 6$, then $d \leq X(\mathbf{W}) \leq 2^{d-1}$.*

Corollary 3.7 *If $\mathbf{W}$ is an arbitrary convex body of constant width in $\mathbb{E}^3$, then $4 \leq I(\mathbf{W}) \leq 6$. If $\mathbf{W}$ is any convex body of constant width in $\mathbb{E}^4$, then $5 \leq I(\mathbf{W}) \leq 12$.*

Moreover, if **W** *is a convex body of constant width in* $\mathbb{E}^d$ *with* $d = 5, 6$, *then* $d + 1 \leq I(\mathbf{W}) \leq 2^d$.

It would be interesting to extend the method described in the paper [21] for the next few dimensions (more exactly, for the dimensions $7 \leq d \leq 14$) in particular, because in these dimensions neither the X-ray Conjecture nor the Illumination Conjecture is known to hold for convex bodies of constant width.

From the proof of Theorem 2.4 it follows in a straightforward way that if **K** is a centrally symmetric convex body in $\mathbb{E}^3$, then $X(\mathbf{K}) \leq 4$. On the other hand, very recently Trelford [78] proved the following related result.

Theorem 3.8 *If* **K** *is a convex body symmetric about a plane in* $\mathbb{E}^3$, *then* $X(\mathbf{K}) \leq 6$.

3.3 Other Relatives

3.3.1 t-covering and t-illumination Numbers

In Sect. 1, we found that the least number of smaller positive homothets of a convex body **K** required to cover it equals the minimum number of translates of the interior of **K** needed to cover **K**. Is this number also equal to the the minimum number $t(\mathbf{K})$ of translates of **K** that are different from **K** and are needed to cover **K**?

Despite being a very natural question, the problem of economical translative coverings have not attracted much attention. To our knowledge, the first systematic study of these was carried out quite recently by Lassak et al. [61] who called them t-coverings and also introduced the corresponding illumination concept, called t-illumination, as follows: A boundary point **x** of a convex body **K** of $\mathbb{E}^d$ is *t-illuminated* by a direction $\mathbf{v} \in \mathbb{S}^{d-1}$ if there exists a different point $\mathbf{y} \in \mathbf{K}$ such that the vector $\mathbf{y} - \mathbf{x}$ has the same direction as **v** (i.e., $\mathbf{y} - \mathbf{x} = \lambda \mathbf{v}$, for some $\lambda > 0$). The minimum number $i(\mathbf{K})$ of directions needed to t-illuminate the entire boundary of **K** is called its t-illumination number. The connection between t-covering and t-illumination is summarized in the next result [61]. Note that a convex body **K** is said to be *strictly convex* if for any two points of **K** the open line segment connecting them belongs to the interior of **K**.

Theorem 3.9

(i) If **K** *is a planar convex body, then* $i(\mathbf{K}) = t(\mathbf{K})$.
(ii) If **K** *is a* d*-dimensional strictly convex body,* $d \geq 3$, *then* $i(\mathbf{K}) = t(\mathbf{K})$.
(iii) If **K** *is a* d*-dimensional convex body,* $d \geq 3$, *then* $i(\mathbf{K}) \leq t(\mathbf{K})$, *where the equality does not hold in general.*

Clearly, $t(\mathbf{K}) \leq I(\mathbf{K})$. In the same paper [61], the following results were obtained about the relationship between $I(\cdot)$ and $t(\cdot)$.

Theorem 3.10

(i) *If* $\mathbf{K}$ *is a planar convex body, then* $t(\mathbf{K}) = I(\mathbf{K})$ *if and only if* $\mathbf{K}$ *contains no parallel boundary segments.*
(i) *If* $\mathbf{K}$ *is a strictly convex body of* $\mathbb{E}^d$, $d \geq 3$, *then* $t(\mathbf{K}) = I(\mathbf{K})$.
(iii) *If* $\mathbf{K}$ *is a convex body of* $\mathbb{E}^d$, $d \geq 3$ *that does not have parallel boundary segments, then* $t(\mathbf{K}) = I(\mathbf{K})$.

However, in general the following remains unanswered [61].

Problem 1 *Characterize the convex bodies* $\mathbf{K}$ *for which* $t(\mathbf{K}) = I(\mathbf{K})$.

In [65], the notion of t-illumination was refined into t-central illumination and strict t-illumination and the corresponding illumination numbers were defined. The paper also introduced metric versions of the classical, t-central and strict t-illumination numbers and investigated their properties at length. The interested reader is referred to [65] for details.

3.3.2 Blocking Numbers

The *blocking number* $\beta(\mathbf{K})$ [84] of a convex body $\mathbf{K}$ is defined as the minimum number of nonoverlapping translates of $\mathbf{K}$ that can be brought into contact with the boundary of K so as to block any other translate of $\mathbf{K}$ from touching $\mathbf{K}$. Since $\beta(\mathbf{K}) = \beta(\mathbf{K} - \mathbf{K})$ and $\mathbf{K} - \mathbf{K}$ is $\mathbf{o}$-symmetric, it suffices to consider the blocking numbers of $\mathbf{o}$-symmetric convex bodies only.

For any $\mathbf{o}$-symmetric convex body $\mathbf{K}$, the relation $I(\mathbf{K}) \leq \beta(\mathbf{K})$ holds [84], while no such relationship exists for general convex bodies. Zong [84] conjectured the following.

Conjecture 3.11 *For any* d*-dimensional convex body* $\mathbf{K}$,

$$2d \leq \beta(\mathbf{K}) \leq 2^d,$$

and $\beta(\mathbf{K}) = 2^d$ *if and only if* $\mathbf{K}$ *is a* d*-dimensional cube.*

If true, Zong's Conjecture would imply the Illumination Conjecture for $\mathbf{o}$-symmetric convex bodies. Some of the known values of the blocking number include $\beta(\mathbf{K}) = 2^d$, if $\mathbf{K}$ is a d-dimensional cube; $\beta(\mathbf{K}) = 6$, if $\mathbf{K}$ is a 3-dimensional ball; and $\beta(\mathbf{K}) = 9$, if $\mathbf{K}$ is a 4-dimensional ball [42]. Some other values and estimates are obtained in [83].

Several generalizations of the blocking number have been proposed. The smallest number of non-overlapping translates of $\mathbf{K}$ such that the interior of $\mathbf{K}$ is disjoint from the interiors of the translates and they can block any other translate from touching $\mathbf{K}$ is denoted by $\beta_1(\mathbf{K})$; the smallest number of translates all of which touch $\mathbf{K}$ at its boundary such that they can block any other translate from touching $\mathbf{K}$ is denoted by $\beta_2(\mathbf{K})$; whereas, $\beta_3(\mathbf{K})$ denotes the smallest number of translates all of which are non-overlapping with **K** such that they can block any other translate from touching **K** [82]. If in the original definition of blocking number, translates are replaced by homothets with homothety ratio $\alpha > 0$ we get the *generalized blocking number* $\beta^{\alpha}(\mathbf{K})$ [39], and if we allow the homothets to overlap, we get the *generalized α-blocking number* $\beta_2^{\alpha}(\mathbf{K})$ [81].

Recently, Wu [81] showed that if **K** and **L** are **o**-symmetric convex bodies that are sufficiently close to each other in the Banach–Mazur sense[7] then there exists $\alpha > 0$ (depending on **K**) such that

$$I(\mathbf{K}) \leq \beta_2^{\alpha}(\mathbf{L}).$$

This gives a series of upper bounds on the illumination number of symmetric convex bodies and a possible way to circumvent the lack of lower semicontinuity of $I(\cdot)$ (see Sect. 4.1 for a discussion of the continuity of the illumination number).

3.3.3 Fractional Covering and Illumination

Naszódi [66] introduced the fractional illumination number and Arstein-Avidan with Raz [3] and with Slomka [4] introduced weighted covering numbers. Both formalisms can be used to study a fractional analogue of the illumination problem. In fact, the Fractional Illumination Conjecture for **o**-symmetric convex bodies was proved in [66], while the case of equality was characterized in [4]. We omit the details as it would lead to a lengthy diversion from the main subject matter.

4 Quantifying Illumination and Covering

4.1 The Illumination and Covering Parameters

It can be seen that in the definition of illumination number $I(\mathbf{K})$, the distance of light sources from **K** plays no role whatsoever. Starting with a relatively small number of light sources, it makes sense to quantify how far they need to be from **K** in order to illuminate it. This is the idea behind the *illumination parameter* as defined by the first author [11].

Let **K** be an **o**-symmetric convex body. Then the norm of $x \in \mathbb{E}^d$ generated by K is defined as

$$\|\mathbf{x}\|_{\mathbf{K}} = \inf\{\lambda > 0 : \mathbf{x} \in \lambda\mathbf{K}\}$$

and provides a good estimate of how far a point **x** is from **K**.

[7] See relation (6) and the discussion preceding it in Sect. 4.1 for an introduction to the Banach–Mazur distance of convex bodies. Note that Wu uses the Hausdorff distance between convex bodies to state his result. However, it can be shown that **K** and **L** are close to each other in the Banach–Mazur sense if and only if there exist affine images of them that are close in the Hausdorff sense. Since the illumination and blocking numbers are affine invariants, we can restate Wu's results in the language of Banach–Mazur distance.

The illumination parameter $\mathrm{ill}(\mathbf{K})$ of an **o**-symmetric convex body **K** estimates how well **K** can be illuminated by relatively few point sources lying as close to **K** on average as possible.

$$\mathrm{ill}(\mathbf{K}) = \inf\left\{\sum_i \|\mathbf{p_i}\|_K : \{\mathbf{p_i}\} \text{ illuminates } \mathbf{K}, \mathbf{p_i} \in \mathbb{E}^d\right\},$$

Clearly, $I(\mathbf{K}) \leq \mathrm{ill}(\mathbf{K})$ holds for any **o**-symmetric convex body **K**. In the papers [15, 56], the illumination parameters of **o**-symmetric Platonic solids have been determined. In [11] a tight upper bound was obtained for the illumination parameter of planar **o**-symmetric convex bodies.

Theorem 4.1 *If* **K** *is an* **o**-*symmetric planar convex body, then* $\mathrm{ill}(\mathbf{K}) \leq 6$ *with equality for any affine regular convex hexagon.*

The corresponding problem in dimension 3 and higher is wide open. The following conjecture is due to Kiss and de Wet [56].

Conjecture 4.2 *The illumination parameter of any* **o**-*symmetric 3-dimensional convex body is at most 12.*

However, for smooth **o**-symmetric convex bodies in any dimension $d \geq 2$, the first named author and Litvak [19] found an upper bound, which was later improved to the following asymptotically sharp bound by Gluskin and Litvak [48].

Theorem 4.3 *For any smooth* **o**-*symmetric d-dimensional convex body* **K**,

$$\mathrm{ill}(\mathbf{K}) \leq 24d^{3/2}.$$

Translating the above quantification ideas from illumination into the setting of covering, Swanepoel [76] introduced the *covering parameter* of a convex body as follows.

$$C(\mathbf{K}) = \inf\left\{\sum_i (1-\lambda_i)^{-1} : \mathbf{K} \subseteq \bigcup_i (\lambda_i \mathbf{K} + \mathbf{t_i}), 0 < \lambda_i < 1, \mathbf{t_i} \in \mathbb{E}^d\right\}.$$

Thus large homothets are penalized in the same way as the far off light sources are penalized in the definition of illumination parameter. Note that here **K** need not be **o**-symmetric. In the same paper, Swanepoel obtained the following Rogers-type upper bounds on $C(\mathbf{K})$ when $d \geq 2$.

Theorem 4.4

$$C(\mathbf{K}) < \begin{cases} e2^d d(d+1)(\ln d + \ln\ln d + 5) = O(2^d d^2 \ln d), & \text{if } \mathbf{K} \text{ is } o\text{-symmetric}, \\ e\binom{2d}{d} d(d+1)(\ln d + \ln\ln d + 5) = O(4^d d^{3/2} \ln d), & \text{otherwise.} \end{cases} \quad (4)$$

He further showed that if K is o-symmetric, then

$$\mathrm{ill}(\mathbf{K}) \leq 2C(\mathbf{K}), \tag{5}$$

and therefore, $\mathrm{ill}(\mathbf{K}) = O(2^d d^2 \ln d)$.

Based on the above results, it is natural to study the following quantitative analogue of the illumination conjecture that was proposed by Swanepoel [76].

Conjecture 4.5 (Quantitative Illumination Conjecture) *For any* **o**-*symmetric d-dimensional convex body* **K**, $\mathrm{ill}(\mathbf{K}) = O(2^d)$.

Before proceeding further, we introduce some terminology and notations. Let us use $\mathcal{K}^d$ and $\mathcal{C}^d$ respectively to denote the set of all d-dimensional convex bodies and the set of all such bodies that are **o**-symmetric. In this section, we consider some of the important properties of the illumination number and the covering parameter as functionals defined on $\mathcal{K}^d$ and the illumination parameter as a functional on $\mathcal{C}^d$. The first observation is that the three quantities are affine invariants (as are several other quantities dealing with the covering and illumination of convex bodies). That is, if $A : \mathbb{E}^d \to \mathbb{E}^d$ is an affine transformation and **K** is any d-dimensional convex body, then $I(\mathbf{K}) = I(A(\mathbf{K}))$, $\mathrm{ill}(\mathbf{K}) = \mathrm{ill}(A(\mathbf{K}))$ and $C(\mathbf{K}) = C(A(\mathbf{K}))$.

Due to this affine invariance, whenever we refer to a convex body **K**, whatever we say about the covering and illumination of **K** is true for all affine images of **K**. In the sequel, $\mathbf{B}^d$ denotes a d-dimensional unit ball[8], $\mathbf{C}^d$ a d-dimensional cube and ℓ a line segment (which is a convex body in $\mathcal{K}^1$) up to an affine transformation.

The Banach–Mazur distance d_{BM} provides a multiplicative metric[9] on $\mathcal{K}^d$ and is used to study the continuity properties of affine invariant functionals on $\mathcal{K}^d$. For $\mathbf{K}, \mathbf{L} \in \mathcal{K}^d$, it is given by

$$d_{BM}(\mathbf{K}, \mathbf{L}) = \inf \{\delta \geq 1 : \mathbf{L} - \mathbf{b} \subseteq T(\mathbf{K} - \mathbf{a}) \subseteq \delta(\mathbf{L} - \mathbf{b}), \mathbf{a} \in \mathbf{K}, \mathbf{b} \in \mathbf{L}\}, \tag{6}$$

where the infimum is taken over all invertible linear operators $T : \mathbb{E}^d \to \mathbb{E}^d$ [72, p. 589].

In the remainder of this paper, $\mathcal{K}^d$ (resp., $\mathcal{C}^d$) is considered as a metric space under the Banach–Mazur distance. Since continuity of a functional can provide valuable insight into its behaviour, it is of considerable interest to check the continuity of $I(\cdot)$, $\mathrm{ill}(\cdot)$ and $C(\cdot)$. Unfortunately, by Example 4.6, the first two quantities are known to be discontinuous, while nothing is known about the continuity of the third.

Example 4.6 (*Smoothed cubes and spiky balls*) In $\mathcal{K}^d$, consider a sequence $(\mathbf{C_n})_{n \in \mathbb{N}}$ of 'smoothed' d-dimensional cubes that approaches $\mathbf{C}^d$ in the Banach–Mazur sense.

---

[8]Without loss of generality, we can assume $\mathbf{B}^d$ to be a unit ball centred at the origin. In what follows, we use the symbol $\mathbf{B}^d$ to denote a d-dimensional unit ball as well as its affine images called ellipsoids.

[9]One can turn the Banach–Mazur distance into an additive metric by applying $\ln(\cdot)$. However, we make no attempt to do that.

Since the smoothed cubes are smooth convex bodies, all the terms of the sequence have illumination number $d + 1$. However, $I(\mathbf{C}^d) = 2^d$, which shows that $I(\cdot)$ is not continuous.

Recently, Naszódi [67] constructed a class of d-dimensional $\mathbf{o}$-symmetric bodies, that he refers to as 'spiky balls'. Pick N points $\mathbf{x}_1, \ldots, \mathbf{x}_N$ independently and uniformly with respect to the Haar probability measure on the $(d-1)$-dimensional unit sphere $\mathbb{S}^{d-1}$ centered at the origin $\mathbf{o}$. Then a spiky ball corresponding to a real number $D > 1$ is defined as

$$\mathbf{K} = \operatorname{conv}\left(\{\pm\mathbf{x}_i : i = 1, \ldots, N\} \cup \frac{1}{D}\mathbf{B}^d\right).$$

Straightaway we observe that K is $\mathbf{o}$-symmetric and satisfies $d_{BM}(\mathbf{K}, \mathbf{B}^d) < D$. Naszódi showed that $I(\mathbf{K}) \geq c^d$, where $c > 1$ is a constant depending on d and D. Thus we have a sequence of spiky balls approaching $\mathbf{B}^d$ in Banach–Mazur distance such that each spiky ball has an exponential illumination number. Since by Theorem 4.3, $\mathrm{ill}(\mathbf{B}^d) = O(d^{3/2})$ and $\mathrm{ill}(\mathbf{K}) \geq I(\mathbf{K})$ we see that $\mathrm{ill}(\cdot)$ is not continuous.

We can state the following about the continuity of $I(\cdot)$ [35].

Theorem 4.7 *The functional $I(\cdot)$ is upper semicontinuous*[10] *on $\mathbb{E}^d$, for all $d \geq 2$.*

Despite the usefulness of the covering parameter, not much is known about it. For instance, we do not know whether $C(\cdot)$ is lower or upper semicontinuous on $\mathcal{K}^d$ and the only known exact value is $C(\mathbf{C}^d) = 2^{d+1}$. Thus there is a need to propose a more refined quantitative version of homothetic covering for convex bodies. Section 4.2 describes how we address this need.

4.2 The Covering Index

The concepts and results presented in this section appear in our recent paper [24]. As stated at the end of Sect. 4.1, the aim here is to come up with a more refined quantification of covering in terms of the covering index with the Covering Conjecture as the eventual goal. The *covering index* of a convex body $\mathbf{K}$ in $\mathbb{E}^d$ combines the notions of the covering parameter $C(\mathbf{K})$ and the m-covering number $\gamma_m(\mathbf{K})$ under the unusual, but highly useful, constraint $\gamma_m(\mathbf{K}) \leq 1/2$, where $\gamma_m(\mathbf{K}) = \inf\left\{\lambda > 0 : \mathbf{K} \subseteq \bigcup_{i=1}^{m}(\lambda\mathbf{K} + \mathbf{t}_i), \mathbf{t}_i \in \mathbb{E}^d, i = 1, \ldots, m\right\}$ is the smallest positive homothety ratio needed to cover $\mathbf{K}$ by m positive homothets. (See Sect. 5 for a detailed discussion of $\gamma_m(\cdot)$.)

[10] Again, the original statement of this result is in terms of Hausdorff distance. However, based on the discussion in footnote 7, we can use the Banach–Mazur distance instead.

Definition 1 (*Covering index*) Let $\mathbf{K}$ be a d-dimensional convex body. We write $N_\lambda(\mathbf{K})$ to denote the covering number $N(\mathbf{K}, \lambda\mathbf{K})$, for any $0 < \lambda \leq 1$. We define the *covering index* of $\mathbf{K}$ as

$$\begin{aligned}\operatorname{coin}(\mathbf{K}) &= \inf\left\{\frac{m}{1-\gamma_m(\mathbf{K})} : \gamma_m(\mathbf{K}) \leq 1/2,\, m \in \mathbb{N}\right\}\\ &= \inf\left\{\frac{N_\lambda(\mathbf{K})}{1-\lambda} : 0 < \lambda \leq 1/2\right\}.\end{aligned}$$

Intuitively, $\operatorname{coin}(\mathbf{K})$ measures how $\mathbf{K}$ can be covered by a relatively small number of positive homothets all corresponding to the same relatively small homothety ratio. The reader may be a bit surprised to see the restriction $\gamma_m(\mathbf{K}) \leq 1/2$. In [24], it was observed that if we start with $\gamma_m(\mathbf{K}) \leq \lambda < 1$, for some λ close to 1 in the definition of $\operatorname{coin}(\mathbf{K})$ and then decrease λ, the properties of the resulting quantity significantly change when $\lambda = 1/2$, at which point we can say a lot about the continuity and maximum and minimum values of the quantity. It was also observed that decreasing λ further does not change these characteristics. Thus $1/2$ can be thought of as a threshold at which the characteristics of the covering problem change.

Note[11] that for $\mathbf{K} \in \mathcal{C}^d$,

$$I(\mathbf{K}) \leq \operatorname{ill}(\mathbf{K}) \leq 2C(\mathbf{K}) \leq 2\operatorname{coin}(\mathbf{K}),$$

and in general for $\mathbf{K} \in \mathcal{K}^d$,

$$I(\mathbf{K}) \leq C(\mathbf{K}) \leq \operatorname{coin}(\mathbf{K}).$$

The following result shows that a lot can be said about the Banach–Mazur continuity of $\operatorname{coin}(\cdot)$. Based on this, $\operatorname{coin}(\cdot)$ seems to be the 'nicest' of all the functionals of covering and illumination of convex bodies discussed here.

Theorem 4.8 *Let d be any positive integer.*

(i) Define $I_{\mathbf{K}} = \{i : \gamma_i(\mathbf{K}) \leq 1/2\} = \{i : \mathbf{K} \in \mathcal{K}_i^d\}$, for any d-dimensional convex body $\mathbf{K}$. If $I_{\mathbf{L}} \subseteq I_{\mathbf{K}}$, for some $\mathbf{K}, \mathbf{L} \in \mathcal{K}^d$, then $\operatorname{coin}(\mathbf{K}) \leq \frac{2d_{BM}(\mathbf{K},\mathbf{L})-1}{d_{BM}(\mathbf{K},\mathbf{L})}\operatorname{coin}(\mathbf{L}) \leq d_{BM}(\mathbf{K},\mathbf{L})\operatorname{coin}(\mathbf{L})$.

(ii) The functional $\operatorname{coin} : \mathcal{K}^d \to \mathbb{R}$ is lower semicontinuous for all d.

(iii) Define $\mathcal{K}^{d*} := \{\mathbf{K} \in \mathcal{K}^d : \gamma_m(\mathbf{K}) \neq 1/2,\, m \in \mathbb{N}\}$. Then the functional $\operatorname{coin} : \mathcal{K}^{d*} \to \mathbb{R}$ is continuous for all d.

We now present some results showing that $\operatorname{coin}(\cdot)$ behaves very nicely with forming direct sums, Minkowski sums and cylinders of convex bodies, making it possible to compute the exact values and estimates of $\operatorname{coin}(\cdot)$ for higher dimensional convex bodies from the covering indices of lower dimensional convex bodies.

[11] In fact, one can obtain $\operatorname{ill}(\mathbf{K}) \leq \frac{3}{2}\operatorname{coin}(\mathbf{K})$ by suitably modifying the proof of (5) which appears in [76].

Theorem 4.9

(i) *Let* $\mathbb{E}^d = \mathbb{L}_1 \oplus \cdots \oplus \mathbb{L}_n$ *be a decomposition of* $\mathbb{E}^d$ *into the direct vector sum of its linear subspaces* $\mathbb{L}_i$ *and let* $\mathbf{K_i} \subseteq \mathbb{L}_i$ *be convex bodies such that* $\Gamma = \max\{\gamma_{m_i}(\mathbf{K_i}) : 1 \leq i \leq n\}$. *Then*

$$\max_{1\leq i\leq n}\{\mathrm{coin}(\mathbf{K_i})\} \leq \mathrm{coin}(\mathbf{K_1} \oplus \cdots \oplus \mathbf{K_n}) \leq \inf_{\lambda\leq\frac{1}{2}} \frac{\prod_{i=1}^n N_\lambda(\mathbf{K_i})}{1-\lambda} \leq \frac{\prod_{i=1}^n N_\Gamma(\mathbf{K_i})}{1-\Gamma} < \prod_{i=1}^n \mathrm{coin}(\mathbf{K_i}). \tag{7}$$

(ii) *The first two upper bounds in* (7) *are tight. Moreover, the second inequality in* (7) *becomes an equality if any* $n-1$ *of the* $\mathbf{K_i}$*'s are tightly covered*

(iii) *Recall that* $\ell \in \mathcal{K}^1$ *denotes a line segment. If* $\mathbf{K}$ *is any convex body, then* $\mathrm{coin}(\mathbf{K} \oplus \ell) = 4N_{1/2}(\mathbf{K})$.

(iv) *Let the convex body K be the Minkowski sum of the convex bodies* $\mathbf{K_1}, \ldots, \mathbf{K_n} \in \mathcal{K}^d$ *and* Γ *be as in part (i). Then*

$$\mathrm{coin}(\mathbf{K}) \leq \inf_{\lambda\leq\frac{1}{2}} \frac{\prod_{i=1}^n N_\lambda(\mathbf{K_i})}{1-\lambda} \leq \frac{\prod_{i=1}^n N_\Gamma(\mathbf{K_i})}{1-\Gamma} < \prod_{i=1}^n \mathrm{coin}(\mathbf{K_i}). \tag{8}$$

The notion of *tightly covered convex bodies* introduced in [24] plays a critical role in Theorem 4.9(ii)–(iii).

Definition 2 We say that a convex body $\mathbf{K} \in \mathcal{K}^d$ is *tightly covered* if for any $0 < \lambda < 1$, $\mathbf{K}$ contains $N_\lambda(\mathbf{K})$ points no two of which belong to the same homothet of $\mathbf{K}$ with homothety ratio λ.

In [24], it was noted that not all convex bodies are tightly covered (e.g., $\mathbf{B}^2$ is not), $\ell \in \mathcal{K}^1$ is tightly covered and so is the d-dimensional cube $\mathbf{C}^d$, for any $d \geq 2$. Do other examples exist?

Problem 2 *For some* $d \geq 2$, *find a tightly covered convex body* $\mathbf{K} \in \mathcal{K}^d$ *other than* $\mathbf{C}^d$ *or show that no such convex body exists.*

Since coin is a lower semicontinuous functional defined on the compact space $\mathcal{K}^d$, it is guaranteed to achieve its infimum over $\mathcal{K}^d$. It turns out that in addition to determining minimizers in all dimensions, we can also find a maximizer in the planar case.

Theorem 4.10

(i) *Let* d *be any positive integer and* $\mathbf{K} \in \mathcal{K}^d$. *Then* $\mathrm{coin}(\mathbf{C}^d) = 2^{d+1} \leq \mathrm{coin}(\mathbf{K})$ *and thus* d*-cubes minimize the covering index in all dimensions.*

(ii) *If* $\mathbf{K}$ *is a planar convex body then* $\mathrm{coin}(\mathbf{K}) \leq \mathrm{coin}(\mathbf{B}^2) = 14$.

Since $\mathbf{B}^2$ maximizes the covering index in the plane, it can be asked if the same is true for $\mathbf{B}^d$ in higher dimensions.

Problem 3 *For any d-dimensional convex body* $\mathbf{K}$, *prove or disprove that* $\operatorname{coin}(\mathbf{K}) \leq \operatorname{coin}(\mathbf{B}^d)$ *holds.*

Since $\operatorname{coin}(\mathbf{B}^d) = O(2^d d^{3/2} \ln d)$ [24], a positive answer to Problem 3 would considerably improve the best known upper bound on the illumination number $I(\mathbf{K}) = O(4^d \sqrt{d} \ln d)$ when $\mathbf{K}$ is a general d-dimensional convex body to within a factor $\sqrt{d}$ of the bound $I(\mathbf{K}) = O(2^d d \ln d)$ when $\mathbf{K}$ is $\mathbf{o}$-symmetric. This gives us a way to closing in on the Illumination Conjecture for general convex bodies.

If we replace the restriction $\gamma_m(\mathbf{K}) \leq 1/2$ from the definition of the covering index with the more usual $\gamma_m(\mathbf{K}) < 1$, the resulting quantity is called the *weak covering index*, denoted by $\operatorname{coin}_{\mathrm{w}}(\mathbf{K})$. As the name suggests, the weak covering index loses some of the most important properties of the covering index. For instance, no suitable analogue of Theorem 4.9 (iii) exists for $\operatorname{coin}_{\mathrm{w}}(\cdot)$. As a result, we can only estimate the weak covering index of cylinders. Also the discussed aspects of continuity of the covering index seem to be lost for the weak covering index. Last, but not the least, unlike the covering index we cannot say much at all about the maximizers and minimizers of the weak covering index.

In the end, we would like to mention that fractional analogues of the covering index and the weak covering index were introduced in [25]. Just like fractional illumination number, we do not discuss these here due to limitation of space.

4.3 Cylindrical Covering Parameters

So far in Sects. 4.1 and 4.2, we have discussed some quantitative versions of the illumination number. The aim of this section is to introduce a quantification of the X-ray number. This quantification has the added advantage of connecting the X-ray problem with the Tarski's plank problem and its relatives (see [22, Chap. 4]).

Given a linear subspace $E \subseteq \mathbb{E}^d$ we denote the orthogonal projection on E by P_E and the orthogonal complement of E by $E^{\perp}$. Given $0 < k < d$, define a k-*codimensional cylinder* $\mathbf{C}$ as a set, which can be presented in the form $\mathbf{C} = B + H$, where H is a k-dimensional linear subspace of $\mathbb{E}^d$ and B is a measurable set in $E := H^{\perp}$. Given a convex body $\mathbf{K}$ and a k-codimensional cylinder $\mathbf{C} = B + H$ denote the cross-sectional volume of $\mathbf{C}$ with respect to $\mathbf{K}$ by

$$\operatorname{crv}_{\mathbf{K}}(\mathbf{C}) := \frac{\operatorname{vol}_{d-k}(\mathbf{C} \cap E)}{\operatorname{vol}_{d-k}(P_E \mathbf{K})} = \frac{\operatorname{vol}_{d-k}(P_E \mathbf{C})}{\operatorname{vol}_{d-k}(P_E \mathbf{K})} = \frac{\operatorname{vol}_{d-k}(B)}{\operatorname{vol}_{d-k}(P_E \mathbf{K})}.$$

We note that if $T : \mathbb{E}^d \to \mathbb{E}^d$ is an invertible affine map, then $\operatorname{crv}_{\mathbf{K}}(\mathbf{C}) = \operatorname{crv}_{T(\mathbf{K})}(T(\mathbf{C}))$. Now we introduce the following.

Definition 3 (*k-th Cylindrical Covering Parameter*). Let $0 < k < d$ and $\mathbf{K}$ be a convex body in $\mathbb{E}^d$. Then the k-th cylindrical covering parameter of $\mathbf{K}$ is labelled by $\mathrm{cyl}_k(\mathbf{K})$ and it is defined as follows:

$$\mathrm{cyl}_k(\mathbf{K}) = \inf_{\bigcup_{i=1}^n \mathbf{C}_i} \left\{ \sum_{i=1}^n \mathrm{crv}_{\mathbf{K}}(\mathbf{C}_i) \ : \ \mathbf{K} \subseteq \bigcup_{i=1}^n \mathbf{C}_i, \ \mathbf{C}_i \text{ is a } k \right.$$
$$\left. - \text{codimensional cylinder}, \ i = 1, \ldots, n \right\}.$$

We observe that if $T : \mathbb{E}^d \to \mathbb{E}^d$ is an invertible affine map, then $\mathrm{cyl}_k(\mathbf{K}) = \mathrm{cyl}_k(T(\mathbf{K}))$. Furthermore, it is clear that $\mathrm{cyl}_k(\mathbf{K}) \leq 1$ holds for any convex body $\mathbf{K}$ in $\mathbb{E}^d$ and for any $0 < k < d$. In terms of X-raying, one can think of $\mathrm{cyl}_k(\mathbf{K})$ as the minimum of the 'sum of sizes' of $(d-k)$-dimensional X-raying windows needed to X-ray $\mathbf{K}$.

Recall that a $(d-1)$-codimensional cylinder of $\mathbb{E}^d$ is also called a *plank* for the reason that it is the set of points lying between two parallel hyperplanes in $\mathbb{E}^d$. The width of a plank is simply the distance between the two parallel hyperplanes. In a remarkable paper [6], Bang has given an elegant proof of the Plank Conjecture of Tarski showing that if a convex body is covered by finitely many planks in $\mathbb{E}^d$, then the sum of the widths of the planks is at least as large as the minimal width of the body, which is the smallest distance between two parallel supporting hyperplanes of the given convex body. A celebrated extension of Bang's theorem to d-dimensional normed spaces has been given by Ball in [5]. In his paper [6], Bang raises his so-called Affine Plank Conjecture, which in terms of our notation can be phrased as follows.

Conjecture 4.11 (Affine Plank Conjecture) *If* $\mathbf{K}$ *is a convex body in* $\mathbb{E}^d$, *then* $\mathrm{cyl}_{d-1}(\mathbf{K}) = 1$.

Now, Ball's celebrated Plank theorem [5] can be stated as follows.

Theorem 4.12 *If* $\mathbf{K}$ *is an* $\mathbf{o}$*-symmetric convex body in* $\mathbb{E}^d$, *then* $\mathrm{cyl}_{d-1}(\mathbf{K}) = 1$.

Bang [6] also raised the important related question of whether the sum of the base areas of finitely many (1-codimensional) cylinders covering a 3-dimensional convex body is at least half of the minimum area of a 2-dimensional projection of the body. This, in terms of our terminology, reads as follows.

Conjecture 4.13 1-(Codimensional Cylinder Covering Conjecture) *If* $\mathbf{K}$ *is a convex body in* $\mathbb{E}^3$, *then* $\mathrm{cyl}_1(\mathbf{K}) \geq \frac{1}{2}$.

If true, then Bang's estimate is sharp due to a covering of a regular tetrahedron by two cylinders described in [6]. In connection with Conjecture 4.13 the first named author and Litvak have proved the following general estimates in [20].

Theorem 4.14 *Let* $0 < k < d$ *and* $\mathbf{K}$ *be a convex body in* $\mathbb{E}^d$. *Then* $\mathrm{cyl}_k(\mathbf{K}) \geq \frac{1}{\binom{d}{k}}$.

Furthermore, it is proved in [20] that if $\mathbf{K}$ is an ellipsoid in $\mathbb{E}^d$, then $\text{cyl}_1(\mathbf{K}) = 1$. Akopyan et al. [2] have recently proved that if $\mathbf{K}$ is an ellipsoid in $\mathbb{E}^d$, then $\text{cyl}_2(\mathbf{K}) = 1$. They have put forward:

Conjecture 4.15 (Ellipsoid Conjecture) *If* $\mathbf{K}$ *is an ellipsoid in* $\mathbb{E}^d$, *then* $\text{cyl}_k(\mathbf{K}) = 1$ *for all* $2 < k < d$.

5 A Computer-Based Approach

Given a positive integer m, Lassak [58] introduced the *m-covering number* of a convex body $\mathbf{K}$ as the minimal positive homothety ratio needed to cover $\mathbf{K}$ by m positive homothets. That is,

$$\gamma_m(\mathbf{K}) = \inf\left\{\lambda > 0 : \mathbf{K} \subseteq \bigcup_{i=1}^{m}(\lambda\mathbf{K} + \mathbf{t_i}), \mathbf{t_i} \in \mathbb{E}^d, i = 1, \ldots, m\right\}.$$

Lassak showed that the m-covering number is well-defined and studied the special case $m = 4$ for planar convex bodies. It should be noted that special values of this quantity had been considered by several authors in the past. For instance, in the 70's and 80's the first named author showed that $\gamma_5(\mathbf{B}^2) = 0.609382\ldots$[12] and $\gamma_6(\mathbf{B}^2) = 0.555905\ldots$ [7, 8].

Zong [85] studied $\gamma_m : \mathcal{K}^d \to \mathbb{R}$ as a functional and proved it to be uniformly continuous for all m and d. He did not use the term m-covering number for $\gamma_m(\mathbf{K})$ and simply referred to it as the smallest positive homothety ratio. In [24], we proved the following stronger result.

Theorem 5.1 *For any* $K, L \in \mathcal{K}^d$, $\gamma_m(K) \leq d_{BM}(K, L)\gamma_m(L)$ *holds and so* γ_m *is Lipschitz continuous on* $\mathcal{K}^d$ *with* $\frac{d^2-1}{2\ln d}$ *as a Lipschitz constant and*

$$|\gamma_m(K) - \gamma_m(L)| \leq d_{BM}(K, L) - 1 \leq \frac{d^2 - 1}{2\ln d}\ln\left(d_{BM}(K, L)\right),$$

for all $K, L \in \mathcal{K}^d$.

Further properties and some variants of $\gamma_m(\cdot)$ are discussed in the recent papers [53, 80]. For instance, it has been shown in [80] that for any d-dimensional convex polytope $\mathbf{P}$ with m vertices, we have

$$\gamma_m(\mathbf{K}) \leq \frac{d-1}{d}.$$

[12]*Cover the Spot* is a popular carnival game in the United States. The objective is to cover a given circular spot by 5 circular disks of smaller radius. It seems that by determining $\gamma_5(\mathbf{B}^2)$, the first named author was unwittingly providing the optimal solution for *Cover the Spot*!

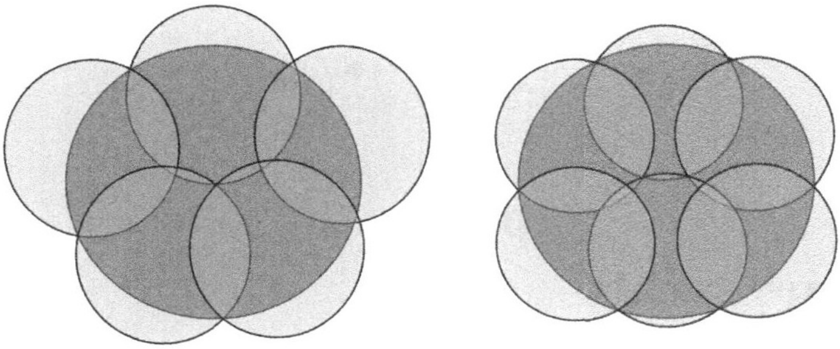

Fig. 6 Optimal configurations that demonstrate $\gamma_5(\mathbf{B}^2) = 0.609382\ldots$ and $\gamma_6(\mathbf{B}^2) = 0.555905\ldots$

Obviously, any $\mathbf{K} \in \mathcal{K}^d$ can be covered by 2^d smaller positive homothets if and only if $\gamma_{2^d}(\mathbf{K}) < 1$. Zong used these ideas to propose a possible computer-based approach to attack the Covering Conjecture [85] (Fig. 6).

Recall that in a metric space, such as $\mathcal{K}^d$, an ϵ-*net* ξ is a finite or infinite subset of $\mathcal{K}^d$ such that the union of closed balls of radius ϵ centered at elements of ξ covers the whole space. Thus if an ϵ-net exists, any element of $\mathcal{K}^d$ is within Banach–Mazur distance ϵ of some element of the cover. The key idea of the procedure proposed by Zong is the construction of a finite ϵ-net of $\mathcal{K}^d$ whose elements are convex polytopes, for every real number[13] $\epsilon > 1$ and positive integer d. Here we describe the construction briefly.

We first take an affine image of a d-dimensional convex body $\mathbf{K}$ that is sandwiched between the unit ball $\mathbf{B^d}$ centered at the origin and the ball $d\mathbf{B^d}$ with radius d. Such an image always exists by John's ellipsoid theorem. Then we take a covering $\{C_1, \ldots, C_m\}$ of the boundary of $d\mathbf{B^d}$ with spherical caps C_i as shown in Fig. 7. The centers of the caps C_i are joined to the origin by lines $\{L_1, \ldots, L_m\}$ and a large number of equidistant points are taken on the lines L_i. We denote by $\mathbf{p_i}$ the point lying in $\mathbf{K} \cap L_i$ that is farthest from the origin. Then the convex hull $\mathbf{P} = \mathrm{conv}\{\mathbf{p_i}\}$ is the required element of our ϵ-net. Zong [85] showed that by taking m large enough and increasing the number of points on L_i we can ensure $d_{BM}(\mathbf{K}, \mathbf{P}) \leq \epsilon$.

He then notes that if we manage to construct a finite ϵ-net $\xi = \{\mathbf{P_i} : i = 1, \ldots, j\}$ of $\mathcal{K}^d$, satisfying $\gamma_{2^d}(\mathbf{P_i}) \leq c_d$ for some $c_d < 1$ and sufficiently small ϵ, then $\gamma_{2^d}(\mathbf{K}) < 1$ would hold for every $\mathbf{K} \in \mathcal{K}^d$. This would imply that the Covering Conjecture is true in dimension d.

The following is a four step approach suggested by Zong [85].

Zong's Program:

1. For a given dimension such as $d = 3$, investigate (with the assistance of a computer) $\gamma_{2^d}(\mathbf{K})$ for some particular convex bodies $\mathbf{K}$ and choose a candidate constant c_d.
2. Choose a suitable ϵ.

[13]Recall that the Banach–Mazur distance is a multiplicative metric and so the condition $\epsilon > 0$ is replaced by the equivalent $\epsilon > 1$ condition.

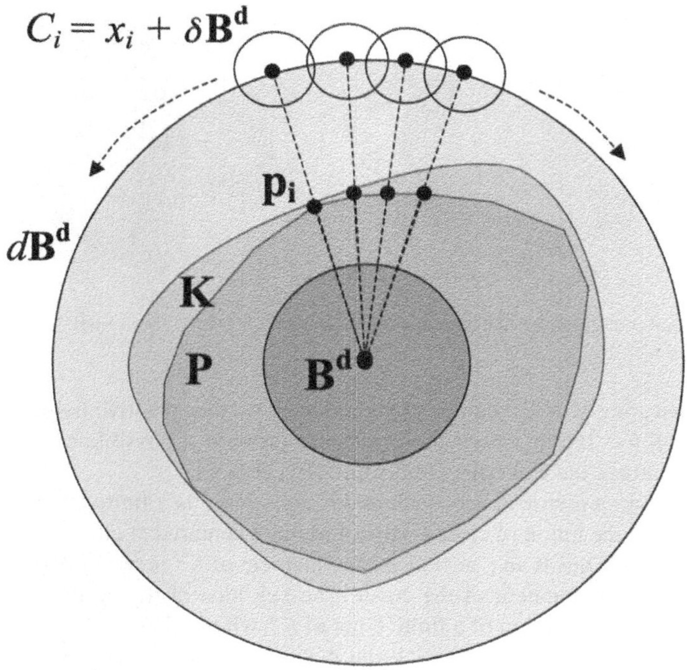

Fig. 7 Construction of an ϵ-net of $\mathcal{K}^d$

3. Construct an ϵ-net ξ of sufficiently small cardinality.
4. Check (with the assistance of a computer) that the minimal γ_{2^d}-value over all elements of ξ is bounded above by c_d.

Indeed this approach appears to be promising and, to the authors' knowledge, is a first attempt at a computer-based resolution of the Covering Conjecture. However, Zong's program is not without its pitfalls. For one, it would take an extensive computational experiment to come up with a good candidate constant c_d. Secondly, Zong's ϵ-net construction leads to a net with exponentially large number of elements. In fact, using Böröczky and Wintsche's estimate [40] on the number of caps in a spherical cap covering, Zong [85] showed that

$$|\xi| \leq \left\lfloor \frac{7d}{\ln \epsilon} \right\rfloor^{\alpha 14^d d^{2d+3} (\ln \epsilon)^{-d}}, \tag{9}$$

where c is an absolute constant. Since Zong's construction does not provide much room for improving the above estimate, better constructions are needed to reduce the size of ξ, while at the same time keeping ϵ sufficiently small.

Problem 4 *Develop a computationally efficient procedure for constructing ϵ-nets of $\mathcal{K}^d$ of small cardinality.*

Addressing the above problem would be a critical first step in implementing Zong's program. Wu [80] (also see [53]) has recently proposed two variants of $\gamma_m(\cdot)$ that can be used in Zong's program instead. However, the challenges and implementation issues remain the same.

Acknowledgements The first author is partially supported by a Natural Sciences and Engineering Research Council of Canada Discovery Grant. The second author is supported by a Vanier Canada Graduate Scholarship (NSERC) and Alberta Innovates Technology Futures (AITF).

References

1. M. Aigner, G.M. Ziegler, *Proofs from the Book*, 4th edn. (Springer, Berlin, 2010)
2. A. Akopyan, R. Karasev, F. Petrov, Bang's problem and symplectic invariants. 1–15 (2014). arXiv:1404.0871v1 [math.MG]
3. S. Artstein-Avidan, O. Raz, Weighted covering numbers of convex sets. Adv. Math. **227**, 730–744 (2011)
4. S. Artstein-Avidan, B.A. Slomka, On weighted covering numbers and the Levi-Hadwiger conjecture. Israel J. Math. **209**(1), 125–155 (2015)
5. K. Ball, The plank problem for symmetric bodies. Invent. Math. **104**, 535–543 (1991)
6. T. Bang, A solution of the "Plank problem". Proc. Am. Math. Soc. **2**, 990–993 (1951)
7. K. Bezdek, *Körök Optimális Fedései*, Ph.D. thesis, Eötvös Loránd University, Budapest, 1979
8. K. Bezdek, Über einige Kreisüberdeckungen. Beiträge Algebra Geom. **14**, 7–13 (1983)
9. K. Bezdek, The problem of illumination of the boundary of a convex body by affine subspaces. Mathematika **38**, 362–375 (1991)
10. K. Bezdek, Hadwiger's covering conjecture and its relatives. Am. Math. Monthly **99**, 954–956 (1992)
11. K. Bezdek, Research problem 46. Period. Math. Hungar. **24**, 119–121 (1992)
12. K. Bezdek, G. Kiss, M. Mollard, An illumination problem for zonoids. Israel J. Math. **81**, 265–272 (1993)
13. K. Bezdek, On affine subspaces that illuminate a convex set. Beiträge Alg. Geom. **35**, 131–139 (1994)
14. K. Bezdek, T. Bisztriczky, A proof of Hadwiger's covering conjecture for dual cyclic polytopes. Geom. Dedicata **68**, 29–41 (1997)
15. K. Bezdek, K. Böroczky, G. Kiss, On the successive illumination parameters of convex bodies. Period. Math. Hungar. **53**(1–2), 71–82 (2006)
16. K. Bezdek, Hadwiger-Levi's covering problem revisited, in *New Trends in Discrete and Computational Geometry*, ed. by J. Pach (Springer, New York, 1993), pp. 199–233
17. K. Bezdek, T. Zamfirescu, A characterization of 3-dimensional convex sets with an infinite X-ray number, in *Intuitive Geometry, Colloquia Mathematica Societatis Janos Bolyai*, vol. 63 (North-Holland, Amsterdam, 1994) pp. 33–38
18. K. Bezdek, Z. Lángi, M. Naszódi, P. Papez, Ball-polyhedra. Discrete Comput. Geom. **38**(2), 201–230 (2007)
19. K. Bezdek, A.E. Litvak, On the vertex index of convex bodies. Adv. Math. **215**(2), 626–641 (2007)
20. K. Bezdek, A.E. Litvak, Covering convex bodies by cylinders and lattice points by flats. J. Geom. Anal. **19**(2), 233–243 (2009)
21. K. Bezdek, G. Kiss, On the X-ray number of almost smooth convex bodies and of convex bodies of constant width. Can. Math. Bull. **52/3**, 342–348 (2009)
22. K. Bezdek, *Classical Topics in Discrete Geometry* (CMS Books in Mathematics, Springer, New York, 2010)

23. K. Bezdek, Illuminating spindle convex bodies and minimizing the volume of spherical sets of constant width. Discrete Comput. Geom. **47**(2), 275–287 (2012)
24. K. Bezdek, M.A. Khan, On the covering index of convex bodies. Aequationes Math. **90**(5), 879–903 (2016)
25. K. Bezdek, M.A. Khan, Quantitative covering of convex bodies, in *International Workshop on Geometry and Symmetry (GeoSym)*, Veszprém, Hungary, 29 June–3 July, 2015, `http://geosym.mik.uni-pannon.hu/images/presentations/khan.pdf`
26. T. Bisztriczky, Separation in neighbourly 4-polytopes. Studia Sci. Math. Hungar. **39**(3-4), 277–289 (2002)
27. T. Bisztriczky, F. Fodor, A separation theorem for totally-sewn 4-polytopes. Studia Sci. Math. Hungar. **52**(3), 386–422 (2015)
28. B. Bollobás, P.A. Catalin, P. Erdős, Hadwiger's conjecture is true for almost every graph. Eur. J. Combin. **1**(3), 195–199 (1980)
29. V. Boltyanski, The problem of illuminating the boundary of a convex body. Izv. Mold. Fil. AN SSSR **76**, 77–84 (1960)
30. V. Boltyanski, Helly's theorem for H-convex sets. Sov. Math. Dokl. **17**(1), 78–81 (1976)
31. V. Boltyanski, I. Gohberg, *Results and Problems in Combinatorial Geometry* (Cambridge University Press, Cambridge, 1985)
32. V. Boltyanski, I. Gohberg, Stories about covering and illuminating of convex bodies. Nieuw Arch. Wisk. **13**(1), 1–26 (1995)
33. V. Boltyanski, H. Martini, Combinatorial geometry of belt bodies. Results Math. **28**(3–4), 224–249 (1995)
34. V. Boltyanski, A solution of the illumination problem for belt bodies. Mat. Zametki **58**, 505–511 (1996)
35. V. Boltyanski, H. Martini, P.S. Soltan, *Excursions into Combinatorial Geometry* (Springer, New York, 1997)
36. V. Boltyanski, Solution of the illumination problem for three dimensional convex bodies. Dokl. Akad. Nauk **375**, 298–301 (2000)
37. V. Boltyanski, Solution of the illumination problem for bodies with md$M = 2$. Discrete Comput. Geom. **26**, 527–541 (2001)
38. V. Boltyanski, H. Martini, Covering belt bodies by smaller homothetical copies. Beiträge Alg. Geom. **42**, 313–324 (2001)
39. K. Böröczky Jr., D.G. Larman, S. Sezgin, C.M. Zong, On generalized kissing numbers and blocking numbers. Rendiconti del Circolo Mathematico di Palermo, Serie **II**(Suppl. 65), 39–57 (2000)
40. K. Böröczky Jr., G. Wintsche, Covering the sphere by equal spherical balls, in *The Goodman-Pollack Festschrift* (Springer, Berlin, 2003), pp. 237–253
41. K. Borsuk, Drei Sätze über die n-dimensionale euklidische Sphäre. Fund. Math. **20**, 177–190 (1933)
42. L. Dalla, D.G. Larman, P. Mani-Levitska, C. Zong, The blocking numbers of convex bodies. Discrete Comput. Geom. **24**(2-3), 267–277 (2000)
43. B.V. Dekster, Each convex body in $\mathbb{E}^3$ symmetric about a plane can be illuminated by 8 directions. J. Geom. **69**(1-2), 37–50 (2000)
44. P. Erdős, C.A. Rogers, Covering space with convex bodies. Acta Arith. **7**, 281–285 (1962)
45. P. Erdős, C.A. Rogers, The star number of coverings of space with convex bodies. Acta Arith. **9**, 41–45 (1964)
46. G. Fejes Tóth, A note on covering by convex bodies. Canad. Math. Bull. **52**(3), 361–365 (2009)
47. R.J. Gardner, Geometric tomography, in *Encyclopedia of Mathematics and its Applications*, vol. 58 (Cambridge University Press, 2006)
48. E.D. Gluskin, A.E. Litvak, A remark on vertex index of the convex bodies, in *Geometric Aspects of Functional Analysis*, Lecture Notes in Mathematics, vol. 2050, pp. 255–265
49. I.T. Gohberg, A.S. Markus, A certain problem about the covering of convex sets with homothetic ones, in *Izvestiya Moldavskogo Filiala Akademii Nauk SSSR*, vol. 10/76 (1960), pp. 87–90 (In Russian)

50. B. Grünbaum, Borsuk's problem and related questions, in *Proceedings of Symposia in Pure Mathematics*, vol. 7 (American Mathematical Society, Providence, RI, 1963)
51. H. Hadwiger, Ungelöste Probleme Nr. 20. Elem. der Math.**12**, 121 (1957)
52. H. Hadwiger, Ungelöste Probleme Nr. 38. Elem. der Math. **15**, 130–131 (1960)
53. H. He, H. Martini, S. Wu, On covering functionals of convex bodies. J. Math. Anal. Appl. **437**(2), 1236–1256 (2016)
54. J. Kahn, G. Kalai, A counterexample to Borsuk's conjecture. Bull. Am. Math. Soc. **29**, 60–62 (1993)
55. G. Kiss, Illumination problems and codes. Period. Math. Hungar. **39**, 65–71 (1999)
56. G. Kiss, P.O. de Wet, Notes on the illumination parameters of convex polytopes. Contrib. Discrete Math.**7/1**, 58–67 (2009)
57. M. Lassak, Solution of Hadwiger's covering problem for centrally symmetric convex bodies in $\mathbb{E}^3$. J. Lond. Math. Soc. **30**, 501–511 (1984)
58. M. Lassak, Covering a plane convex body by four homothetical copies with the smallest positive ratio. Geom. Dedicata **21**, 157–167 (1986)
59. M. Lassak, Covering the boundary of a convex set by tiles. Proc. Am. Math. Soc. **104**, 269–272 (1988)
60. M. Lassak, Illumination of three-dimensional convex bodies of constant width, in *Proceedings of the 4th International Congress of Geometry* (Aristotle University of Thessaloniki, 1997), pp. 246–250
61. M. Lassak, H. Martini, M. Spirova, On translative coverings of convex bodies. Rocky Mountain J. Math. **44**(4), 1281–1299 (2014)
62. F.W. Levi, Überdeckung eines Eibereiches durch Parallelverschiebungen seines offenen Kerns. Arch. Math. **6**(5), 369–370 (1955)
63. H. Martini, Some results and problems around zonotopes, in *Colloquia Mathematica Societatis Janos Bolyai, Intuitive Geometry, Sioófok 1985*, vol. 48 (North Holland, 1987), pp. 383–418
64. H. Martini, V. Soltan, Combinatorial problems on the illumination of convex bodies. Aequationes Math. **57**, 121–152 (1999)
65. H. Martini, C. Richter, M. Spirova, Illuminating and covering convex bodies. Discrete Math. **337**, 106–118 (2014)
66. M. Naszódi, Fractional illumination of convex bodies. Contrib. Discrete Math. **4**(2), 83–88 (2009)
67. M. Naszódi, A spiky ball. Mathematika **62**(2), 630–636 (2016)
68. I. Papadoperakis, An estimate for the problem of illumination of the boundary of a convex body in $\mathbb{E}^3$. Geom. Dedicata **75**, 275–285 (1999)
69. C.A. Rogers, A note on coverings. Mathematika **4**, 1–6 (1957)
70. C.A. Rogers, G.C. Shephard, The difference body of a convex body. Arch. Math. **8**, 220–233 (1957)
71. C.A. Rogers, C. Zong, Covering convex bodies by translates of convex bodies. Mathematika **44**, 215–218 (1997)
72. R. Schneider, *Convex Bodies: The Brun-Minkowski Theory, Encyclopedia of Mathematics and Its Applications*, vol. 151, 2nd edn. (Cambridge University Press, Cambridge, UK, 2014)
73. O. Schramm, Illuminating sets of constant width. Mathematika **35**, 180–189 (1988)
74. P. Soltan, Helly's theorem for d-convex sets (in Russian). Dokl. Akad. Nauk. SSSR **205/3**, 537–539 (1972)
75. P. Soltan, V. Soltan, Illumination through convex bodies. Dokl. Akad. Nauk. SSSR **286**, 50–53 (1986)
76. K.J. Swanepoel, Quantitative illumination of convex bodies and vertex degrees of geometric Steiner minimal trees. Mathematika **52**(1-2), 47–52 (2005)
77. I. Talata, Solution of Hadwiger-Levi's covering problem for duals of cyclic $2k$-polytopes. Geom. Dedicata **74**, 61–71 (1999)
78. R.G. Trelford, Separation in and X-raying of convex bodies, Ph.D. thesis, University of Calgary, 2014, pp. 1–126
79. B. Weissbach, Invariante Beleuchtung konvexer Körper. Beiträge Alg. Geom. **37**, 9–15 (1996)

80. S. Wu, Two equivalent forms of Hadwiger's covering conjecture. Sci. Sin. Math. **44**(3), 275–285 (2014). (in Chinese)
81. S. Wu, Upper bounds for the covering number of centrally symmetric convex bodies in $\mathbb{R}^n$. Math. Inequal. Appl. **17**(4), 1281–1298 (2014)
82. L. Yu, C. Zong, On the blocking number and the covering number of a convex body. Adv. Geom. **9**(1), 13–29 (2009)
83. L. Yu, Blocking numbers and fixing numbers of convex bodies. Discrete Math. **309**, 6544–6554 (2009)
84. C. Zong, Some remarks concerning kissing numbers, blocking numbers and covering numbers. Period. Math. Hungar. **30**(3), 233–238 (1995)
85. C. Zong, A quantitative program for Hadwigers covering conjecture. Sci. China Math. **53**(9), 2551–2560 (2010)

Stability of the Simplex Bound for Packings by Equal Spherical Caps Determined by Simplicial Regular Polytopes

Károly Böröczky, Károly J. Böröczky, Alexey Glazyrin and Ágnes Kovács

Abstract It is well known that the vertices of any simplicial regular polytope in $\mathbb{R}^d$ determine an optimal packing of equal spherical balls in S^{d-1}. We prove a stability version of optimal order of this result.

Keywords Simplex bound · Packing of equal balls · Spherical space · Simplicial Polytopes · stability

MSC2010 Subject class: 52C17

Research is supported in parts by NKFIH grants 109789, 121649 and 116451.
Research is partially supported by NSF grant DMS-1400876

K. Böröczky · Á. Kovács
Roland Eötvös University,
Pázmány Péter sétány 1/C, H-1117 Budapest, Hungary
e-mail: boroczky@cs.elte.hu

Á. Kovács
e-mail: kovacsa@cs.elte.hu

K. J. Böröczky (✉)
Alfréd Rényi Institute of Mathematics, Hungarian Academy of Sciences,
Reáltanoda u. 13-15, H-1053 Budapest, Hungary
e-mail: carlos@renyi.hu; boroczky.karoly.j@renyi.mta.hu

K. J. Böröczky
Central European University, Nádor u. 9, H-1051 Budapest, Hungary

A. Glazyrin
School of Mathematical & Statistical Sciences,
The University of Texas Rio Grande Valley, One West University Blvd,
Brownsville TX, USA
e-mail: Alexey.Glazyrin@utrgv.edu

M. D. E. Conder et al. (eds.), *Discrete Geometry and Symmetry*,
Springer Proceedings in Mathematics & Statistics 234,
https://doi.org/10.1007/978-3-319-78434-2_2

1 Introduction

Euclidean regular polytopes are in the center of scientific studies since the Antiquity (see [18] or [9]). Packings of equal balls in spaces of constant curvature have been investigated rather intensively since the middle of the 20th century (see [3], [7], [12] and [21]). In this paper, we focus on packings of equal spherical balls (see [8], [11] and [19]) that are related to some Euclidean simplicial regular polytope P with its $f_0(P)$ vertices being on S^{d-1}, $d \geq 3$. We write φ_P to denote the acute angle satisfying that edge length of P is $2\sin\varphi_P$. We note that the simplicial regular polytopes in $\mathbb{R}^d$, $d \geq 3$, are the regular simplex and crosspolytope in all dimensions, and in addition the icosahedron in $\mathbb{R}^3$ and the 600-cell in $\mathbb{R}^4$ (the latter has Schläfli symbol (3, 3, 5)). The corresponding data is summarized in the following table.

Regular Polytope P	$f_0(P)$	φ_P
Simplex in $\mathbb{R}^d$	$d+1$	$\frac{1}{2}\arccos\frac{-1}{d}$
Crosspolytope in $\mathbb{R}^d$	$2d$	$\frac{\pi}{4}$
Icosahedron in $\mathbb{R}^3$	12	$\frac{1}{2}\arccos\frac{1}{\sqrt{5}}$
600-cell in $\mathbb{R}^4$	120	$\frac{\pi}{10}$

Theorem A *If P is a simplicial regular polytope in $\mathbb{R}^d$ having its vertices on S^{d-1}, $d \geq 3$, then the vertices are centers of an optimal packing of equal spherical balls of radius φ_P on S^{d-1}.*

Theorem A is due to Jung [17] if P is a regular simplex. For the case of a regular crosspolytope, the statement of Theorem A was proposed as a problem by Davenport and Hajós [10]. Numerous solutions arrived in a relatively short time; namely, the ones by Aczél [1] and by Szele [22] and the unpublished ones due to M. Bognár, Á. Császár, T. Kővári and I. Vincze. Independently, Rankin [20] solved the case of crosspolytopes. There are two more simplical regular polytopes. The case of icosahedron was handled by Fejes Tóth [13] (see, say, [15] or [16]), and the case of the 600-cell is due to Böröczky [4]. All these arguments yield (explicitly or hidden) also the uniqueness of the optimal configuration up to orthogonal transformations. For the case of the 600-cell, Andreev [2] provided an argument for optimality based on the linear programming bound in coding theory. The proof of uniqueness via the linear programming bound was given by Boyvalenkov and Danev [6].

In this paper, we provide a stability version of Theorem A of optimal order. For $u, v \in S^{d-1}$, we write $\delta(u, v) \in [0, \pi]$ to denote the spherical (geodesic) distance of u and v, which is just their angle as vectors in $\mathbb{R}^d$.

Theorem 1.1 *Let P be a simplicial regular polytope in $\mathbb{R}^d$ having its vertices on S^{d-1}, $d \geq 3$. For suitable $\varepsilon_P, c_P > 0$, if $x_1, \ldots, x_k \in S^{d-1}$ are centers of non-overlapping spherical balls of radius at least $\varphi_P - \varepsilon$ for $\varepsilon \in [0, \varepsilon_P)$ and $k \geq f_0(P)$,* *then* $k = f_0(P)$, *and there exists a* $\Phi \in O(d)$, *such that for any* x_i *one finds a vertex* v *of* P *satisfying* $\delta(x_i, \Phi v) \leq c_P \varepsilon$.

We even provide explicit expressions for ε_P and c_P. If P is a d-simplex or a d-crosspolytope, then c_P is of polynomial growth in d ($c_P = 9d^{3.5}$ if P is a d-simplex, and $c_P = 96d^3$ if P is a d-crosspolytope).

Concerning notation, if $p \in S^{d-1}$ and $\varphi \in (0, \pi/2)$, then we write $B(p, \varphi)$ for the spherical ball of center p and radius φ. When working in $\mathbb{R}^d$, we write either $|X|$ or $\mathcal{H}^{d-1}(X)$ to denote the $(d-1)$-dimensional Hausdorff-measure of X. For $x_1, \ldots, x_k \in \mathbb{R}^d$, their convex hull, linear hull and affine hull in $\mathbb{R}^d$ are denoted by $[x_1, \ldots, x_k]$, $\text{lin}\{x_1, \ldots, x_k\}$ and $\text{aff}\{x_1, \ldots, x_k\}$, respectively. For $x, y \in \mathbb{R}^d$, we write $\langle x, y\rangle$ to denote the scalar product, and $\|x\|$ to denote the Euclidean norm. As usual, $\text{int}\, K$ stands for the interior of $K \subset \mathbb{R}^d$.

The paper uses various tools to establish Theorem 1.1. Only elementary linear algebra is needed for the case of a simplex, the linear programming bound is used for the case of a crosspolytope, and the simplex bound is applied to the icosahedron and the 600-cell.

Concerning the structure of the paper, Sects. 3 and 5 handle the cases of the simplex and the crosspolytope, respectively, and Sect. 4 in between reviews the linear programming bound used for the case of crosspolytopes. Results in these sections will be used also to handle the cases of the icosahedron in Sect. 8 and the 600-cell in Sect. 9, as well. After reviewing the Delone and Dirichlet-Voronoi cell decompositions and the corresponding simplex bound in Sect. 6, and verifying some volume estimates in Sect. 7, Theorem 1.1 is proved in Sects. 8 and 9 in the cases of the icosahedron and the 600-cell, respectively.

2 Some Simple Preparatory Statements

The following statement will play a key role in the arguments for the cases of simplices and crosspolytopes of Theorem 1.1.

Lemma 2.1 *Let* $n \geq 2$ *and* $0 \leq \eta < \frac{1}{n-1}$. *If* $u_1, \ldots, u_n \in S^{n-1}$ *satisfy that* $|\langle u_i, u_j\rangle| \leq \eta$ *for* $i \neq j$, *then there exists an orthonormal basis* $v_1, \ldots, v_n$ *of* $\mathbb{R}^n$ *such that* $\text{lin}\{u_i, \ldots, u_n\} = \text{lin}\{v_i, \ldots, v_n\}$ *and* $\langle u_i, v_i\rangle > 0$ *for* $i = 1, \ldots, n$, *and*

$$|\langle u_i, v_j\rangle| \leq \frac{\eta}{1-(n-2)\eta} \text{ for } i \neq j. \tag{1}$$

Moreover, $\delta(u_i, v_i) \leq 2n\eta$ *holds for* $i = 1, \ldots, n$ *provided that* $\eta < \frac{1}{2n}$.

Proof We prove the lemma by induction on n where the case $n = 2$ readily holds. Therefore, we assume that $n \geq 3$, and the lemma holds in $\mathbb{R}^{n-1}$.

Let $v_n = u_n$. For $i = 1, \ldots, n-1$, let $u_i = w_i + t_i v_n$ for $w_i \in v_n^{\perp}$ and $t_i \in \mathbb{R}$. It follows that $|t_i| \leq \eta$ and $\|w_i\| = (1-t_i^2)^{\frac{1}{2}} \geq (1-\eta^2)^{\frac{1}{2}}$ for $i = 1, \ldots, n-1$, and we define $\bar{w}_i = w_i/\|w_i\| \in S^{n-1}$. We observe that if $1 \le i < j \le n-1$, then

$$|\langle \bar{w}_i, \bar{w}_j \rangle| = \frac{|\langle w_i, w_j \rangle|}{(1-t_i^2)^{\frac{1}{2}}(1-t_j^2)^{\frac{1}{2}}} \le \frac{|\langle u_i, u_j \rangle| + |t_i t_j|}{1-\eta^2} \le \frac{\eta+\eta^2}{1-\eta^2} = \frac{\eta}{1-\eta}.$$

As $\bar{\eta} = \frac{\eta}{1-\eta} < \frac{1}{n-2}$ follows from $\eta < \frac{1}{n-1}$, we may apply the induction hypothesis to $\bar{w}_1, \ldots, \bar{w}_{n-1}$ and $\bar{\eta}$. We obtain an orthonormal basis $v_1, \ldots, v_{d-1}$ for $v_n^\perp$ such that $\text{lin}\{\bar{w}_i, \ldots, \bar{w}_{n-1}\} = \text{lin}\{v_i, \ldots, v_{n-1}\}$ and $\langle \bar{w}_i, v_i \rangle > 0$ for $i = 1, \ldots, n-1$, and

$$|\langle \bar{w}_i, v_j \rangle| \le \frac{\bar{\eta}}{1-(n-3)\bar{\eta}} = \frac{\eta}{1-(n-2)\eta} \text{ for } i \ne j.$$

If $1 \le i \le n-1$ then $\langle u_n, v_i \rangle = \langle v_n, v_i \rangle = 0$ and $|\langle u_i, v_n \rangle| = |t_i| \le \eta$. However if $i \ne j$ for $i, j \in \{1, \ldots, n-1\}$, then

$$|\langle u_i, v_j \rangle| = |\langle (1-t_i^2)^{\frac{1}{2}} \bar{w}_i + t_i v_n, v_j \rangle| \le |\langle \bar{w}_i, v_j \rangle| \le \frac{\eta}{1-(n-2)\eta}.$$

Therefore, we have verified (1), and we readily have $\text{lin}\{u_i, \ldots, u_n\} = \text{lin}\{v_i, \ldots, v_n\}$ for $i = 1, \ldots, n$ by construction.

Finally, for the estimate $\delta(u_i, v_i)$ if $\eta < \frac{1}{2n}$ and $i = 1, \ldots, n$, we observe that $|\langle u_i, v_j \rangle| < 2\eta$ provided $j \ne i$. It follows from $\|u_i\| = 1$ and $\langle u_i, v_i \rangle > 0$ that

$$0 \le \langle v_i - u_i, v_i \rangle = 1 - \sqrt{1 - \sum_{j \ne i} \langle u_i, v_j \rangle^2} \le \sum_{j \ne i} \langle u_i, v_j \rangle^2 \le (n-1)4\eta^2 < 2\eta.$$

In particular,

$$\|v_i - u_i\| = \sqrt{\sum_{j=1}^{n} \langle v_i - u_i, v_j \rangle^2} < \sqrt{n4\eta^2} = 2\sqrt{n}\,\eta,$$

and hence $\delta(u_i, v_i) < 2n\eta$. **Q.E.D.**

The following Lemma 2.2 and its consequence Corollary 2.3 are due to Rankin [20], and will be used, say, for simplices.

Lemma 2.2 *If $u_1, \ldots, u_{d+1} \in S^{d-1}$, $d \ge 2$, are contained in a closed hemisphere, then there exist i and j, $1 \le i < j \le d+1$, such that $\langle u_i, u_j \rangle \ge 0$.*

Proof We prove the statement by induction on d where the case $d = 2$ readily holds. If $d \ge 3$, then we may assume that $\langle u_i, u_j \rangle \le 0$ if $1 \le i < j \le d+1$. Let $v \in S^{n-1}$ such that $\langle v, u_i \rangle \ge 0$ for $i = 1, \ldots, d+1$, and hence $u_i = w_i + \lambda_i v$ for $i = 1, \ldots, d+1$ where $w_i \in v^\perp$ and $\lambda_i \ge 0$. If $u_i = v$ for some $i \in \{1, \ldots, d+1\}$, then $\langle u_j, u_i \rangle = 0$ for $j \ne i$, thus we are done. Otherwise $w_i \ne o$ for $i = 1, \ldots, d+1$. If $i = 1, \ldots, d$, then

$$0 \geq \langle u_{d+1}, u_i \rangle = \langle w_{d+1}, w_i \rangle + \lambda_{d+1} \cdot \lambda_i \geq \langle w_{d+1}, w_i \rangle,$$

therefore, the induction hypothesis applied to $\frac{w_1}{\|w_1\|}, \ldots, \frac{w_d}{\|w_d\|} \in v^{\perp} \cap S^{d-1}$ yields $\langle w_i, w_j \rangle \geq 0$ for some $1 \leq i < j \leq d$, and hence $\langle u_i, u_j \rangle \geq 0$. **Q.E.D.**

Corollary 2.3 *If $k \geq d+2$, $d \geq 2$, and $u_1, \ldots, u_k \in S^{d-1}$, then there exist i and j, $1 \leq i < j \leq d+1$, such that $\langle u_i, u_j \rangle \geq 0$.*

3 The Proof of Theorem 1.1 in the Case of Simplices

Theorem 3.1 covers the case of regular simplex of Theorem 1.1.

Theorem 3.1 *If $u_0, \ldots, u_d \in S^{d-1}$ satisfy $\delta(u_i, u_j) \geq \arccos \frac{-1}{d} - 2\varepsilon$ for $\varepsilon \in [0, \varepsilon_d)$ and $0 \leq i < j \leq d$, $d \geq 2$, then there exists a regular simplex $[v_0, \ldots, v_d]$ with $v_0, \ldots, v_d \in S^{d-1}$ such that $\delta(u_i, v_i) \leq c_d \varepsilon$ for $i = 0, \ldots, d$ where $c_d = 9d^{3.5}$ and $\varepsilon_d = 1/c_d$.*

Remark If $d = 2$, then one may even choose $c_2 = 3$ and $\varepsilon_2 = \frac{\pi}{12}$.

Proof We first handle the case $d = 2$, because this case is much more elementary. We define ε_2 to be $\frac{\pi}{12} = \frac{1}{2}(\frac{2\pi}{3} - \frac{\pi}{2})$. Thus $\arccos \frac{-1}{2} = \frac{2\pi}{3}$ and $\varepsilon < \varepsilon_2$ yield that no closed semicircle contains u_0, u_1, u_2, and hence the sum of the three angles of type $\delta(u_i, u_j)$ is 2π. We may assume that $\delta(u_0, u_1) \leq \delta(u_0, u_2) \leq \delta(u_1, u_2)$, and hence

$$\frac{2\pi}{3} - 2\varepsilon \leq \delta(u_0, u_1) \leq \frac{2\pi}{3} \leq \delta(u_1, u_2) \leq \frac{2\pi}{3} + 4\varepsilon. \tag{2}$$

We choose $v_1, v_2, v_3 \in S^1$ that are vertices of a regular triangle, and

$$\delta(u_0, v_0) = \delta(u_1, v_1) \leq \varepsilon.$$

We deduce from (2) that $\delta(u_2, v_2) \leq 3\varepsilon$, thus we may choose c_2 to be 3.

Turning to the case $d \geq 3$, let

$$0 < \varepsilon < \frac{1}{9d^{3.5}}.$$

If $0 \leq i < j \leq d$, then we have

$$\begin{aligned} \|u_i - u_j\|^2 = 2 - 2\cos\delta(u_i, u_j) &\geq 2 + 2\left(\frac{\cos 2\varepsilon}{d} - \frac{\sqrt{d^2-1}}{d} \cdot \sin 2\varepsilon\right) \\ &> 2 + 2\left(\frac{1-2\varepsilon}{d} - 2\varepsilon\right) > \frac{2(d+1)}{d} - 6\varepsilon. \end{aligned} \tag{3}$$

Using (3) and the estimate

$$(d+1)^2 = \left\| \sum_{i=0}^{d} u_i \right\|^2 + \sum_{0 \le i < j \le d} \|u_i - u_j\|^2 \ge \sum_{0 \le i < j \le d} \|u_i - u_j\|^2,$$

we deduce for any $i < j$ the upper bound

$$\|u_i - u_j\|^2 < \frac{2(d+1)}{d} + 3d(d+1)\varepsilon.$$

In particular, if $i < j$, then

$$\frac{-1}{d} - \frac{3}{2} d(d+1)\varepsilon \le \langle u_i, u_j \rangle \le \frac{-1}{d} + 2\varepsilon. \tag{4}$$

We embed $\mathbb{R}^d$ into $\mathbb{R}^{d+1}$ as $\mathbb{R}^d = e^\perp$ for suitable $e \in S^d \subset \mathbb{R}^{d+1}$. For $i = 0, \dots, d$, we define

$$w_i = \sqrt{\frac{1}{d+1}}\, e + \sqrt{\frac{d}{d+1}}\, u_i \in S^d,$$

and hence (4) yields that if $i \ne j$, then

$$|\langle w_i, w_j \rangle| = \left| \frac{1}{d+1} + \frac{d}{d+1} \langle u_i, u_j \rangle \right| = \frac{d}{d+1} \left| \frac{1}{d} + \langle u_i, u_j \rangle \right| \le \frac{3}{2} d^2 \varepsilon.$$

Since $\frac{3}{2} d^2 \varepsilon < \frac{1}{2(d+1)}$, Lemma 2.1 can be applied, and hence there exists an orthonormal basis $q_0, \dots, q_d$ of $\mathbb{R}^{d+1}$ such that $\delta(w_i, q_i) \le 3(d+1)d^2\varepsilon$ holds for $i = 0, \dots, d$. We define $q = \sum_{i=0}^{d} \frac{1}{\sqrt{d+1}} q_i$ and deduce that $q \in S^d$.

Since for any $i = 0, \dots, d$, we have $\langle e, w_i \rangle = \frac{1}{\sqrt{d+1}}$ and $\delta(w_i, q_i) \le 3(d+1)d^2\varepsilon$, it follows from $|\cos(\alpha + \beta) - \cos\alpha| \le |\beta|$ for $\alpha, \beta \in \mathbb{R}$ that $\left| \langle e, q_i \rangle - \frac{1}{\sqrt{d+1}} \right| \le 3(d+1)d^2\varepsilon$, and hence $|\langle e - q, q_i \rangle| \le 3(d+1)d^2\varepsilon$. We deduce that

$$\|e - q\| \le 3(d+1)^{\frac{3}{2}} d^2 \varepsilon.$$

Let $A \in \mathrm{O}(d+1)$ be the identity if $e = q$, and be the rotation around the linear $(d-1)$-space of $\mathbb{R}^{d+1}$ orthogonal to $\mathrm{lin}\{e, q\}$ with $Aq = e$ if $e \ne q$. It follows that $\|Au - u\| \le \|e - q\|$ for $u \in S^d$. For each $i = 0, \dots, q$, $\bar{q}_i = Aq_i \in S^d$ satisfies $\|\bar{q}_i - q_i\| \le \|e - q\| \le 3(d+1)^{\frac{3}{2}} d^2 \varepsilon$ and combining the last estimate with $\delta(w_i, q_i) \le 3(d+1)d^2\varepsilon \le \frac{3}{2}(d+1)^{\frac{3}{2}} d^2 \varepsilon$ yields

$$\|w_i - \bar{q}_i\| \le \frac{9}{2}(d+1)^{\frac{3}{2}} d^2 \varepsilon. \tag{5}$$

As $Aq = e$, we also have that $\langle \bar{q}_i, e\rangle = \sqrt{\frac{1}{d+1}} = \langle w_i, e\rangle$ for $i = 0, \dots, q$. Therefore,

$$v_i = \sqrt{\frac{d+1}{d}}\left(\bar{q}_i - \sqrt{\frac{1}{d+1}}\,e\right) \in e^{\perp} \cap S^d = S^{d-1}$$

for $i = 0, \dots, q$, $[v_0, \dots, v_d]$ is a regular d-simplex, and

$$\|v_i - u_i\| = \sqrt{\frac{d+1}{d}} \cdot \|\bar{q}_i - w_i\| \leq \frac{9}{2}(d+1)^2 d^{\frac{3}{2}}\varepsilon \leq 8d^{3.5}\varepsilon \leq \frac{8}{9}$$

for $i = 0, \dots, q$ where we used $d \geq 3$ at the last estimate. Using that $2\arcsin\frac{t}{2} \leq \frac{9}{8}t$ for any $t \in [0, \frac{8}{9}]$, we conclude that $\delta(v_i, u_i) = 2\arcsin\frac{\|v_i - u_i\|}{2} \leq \frac{9}{8}\|v_i - u_i\| \leq 9d^{3.5}\varepsilon$ for $i = 0, \dots, q$. **Q.E.D.**

4 The Linear Programming Bound

Let $d \geq 2$. The presentation about the linear programming bound for sphere packings on S^{d-1} in this section is based on Ericson and Zinoviev [11, Chap. 2]. A central role in the theory is played by certain real Gegenbauer polynomials Q_i, $i \in \mathbb{N}$, in one variable where each Q_i is of degree i, and satisfies the following recursion:

$$\begin{aligned}
Q_0(t) &= 1\\
Q_1(t) &= t\\
Q_2(t) &= \frac{dt^2 - 1}{d-1}\\
(i + d - 2)Q_{i+1}(t) &= (2i + d - 2)tQ_i(t) - iQ_{i-1}(t) \text{ for } i \geq 2.
\end{aligned}$$

We do not signal the dependence of Q_i on d because the original notation for the Gegenbaur polynomial is $Q_i = Q_i^{(\alpha)}$ for $\alpha = \frac{d-2}{2}$ as

$$\int_{-1}^{1} Q_i(t)Q_j(t)(1-t^2)^{\frac{d-3}{2}}\,dt = 0 \text{ if } i \neq j.$$

Actually, Q_i is normalized in a way such that $Q_i(1) = 1$ for $i \in \mathbb{N}$.

The basis of our considerations is the following version of the linear programming bound, which is contained in the proof of Theorem 2.3.1 in [11]. We write $|X|$ to denote the cardinality of a finite set X.

Theorem 4.1 *For $d \geq 2$, if $f = f_0Q_0 + f_1Q_1 + \ldots + f_kQ_k$ for $k \geq 1$, $f_0 > 0$ and $f_1, \dots, f_k \geq 0$, then any finite $X \subset S^{d-1}$ satisfies*

$$|X|f(1) + \sum_{\substack{x,y\in X\\ x\neq y}} f(\langle x, y\rangle) \geq |X|^2 f_0. \tag{6}$$

Remark The classical linear programming bound is a consequence; namely, if in addition, $f(t) \leq 0$ for fixed $s \in (-1, 1)$ and variable $t \in [-1, s]$, then

$$|X| \leq f(1)/f_0. \tag{7}$$

If we have equality in (7), then (6) shows that all values $\langle x, y\rangle$ for $x \neq y$, $x, y \in X$ are roots of f.

As an example, let $X \subset S^{d-1}$ be the centers for a packing of spherical balls of radius $\frac{\pi}{4}$, and hence $\langle x, y\rangle \leq 0$ for $x, y \in X$ with $x \neq y$. The polynomial

$$f(t) = t(t+1) = f_0 Q_0 + f_1 Q_1 + f_2 Q_2$$

satisfies $f(t) \leq 0$ for $t \in [-1, 0]$ and

$$f_0 = \frac{1}{d}, \quad f_1 = 1, \quad f_2 = 1 - \frac{1}{d}, \quad f(1) = 2,$$

therefore, (7) yields $|X| \leq 2d$.

Next we quantify the obvious statement that for any packing of m spherical balls of radius r on S^{n-1}, if r is close to $\frac{\pi}{4}$ then $m \leq 2n$.

Lemma 4.2 *If $Y \subset S^{n-1}$, $n \geq 2$, satisfies that $\langle x, y\rangle < \frac{1}{2n^2-n}$ for $x, y \in Y$ with $x \neq y$, then $|Y| \leq 2n$.*

Proof Let $s = \max\{\langle x, y\rangle : x, y \in Y \text{ and } x \neq y\} < \frac{1}{2n^2-n}$. We consider the polynomial

$$f(t) = (t+1)(t-s) = f_0 Q_0 + f_1 Q_1 + f_2 Q_2$$

where $f(t) \leq 0$ for $t \in [-1, s]$ and

$$f_0 = \frac{1}{n} - s, \quad f_1 = 1 - s, \quad f_2 = 1 - \frac{1}{n}, \quad f(1) = 2(1-s).$$

We deduce from the linear programming bound (7) and $s < \frac{1}{2n^2-n}$ that

$$|Y| \leq \frac{2n(1-s)}{1-ns} = 2n + \frac{2n(n-1)s}{1-ns} < 2n + 1.$$ **Q.E.D.**

The linear programming bound could have been used in the case of simplex to prove (4). However, this could be proved easily by elementary arguments, as well.

The linear programming bound can be also used to prove the optimality of the icosahedron and the 600-cell however the corresponding polynomials are more complicated. Say, in the case of 600-cell, the polynomial is of degree 17 and $f_{12} = f_{13} = 0$ according to Andreev [2]. Therefore we use volume estimates to handle the cases of the icosahedron and the 600-cell.

5 The Proof of Theorem 1.1 in the Case of Crosspolytopes

Let $X \subset S^{d-1}$ be the centers for a packing of at least $2d$ spherical balls of radius $\frac{\pi}{4} - \varepsilon$, $0 < \varepsilon < \frac{1}{64d^4}$, and hence $\langle x, y\rangle \leq s$ for $x, y \in X$ with $x \neq y$ and

$$s = \sin 2\varepsilon < 2\varepsilon < \frac{1}{32d^4}.$$

We deduce from Lemma 4.2 that

$$|X| = 2d.$$

We consider the polynomial

$$f(t) = (t+1)(t-s) = f_0 Q_0 + f_1 Q_1 + f_2 Q_2$$

where $f(t) \leq 0$ for $t \in [-1, s]$ and

$$f_0 = \frac{1}{d} - s, \quad f_1 = 1 - s, \quad f_2 = 1 - \frac{1}{d}, \quad f(1) = 2(1-s).$$

It follows from (6) and $f(t) \leq 0$ for $t \in [-1, s]$ that if $x, y \in X$ with $x \neq y$, then

$$f(\langle x, y\rangle) \geq |X|^2 f_0 - |X| f(1) = 4d^2\left(\frac{1}{d} - s\right) - 4d(1-s) = -4d(d-1)s. \tag{8}$$

Since $t - s \leq \frac{-1}{2}$ if $t \leq \frac{-1}{2}$ and $t + 1 \geq \frac{1}{2}$ if $t \geq \frac{-1}{2}$, we have

$$f(t) \leq -\frac{1}{2} \min\{|t+1|, |t-s|\} \text{ for } t \in [-1, s].$$

We deduce from (8) that if $x, y \in X$ with $x \neq y$, then

$$\min\{\langle x, y\rangle + 1, s - \langle x, y\rangle\} \leq 8d(d-1)s,$$

or in other words,

$$\begin{array}{ll}\text{either} & -1 \le \langle x, y\rangle \le -1 + \frac{1}{4d^2} < \frac{-3}{4} \\ \text{or} & -8d(d-1)s \le \langle x, y\rangle \le s < \frac{1}{32}.\end{array} \tag{9}$$

We define

$$\eta = 8d(d-1)s < \frac{1}{4d^2}. \tag{10}$$

We claim that for every $x \in X$

$$\text{there exists a unique } y \in X \text{ such that } \langle x, y\rangle \le \frac{-3}{4}, \tag{11}$$

which we call the element of X opposite to x. For any $y \in X$, we write $\bar{y}$ to denote its projection into $x^{\perp}$, and if $y \ne \pm x$, then we set $y^* = \bar{y}/\|\bar{y}\|$.

The first step towards (11) is to show that if $y, z \in X$, then

$$\langle x, y\rangle \le \frac{-3}{4} \text{ and } \langle x, z\rangle \le \frac{-3}{4} \text{ yield } y = z. \tag{12}$$

Since $\|\bar{y}\| = \sqrt{1 - \langle x, y\rangle^2} < \sqrt{\frac{1}{2}}$ and similarly $\|\bar{z}\| < \sqrt{\frac{1}{2}}$, we have

$$\langle y, z\rangle = \langle x, y\rangle\langle x, z\rangle + \langle \bar{y}, \bar{z}\rangle > \frac{9}{16} - \frac{1}{2} = \frac{1}{16},$$

which proves $\langle y, z\rangle = 1$ by (9), and in turn verifies (12).

Next, set $\widetilde{X} = \{y \in X : |\langle x, y\rangle| \le \eta\}$. For (11), it is sufficient to verify that

$$|\widetilde{X}| \le 2(d-1). \tag{13}$$

For $y_1, y_2 \in \widetilde{X}$, we have $y_i = \bar{y}_i + p_i x$ for $i = 1, 2$ where $p_i \in [-\eta, \eta]$. In particular, $\|\bar{y}_i\| = (1 - p_i^2)^{\frac{1}{2}} \ge (1 - \eta^2)^{\frac{1}{2}}$, and hence

$$\langle y_1^*, y_2^*\rangle = \frac{\langle \bar{y}_1, \bar{y}_2\rangle}{(1-p_1^2)^{\frac{1}{2}}(1-p_2^2)^{\frac{1}{2}}} = \frac{\langle y_1, y_2\rangle - p_1 p_2}{(1-p_1^2)^{\frac{1}{2}}(1-p_2^2)^{\frac{1}{2}}} \le \frac{\eta + \eta^2}{1 - \eta^2} = \frac{\eta}{1 - \eta} < 2\eta.$$

Since $2\eta < \frac{1}{2d^2}$, Lemma 4.2 with $n = d - 1$ yields (13), and in turn (11).

We deduce from (11) that X can be divided into d pairs of opposite vectors. Choosing one unit vector from each pair, we obtain $x_1, \dots, x_d \in X$ such that $|\langle x_i, x_j\rangle| \le \eta$ for $i \ne j$. It follows from Lemma 2.1 that for every such d-tuple $x_1, \dots, x_d \in X$ there exists an orthonormal basis $v_1, \dots, v_d$ of $\mathbb{R}^d$ such that $\text{lin}\{x_i, \dots, x_d\} = \text{lin}\{v_i, \dots, v_d\}$ and $\delta(x_i, v_i) \le 2d\eta$ for $i = 1, \dots, d$.

We claim that if $x, y \in X$ are opposite vectors, then

$$\delta(y, -x) \le 4d\eta. \tag{14}$$

We choose $x_2, \ldots, x_d \in X$ representatives from the other $d-1$ opposite pairs, and let v be the unit vector orthogonal to $\mathrm{lin}\{x_2, \ldots, x_d\}$ with $\langle x, v\rangle > 0$. Taking $x = x_1$ and considering the approximating orthonormal basis $v_1, \ldots, v_d$ for this $x_1, \ldots, x_d$, we deduce that $v = v_1$, and hence $\delta(x, v) \leq 2d\eta$. Similarly, taking $y = x_1$, we have $v_1 = -v$ for the approximating orthonormal basis, thus $\delta(y, -v) \leq 2d\eta$. In turn, we conclude (14) by the triangle inequality.

Finally, we fix representatives $u_1, \ldots, u_d$ from each of the d pairs of opposite vectors, and hence there exists an orthonormal basis $w_1, \ldots, w_d$ of $\mathbb{R}^d$ such that $\delta(u_i, w_i) \leq 2d\eta$ for $i = 1, \ldots, d$. We write u_{i+d} to denote the vector of X opposite to u_i, $i = 1, \ldots, d$, and hence $\delta(u_{i+d}, -u_i) \leq 4d\eta$ according to (14). Therefore,

$$\delta(u_{i+d}, -w_i) \leq \delta(u_{i+d}, -u_i) + \delta(-u_i, -w_i) \leq 4d\eta + 2d\eta = 6d\eta \leq 48d^3 s \leq 96d^3\varepsilon.$$

Therefore, $c_d = 96d^3$ can be chosen for Theorem 1.1 in the case of crosspolytopes.

6 Spherical Dirichlet-Voronoi and Delone Cell Decomposition

For $v \in S^{d-1}$ and acute angle θ, we write $B(v, \theta)$ to denote the spherical ball of center v and radius θ. For $u, v \in S^{d-1}$, $u \neq -v$, we write $\overline{uv}$ to denote the smaller geodesic arc connecting u and v. We will frequently use the Spherical Law of Cosines: If a, b, c are side lengths of a spherical triangle contained in an open hemisphere, and the opposite angles are α, β, γ, respectively, then

$$\cos c = \cos a \cdot \cos b + \sin a \cdot \sin b \cdot \cos \gamma. \tag{15}$$

A set $C \subset \mathbb{R}^d$ is a convex cone if it is closed and $\alpha x + \beta y \in C$ for $\alpha, \beta \geq 0$ and $x, y \in C$. If C contains a half-line, then $M = C \cap S^{d-1}$ is called a spherically convex set whose dimension is one less than the Euclidean dimension of C. The relative interior of M is the intersection of S^{d-1} and the relative interior of C with respect to $\mathrm{lin}\, C$. If the origin is a face of C and C is a polyhedron (namely, intersection of finitely many half-spaces) then M is called a spherical polytope. In this case, the faces of M are intersections of S^{d-1} with the faces of C different from the origin.

Let $x_1, \ldots, x_k \in S^{d-1}$ satisfy that each open hemisphere contains some of $x_1, \ldots, x_k$, and hence $o \in \mathrm{int}\, P$ for $P = [x_1, \ldots, x_k]$. The radial projections of the facets of P onto S^{d-1} form the Delone (or Delaunay) cell decomposition of S^{d-1}. We observe that if the distance of o from aff F is ϱ for a facet F, then $\arccos \varrho$ is the spherical radius of the spherical cap cut off by aff F. We call $\arccos \varrho$ the spherical circumradius of the corresponding Delone cell.

To define the other classical decomposition of S^{d-1} corresponding to $x_1, \ldots, x_k$, let

$$D_i = \{u \in S^{d-1} : \delta(u, x_i) \leq \delta(u, x_j) \text{ for } j = 1, \ldots, k\}$$

for $i = 1, \dots, k$, which is the Dirichlet-Voronoi cell of x_i. The Dirichlet-Voronoi cells also form a cell decomposition of S^{n-1} that is dual to the Delone cell decomposition by providing the following bijective correspondence between vertices of Dirichlet cells and Delone cells. If v is a vertex of $D_i, i \in \{1, \dots, k\}$, and $\delta(v, x_i) = \theta$, then $\delta(v, x_j) \geq \theta$ for all $j = 1, \dots, k$, and points x_j with $\delta(v, x_j) = \theta$ form the vertex set of a Delone cell (see, say, Böröczky [5]). In addition, if F is an m-dimensional face of some D_i, and p is the closest point of the m-dimensional great sphere Σ of F, then there exists a $(d-1-m)$-dimensional face G of the Delone cell complex contained in the $(d-1-m)$-dimensional great sphere Σ' orthogonal to Σ at p whose vertices are all of distance $\delta(p, x_i)$ from p.

A simplex with ordered vertices $p_0, \dots, p_{d-1}$ on S^{d-1} is called an orthoscheme if for $i = 1, \dots, d-2$, the i-dimensional great sphere through $p_0, \dots, p_i$ is orthogonal to the $(d-1-i)$-dimensional great sphere through $p_i, \dots, p_{d-1}$.

For any face F of a Dirichlet-Voronoi cell D_i, we write $q_i(F)$ to denote the point of F closest to x_i. It follows from the convexity of F and the Spherical Law of Cosines that if $x \in F \backslash q_i(F)$, then

(a) the angle between the arcs $\overline{q_i(F), x_i}$ and $\overline{q_i(F), x}$ is at least $\frac{\pi}{2}$,
(b) and is actually exactly $\frac{\pi}{2}$ if $q_i(F)$ lies in the relative interior of F.

For a Dirichlet-Voronoi cell D_i, we say that a sequence $(F_0, \dots, F_{d-2})$ is a tower, if F_j is a j-face of D_i, $j = 0, \dots, d-2$, and $F_j \subset F_l$ if $j < l$. In addition, $(F_0, \dots, F_{d-2})$ is a proper tower, if $q_i(F_j) \neq q_i(F_l)$ for $j < l$, and, in this case, we call the simplex Ξ with ordered vertices $x_i, q_i(F_{d-2}), \dots, q_i(F_0)$, a quasi-orthoscheme. We observe that according to (b), a quasi-orthoscheme is an orthoscheme if each $q_i(F_j)$, $j = 1, \dots, d-2$, lies in the relative interior of F_j. Moreover, (a) yields that quasi-orthoschemes provide a triangulation of S^{d-1} refining the Dirichlet-Voronoi cell decomposition.

For any $\varphi \in (0, \frac{\pi}{2})$ and $i \geq 1$, we write $r_i(\varphi) \in (0, \frac{\pi}{2})$ to denote the circumradius of the i-dimensional spherical regular simplex of edge length 2φ. In particular, there exists a spherical triangle with equal sides $r_i(\varphi)$ enclosing the angle $\arccos \frac{-1}{i}$ where the third side of the triangle is 2φ. In addition, we define $r_\infty(\varphi) \in (0, \frac{\pi}{2})$ in a way such that there exists a spherical triangle with equal sides $r_\infty(\varphi)$ enclosing the right angle where the third side of the triangle is 2φ. We have

$$\varphi = r_1(\varphi) < \cdots < r_{d-1}(\varphi) < r_\infty(\varphi).$$

It follows from (15) that if $j = 1, \dots, d-1$, then

$$\cos 2\varphi = \cos^2 r_j(\varepsilon) - \frac{\sin^2 r_j(\varepsilon)}{j} \quad \text{and} \quad \cos 2\varphi = \cos^2 r_\infty(\varepsilon), \tag{16}$$

which in turn yields that

$$\sin r_j(\varphi) = \sqrt{\frac{2j}{j+1}} \sin \varphi \quad \text{and} \quad \sin r_\infty(\varphi) = \sqrt{2} \sin \varphi. \tag{17}$$

The following lemma is due to Boroczky [4]. We include the argument because the second statement is only implicit in [4].

Lemma 6.1 *Let $\varphi \in (0, \frac{\pi}{2})$, and let $x_1, \ldots, x_k \in S^{d-1}$ satisfy that each open hemisphere contains some of $x_1, \ldots, x_k$, and $\delta(x_i, x_j) \geq 2\varphi$ for $i \neq j$, and let D_j be the Dirichlet-Voronoi cell of x_j. If F is an m-dimensional face of certain D_i, then*

(i) $\delta(x_i, q_i(F)) \geq r_{d-1-m}(\varphi)$;
(ii) and even $\delta(x_i, q_i(F)) \geq r_\infty(\varphi)$ if $q_i(F)$ is not contained in the relative interior of F.

Proof Let p be the closest to x_i point of the m-dimensional great subsphere Σ containing F, and let I be the set of all indices j such that F is a face of D_j. In particular, all x_j with $j \in I$ span the $(d-1-m)$-dimensional great subsphere Σ' passing through p and perpendicular to Σ, and hence the cardinality of I is at least $d-m$. It follows that for $\theta = \delta(x_i, p) \leq \delta(x_i, q_i(F))$, we have $\theta = \delta(x_j, p)$ for $j \in I$. For $j \in I$, let u_j be a unit vector tangent to the arc $\overline{p, x_j}$ at p, and hence all u_j, $j \in I$, span the $(d-1-m)$-dimensional linear subspace L' tangent to Σ' at p. According to Jung's theorem (see also Lemma 3.1), there exist different $l, j \in I$ such that $\delta(u_l, u_j) \leq \arccos \frac{-1}{d-1-m}$. Since $\delta(x_l, p) = \delta(x_j, p) = \theta$, we deduce (i) from the Spherical Law of Cosines (15).

Turning to (ii), we assume that p is not contained in the relative interior of F. In this case, there exists an $x_g \in S^{d-1} \backslash \Sigma'$ such that $0 < \delta(x_g, p) \leq \theta$. Let $u_g \in S^{d-1}$ be a unit vector tangent to the arc $\overline{p, x_g}$ at p. We claim that there exist different $j, l \in I \cup \{g\}$ such that

$$\langle u_j, u_l \rangle \geq 0. \tag{18}$$

Let L be the m-dimensional linear subspace L tangent to Σ at p, which is the orthogonal complement of L' inside the tangent space to S^{d-1} at p. Therefore, there exist unit vectors $v \in L$ and $v' \in L'$ and a real number $t \in [0, \frac{\pi}{2}]$ such that $u_g = v \cos t + v' \sin t$. If $\langle v', u_j \rangle < 0$ for all $j \in I$, then Lemma 2.2 yields different $j, l \in I$ such that $\langle u_j, u_l \rangle \geq 0$. Otherwise there exists $j \in I$ such that $\langle v', u_j \rangle \geq 0$, and hence $\langle u_g, u_j \rangle \geq 0$, as well.

Using these u_j and u_l in (18), we apply the Spherical Law of Cosines (15) to the triangle with vertices p, x_j, x_l to obtain

$$\cos 2\varphi \geq \cos \delta(x_j, x_l) \geq \cos \delta(p, x_j) \cdot \cos \delta(p, x_l) \geq \cos^2 \theta.$$

Therefore, $\theta \geq r_\infty(\varphi)$ by (16). **Q.E.D.**

We fix a point $z_0 \in S^{d-1}$, and for $0 < t_1 < \cdots < t_{d-1} < \frac{\pi}{2}$, we write $\Theta(t_1, \ldots, t_{d-1})$ to denote an orthoscheme with ordered vertices $z_0, z_1, \ldots, z_{d-1}$ such that $\delta(z_0, z_i) = t_i$ for $i = 1, \ldots, d-1$. We observe that the (spherical) diameter of $\Theta(t_1, \ldots, t_{d-1})$ is t_{d-1}. For any $\varphi \in (0, t_1]$, we define

$$\Delta(t_1, \ldots, t_{d-1}) = \frac{|\Theta(t_1, \ldots, t_{d-1}) \cap B(z_0, \varphi)|}{|\Theta(t_1, \ldots, t_{d-1})| \cdot |B(z_0, \varphi)|},$$

whose value does not depend on the choice of $\varphi \in (0, t_1]$. If $\Psi \subset z_0^\perp$ is the Euclidean convex polyhedral cone generated by the rays tangent to the arcs $\overline{z_0, z_i}$ at z_0, $i = 1, \dots, d-1$, then

$$\Delta(t_1, \dots, t_{d-1}) = \frac{\mathcal{H}^{d-2}(\Psi \cap S^{d-1})}{|\Theta(t_1, \dots, t_{d-1})| \cdot \mathcal{H}^{d-2}(S^{d-2})}.$$

According to one of the core results of Boroczky [4], if $s_1 < \cdots s_{d-1} < \frac{\pi}{2}$, and $t_i \leq s_i$ for $i = 1, \dots, d-1$, then

$$\Delta(t_1, \dots, t_{d-1}) \geq \Delta(s_1, \dots, s_{d-1}). \tag{19}$$

We deduce from Lemma 6.1 and (19) the following estimate.

Lemma 6.2 *Let $\sigma \in (0, \frac{\pi}{2})$, and let $x_1, \dots, x_k \in S^{d-1}$, $d \geq 3$, satisfy that each open hemisphere contains some of $x_1, \dots, x_k$, and $\delta(x_i, x_j) \geq 2\sigma$ for $i \neq j$, and let D_i be the Dirichlet-Voronoi cell of x_i. If Ξ is a quasi-orthoscheme associated to some D_i and it is known that Ξ is an orthoscheme, and the diameter of Ξ is R, then*

$$\frac{|\Xi \cap B(x_i, \sigma)|}{|\Xi| \cdot |B(x_i, \sigma)|} \leq \Delta(r_1(\sigma), \dots, r_{d-2}(\sigma), R) \tag{20}$$

$$\leq \Delta(r_1(\sigma), \dots, r_{d-2}(\sigma), r_{d-1}(\sigma)). \tag{21}$$

We note that the ideas in Boroczky [4] yield (21) even if the quasi-orthoscheme Ξ is not an orthoscheme, but they actually even imply the following stronger bound in the low dimensions we are interested in.

Lemma 6.3 *Let $\sigma \in (0, \frac{\pi}{2})$, and let $x_1, \dots, x_k \in S^{d-1}$, $d = 3, 4$, satisfy that each open hemisphere contains some of $x_1, \dots, x_k$, and $\delta(x_i, x_j) \geq 2\sigma$ for $i \neq j$, and let D_i be the Dirichlet-Voronoi cell of x_i. If Ξ is a quasi-orthoscheme associated to some D_i and it is known that Ξ is not an orthoscheme, then*

$$\frac{|\Xi \cap B(x_i, \sigma)|}{|\Xi| \cdot |B(x_i, \sigma)|} \leq \Delta(r_1(\sigma), \dots, r_{d-2}(\sigma), r_\infty(\sigma)).$$

Proof Let $F_0 \subset \cdots \subset F_{d-2}$ be the proper tower of faces of D_i associated to Ξ. If $\delta(x_i, q_i(F_{d-2})) \geq r_\infty(\sigma)$, then F_{d-2} does not intersect the interior of $B(x_i, r_\infty(\sigma))$, and hence Lemma 6.1 yields

$$\frac{|\Xi \cap B(x_i, \sigma)|}{|\Xi|} \leq \frac{|\Xi \cap B(x_i, \sigma)|}{|\Xi \cap B(x_i, r_\infty(\sigma))|} = \frac{|B(x_i, \sigma)|}{|B(x_i, r_\infty(\sigma))|}.$$

Since $\Theta(r_1(\sigma), \dots, r_{d-2}(\sigma), r_\infty(\sigma)) \subset B(z_0, r_\infty(\sigma))$, we have

$$\frac{|\Theta(r_1(\sigma), \dots, r_{d-2}(\sigma), r_\infty(\sigma)) \cap B(z_0, \sigma)|}{|\Theta(r_1(\sigma), \dots, r_{d-2}(\sigma), r_\infty(\sigma))|} \geq \frac{|B(z_0, \sigma))}{|B(z_0, r_\infty(\sigma))|}.$$

we conclude the lemma in this case.

This covers the case $d = 3$ completely because the condition $\delta(x_i, q_i(F_1)) < r_\infty(\sigma)$ implies by Lemma 6.1 that Ξ is an orthoscheme. The only case left open is when $d = 4$, $\delta(x_i, q_i(F_2)) < r_\infty(\sigma)$, and hence $q_i(F_2)$ is contained in the relative interior of F_2, but $q_i(F_1)$ is not contained in the relative interior of F_1 because otherwise Ξ is an orthoscheme. Then there exists $p \in \overline{q_i(F_2), q_i(F_1)}$ such that $\delta(x_i, p) = r_\infty(\varphi)$. We consider the spherical cone C obtained by rotating the triangle with vertices $x_i, q_2(F_2), p$ around $\overline{x_i, q_2(F_2)}$. Since $F_2 \backslash C$ does not intersect $B(x_i, r_\infty(\varphi))$, the argument as above leads to

$$\frac{|(\Xi \backslash C) \cap B(x_i, \sigma)|}{|(\Xi \backslash C)| \cdot |B(x_i, \sigma)|} \leq \Delta(r_1(\sigma), r_2(\sigma), r_\infty(\sigma)). \tag{22}$$

In addition, (19) and the argument of K. Boroczky [4] yield

$$\begin{aligned} \frac{|C \cap B(x_i, \sigma)|}{|C| \cdot |B(x_i, \sigma)|} &= \lim_{s \to 0^+} \Delta(r_1(\sigma), r_\infty(\sigma) - s, r_\infty(\sigma)) \\ &\leq \Delta(r_1(\sigma), r_2(\sigma), r_\infty(\sigma)). \end{aligned} \tag{23}$$

Combining (22) and (23) proves Lemma 6.3. **Q.E.D.**

Actually, the argument in Boroczky [4] shows that Lemma 6.3 holds in any dimension. More precisely, [4] proved the so-called *simplex bound*; namely, if $\sigma \in (0, \frac{\pi}{2})$, and there exist k non-overlapping spherical balls of radius σ on S^{d-1}, then

$$k \leq \Delta(r_1(\sigma), \ldots, r_{d-1}(\sigma)) \cdot \mathcal{H}^{d-1}(S^{d-1}), \tag{24}$$

and equality holds in the simplex bound if and only if the centers are vertices of a regular simplicial polytope P with edge length $2 \sin \sigma$.

The following statement shows in a qualitative way that if for an acute angle φ, all simplices in a Delone triangulation of S^{d-1} are close to be regular with spherical edge length 2φ, then the whole Delone triangulation is close to a one induced by a simplicial regular polytope.

Lemma 6.4 *Let $\varphi \in (0, \pi/4]$, let $u_0, \ldots, u_d \in S^{d-1}$, $d \geq 3$ be such that $u_1, \ldots, u_{d-1}$ determines a unique $(d-2)$-dimensional great subsphere that separates u_0 and u_d, and let $\varepsilon \in (0, \varepsilon_0)$ for $\varepsilon_0 = \frac{\sin \varphi}{16\sqrt{d-1}}$. If there exist two spherical regular simplices of edge length φ with vertices $v_0, \ldots, v_{d-1}$ and $w_1, \ldots, w_d$ such that $\delta(u_i, v_i) \leq \varepsilon$ for $i = 0, \ldots, d-1$, and $\delta(u_i, w_i) \leq \varepsilon$ for $i = 1, \ldots, d$, then $\delta(u_d, v_d) \leq c\varepsilon$, where $v_1, \ldots, v_d$ are vertices of a regular simplex, $v_d \neq v_0$ and $c = \frac{16\sqrt{d-1}}{\sin \varphi}$.*

Proof It is sufficient to prove that $\delta(v_d, w_d) \leq (c-1)\varepsilon$. Using $\delta(v_d, w_d) = 2 \arcsin \frac{\|v_d - w_d\|}{2} \leq 2\|v_d - w_d\|$ given $\|v_d - w_d\| \leq 1$, it is sufficient to show

$$\|v_d - w_d\| \leq \frac{c-1}{2} \cdot \varepsilon. \tag{25}$$

We will use that if $x_1, \ldots, x_k, y_1, \ldots, y_k \in \mathbb{R}^k$, $\|x_i - y_i\| \leq \eta$ for all $i = 1, \ldots, k$, and $\lambda_1, \ldots, \lambda_k \geq 0$, then the triangle inequality yields

$$\|(\lambda_1 x_1 + \ldots + \lambda_k x_k) - (\lambda_1 y_1 + \ldots + \lambda_k y_k)\| \leq (\lambda_1 + \ldots + \lambda_k)\eta. \tag{26}$$

We have $\delta(v_i, w_i) \leq 2\varepsilon$ for $i = 1, \ldots, d-1$, thus $\|v_i - w_i\| \leq 2\varepsilon$ for $i = 1, \ldots, d-1$. We deduce from (26) that $\|p - p'\| \leq 2\varepsilon$ holds for the centroids

$$p = \frac{1}{d-1}(v_1 + \cdots + v_{d-1}) \text{ and } p' = \frac{1}{d-1}(w_1 + \cdots + w_{d-1})$$

of the $(d-2)$-dimensional regular Euclidean simplices $[v_1, \ldots, v_{d-1}]$ and $[w_1, \ldots, w_{d-1}]$. We consider $\alpha > \beta > 0$, and an orthonormal basis $\tilde{v}_1, \ldots, \tilde{v}_d$ such that $v_d, \tilde{v}_d$ lie in the same half-space with respect to $\operatorname{lin}\{v_1, \ldots, v_{d-1}\} = \operatorname{lin}\{\tilde{v}_1, \ldots, \tilde{v}_{d-1}\}$ and satisfy

$$v_i = \alpha \tilde{v}_i + \sum_{\substack{j \neq i \\ j \in \{1, \ldots, d-1\}}} \beta \tilde{v}_j \text{ for } i = 1, \ldots, d-1 \tag{27}$$

Then α, β satisfy

$$\begin{aligned} 1 = \langle v_1, v_1 \rangle &= \alpha^2 + (d-2)\beta^2 \\ \cos 2\varphi = \langle v_1, v_2 \rangle &= 2\alpha\beta + (d-3)\beta^2, \end{aligned}$$

therefore taking the difference leads to

$$\frac{(\alpha - \beta)^2}{2} = \frac{1 - \cos 2\varphi}{2} = \sin^2 \varphi. \tag{28}$$

Similarly, we define an orthonormal basis $\tilde{w}_1, \ldots, \tilde{w}_d$ of $\mathbb{R}^d$ such that $w_d, \tilde{w}_d$ lie in the same half-space with respect to $\operatorname{lin}\{w_1, \ldots, w_{d-1}\} = \operatorname{lin}\{\tilde{w}_1, \ldots, \tilde{w}_{d-1}\}$ and satisfy

$$w_i = \alpha \tilde{w}_i + \sum_{\substack{j \neq i \\ j \in \{1, \ldots, d-1\}}} \beta \tilde{w}_j \text{ for } i = 1, \ldots, d-1.$$

This basis exists when α, β satisfy the conditions derived above.

According to (27), the $(d-1) \times (d-1)$ symmetric matrix M whose main diagonals are α, and the rest of the entries are β, satisfies that $M\tilde{v}_i = v_i$ $i = 1, \ldots, d-1$. One of the eigenvectors of M in $\tilde{v}_d^{\perp}$ is $v_* = \sum_{j=1}^{d-1} \tilde{v}_j$ with eigenvalue $\alpha + (d-2)\beta$. Any vector in $\tilde{v}_d^{\perp}$ orthogonal to v_* is an eigenvector with eigenvalue $\alpha - \beta$. We deduce with help of (28) that if $v \in \tilde{v}_d^{\perp}$, then

$$\|M^{-1}v\| \leq (\alpha - \beta)^{-1}\|v\| = \frac{\|v\|}{\sqrt{2}\sin\varphi}. \tag{29}$$

For $i = 1, \dots, d-1$, we have $\langle \tilde{w}_d, w_i \rangle = 0$ and $\|v_i - w_i\| \le 2\varepsilon$, therefore,

$$2\varepsilon \ge |\langle \tilde{w}_d, v_i \rangle| = \left\| \alpha \langle \tilde{w}_d, \tilde{v}_i \rangle + \sum_{\substack{j \ne i \\ j \in \{1,\dots,d-1\}}} \beta \langle \tilde{w}_d, \tilde{v}_j \rangle \right\|.$$

In particular, the length of the vector $v = \langle \tilde{w}_d, v_1 \rangle \tilde{v}_1 + \cdots + \langle \tilde{w}_d, v_{d-1} \rangle \tilde{v}_{d-1}$ is at most $2\varepsilon\sqrt{d-1}$, thus (29) implies that

$$\|M^{-1}v\| = \sqrt{\sum_{j=1}^{d-1} \langle \tilde{w}_d, \tilde{v}_j \rangle^2} \le \frac{2\varepsilon\sqrt{d-1}}{\sqrt{2}\sin\varphi}.$$

In other words, the projection of the unit vector $\tilde{w}_d$ into $\tilde{v}_d^{\perp}$ is of length at most $\frac{2\varepsilon\sqrt{d-1}}{\sqrt{2}\sin\varphi}$, therefore, possibly after exchanging $\tilde{w}_d$ by $-\tilde{w}_d$, we have

$$\|\tilde{v}_d - \tilde{w}_d\| \le \frac{2\varepsilon\sqrt{d-1}}{\sqrt{2}\sin\varphi}\sqrt{2} = \frac{2\varepsilon\sqrt{d-1}}{\sin\varphi}.$$

Now the orthogonal projection of the origin o into $\mathrm{aff}\{v_1, \dots, v_d\}$ lies inside $[p, v_d]$, thus the angle of the triangle $[o, p, v_d]$ at p is acute. In addition, the angle of p and v_d is also acute by $\varphi \le \frac{\pi}{4}$. Therefore, there exist $t, s \in (0, 1)$ such that $v_d = tp + s\tilde{v}_d$, and hence also $w_d = tp' + s\tilde{w}_d$. We deduce from $\|p - p'\| \le 2\varepsilon \le \frac{2\varepsilon\sqrt{d-1}}{\sin\varphi}$ and (26) that

$$\|v_d - w_d\| \le (t+s)\frac{2\varepsilon\sqrt{d-1}}{\sin\varphi} \le \frac{4\varepsilon\sqrt{d-1}}{\sin\varphi}.$$

According to (25), we may choose $c = \frac{16\sqrt{d-1}}{\sin\varphi}$. **Q.E.D.**

We note that the lengthy calculations in the rest of paper (say, Sect. 7) are mostly aiming at providing upper estimates for the derivatives of $\Delta(\varphi_I - \varepsilon, r_2(\varphi_I - \varepsilon))$ (see (34)), $\Delta(\varphi_I - \varepsilon, r_2(\varphi_I) + \gamma_2\varepsilon)$ (see Lemma 8.1), $\Delta(\varphi_Q - \varepsilon, r_2(\varphi_Q - \varepsilon), r_3(\varphi_Q - \varepsilon))$ [see (43)] and $\Delta(\varphi_Q - \varepsilon, r_2(\varphi_Q - \varepsilon), r_3(\varphi_Q) + \gamma_3\varepsilon)$ (see Lemma 9.1) as a function of small $\varepsilon > 0$ where γ_2 and γ_3 are suitable large constants. These estimates can be obtained by some math computer packages based on formulas in Fejes Tóth [14, 15] and . However, we preferred a more theoretical approach, because the ideas can be used in any dimension for similar problems.

7 Volume Estimates Related to the Simplex Bound

To calculate or estimate $(d-1)$-volume of a compact $X \subset S^{d-1}$, we use Lemmas 7.1 and 7.2.

Lemma 7.1 *If $t \in (0, 1)$, and $X \subset S^{d-1}$, $d \geq 3$, is spherically convex that for some $v \in X$ satisfies $\langle u, v\rangle \geq t$ for all $u \in X$, then $\mathcal{H}^{d-1}(X) \geq \mathcal{H}^{d-1}(X')$ holds for the radial projection X' of X into $tv + v^{\perp}$.*

Proof The statement follows from the fact that the orthogonal projection of X into $tv + v^{\perp}$ covers X'. **Q.E.D.**

Lemma 7.2 *If $v \in S^{d-1}$, $d \geq 3$, and $X \subset S^{d-1}$ is compact and satisfies $\delta(u, v) \leq \Theta$, $\Theta < \frac{\pi}{2}$, for all $u \in X$, and $\widetilde{X}$ is the radial projection of X into the tangent hyperplane to S^{d-1} at v, then*

$$\mathcal{H}^{d-1}(X) = \int_{\widetilde{X}} (1 + \|y - v\|^2)^{-d/2}\, d\mathcal{H}^{d-1}(y) \geq \cos^d \Theta \cdot \mathcal{H}^{d-1}(\widetilde{X}).$$

Proof The statement follows from the facts that if $y \in \widetilde{X}$, then $\|y\| = (1 + \|y - v\|^2)^{1/2}$ and $u = y/\|y\|$ satisfies $\langle u, v\rangle = (1 + \|y - v\|^2)^{-1/2} \geq \cos\Theta$. **Q.E.D.**

The main results of this section are Lemma 7.3, its Corollary 7.4, and Lemma 7.5, which provide estimates when we slightly deform the "regular" orthoscheme $\Theta(r_1(\varphi), \ldots, r_{d-1}(\varphi))$.

Lemma 7.3 *For $\varphi \in \left(0, \arcsin\sqrt{\frac{d}{4(d-1)}}\right)$, if $\varepsilon \in (0, \varphi)$, then*

$$|\Theta(r_1(\varphi - \varepsilon), \ldots, r_{d-1}(\varphi - \varepsilon))| > |\Theta(r_1(\varphi), \ldots, r_{d-1}(\varphi))|(1 - \aleph \cdot \varepsilon),$$

where $\aleph = d\, 2^{(d+3)/2}/\sin r_{d-1}(\varphi)$.

Proof We deduce from (17) that $r_{d-1}(\varphi) < \pi/4$. Let $v \in S^{d-1}$, let $H = v + v^{\perp}$ be the hyperplane tangent to S^{d-1} at v, and let σ be a spherical arc of length $\pi/4$ starting from v. For $\varepsilon \in (0, \varphi)$, we consider the spherical regular simplex $T(\varepsilon)$ whose spherical circumscribed ball is of center v and radius $r_{d-1}(\varphi - \varepsilon)$, and one vertex of $T(\varepsilon)$ is contained in σ. In particular,

$$|\Theta(r_1(\varphi - \varepsilon), \ldots, r_{d-1}(\varphi - \varepsilon))| = |T(\varepsilon)|/d!.$$

We write $\widetilde{T}(\varepsilon)$ to denote the radial projection of $T(\varepsilon)$ into H, which is a Euclidean regular simplex of circumradius $R(\varepsilon) = \tan r_{d-1}(\varphi - \varepsilon) < 1$. Bounding $\mathcal{H}^{d-1}(\widetilde{T}(0)) \leq 2^{\frac{d}{2}}|T(0)|$ by Lemma 7.2 we deduce that

$$\begin{aligned}
|T(0)| - |T(\varepsilon)| &\leq |\widetilde{T}(0)\backslash\widetilde{T}(\varepsilon)| \\
&= \left(1 - \frac{R(\varepsilon)^{d-1}}{R(0)^{d-1}}\right)\mathcal{H}^{d-1}(\widetilde{T}(0)) \\
&\leq \left(1 - \left(1 - \frac{R(0) - R(\varepsilon)}{R(0)}\right)^{d-1}\right) 2^{d/2}|T(0)| \\
&\leq \frac{R(0) - R(\varepsilon)}{R(0)} \cdot d\, 2^{d/2}|T(0)|. \qquad (30)
\end{aligned}$$

For $r(\varepsilon) = r_{d-1}(\varphi - \varepsilon)$, we deduce from (17) that $r'(\varepsilon) = -\frac{\cos(\varphi-\varepsilon)}{\cos r(\varepsilon)}\sqrt{\frac{2(d-1)}{d}}$, therefore,

$$R'(\varepsilon) = (1 + R(\varepsilon)^2)r'(\varepsilon) \geq -\frac{\sqrt{2}(1 + R(\varepsilon)^2)}{\cos r(0)} \geq -\frac{2^{3/2}}{\cos r(0)}.$$

Using (30) and $R(0) \cdot \cos r(0) = \sin r_{d-1}(\varphi)$,

$$\frac{|T(0)| - |T(\varepsilon)|}{|T(0)|} \leq \frac{2^{3/2}\varepsilon}{R(0) \cdot \cos r(0)} \cdot d\, 2^{d/2} = \frac{d\, 2^{(d+3)/2}}{\sin r_{d-1}(\varphi)}\varepsilon.$$

Q.E.D.

Corollary 7.4 *For* $\varphi \in \left(0, \arcsin\sqrt{\frac{d}{4(d-1)}}\right)$, *if* $\varepsilon \in (0, \frac{1}{2\aleph})$ *for the* $\aleph$ *of Lemma* 7.3, *then*

$$\Delta(r_1(\varphi - \varepsilon), \ldots, r_{d-1}(\varphi - \varepsilon)) \leq \Delta(r_1(\varphi), \ldots, r_{d-1}(\varphi))(1 + 2\aleph \cdot \varepsilon).$$

Proof $1 + 2\aleph\varepsilon \geq 1/(1 - \aleph\varepsilon)$ so, according to Lemma 7.3, it is sufficient to prove that if $0 < s < \varphi$, then, for any $\tau < r_1(s)$,

$$|B(z_0, \tau) \cap \Theta(r_1(s), \ldots, r_{d-1}(s))| \leq |B(z_0, \tau) \cap \Theta(r_1(\varphi), \ldots, r_{d-1}(\varphi))|. \quad (31)$$

Essentially, this statement means that the angle measure at a vertex of a regular spherical simplex increases when the side length of the simplex increases. For the sake of completeness we give an argument for this statement.

Consider two regular spherical simplices T' and T with side lengths $2s$ and 2φ respectively such that they share a common center v and each vertex z_i' of T' belongs to the arc $\overline{z_i, v}$. Triangle $[z_1', z_2', v]$ is inside $[z_1, z_2, v]$ so the area of $[z_1', z_2', v]$ is less than the area of $[z_1, z_2, v]$. Since the area of a spherical triangle is the sum of its angles minus π, the angle between $\overline{z_1', z_2'}$ and $\overline{z_1', z_v'}$ is less than the angle between $\overline{z_1, z_2}$ and $\overline{z_1, v}$.

Now we consider two regular simplices T' of side length $2s$ with vertices $z_0, z_1', \ldots, z_{d-1}'$ and T of side length 2φ with vertices $z_0, z_1, \ldots, z_{d-1}$ such that the center v' of T' belongs to the arc $\overline{v, z_0}$, where v is the center of T, and all triangles $[z_0, v, z_i]$ and $[z_0, v', z_i']$ overlap. Then all arcs $\overline{z_0, z_i}$ belong to the cone formed by T at z_0 because all corresponding 2-dimensional angles in T' are smaller than those in T. Therefore, the angle measure for T' is smaller than the one for T. **Q.E.D.**

We set up a notation for Lemma 7.5. For $\varphi \in (0, \frac{\pi}{4})$, let $z_0 = z_0(\varphi), z_1(\varphi), \ldots, z_{d-1}(\varphi)$ be the vertices of $\Theta(r_1(\varphi), \ldots, r_{d-1}(\varphi))$. For $t \in [r_{d-1}(\varphi), \frac{\pi}{2})$, we set

$$\widetilde{\Theta}(\varphi, t) = \Theta(r_1(\varphi), \ldots, r_{d-2}(\varphi), t),$$

and we may assume that $z_0(\varphi), \ldots, z_{d-2}(\varphi)$ are vertices of $\widetilde{\Theta}(\varphi, t)$, and its d-th vertex $z_{d-1}(\varphi, t)$ satisfies $z_{d-1}(\varphi) \in \overline{z_{d-2}(\varphi), z_{d-1}(\varphi, t)}$.

Lemma 7.5 *If* $\varphi \in \left(0, \arcsin\sqrt{\frac{d}{4(d-1)}}\right)$ *and* $t \in (\varphi, \frac{\pi}{3})$, *then*

$$\left|\widetilde{\Theta}(\varphi, t)\backslash\widetilde{\Theta}(\varphi, r_{d-1}(\varphi))\right| \geq \frac{t - r_{d-1}(\varphi)}{2^d} \cdot \left|\widetilde{\Theta}(\varphi, r_{d-1}(\varphi))\right|.$$

Proof For brevity, we set $z_i = z_i(\varphi)$ for $i = 0, \ldots, d-1$, and $r_{d-1} = r_{d-1}(\varphi)$. The condition on φ yields that $r_{d-1} \leq \frac{\pi}{4}$.

Let s be the length of the arc $\overline{z_{d-1}, z_{d-1}(\varphi, t)}$. Since the length of the arc $\overline{z_{d-1}, z_0}$ is r_{d-1}, and the angle of these two arcs is $\arccos \frac{-1}{d}$, the Law of Cosines (15) yields

$$\cos t = \cos r_{d-1} \cos s - (\sin r_{d-1} \sin s)/d,$$

we deduce from $\sin t \geq \sin r_{d-1}$ that

$$\frac{dt}{ds} = \frac{\cos r_{d-1} \sin s + (\sin r_{d-1} \cos s)/d}{\sin t} \leq \frac{1}{\sin r_{d-1}},$$

therefore,

$$s \geq (t - r_{d-1}) \sin r_{d-1}. \tag{32}$$

We set $\widetilde{\Theta} = \widetilde{\Theta}(\varphi, r_{d-1}(\varphi))$, and observe that the closure of $\widetilde{\Theta}(\varphi, t)\backslash\widetilde{\Theta}$ is the spherical simplex T with vertices $z_0, \ldots, z_{d-3}, z_{d-1}, z_{d-1}(\varphi, t)$. Let H be the hyperplane tangent to S^{d-1} at z_{d-1}, and we write X' to denote the radial projection of some $X \subset S^{d-1}$ in H. It follows that $\widetilde{\Theta}'$ is the Euclidean orthoscheme such that $d!$ of its copies tile the Euclidean regular simplex of circumradius $\tan r_{d-1} \leq 1$, and hence $\|z'_{d-2} - z'_{d-1}\| = (\tan r_{d-1})/(d-1)$. We deduce from Lemma 7.2 and (32) that

$$\begin{aligned}|T| &\geq \frac{|T'|}{2^d} = \frac{|\widetilde{\Theta}'| \tan s}{2^d \|z'_{d-2} - z'_{d-1}\|} \geq \frac{|\widetilde{\Theta}'|(t - r_{d-1}) \sin r_{d-1}}{2^d (\tan r_{d-1})/(d-1)} \\ &\geq \frac{|\widetilde{\Theta}'|(t - r_{d-1})}{2^d} \geq \frac{|\widetilde{\Theta}|(t - r_{d-1})}{2^d}.\end{aligned}$$

Q.E.D.

8 The Case of the Icosahedron

In this section, we write I to denote the regular icosahedron with vertices on S^2. In particular,

$$\varphi_I = \frac{1}{2} \arccos \frac{1}{\sqrt{5}} < \arcsin\sqrt{\frac{3}{8}}, \tag{33}$$

thus Corollary 7.4 and Lemma 7.5 can be applied with $\varphi = \varphi_I$. Since S^2 can be dissected into 120 congruent copies of $\Theta(\varphi_I, r_2(\varphi_I))$, we have

$$|\Theta(\varphi_I, r_2(\varphi_I))| = \frac{\pi}{30},$$

and it follows from (24) that

$$\Delta(\varphi_I, r_2(\varphi_I)) = \frac{3}{\pi}.$$

According to (17), we have $\sin r_2(\varphi_I) = \frac{2}{\sqrt{3}} \sin \varphi_I$, thus the constant $\aleph$ of Lemma 7.3 satisfies $\aleph = \frac{3 \cdot 2^3}{\sin r_2(\varphi_I)} < 40$. In particular, Corollary 7.4 yields that if $\varepsilon \in (0, 0.01)$, then

$$\Delta(\varphi_I - \varepsilon, r_2(\varphi_I - \varepsilon)) < \frac{3}{\pi}(1 + 80\varepsilon) < \frac{3}{\pi} + 80\varepsilon. \tag{34}$$

We also note that if $v \in S^2$ and $\eta \in (0, \frac{\pi}{2})$, then

$$|B(v, \eta)| = 2\pi(1 - \cos \eta). \tag{35}$$

Lemma 8.1 *For $\gamma \geq 10^4$ and $\varepsilon \in (0, \frac{1}{100\gamma})$, we have*

$$\Delta(\varphi_I - \varepsilon, r_2(\varphi_I) + \gamma\varepsilon) \leq \Delta(\varphi_I, r_2(\varphi_I)) - \frac{\gamma\varepsilon}{200}.$$

Proof To simplify the notation, we write $\varphi = \varphi_I$ and $r_2 = r_2(\varphi) = \arcsin \frac{2 \sin \varphi}{\sqrt{3}}$, which satisfy $r_2 + \gamma\varepsilon < \frac{\pi}{3}$ (in order to apply Lemma 7.5). We may assume that $\Theta(\varphi - \varepsilon, r_2(\varphi - \varepsilon))$ and $\Theta(\varphi - \varepsilon, r_2 + \gamma\varepsilon)$ share a side of length $\varphi - \varepsilon$.

We deduce from $r_2(\varphi - \varepsilon) \leq r_2$ that $(r_2 + \gamma\varepsilon) - r_2(\varphi - \varepsilon) \geq \gamma\varepsilon$.

We set T to be the closure of

$$\Theta(\varphi - \varepsilon, r_2 + \gamma\varepsilon) \backslash \Theta(\varphi - \varepsilon, r_2(\varphi - \varepsilon)),$$

thus Lemma 7.5 yields

$$|T| \geq \frac{\gamma\varepsilon}{8} \cdot |\Theta(\varphi - \varepsilon, r_2(\varphi - \varepsilon))|. \tag{36}$$

In addition, if $\sigma \in (0, \varphi - \varepsilon)$, then we deduce from $\varepsilon < 10^{-6}$, that

$$\begin{aligned} \frac{|T \cap B(z_0, \sigma)|}{|B(z_0, \sigma)| \cdot |T|} &< \frac{|T \cap B(z_0, \sigma)|}{|B(z_0, \sigma)| \cdot |T \cap B(z_0, r_2(\varphi - \varepsilon))|} = \frac{|B(z_0, \sigma)|}{|B(z_0, \sigma)| \cdot |B(z_0, r_2(\varphi - \varepsilon))|} \\ &\leq \frac{1}{|B(z_0, r_2(\varphi - 10^{-6}))|} = \Delta_0 < \frac{3}{\pi} - 0.175, \end{aligned}$$

because $\Delta_0 \approx 0.7751$ and $\frac{3}{\pi} - 0.175 \approx 0.7799$.

Therefore $\gamma \geq 10^4$ yields

$$\begin{aligned}\Delta(\varphi-\varepsilon, r_2+\gamma\varepsilon) &\leq \frac{(\frac{3}{\pi}+80\varepsilon)|\Theta(\varphi-\varepsilon, r_2(\varphi-\varepsilon))|+\Delta_0|T|}{|\Theta(\varphi-\varepsilon, r_2(\varphi-\varepsilon))|+|T|}\\ &\leq \frac{3}{\pi}+80\varepsilon-\left(\frac{3}{\pi}+80\varepsilon-\Delta_0\right)\frac{\gamma\varepsilon/8}{1+\frac{\gamma\varepsilon}{8}}\\ &= \frac{3}{\pi}+\gamma\varepsilon\left(\frac{80}{\gamma}-\frac{\frac{3}{\pi}+80\varepsilon-\Delta_0}{8+\gamma\varepsilon}\right)\\ &\leq \frac{3}{\pi}+\gamma\varepsilon\left(10^{-2}-\frac{\frac{3}{\pi}-\Delta_0}{10}\right)\leq\frac{3}{\pi}-\frac{\gamma\varepsilon}{200}.\end{aligned}$$

Q.E.D.

The following two simple statements are useful tools in the case of the 600-cell as well.

Lemma 8.2 *If $T \subset \mathbb{R}^2$ is a triangle such that all sides are of length at least a, and the center of the circle passing through the vertices lies in T, then $|T| \geq \frac{\sqrt{3}}{4}a^2$.*

Proof The largest angle α of T satisfies $\frac{\pi}{3} \leq \alpha \leq \frac{\pi}{2}$. **Q.E.D.**

Lemma 8.3 *For $x, y, v \in S^2$, let $\delta(x, y) \geq 2\psi$, and let $\delta(x, v) = \delta(y, v) = R$ for $0 < \psi < R < \frac{\pi}{2}$. If the angle between $\overline{v, x}$ and $\overline{v, y}$ is ω, then*

(i) $\cos\omega \leq 1 - \frac{2\sin^2\psi}{\sin^2 R}$;
(ii) *If $\psi = \varphi - \varepsilon$ and $R \leq r + \gamma\varepsilon$ where $\psi < \varphi < r < \frac{\pi}{2} - \gamma\varepsilon$ and $\gamma > 1$, then* $\cos\omega \leq 1 - \frac{2\sin^2\varphi}{\sin^2 r} + \frac{4\gamma\varepsilon}{\sin^2 r}$.

Proof For (i), the Spherical Law of Cosines (15) yields

$$1-2\sin^2\psi = \cos 2\psi \geq \cos^2 R + (\sin^2 R)\cos\omega = 1-(1-\cos\omega)\sin^2 R.$$

Turning to (ii), we deduce from $\frac{d}{dt}\sin^2 t = \sin 2t \leq 1$ that

$$\frac{2\sin^2(\varphi-\varepsilon)}{\sin^2(r+\gamma\varepsilon)} \geq \frac{2(\sin^2\varphi-\varepsilon)}{\sin^2 r+\gamma\varepsilon} = \frac{(1-\frac{\varepsilon}{\sin^2\varphi})2\sin^2\varphi}{(1+\frac{\gamma\varepsilon}{\sin^2\varphi})\sin^2 r} \geq \frac{\left(1-\frac{(\gamma+1)\varepsilon}{\sin^2\varphi}\right)2\sin^2\varphi}{\sin^2 r},$$

and hence (i) implies (ii). **Q.E.D.**

Proof of Theorem 1.1 in the case of the icosahedron Let I be the icosahedron with vertices on S^2, therefore, the vertices determine the optimal packing of 12 spherical circular discs of radius $\varphi_I = \frac{1}{2}\arccos\frac{1}{\sqrt{5}}$. We set $\varphi = \varphi_I$, $r_2 = r_2(\varphi)$ and $r_\infty = r_\infty(\varphi)$. For $\varepsilon_0 = 10^{-9}$ and $\eta = 0.11$, we observe that

$$r_2 + 10^7\varepsilon_0 < r_2 + \eta < r_\infty - \eta. \tag{37}$$

Let $\varepsilon \in (0, \varepsilon_0)$, and let $x_1, \ldots, x_k \in S^2$ satisfy that $k \geq 12$, and $\delta(x_i, x_j) \geq 2(\varphi - \varepsilon)$ for $i \neq j$. We may assume that for any $x \in S^2$ there exists x_i such that $\delta(x_i, x) < 2(\varphi - \varepsilon)$. Let $P = [x_1, \ldots, x_k]$, and hence $o \in \text{int}\, P$. We prove Theorem 1.1 for the icosahedron in two steps.

Step 1 *Proving that all Delone cells are of circumradius at most $r_2 + 10^7\varepsilon$*

We suppose that there exists a Delone cell of spherical circumradius at least $r_2 + 10^7\varepsilon$, and seek a contradiction. Let us consider the triangulation of S^2 by all quasi-orthoschemes associated to the Dirichlet cell decomposition induced by $x_1, \ldots, x_k$. Among them, let $\mathcal{O}$ and $\mathcal{Q}$ denote the family of the ones with diameter less than $r_2 + 10^7\varepsilon$, and with diameter at least $r_2 + 10^7\varepsilon$, respectively. We claim that

$$\sum_{\Xi \in \mathcal{Q}} |\Xi| \geq 2\pi(1 - \cos\eta) > 0.03. \tag{38}$$

Let $\varrho > 0$ be the largest number such that $\varrho B^3 \subset P$, and let $R = \arccos \varrho$. Then ϱB^3 touches ∂P at a point $y \in \partial P$ in the relative interior of a two-dimensional face F of P, R is the spherical circumradius of the corresponding Delone cell, and $R \geq r_2 + 10^7\varepsilon$. By construction, R is the maximal circumradius among all Delone cells.

We may assume that x_1, x_2, x_3 are vertices of F such that $y \in [x_1, x_2, x_3] = T$. Let $v = y/\|y\|$, and let $\widetilde{T}$ be the radial projection of T into S^2, that is the associated spherical "Delone triangle", and satisfies $v \in \widetilde{T}$. If $R < r_\infty$, then all quasi-orthoschemes having vertex v are actual orthoschemes by Lemma 6.1, and hence their union is $\widetilde{T}$. In particular, Lemmas 7.1 and 8.2 yield that

$$\sum_{\Xi \in \mathcal{Q}} |\Xi| \geq |\widetilde{T}| \geq |T| \geq \frac{\sqrt{3}}{4}(2\sin(\varphi - \varepsilon_0))^2 > 0.4.$$

However, if $R \geq r_\infty$ and $x \in B(v, \eta)$, then $\delta(x, x_i) \geq r_2 + \eta$ for all $i = 1, \ldots, k$, thus any quasi-orthoscheme Ξ containing x has a diameter at least $r_2 + 10^7\varepsilon$ by (37). Therefore,

$$\sum_{\Xi \in \mathcal{Q}} |\Xi| \geq |B(v, \eta)| = 2\pi(1 - \cos\eta)$$

in this case, proving (38).

We note that $12 = \frac{3}{\pi} \cdot |S^2|$ according to the equality case of the simplex bound (24). We deduce from (34), Lemma 8.1 with $\gamma = 10^7$ and (38) that

$$\begin{aligned} k &\leq \sum_{\Xi \in \mathcal{O}} |\Xi| \frac{3}{\pi} \cdot (1 + 80\varepsilon) + \sum_{\Xi \in \mathcal{Q}} |\Xi| \left(\frac{3}{\pi} - 50{,}000\varepsilon\right) \\ &\leq 12 + \frac{3}{\pi}[4\pi \cdot 80\varepsilon - 0.03 \cdot 50{,}000 \cdot \varepsilon] < 12. \end{aligned}$$

This contradiction completes the proof of Step 1.

Step 2 *Assuming all Delone cells are of circumradius at most* $r_2 + 10^7\varepsilon$

It follows from (24) and (34) that $k = 12$.

We set $\gamma = 10^7$. Let Ω be a Delone cell, and let v be the center of the circumcircle of radius R. We claim that Ω is a triangle, and there exists a regular spherical triangle Ω_0 of side length 2φ, such that for any vertex x_i of Ω there exists a vertex w of Ω_0 with

$$\delta(x_i, w) \leq 25\gamma\varepsilon. \tag{39}$$

If $x_i \neq x_j$ are the vertices of Ω, and the angle between $\overline{v, x_i}$ and $\overline{v, x_j}$ is ω_{ij}, then Lemma 8.3, $\sin\varphi / \sin r_2 = \sqrt{3}/2$ and $\gamma\varepsilon < 10^{-2}$ yield

$$\cos\omega_{ij} \leq 1 - \frac{2\sin^2\varphi}{\sin^2 r_2} + \frac{4\gamma\varepsilon}{\sin^2 r_2} \leq \frac{-1}{2} + 12\gamma\varepsilon < 0.$$

In particular, Ω is a triangle by Corollary 2.3. Since $(\cos t)' = -\sin t$ is at most $\frac{-3}{4}$ if $t \in [\frac{\pi}{2}, \frac{2\pi}{3}]$, we have

$$\omega_{ij} \geq \frac{2\pi}{3} - 16\gamma\varepsilon. \tag{40}$$

We deduce from the Remark after Theorem 3.1 that one may find a regular spherical triangle Ω' with vertices on the spherical circle with center v and radius R such that for any vertex x_i of Ω there exists a vertex w' of Ω' such that the angle between $\overline{x_i, v}$ and $\overline{w', v}$ is at most $24\gamma\varepsilon$, and hence $\delta(x_i, w') \leq 24\gamma\varepsilon$. We take Ω_0 with the circumcenter v so that for any vertex w of Ω_0 there exists a vertex w' of Ω' such that $w \in \overline{w', v}$ or $w' \in \overline{w, v}$. As $R \leq r_2 + \gamma\varepsilon$ by the condition of Step 2, and $R \geq r_2(\varphi - \varepsilon) \geq r_2 - \gamma\varepsilon$, we conclude (39) by the triangle inequality.

Now we fix a Delone cell Θ and let Θ_0 be the spherical regular triangle provided by (39). We observe that $c < 44$ for the constant of Lemma 6.4 in our case. We may assume that the vertices of Θ_0 are vertices of the face F_0 of the icosahedron I. There exist nine more faces $F_1, \ldots, F_9$ of I, such that $F_i \cap F_{i-1}$ is a common edge for $i = 1, \ldots, 9$, and any vertex of I is a vertex of some F_i, $i \leq 9$. Attaching the corresponding nine more Delone cells to Θ, we conclude from Lemma 6.4 that we may choose $c_I = 44^9 \cdot 25\gamma$. **Q.E.D.**

9 The Case of the 600-Cell

In this section, by Q we denote the regular 600-cell with vertices on S^2. In particular,

$$\varphi_Q = \frac{\pi}{10} < \arcsin\sqrt{\frac{1}{3}} \tag{41}$$

thus Corollary 7.4 and Lemma 7.5 can be applied with $\varphi = \varphi_Q$. Since S^3 can be dissected into 14400 congruent copies of $\Theta(\varphi_Q, r_2(\varphi_Q), r_3(\varphi_Q))$, we have

$$|\Theta(\varphi_Q, r_2(\varphi_Q), r_3(\varphi_Q))| = \frac{|S^3|}{14400} = \frac{\pi^2}{7200},$$

and it follows from (24) that

$$\Delta(\varphi_Q, r_2(\varphi_Q), r_3(\varphi_Q)) = \frac{60}{\pi^2}. \tag{42}$$

The main idea of the argument in the case of the 600-cell will be similar to the one for the icosahedron. According to (17), we have $\sin r_3(\varphi_Q) = \sqrt{\frac{3}{2}} \sin \varphi_Q$, thus the constant $\aleph$ of Lemma 7.3 satisfies $\aleph = \frac{4 \cdot 2^{7/2}}{\sin r_3(\varphi_Q)} < 120$. In particular, Corollary 7.4 yields that if $\varepsilon \in (0, 0.004)$, then

$$\Delta(\varphi_Q - \varepsilon, r_2(\varphi_Q - \varepsilon), r_3(\varphi_Q - \varepsilon)) < \frac{60}{\pi^2}(1 + 240\varepsilon) < \frac{60}{\pi^2} + 1500\varepsilon. \tag{43}$$

Next Lemma 9.1 estimates $\Delta(\varphi_Q - \varepsilon, r_2(\varphi_Q - \varepsilon), r_3(\varphi_Q) + \gamma\varepsilon)$ for large γ and small $\varepsilon > 0$, and Lemma 9.2 estimates the volume of a tetrahedron.

Lemma 9.1 *For $\gamma \geq 10^6$ and $\varepsilon \in (0, \frac{1}{100\gamma})$, we have*

$$\Delta(\varphi_Q - \varepsilon, r_2(\varphi_Q - \varepsilon), r_3(\varphi_Q) + \gamma\varepsilon) \leq \Delta(\varphi_Q, r_2(\varphi_Q), r_3(\varphi_Q)) - \frac{\gamma\varepsilon}{100}.$$

Proof To simplify notation, we write $\varphi = \varphi_Q$ and $r_3 = r_3(\varphi) = \arcsin \frac{3 \sin \varphi}{2}$, and use the notation set up before Lemma 7.5.

We deduce from $r_3(\varphi - \varepsilon) \leq r_3$ that $(r_3 + \gamma\varepsilon) - r_3(\varphi - \varepsilon) \geq \gamma\varepsilon$.

For the closure T of

$$\widetilde{\Theta}(\varphi - \varepsilon, r_3 + \gamma\varepsilon) \backslash \widetilde{\Theta}(\varphi - \varepsilon, r_3(\varphi - \varepsilon)),$$

Lemma 7.5 yields

$$|T| \geq \frac{\gamma\varepsilon}{16} \cdot |\widetilde{\Theta}(\varphi - \varepsilon, r_3(\varphi - \varepsilon))|. \tag{44}$$

Let $\sigma \in (0, \varphi - \varepsilon_0)$. We consider two spherical cones C and C_0, where C is obtained by rotating the triangle with vertices $z_0, z_1(\varphi - \varepsilon), z_3(\varphi - \varepsilon)$ around $z_0, z_1(\varphi - \varepsilon)$, and C_0 is obtained by rotating the triangle with vertices $z_0, z_1(\varphi - \varepsilon_0), z_3(\varphi - \varepsilon_0)$ around $z_0, z_1(\varphi - \varepsilon_0)$. For the two-face F of T opposite to z_0, $F \backslash C$ is disjoint from $B(z_0, r_3(\varphi - \varepsilon))$, which in turn contains C, and hence we have the density estimates

$$\frac{|(T\backslash C)\cap B(z_0,\sigma)|}{|T\backslash C|\cdot|B(z_0,\sigma)|}\leq\frac{|B(z_0,\sigma)|}{|B(z_0,r_3(\varphi-\varepsilon))|\cdot|B(z_0,\sigma)|}\leq\frac{|C\cap B(z_0,\sigma)|}{|C|\cdot|B(z_0,\sigma)|}.$$

Since the density of $B(z_0,\sigma)$ in $C\cap T$ is $\frac{|C\cap B(z_0,\sigma)|}{|C|}$, and in $T\backslash C$ the density is at most $\frac{|C\cap B(z_0,\sigma)|}{|C|}$, we deduce using (19) and the argument of Boroczky [4] that

$$\begin{aligned}\frac{|T\cap B(z_0,\sigma)|}{|T|\cdot|B(z_0,\sigma)|}&\leq\frac{|C\cap B(z_0,\sigma)|}{|C|\cdot|B(z_0,\sigma)|}=\lim_{s\to 0^+}\Delta(\varphi-\varepsilon,r_3(\varphi-\varepsilon)-s,r_3(\varphi-\varepsilon))\\&\leq\lim_{s\to 0^+}\Delta(\varphi-\varepsilon_0,r_3(\varphi-\varepsilon_0)-s,r_3(\varphi-\varepsilon_0))\\&\leq\frac{|C_0\cap B(z_0,\sigma)|}{|C_0|\cdot|B(z_0,\sigma)|}=\Delta_0.\end{aligned}\tag{45}$$

Now C_0 is a spherical cone whose base is a circular disc of radius $\xi=\arccos\frac{\cos r_3(\varphi-\varepsilon_0)}{\cos(\varphi-\varepsilon_0)}$, center $z_1(\varphi-\varepsilon_0)$ and height $\varphi-\varepsilon_0$. Let $H\subset\mathbb{R}^4$ be the hyperplane tangent to S^3 at $z_1(\varphi-\varepsilon_0)$, let C_0' be the radial projection of C_0 into H, which is a Euclidean cone whose base is a circular disc of radius $\varrho=\tan\xi$, center $z_1(\varphi-\varepsilon_0)$ and height $h=\tan(\varphi-\varepsilon_0)$. Therefore, Lemma 7.2 yields

$$\begin{aligned}|C_0|&=\int_{C_0'}(1+\|x-z_1(\varphi-\varepsilon_0)\|^2)^{-2}\,dx\\&=\int_0^h\int_0^{\varrho-\frac{\varrho t}{h}}(1+t^2+r^2)^{-2}\cdot 2\pi r\,drdt.\end{aligned}$$

In addition, if the angle between the arcs $\overline{z_0,z_1(\varphi-\varepsilon_0)}$ and $\overline{z_0,z_3(\varphi-\varepsilon_0)}$ is α, then $\cos\alpha=\frac{\tan(\varphi-\varepsilon_0)}{\tan r_3(\varphi-\varepsilon_0)}$. Therefore, (35) yields

$$\Delta_0=\frac{1-\cos\alpha}{2|C_0|}<\frac{60}{\pi^2}-0.3.$$

For $\Delta=\Delta(\varphi-\varepsilon,r_2(\varphi-\varepsilon),r_3+\gamma\varepsilon)$, $\gamma\geq 10^6$ yields

$$\begin{aligned}\Delta&\leq\frac{(\frac{60}{\pi^2}+1500\varepsilon)\widetilde{\Theta}(\varphi-\varepsilon,r_3(\varphi-\varepsilon))|+\Delta_0|T|}{|\widetilde{\Theta}(\varphi-\varepsilon,r_2(\varphi-\varepsilon))|+|T|}\\&\leq\frac{60}{\pi^2}+1500\varepsilon-\left(\frac{60}{\pi^2}+1500\varepsilon-\Delta_0\right)\frac{\gamma\varepsilon/16}{1+\frac{\gamma\varepsilon}{16}}\\&=\frac{60}{\pi^2}+\gamma\varepsilon\left(\frac{1500}{\gamma}-\frac{\frac{60}{\pi^2}+1500\varepsilon-\Delta_0}{16+\gamma\varepsilon}\right)\\&\leq\frac{60}{\pi^2}+\gamma\varepsilon\left(2\cdot 10^{-3}-\frac{\frac{60}{\pi^2}-\Delta_0}{20}\right)\leq\frac{60}{\pi^2}-\frac{\gamma\varepsilon}{100}.\end{aligned}$$

Q.E.D.

Lemma 9.2 *If* $\theta \in (0, \frac{1}{3})$*, and* $u_1, u_2, u_3, u_4 \in S^2$ *satisfy that* $\langle u_i, u_j\rangle \le -\theta$ *for* $i \neq j$*, then*

$$\mathcal{H}^3([u_1, u_2, u_3, u_4]) \ge \sqrt{\theta}/4.$$

Proof For $T = [u_1, u_2, u_3, u_4]$, we have $o \in \text{int}\, T$ by Lemma 2.2. Let $r > 0$ be the maximal number such that $rB^3 \subset T$, and hence $r \le \frac{1}{3}$ (see, say, Boroczky [5], Section 6.5). We may assume that rB^3 touches ∂T in a point y of $F = [u_1, u_2, u_3]$, which lies in the relative interior of F. We set $u = y/r \in S^2$, and $v_i = (u_i - y)/\sqrt{1-r^2} \in S^2$ for $i = 1, 2, 3$. We have $\alpha \in [\arccos\frac{1}{3}, \frac{\pi}{2})$ and $\beta \in (\frac{\pi}{2}, \pi]$ such that $\delta(u_i, u) = \alpha$ for $i = 1, 2, 3$, $\delta(u_4, u) = \beta$. Thus $u_i = u\cos\alpha + v_i \sin\alpha$ for $i = 1, 2, 3$, and $u_4 = -u|\cos\beta| + w\sin\beta$ for some $w \in u^{\perp} \cap S^2$.

Since $\langle u_i, u_j\rangle < 0$ for $1 \le i < j \le 3$, we have $\langle v_i, v_j\rangle = \langle u_i, u_j\rangle - \cos\alpha\cos\alpha < 0$ for $1 \le i < j \le 3$. We deduce that $\|u_i - u_j\| \ge \sqrt{2(1-r^2)}$ for $1 \le i < j \le 3$, and there exists $l \in \{1, 2, 3\}$ such that $\langle v_l, w\rangle > 0$. In particular, we have

$$-\theta \ge \langle u_4, u_l\rangle \ge -|\cos\beta| \cdot \cos\alpha.$$

It follows from Lemma 8.2 and $1 - r^2 \ge \frac{8}{9}$ that

$$\mathcal{H}^3(T) = \frac{|\cos\beta| + \cos\alpha}{4} \cdot \mathcal{H}^2(F) \ge \frac{\sqrt{|\cos\beta| \cdot \cos\alpha}}{2} \cdot \frac{\sqrt{3}(1-r^2)}{2} > \frac{\sqrt{\theta}}{4}.$$

Q.E.D.

It is not hard to see that the lower bound $\sqrt{\theta}/4$ in Lemma 9.2 can't be replaced by, say, $2\sqrt{\theta}$.

Proof of Theorem 1.1 **in the case of the 600 -cell** Let Q be an 600-cell with vertices on S^3, therefore, its vertices determine the optimal packing of 120 spherical circular discs of radius $\varphi_Q = \frac{\pi}{10}$. We set $\varphi = \varphi_Q$, $r_2 = r_2(\varphi)$, $r_3 = r_3(\varphi)$ and $r_\infty = r_\infty(\varphi)$. For $\gamma = 10^{12}$, $\varepsilon_0 = 10^{-14}$ and $\eta = 0.02$, we observe that

$$r_3 + \gamma\varepsilon_0 < r_3 + \eta < r_\infty - 2\eta. \tag{46}$$

Let $\varepsilon \in (0, \varepsilon_0)$, and let $x_1, \ldots, x_k \in S^2$ satisfy that $k \ge 120$, and $\delta(x_i, x_j) \ge 2(\varphi - \varepsilon)$ for $i \neq j$. We may assume that for any $x \in S^3$, there exists x_i such that $\delta(x_i, x) < 2(\varphi - \varepsilon)$. Let $P = [x_1, \ldots, x_k]$, and hence $o \in \text{int}\, P$. We prove Theorem 1.1 for the 600-cell in two steps.

Step 1 *Proving that all Delone cells are of circumradius at most* $r_3 + \gamma\varepsilon$

We suppose that there exists a Delone cell of spherical circumradius at least $r_3 + \gamma\varepsilon$ and seek a contradiction. Let us consider the triangulation of S^3 by all quasi-orthoschemes associated to the Dirichlet cell decomposition induced by $x_1, \ldots, x_k$. Among them, let $\mathcal{O}$ and $\mathcal{Q}$ denote the family of the ones with diameter less than $r_3 + \gamma\varepsilon$, and with diameter at least $r_3 + \gamma\varepsilon$, respectively. We claim that

$$\sum_{\Xi\in\mathcal{Q}} |\Xi| > (4\pi/3)\sin^3\eta > 10^{-5}. \tag{47}$$

Let $\varrho > 0$ be the largest number such that $\varrho B^4 \subset P$ and let $R = \arccos\varrho$. Then ϱB^4 touches ∂P at a point $y \in \partial P$ in the relative interior of a three-dimensional face F of P, R is the spherical circumradius of the corresponding Delone cell, and $R \geq r_3 + \gamma\varepsilon$.

We may assume that x_1, x_2, x_3, x_4 are vertices of F in a way such that $y \in [x_1, x_2, x_3, x_4] = T$. Let $v = y/\|y\|$, and let $\widetilde{T}$ be the radial projection of T into S^3, that is the associated spherical "Delone simplex", and satisfies $v \in \widetilde{T}$. If $R < r_3 + 2\eta$, then all quasi-orthoschemes having vertex v are actual orthoschemes by Lemma 6.1, and hence their union is $\widetilde{T}$. If for some $\{i, j\} \subset \{1, 2, 3, 4\}$, the angle between $\overline{v, x_i}$ and $\overline{v, x_j}$ is ω_{ij}, then Lemma 8.3 yields

$$\cos\omega_{ij} \leq 1 - \frac{2\sin^2(\varphi - \varepsilon)}{\sin^2 R} < 1 - \frac{2\sin^2(\varphi - \varepsilon_0)}{\sin^2(r_3 + 2\eta)} < -0.1.$$

In particular, Lemmas 7.1 and 9.2 yield that

$$\sum_{\Xi\in\mathcal{Q}} |\Xi| \geq |\widetilde{T}| \geq |T| \geq \sqrt{0.1}/4 > 0.07.$$

However, if $R \geq r_3 + 2\eta$ and $x \in B(v, \eta)$, then $\delta(x, x_i) \geq r_3 + \eta$ for all $i = 1, \ldots, k$, thus any quasi-orthoscheme Ξ containing x has diameter at least $r_3 + \gamma\varepsilon$ by (46). We deduce from Lemma 7.1 that

$$\sum_{\Xi\in\mathcal{Q}} |\Xi| \geq |B(v, \eta)| = (4\pi/3)\sin^3\eta$$

in this case, proving (47).

We note that $120 = \frac{60}{\pi^2} \cdot |S^3|$ according to the equality case of the simplex bound (24). We deduce from (34), Lemma 8.1 with $\gamma = 10^{12}$ and (38) that

$$\begin{aligned} k &\leq \sum_{\Xi\in\mathcal{O}} |\Xi| \frac{60}{\pi^2} \cdot (1 + 1500\varepsilon) + \sum_{\Xi\in\mathcal{Q}} |\Xi| \frac{60}{\pi^2} \cdot (1 - 10^{10} \cdot \varepsilon) \\ &\leq 12 + \frac{60}{\pi^2}[2\pi^2 \cdot 1500\varepsilon - 10^{-5} \times 10^{10} \cdot \varepsilon] < 12. \end{aligned}$$

This contradiction completes the proof of Step 1.

Step 2 *Assuming all Delone cells are of circumradius at most* $r_3 + \gamma\varepsilon$

It follows from (24) and (43) that $k = 120$.

Let Ω be a Delone cell, and let v be the center of the circumscribed spherical ball of radius R. We claim that Ω is a spherical tetrahedron and there exists a regular spherical tetrahedron Ω_0 of side length 2φ such that for any vertex x_i of Ω there exists a vertex w of Ω_0 with

$$\delta(x_i, w) \leq 10{,}000\gamma\varepsilon. \tag{48}$$

If $x_i \neq x_j$ are the vertices of Ω, and the angle between $\overline{v, x_i}$ and $\overline{v, x_j}$ is ω_{ij}, then Lemma 8.3, $\sin\varphi/\sin r_3 = \sqrt{2/3}$ and $\gamma\varepsilon < 10^{-2}$ yield

$$\cos\omega_{ij} \leq 1 - \frac{2\sin^2\varphi}{\sin^2 r_2} + \frac{4\gamma\varepsilon}{\sin^2 r_3} \leq \frac{-1}{3} + 30\gamma\varepsilon < 0.$$

In particular, Ω is a tetrahedron by Corollary 2.3. Since $(\cos t)' = -\sin t$ is at most $\frac{-3}{4}$ if $t \in [\frac{\pi}{2}, \frac{2\pi}{3}]$, we have

$$\omega_{ij} \geq \arccos\frac{-1}{3} - 40\gamma\varepsilon. \tag{49}$$

We deduce from Theorem 3.1 that one may find a regular spherical tetrahedron Ω' with vertices on the subsphere with center v and radius R such that for any vertex x_i of Ω there exists a vertex w' of Ω' such that the angle between $\overline{x_i, v}$ and $\overline{w', v}$ is at most $9000\gamma\varepsilon$ and hence $\delta(x_i, w') \leq 9000\gamma\varepsilon$. We take Ω_0 with circumcenter v so that for any vertex w of Ω_0 there exists a vertex w' of Ω' such that $w \in \overline{w', v}$ or $w' \in \overline{w, v}$. As $R \leq r_3 + \gamma\varepsilon$ by the condition of Step 2, and $R \geq r_3(\varphi - \varepsilon) \geq r_3 - \gamma\varepsilon$, we conclude (48) by the triangle inequality.

Now we fix a Delone cell Θ and let Θ_0 be the spherical regular tetrahedron provided by (48). We observe that $c < 90$ for the constant of Lemma 6.4 in our case. We may assume that the vertices of Θ_0 are vertices of the face F_0 of the 600-cell Q. There exist 116 more faces $F_1, \ldots, F_{116}$ of Q, such that $F_i \cap F_{i-1}$ is a common edge for $i = 1, \ldots, 116$, and any vertex of Q is a vertex of some F_i, $i \leq 116$. Attaching the corresponding 116 more Delone cells to Θ, we conclude from Lemma 6.4 that we may choose $c_Q = 90^{116} \cdot 10{,}000\,\gamma$. **Q.E.D.**

References

1. J. Aczél, Solution to problem 35, I. (Hungarian). Mat. Lapok, **3**, 94-95 (1952)
2. N.N. Andreev, A spherical code. Russ. Math. Surv. **54**, 251–253 (1999)
3. K. Bezdek, *Classical Topics in Discrete Geometry* (Springer, 2010)
4. K. Böröczky, Packing of spheres in spaces of constant curvature. Acta Math. Hung. **32**, 243–261 (1978)
5. K. Böröczky Jr., *Finite Packing and Covering* (Cambridge Unversity Press, 2004)
6. P. Boyvalenkov, D. Danev, Uniqueness of the 120-point spherical 11-design in four dimensions. Arch. Math. **77**, 360–268 (2001)
7. S.A. Chepanov, S.S. Ryshkov, N.N. Yakovlev, On the disjointness of point systems. (Russian) Trudy Mat. Inst. Steklov. **196**, 147–155 (1991)
8. J.H. Conway, N.J.A. Sloane, *Sphere Packings, Lattices and Groups* (Springer, Berlin, New York, 1998)

9. H.S.M. Coxeter, *Regular Polytopes* (Originally Published in 1947) (Dover, 1973)
10. H. Davenport, Gy. Hajós, Problem 35 (Hungarian). Mat. Lapok, **2**, 68 (1951)
11. T. Ericson, V. Zinoviev, *Codes on Euclidean Spheres* (North-Holland, 2001)
12. G. Fejes Tóth, Packing and Covering, in *CRC Handbook on Discrete and Computational Geometry*, ed. by E. J. Goodman, J. O'Rourke (CRC Press, 2004)
13. L. Fejes Tóth, On the densest packing of spherical caps. Am. Math. Monthly **56**, 330–331 (1949)
14. L. Fejes Tóth, On the volume of a polyhedron in non-Euclidean spaces. Publ. Math. Debrecen **4**, 256–261 (1956)
15. L. Fejes Tóth, *Regular Figures* (Pergamon Press, Oxford, 1964)
16. L. Fejes Tóth, *Lagerungen in der Ebene, auf der Kugel und im Raum* (Springer, Berlin, 1972)
17. H.W.E. Jung, Über die kleineste Kugel die eine räumliche Figur einschliesst. J. Reine ang. Math. **123**, 241–257 (1901)
18. P. McMullen, E. Schulte, *Abstract Regular Polytopes* (Cambridge University Press. 2002)
19. O. Musin, The kissing number in four dimensions. Ann. Math. **168**, 1–32 (2008)
20. R.A. Rankin, The closest packing of spherical caps in n dimensions. Proc. Glasgow Math. Assoc. **2**, 139–144 (1955)
21. C.A. Rogers, *Packing and Covering* (Cambridge University Press, Cambridge, 1964)
22. T. Szele, Solution to Problem 35, II. (Hungarian). Mat. Lapok, **3**, 95 (1952)

Vertex-Transitive Haar Graphs That Are Not Cayley Graphs

Marston D. E. Conder, István Estélyi and Tomaž Pisanski

Dedicated to Egon Schulte and Károly Bezdek on the occasion of their 60th birthdays

Abstract In a recent paper in *Electron. J. Combin.* 23 (2016), Estélyi and Pisanski raised a question whether there exist vertex-transitive Haar graphs that are not Cayley graphs. In this note we construct an infinite family of trivalent Haar graphs that are vertex-transitive but non-Cayley. The smallest example has 40 vertices and is the well-known Kronecker cover over the dodecahedron graph $G(10, 2)$, occurring as the graph '**40**' in the Foster census of connected symmetric trivalent graphs.

The first author was supported by New Zealand's Marsden Fund (grant no. UOA1626). The second author was supported in part by the Slovenian Research Agency (research program P1-0294, research project J1-7051 and a Young Researchers Grant), and by the Czech Ministry of Education, Youth and Sports (project LO1506). The third author was supported in part by the Slovenian Research Agency (research program P1-0294 and research projects N1-0032, L7-5554 and J1-6720), and in part by the H2020 Teaming InnoRenew CoE.

M. D. E. Conder (✉)
Mathematics Department, University of Auckland,
Private Bag 92019, Auckland 1142, New Zealand
e-mail: m.conder@auckland.ac.nz

I. Estélyi
FMF, University of Ljubljana, Jadranska 19, 1000 Ljubljana, Slovenia
e-mail: istvan.estelyi@student.fmf.uni-lj.si

I. Estélyi · T. Pisanski
IAM, University of Primorska, Muzejski trg 2, 6000 Koper, Slovenia

I. Estélyi
NTIS, University of West Bohemia, Technická 8, 306 14 Plzeň 3, Czech Republic

T. Pisanski
FAMNIT, University of Primorska, Glagoljaška 8, 6000 Koper, Slovenia
e-mail: tomaz.pisanski@upr.si

M. D. E. Conder et al. (eds.), *Discrete Geometry and Symmetry*,
Springer Proceedings in Mathematics & Statistics 234,
https://doi.org/10.1007/978-3-319-78434-2_3

Keywords Haar graph · Cayley graph · Vertex-transitive graph

MSC 2010 05E18 (primary) · 20B25 (secondary)

1 Introduction

Let G be a group, and let S be a subset of G with $1_G \notin S$. Then the *Cayley graph* $\mathrm{Cay}(G, S)$ is the graph with vertex-set G and with edges of the form $\{g, sg\}$ for all $g \in G$ and $s \in S$. Equivalently, since all edges can be written in the form $\{1, s\}g$, this is a covering graph over a single-vertex graph having loops and semi-edges, with voltages taken from S: the order of a voltage over a semi-edge is 2 (corresponding to an involution in S), while the order of a voltage over a loop is greater than 2 (corresponding to a non-involution in S). Note that we may assume $S = S^{-1}$.

A natural generalisation of Cayley graphs are the so called *Haar graphs*, introduced in [16] by Hladnik et al., as follows. A *dipole* is a graph with two vertices, say black and white, and parallel edges (each from the *white* vertex to the *black* vertex), but no loops. Given a group G and an arbitrary subset S of G, the *Haar graph* $H(G, S)$ is the regular G-cover of a dipole with $|S|$ parallel edges, labeled by elements of S. In other words, the vertex-set of $H(G, S)$ is $G \times \{0, 1\}$, and the edges are of the form $\{(g, 0), (sg, 1)\}$ for all $g \in G$ and $s \in S$. If it is not ambiguous, we use the notation $(x, 0) \sim (y, 1)$ to indicate an edge $\{(x, 0), (y, 1)\}$ of $H(G, S)$. The name 'Haar graph' comes from the fact that when G is an abelian group, the Schur norm of the corresponding adjacency matrix can be easily evaluated via the so-called Haar integral on G (see [15]).

Note that the group G acts on $H(G, S)$ as a group of automorphisms, by right multiplication, and moreover, G acts regularly on each of the two parts of $H(G, S)$, namely $\{(g, 0) : g \in G\}$ and $\{(g, 1) : g \in G\}$. Conversely, if Γ is any bipartite graph and its automorphism group $\mathrm{Aut}\,\Gamma$ has a subgroup G that acts regularly on each part of Γ, then Γ is a Haar graph — indeed Γ is isomorphic to $H(G, S)$ where S is determined by the edges incident with a given vertex of Γ.

Haar graphs form a special subclass of the more general class of *bi-Cayley graphs*, which are graphs that admit a semiregular group of automorphisms with two orbits of equal size. Every bi-Cayley graph can be realised as follows. Let L and R be subsets of a group G such that $L = L^{-1}$, $R = R^{-1}$ and $1 \notin L \cup R$, and let S be any subset of G. Now take a dipole with edges labelled by elements of S, and add $|L|$ loops to the white (or 'left') vertex and label these by elements of L, and similarly add $|R|$ loops to the black (or 'right') vertex and label these by elements of R. This is a voltage graph, and the bi-Cayley graph $\mathrm{BCay}(G, L, R, S)$ is its regular G-cover. The vertex-set of $\mathrm{BCay}(G, L, R, S)$ is $G \times \{0, 1\}$, and the edges are of three forms: $\{(g, 0), (lg, 0)\}$ for $l \in L$, $\{(g, 1), (rg, 1)\}$ for $r \in R$, and $\{(g, 0), (sg, 1)\}$ for $s \in S$, for all $g \in G$. Note that the Haar graph $H(G, S)$ is exactly the same as the bi-Cayley graph $\mathrm{BCay}(G, \emptyset, \emptyset, S)$.

Recently bi-Cayley graphs (and Haar graphs in particular) have been investigated by several authors — see [8–10, 16–21, 23–25, 28, 30], for example.

It is known that every Haar graph over an abelian group is a Cayley graph (see [23]). More precisely, if A is an abelian group, then a Haar graph over A is a Cayley graph over the corresponding *generalised dihedral group* $D(A)$, which is the group generated by A and the automorphism of A that inverts every element of A (see [26]). The authors of [16] considered only cyclic Haar graphs — that is, Haar graphs $H(G, S)$ where G is a cyclic group. In [9], the second and third authors of this paper extended the study of Haar graphs to those over non-abelian groups, and found some that are not vertex-transitive, and some others that are Cayley graphs. The existence of Haar graphs that are vertex-transitive but non-Cayley remained open, and led to the following question.

Problem 1 Is there a non-abelian group G and a subset S of G such that the Haar graph $H(G, S)$ is vertex-transitive but non-Cayley?

In this note we give a positive answer to the above question, by exhibiting an infinite family of trivalent examples, coming from a family of double covers of generalised Petersen graphs. These graphs, which we denote by $D(n, r)$ for any integers n and r with $n \geq 3$ and $0 < r < n$, are described in Sect. 2. They have been considered previously by other authors (see later); in particular, by a theorem of Feng and Zhou [30], it is known exactly which of the graphs $D(n, r)$ are vertex-transitive, and which are Cayley. Then in Sect. 3 we determine necessary and sufficient conditions for $D(n, r)$ to be a Haar graph, and this provides the answer to Problem 1 in Sect. 4.

2 The Graphs $D(n, r)$ and Their Properties

Let $G(n, r)$ be the generalised Petersen graph on $2n$ vertices with span r. By $D(n, r)$ we denote a double cover of $G(n, r)$, in which the edges get non-trivial voltage if and only if they belong to the 'inner rim' (see below). This gives a class of graphs that was introduced by Zhou and Feng [29] under the name of *double generalised Petersen graphs*, and studied recently also by Kutnar and Petecki [22]. In both [29] and [22], the notation $DP(n, r)$ was used for the graph $D(n, r)$.

It is easy to define the vertices and edges of the graph $D(n, r)$ explicitly. There are four types of vertices, called u_i, v_i, w_i and z_i (for $i \in \mathbb{Z}_n$), and three types of edges, given by the sets

$$\Omega = \{\{u_i, u_{i+1}\}, \{z_i, z_{i+1}\} : i \in \mathbb{Z}_n\} \qquad \text{(the 'outer' edges),}$$
$$\Sigma = \{\{u_i, v_i\}, \{z_i, w_i\} : i \in \mathbb{Z}_n\} \qquad \text{(the 'spokes'), and}$$
$$I = \{\{v_i, w_{i+r}\}, \{v_i, w_{i-r}\} : i \in \mathbb{Z}_n\} \qquad \text{(the 'inner' edges).}$$

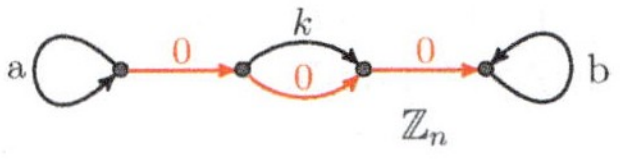

Fig. 1 Voltage graph defining the tetracirculant $\Sigma_0(n, a, k, b)$

This definition makes it easy to see that each $D(n, r)$ is a special tetracirculant [13], which is a cyclic cover $\Sigma_0(n, a, k, b)$ over the voltage graph given in Fig. 1. To see this, simply take $a = b = 1$ and $k = 2r$, and then $D(n, r) \cong \Sigma_0(n, 1, 2r, 1)$.

We now describe some other properties of the graphs $D(n, r)$ which are helpful. Many of these properties are already known, but we explain them here in detail for completeness.

Proposition 1 *Every $D(n, r)$ is connected.*

Proof Clearly all of the u_i lie in the same component as each other, as do all the z_j. Next, all the v_i lie in the same component as the u_i, and similarly, all the w_j lie in the same component as the z_j. Finally, there are edges between the vertices v_i and some of the w_j, and this makes the whole graph connected. □

Proposition 2 *The graph $D(n, r)$ is bipartite if and only if n is even.*

Proof If n is odd, then the vertices u_i lie in a cycle of odd length, and so the graph is not bipartite. On the other hand, if n is even, then the graph is bipartite, with one part containing the vertices u_i and $w_{i\pm r}$ for even i and the vertices v_j and $z_{j\pm r}$ for odd j. □

We now consider automorphisms of the graphs $D(n, r)$. Some automorphisms are apparent from the definition, such as these, which were noted in [22]:

$$\begin{array}{llllll} \alpha: & u_i \mapsto u_{i+1}, & v_i \mapsto v_{i+1}, & w_i \mapsto w_{i+1}, & z_i \mapsto z_{i+1} & \text{(rotation)}, \\ \beta: & u_i \mapsto z_i, & v_i \mapsto w_i, & w_i \mapsto v_i, & z_i \mapsto u_i & \text{(flip symmetry)}, \\ \gamma: & u_i \mapsto u_{-i}, & v_i \mapsto v_{-i}, & w_i \mapsto w_{-i}, & z_i \mapsto z_{-i} & \text{(reflection)}. \end{array}$$

Immediately we obtain the following:

Proposition 3 *The automorphism group of the graph $D(n, r)$ has at most two orbits on vertices, namely the set of all u_i and all z_j, and the set of all v_i and all w_j.*

Note also that α and β commute with each other. In fact, Zhou and Feng [29] proved that $D(n, r)$ is isomorphic to the bi-Cayley graph $\mathrm{BCay}(G, R, L, \{1\})$ over the abelian group $G = \langle \alpha, \beta \rangle \cong \mathbb{Z}_n \times \mathbb{Z}_2$, and $R = \{\alpha, \alpha^{-1}\}$ and $L = \{\alpha^r \beta, \alpha^{-r} \beta\}$.

Next, we consider isomorphisms among the graphs $G(n, r)$ and $D(n, r)$.

Proposition 4 *For every n and r, the graph $D(n, r)$ is isomorphic to $D(n, n-r)$, and $D(2n, r)$ is isomorphic to $D(2n, n-r)$.*

Proof First, $D(n,r) \cong D(n,n-r)$ because $G(n,r)$ is identical to $G(n,n-r)$. For the second part, consider a 180 degree rotation of the two 'inner' layers, namely $w_i \mapsto w_{i+n}$ and $z_i \mapsto z_{i+n}$ for all i. This shows that $D(2n,r)$ is isomorphic to $D(2n,n+r)$, and then applying the first part gives $D(2n,r) \cong D(2n,2n-(n+r)) = D(2n,n-r)$. □

Here we note that it can happen that the graphs $D(n,r)$ and $D(n,s)$ are different when $G(n,r)$ is isomorphic to $G(n,s)$. For instance, $G(7,2)$ is isomorphic to $G(7,3)$ but $D(7,2)$ is not isomorphic to $D(7,3)$, since $D(7,3)$ is planar but $D(7,2)$ is not.

Also we have the following:

Proposition 5 *For every r, the graph $D(2r+1,r)$ is planar, and isomorphic to the generalised Petersen graph $G(4r+2,2)$.*

Proof To see that $D(2r+1,r)$ is planar, first note that since r is coprime to $2r+1$, the edges between the vertices v_i and w_j give a cycle of length $2(2r+1)$, namely $(v_0, w_{-r}, v_1, w_{1-r}, v_2, w_{2-r}, \dots, v_{-2}, w_{r-1}, v_{-1}, w_r)$. Now draw three concentric circles, with the middle one for this $2(2r+1)$-cycle, the inside one for the $(2r+1)$-cycle $(u_0, u_1, \dots, u_{2r})$, and the outside one for the $(2r+1)$-cycle $(z_0, z_1, \dots, z_{2r})$, in a consistent order, and then add the spoke edges $\{u_i, v_i\}$ and $\{w_i, z_i\}$ in the natural way. In the resulting planar drawing of $D(2r+1,r)$, there is an inner face of length $2r+1$ (with the u_i as vertices), then two layers of pentagonal faces (bounded by cycles of the form $(u_i, v_i, w_{i-r}, v_{i+1}, u_{i+1})$ and $(v_j, w_{j+r}, z_{j+r}, z_{j-r}, w_{j-r})$), and an outer face of length $2r+1$ (with the z_j as vertices). After doing this, it is also easy to see that $D(2r+1,r)$ is isomorphic to the generalised Petersen graph $G(4r+2,2)$, with the spoke edges joining vertices of the large $2(2r+1)$-cycle (on the vertices v_i and w_j) to the two $(2r+1)$-cycles (on the vertices u_i and vertices z_j respectively). □

In particular, the graph $D(5,2)$ is isomorphic to the dodecahedral graph $G(10,2)$, and hence $D(5,2)$ is vertex-transitive. But as we will see, it is not a Haar graph.

Finally in this section, we consider the questions of which of the graphs $D(n,r)$ are vertex-transitive, and which are Cayley (or equivalently, which have the property that $\mathrm{Aut}(D(n,r))$ has a subgroup that acts regularly on vertices). Recall that $\mathrm{Aut}(D(n,r))$ has at most two orbits on vertices, and just one when $(n,r)=(5,2)$. The complete picture was determined by Feng and Zhou in [30, Theorem 1.3], as follows:

Theorem 6 *The graph $D(n,r)$ is vertex-transitive if and only if $n=5$ and $r=\pm 2$, or n is even and $r^2 \equiv \pm 1 \bmod \frac{n}{2}$. In the first case, $D(n,r)$ is isomorphic to the dodecahedral graph $G(10,2)$, which is non-Cayley, and in the second case, if $r^2 \equiv 1 \bmod \frac{n}{2}$ then $D(n,r)$ is a Cayley graph, while if $r^2 \equiv -1 \bmod \frac{n}{2}$ then $D(n,r)$ is non-Cayley.*

3 The Graphs $D(n,r)$ as Haar Graphs

Recall that a Haar graph is a regular cover of a dipole, and also a bi-Cayley graph. Also we have the following, proved in a different way in [9, Proposition 5]:

Proposition 7 *A Cayley graph is a Haar graph if and only if it is bipartite.*

Proof Let Γ be a Cayley graph, say for a group K. Then K acts on Γ as a group of automorphisms, and acts regularly on the vertices of Γ. Now if Γ is a Haar graph, then by definition Γ is bipartite. Conversely, suppose Γ is bipartite. Then the subgroup G of K preserving each of the two parts of Γ has index 2 in K, and acts regularly on each part, so Γ is a Haar graph (by the argument given in the third paragraph of the Introduction). □

We can now prove our main theorem:

Theorem 8 *$D(n,r)$ is a Haar graph if and only if it is vertex-transitive and n is even.*

Proof First, we note that $D(n,r)$ is bipartite if and only if n is even, by Proposition 2, and hence we may suppose that n is even, and then show that under that assumption, $D(n,r)$ is a Haar graph if and only if it is vertex-transitive.

One direction is easy. Suppose $\Gamma = D(n,r)$ is a Haar graph, say $H(G,S)$. Then by the definition of a Haar graph given in the Introduction, the subgroup G_R of Aut Γ induced by G has two orbits on vertices, namely the two parts of the bipartition of Γ. On the other hand, by Proposition 3, all the vertices u_i lie in the same orbit of Aut Γ; and then since these vertices lie in both parts of Γ, it follows that Aut Γ has a single orbit on vertices. Thus Γ is vertex-transitive.

For the converse, suppose that $\Gamma = D(n,r)$ is vertex-transitive, and let $m = \frac{n}{2}$. Then by Theorem 6, we know that $r^2 \equiv \pm 1 \bmod m$. Also by Proposition 4 we may suppose that $0 < r < m$, and further, we may suppose that r is odd, because if r is even then m is odd, and then by Proposition 4 we can replace r by $m - r$. We now proceed by considering separately the two cases $r^2 \equiv \pm 1 \bmod m$.

Case (a): Suppose that $r^2 \equiv 1 \bmod m$. Then by Theorem 6, we know that $D(n,r)$ is a Cayley graph, and also since it is bipartite, it follows from Proposition 7 that it is a Haar graph as well.

Case (b): Suppose that $r^2 \equiv -1 \bmod m$. In this case we construct a group of automorphisms of $D(n,r)$ that acts regularly on each part of $D(n,r)$. To do this, we take the automorphism α from the previous section, given by

$$\alpha : u_i \mapsto u_{i+1}, \quad v_i \mapsto v_{i+1}, \quad w_i \mapsto w_{i+1}, \quad z_i \mapsto z_{i+1},$$

and then take an additional automorphism δ, given by

$$\delta : u_i \mapsto v_{ri+1}, \quad v_i \mapsto u_{ri+1}, \quad w_i \mapsto z_{ri+1}, \quad z_i \mapsto w_{ri+1} \text{ if } m \text{ is odd and } i \text{ is even,}$$
$$\delta : u_i \mapsto w_{ri+1}, \quad v_i \mapsto z_{ri+1}, \quad w_i \mapsto u_{ri+1}, \quad z_i \mapsto v_{ri+1} \text{ if } m \text{ is odd and } i \text{ is odd,}$$

or

$$\delta: u_i \mapsto v_{ri+1},\ v_i \mapsto u_{ri+1},\ w_i \mapsto z_{ri+m+1},\ z_i \mapsto w_{ri+m+1} \text{ if } m \text{ is even and } i \text{ is even},$$
$$\delta: u_i \mapsto w_{ri+1},\ v_i \mapsto z_{ri+1},\ w_i \mapsto u_{ri+m+1},\ z_i \mapsto v_{ri+m+1} \text{ if } m \text{ is even and } i \text{ is odd}.$$

It is a straightforward exercise to verify that δ preserves the edge-set $\Omega \cup \Sigma \cup I$ of $D(n, r)$, and also preserves the two parts of $D(n, r)$, given in the proof of Proposition 2. To do the former, it is important to note that $r^2 \equiv 1 \bmod 4$ (because r is odd), and hence that $r^2 \equiv -1 \bmod n$ when m is odd, while $r^2 \equiv m-1 \bmod n$ when m is even. For example, if m and i are even then $\{v_i, w_{i+r}\}^\delta = \{u_{ri+1}, u_{r(i+r)+m+1}\} = \{u_{ri+1}, u_{ri}\}$.

It is also easy to see that conjugation by δ takes α^2 to α^{2r}, and so the subgroup G of $\mathrm{Aut}(D(n, r))$ generated by α^2 and δ is isomorphic to the semi-direct product $\mathbb{Z}_m \rtimes_r \mathbb{Z}_4$. In particular, G has order $4m = 2n$. Also G acts transitively and hence regularly on each of the two parts of $D(n, r)$, and therefore $D(n, r)$ is a Haar graph. □

4 Vertex-Transitive Haar Graphs That Are Not Cayley Graphs

Combining Theorems 6 and 8, we have the following, in answer to Problem 1:

Theorem 9

(a) *If n is odd, or if n is even and $r^2 \not\equiv \pm 1$ mod $\frac{n}{2}$, then $D(n, r)$ is not a Haar graph, and is vertex-transitive only when $(n, r) = (5, \pm 2)$;*

(b) *If n is even and $r^2 \equiv 1$ mod $\frac{n}{2}$, then $D(n, r)$ is a Haar graph and a Cayley graph;*

(c) *If n is even and $r^2 \equiv -1$ mod $\frac{n}{2}$, then $D(n, r)$ is a Haar graph and is vertex-transitive but not a Cayley graph.*

Corollary 10 *If $m > 2$ and $r^2 \equiv -1$ mod m, then $D(2m, r)$ is a Haar graph that is vertex-transitive but non-Cayley. In particular, there are infinitely many such graphs.*

Proof The first part follows immediately from Theorem 9, and the second part follows from a well known fact in number theory, namely that -1 is a square mod m if and only if m or $m/2$ is a product of primes $p \equiv 1 \bmod 4$ (see [14, Chapter 6]), or simply by taking $m = r^2 + 1$ for each integer $r \geq 2$. □

We discovered the first few of these examples during the week of the conference *Geometry and Symmetry*, held in 2015 at Veszprém, Hungary, to celebrate the 60th birthdays of Károly Bezdek and Egon Schulte.

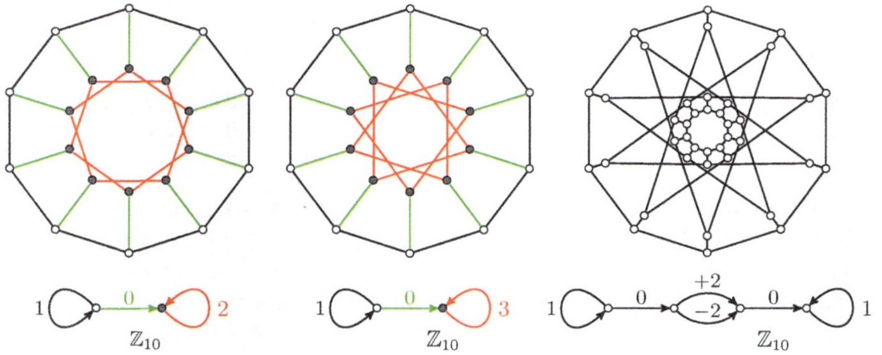

Fig. 2 The dodecahedral graph $G(10, 2)$, the Desargues graph $G(10, 3)$, and the Haar graph $D(10, 2) \cong D(10, 3) \cong$ **F40**

The smallest of our examples is $D(10, 2)$, of order 40, occurring when $m = 5$ and $r \equiv \pm 2$ or ± 3 mod 10 (noting that $m - r = 3$ when $(m, r) = (5, 2)$). We will show that this is also the smallest Haar graph that is vertex-transitive and non-Cayley. It is a Kronecker cover over the dodecahedral graph $G(10, 2)$, and is also a double cover over the Desargues graph $G(10, 3)$. These graphs are illustrated in Fig. 2.

The graph $D(10, 2)$ was known by R.M. Foster as early as the late 1930s, and appears as the graph '**40**' (alternatively known as '**F40**') in the *Foster Census* of connected symmetric trivalent graphs [7]. It was also studied in [27] by Asia Ivić Weiss (the chair of the Veszprém conference), and by Betten, Brinkmann and Pisanski in [1], and Boben, Grünbaum, Pisanski and Žitnik in [2]. It has girth 8, and automorphism group of order 480, and it is not just vertex-transitive, but is also arc-transitive. Moreover, by a very recent theorem of Kutnar and Petecki [22], the graph $D(n, r)$ is arc-transitive only when $(n, r) = (5, 2)$ or $(10, 2)$ or $(10, 3)$. This implies that **F40** is the only example from the family of graphs $D(n, r)$ that is arc-transitive but non-Cayley.

In fact, **F40** is the smallest vertex-transitive non-Cayley Haar graph, in terms of both the graph order and the number of edges. We found this by running a Magma [3] computation to construct all Haar graphs with at most 40 vertices or at most 60 edges, with a check for which of the graphs are vertex-transitive but non-Cayley. Incidentally, this computation shows that there are 60 different examples of order 40, with valencies running between 3 and 17, but just one of valency 3, namely **F40**.

Finally, there are many other examples of vertex-transitive non-Cayley Haar graphs that are not of the form $D(n, r)$, including 3-valent examples of orders 80, 112, 120 and 128, and higher-valent examples of orders 48, 64, 72, 78 and 80. Among the 3-valent examples, many are arc-transitive, including the graphs **F80** and **F640** in the Foster census [7] and its extended version in [5, 6], and others in the first author's complete set of all arc-transitive trivalent graphs of order up to 10000 described on his website [4]. Most of these 'small' examples are abelian regular covers of **F40**, of orders 1280, 2560, 3240, 5000, 5120, 6480, 6720, 9720 and 10000, and are 3-arc-regular, but two others are 2-arc-regular of type 2^2, with orders 6174 and 8064, and these are abelian regular covers of the Pappus graph (**F18**) and the Coxeter graph (**F28**) respectively.

In particular, the graph of order 6174 is a member of an infinite family of 2-arc-regular covers of the Pappus graph, investigated in [11, 12]. Each graph in this family is a 2-arc-regular 3-valent graph of type 2^2 and order $18n^3$ for some $n \geq 7$ for which $\mathbb{Z}_n$ contains a root of the polynomial $x^2 + x + 1$. Also each member of this family is Haar but not Cayley, since the 2-arc-regular subgroup of type 2^2 in the automorphism group of the Pappus graph contains a subgroup of order 9 acting regularly on each of the two parts of the graph, but contains no subgroup of order 18 acting transitively on the vertices. Hence we have another infinite family of examples of vertex-transitive non-Cayley Haar graphs, but of considerably larger orders.

Acknowledgements We acknowledge with gratitude the use of the MAGMA system [3] for helping to find and analyse examples. We would also like to thank Klavdija Kutnar and István Kovács for fruitful conversations and for pointing out several crucial references, and the referees for helpful comments, and in particular, one of the referees for suggesting the family of 2-arc-regular covers of the Pappus graph as another potential source of examples.

References

1. A. Betten, G. Brinkmann, T. Pisanski, Counting symmetric configurations v_3, in Proceedings of the 5th Twente Workshop on Graphs and Combinatorial Optimization (Enschede, 1997), Discrete Appl. Math. **99**(1–3), 331–338 (2000)
2. M. Boben, B. Grünbaum, T. Pisanski, A. Žitnik, Small triangle-free configurations of points and lines. Discrete Comput. Geom. **35**(3), 405–427 (2006)
3. W. Bosma, J. Cannon, C. Playoust, The Magma algebra system I: the user language. J. Symbolic Comput. **24**, 235–265 (1997)
4. M.D.E. Conder, *Trivalent* (cubic) *symmetric graphs on up to 10000 vertices*, available at http://www.math.auckland.ac.nz/~conder/symmcubic10000list.txt
5. M. Conder, P. Dobcsányi, Trivalent symmetric graphs on up to 768 vertices. J. Combin. Math. Combin. Comput. **40**, 41–63 (2002)
6. M. Conder, R. Nedela, A refined classification of symmetric cubic graphs. J. Algebra **322**, 722–740 (2009)
7. I.Z. Bouwer, W.W. Chernoff, B. Monson, Z. Star, *The Foster Census* (Charles Babbage Research Centre, Winnipeg, 1988)
8. S.F. Du, M.Y. Xu, A classification of semi-symmetric graphs of order $2pq$, Comm. Algebra **28**(6), 2685–2715 (2000)
9. I. Estélyi, T. Pisanski, Which Haar graphs are Cayley graphs? Electron. J. Combin. **23**(Paper 3.10), 13pp (2016)
10. G. Exoo, R. Jajcay, On the girth of voltage graph lifts. Eur. J. Combin. **32**, 554–562 (2011)
11. Y.-Q. Feng, K. Kutnar, D. Marušič, D.-W. Yang, On cubic symmetric non-Cayley graphs with solvable automorphism groups, manuscript. arXiv: math/1609703
12. Y.-Q. Feng, C.H. Li, J.-X. Zhou, Symmetric cubic graphs with solvable automorphism groups. Eur. J. Combin. **45**, 1–11 (2015)
13. B. Frelih, K. Kutnar, Classification of symmetric cubic tetracirculants and pentacirculants. Eur. J. Combin. **34**, 169–194 (2013)
14. G.H. Hardy, E.M. Wright, *An Introduction to the Theory of Numbers*, 6th edn. (Oxford University Press, Oxford, 2008)

15. M. Hladnik, Schur norms of bicirculant matrices. Lin. Alg. Appl. **286**, 261–272 (1999)
16. M. Hladnik, D. Marušič, T. Pisanski, Cyclic Haar graphs. Discrete Math. **244**, 137–153 (2002)
17. W. Jin, W. Liu, A classification of nonabelian simple 3-BCI-groups. Eur. J. Combin. **31**, 1257–1264 (2010)
18. H. Koike, I. Kovács, Isomorphic tetravalent cyclic Haar graphs. Ars Math. Contemp. **7**, 215–235 (2014)
19. H. Koike, I. Kovács, T. Pisanski, The number of cyclic configurations of type (v_3) and the isomorphism problem. J. Combin. Des. **22**(5), 216–229 (2014)
20. I. Kovács, K. Kutnar, D. Marušič, Classification of edge-transitive rose window graphs. J. Graph Theory **65**, 216–231 (2010)
21. I. Kovács, A. Malnič, D. Marušič, Š. Miklavič, One-matchng bi-Cayley graphs over abelian groups. Eur. J. Combin. **30**, 602–616 (2009)
22. K. Kutnar, P. Petecki, On automorphisms and structural properties of double generalized Petersen graphs. Discrete Math. **339**, 2861–2870 (2016)
23. Z.P. Lu, On the automorphism groups of bi-Cayley graphs. Acta Sci. Nat. Univ. Peking **39**(1), 1–5 (2003)
24. Z.P. Lu, C.Q. Wang, M.Y. Xu, Semisymmetric cubic graphs constructed from bi-Cayley graphs of A_n. Ars Combin. **80**, 177–187 (2006)
25. T. Pisanski, A classification of cubic bicirculants. Discrete Math. **307**, 567–578 (2007)
26. W.R. Scott, *Group Theory* (Prentice-Hall, Upper Saddle River, 1964)
27. A.I. Weiss, An infinite graph of girth 12. Trans. Amer. Math. Soc. **283**, 575–588 (1984)
28. S. Wilson, Rose window graphs. Ars Math. Contemp. **1**, 7–19 (2008)
29. J.X. Zhou, Y.Q. Feng, Cubic vertex-transitive non-Cayley graphs of order $8p$. Electron. J. Combin. **19**, 53 (2012)
30. J.X. Zhou, Y.Q. Feng, Cubic bi-Cayley graphs over abelian groups. Eur. J. Combin. **36**, 679–693 (2014)

On the Volume of Boolean Expressions of Large Congruent Balls

Balázs Csikós

Dedicated to Károly Bezdek and Egon Schulte on the occasion of their 60th birthdays.

Abstract We consider the volume of a Boolean expression of some congruent balls about a given system of centers in the d-dimensional Euclidean space. When the radius r of the balls is large, this volume can be approximated by a polynomial of r, which will be computed up to an $O(r^{d-3})$ error term. We study how the top coefficients of this polynomial depend on the set of the centers. It is known that in the case of the union of the balls, the top coefficients are some constant multiples of the intrinsic volumes of the convex hull of the centers. Thus, the coefficients in the general case lead to generalizations of the intrinsic volumes, in particular, to a generalization of the mean width of a set. Some known results on the mean width, along with the theorem on its monotonicity under contractions are extended to the "Boolean analogues" of the mean width.

Keywords Volume · Intrinsic volume · Quermassintegral · Unions and intersections of balls

1 Introduction

The long-standing conjecture of Kneser [10] and Poulsen [11] claims that if the points $\mathbf{p}_1, \ldots, \mathbf{p}_N$ and $\mathbf{q}_1, \ldots, \mathbf{q}_N$ of the d-dimensional Euclidean space $\mathbb{R}^d$ satisfy

B. Csikós (✉)
Institute of Mathematics, Eötvös Loránd University, Budapest, Hungary
e-mail: csikos@cs.elte.hu

Present address:
B. Csikós
Pázmány Péter stny. 1/C, Budapest 1117, Hungary

M. D. E. Conder et al. (eds.), *Discrete Geometry and Symmetry*,
Springer Proceedings in Mathematics & Statistics 234,
https://doi.org/10.1007/978-3-319-78434-2_4

the inequalities $\mathrm{d}(\mathbf{p}_i, \mathbf{p}_j) \geq \mathrm{d}(\mathbf{q}_i, \mathbf{q}_j)$ for all $0 \leq i, j \leq N$, then

$$\mathrm{vol}_d\left(\bigcup_{i=1}^{N} B^d(\mathbf{p}_i, r)\right) \geq \mathrm{vol}_d\left(\bigcup_{i=1}^{N} B^d(\mathbf{q}_i, r)\right)$$

for any $r > 0$, where $B^d(\mathbf{p}, r)$ denotes the closed d-dimensional ball of radius r about the point $\mathbf{p}$ and vol_d is the d-dimensional volume. K. Bezdek and Connelly [2] proved the conjecture in the plane, but it is still open in dimensions $d \geq 3$.

Results of Gromov [9], Gordon and Meyer [8] and the author [4, 5] suggest that the Kneser–Poulsen conjecture could be true in a more general form, which we formulate below.

Let $\mathcal{B}_N$ be the free Boolean algebra generated by $N \geq 1$ symbols $x_1, \ldots, x_N$. We denote the greatest element of $\mathcal{B}_N$ by X, and the least element of $\mathcal{B}_N$ by $\emptyset$. Elements of $\mathcal{B}_N$ are equivalence classes of formal expressions built from the symbols $x_1, \ldots, x_N$, X and $\emptyset$, the binary operations $\cup$, $\cap$, and the unary operator $f \mapsto \bar{f}$. Two expressions are called equivalent if and only if we can prove their equality assuming that the operations satisfy the axioms of a Boolean algebra. We shall refer to an element of $\mathcal{B}_N$ by choosing a Boolean expression from its equivalence class, and we write "$=$" between two Boolean expressions if they are equivalent. We shall also use the derived operator $f \setminus g = f \cap \bar{g}$ and the partial ordering $f \subseteq g \overset{\text{def}}{\Longleftrightarrow} f \cup g = g$. We refer to [6] for more details on Boolean algebras.

Take a Boolean expression $f \in \mathcal{B}_N$ which can be represented by a formula built exclusively from the variables $x_1, \ldots, x_N$ and the operations $\cup$, $\cap$, $\setminus$ in such a way that each of the variables occurs in the formula exactly once. For any pair of indices $i \neq j$, $1 \leq i, j \leq N$, evaluate f replacing the variables x_k, $k \notin \{i, j\}$ by X or $\emptyset$ in all possible ways. It can be seen that the results of those evaluations that are not equal to X or $\emptyset$, are all equal to one another and to one of the expressions $x_i \cap x_j$, $x_i \setminus x_j$, $x_j \setminus x_i$, $x_i \cup x_j$. Let the sign ϵ_{ij}^f be -1 if the evaluations not equal to X or $\emptyset$ are equal to $x_i \cap x_j$, and set $\epsilon_{ij}^f = 1$ in the remaining three cases.

The generalization of the Kneser–Poulsen conjecture for Boolean expressions of balls claims that if the Boolean expression $f \in \mathcal{B}_N$ obeys the conditions of the previous paragraph, and the points $\mathbf{p}_1, \ldots, \mathbf{p}_N$ and $\mathbf{q}_1, \ldots, \mathbf{q}_N$ in $\mathbb{R}^d$ satisfy the inequalities $\epsilon_{ij}^f(\mathrm{d}(\mathbf{p}_i, \mathbf{p}_j) - \mathrm{d}(\mathbf{q}_i, \mathbf{q}_j)) \geq 0$ for all $0 \leq i, j \leq N$, then

$$\mathrm{vol}_d\left(f(B^d(\mathbf{p}_1, r_1), \ldots, B^d(\mathbf{p}_N, r_N))\right) \geq \mathrm{vol}_d\left(f(B^d(\mathbf{q}_1, r_1), \ldots, B^d(\mathbf{q}_N, r_N))\right) \tag{1}$$

for any choice of the radii $r_1, \ldots, r_N$.

A suitable modification of the arguments of Bezdek and Connelly [2] shows that this generalization of the Kneser–Poulsen conjecture is also true in the Euclidean plane (see [5]).

As it was pointed out by Capoyleas, Pach [3], and Gorbovickis [7], the original Kneser–Poulsen conjecture for large congruent balls is closely related to the monotonicity of the mean width of a set under contractions. The relation is based on the formula

$$\operatorname{vol}_d\left(\bigcup_{i=1}^{N} B^d(\mathbf{p}_i, r)\right) = \kappa_d r^d + \frac{d\kappa_d}{2}\omega_d(\{\mathbf{p}_1, \ldots, \mathbf{p}_N\})r^{d-1} + O(r^{d-2}), \quad (2)$$

where κ_d is the volume of the unit ball in $\mathbb{R}^d$, $\omega_d(S)$ denotes the mean width of the bounded set $S \subset \mathbb{R}^d$. We remark that the mean width function ω_d depends on the dimension d of the ambient space, but only up to a constant factor. More explicitly, if $\Phi\colon \mathbb{R}^d \to \mathbb{R}^{\tilde{d}}$ is an isometric embedding, then we have $\frac{d\kappa_d}{\kappa_{d-1}}\omega_d(S) = \frac{\tilde{d}\kappa_{\tilde{d}}}{\kappa_{\tilde{d}-1}}\omega_{\tilde{d}}(\Phi(S))$ for any bounded set $S \subset \mathbb{R}^d$. Applying Formula (2) and the fact that the Kneser–Poulsen conjecture is true if the dimension of the space is at least $N-1$ (see [9]), Capoyleas and Pach [3] proved that the mean width of a set cannot increase when the set is contracted. Using rigidity theory, Gorbovickis [7] sharpened this result by proving that if the d-dimensional configurations $(\mathbf{p}_1, \ldots, \mathbf{p}_N)$ and $(\mathbf{q}_1, \ldots, \mathbf{q}_N)$ are not congruent and satisfy the inequalities $\mathrm{d}(\mathbf{p}_i, \mathbf{p}_j) \ge \mathrm{d}(\mathbf{q}_i, \mathbf{q}_j)$ for all $0 \le i, j \le N$, then the strict inequality

$$\omega_d(\{\mathbf{p}_1, \ldots, \mathbf{p}_N\}) > \omega_d(\{\mathbf{q}_1, \ldots, \mathbf{q}_N\})$$

holds. This strict inequality, in return, implies that the Kneser–Poulsen conjecture is true if the radius of the balls is bigger than a constant depending on the configurations of the centers.

Gorbovickis [7] proved also that for the volume of the intersection of large congruent balls we have

$$\operatorname{vol}_d\left(\bigcap_{i=1}^{N} B^d(\mathbf{p}_i, r)\right) = \kappa_d r^d - \frac{d\kappa_d}{2}\omega_d(\{\mathbf{p}_1, \ldots, \mathbf{p}_N\})r^{d-1} + O(r^{d-2}), \quad (3)$$

thus, as a consequence of the strict monotonicity of the mean width, the above mentioned generalization of the Kneser–Poulsen conjecture is true also for the intersections of congruent balls if the radius of the balls is greater than a constant depending on the configurations of the centers.

In 2013 K. Bezdek [1] posed the problem of finding a suitable generalization of Eqs. (2) and (3) for the volume of an arbitrary Boolean expression of large congruent balls, and suggested to explore the interplay between the generalized Kneser–Poulsen conjecture and the monotonicity properties of the coefficient of r^{d-1} in the general formula. In the present paper, we summarize the results of the research initiated by these questions.

The outline of the paper is the following. In Sect. 2, we sharpen Eq. (2), expressing the volume of the union of some large congruent balls with an error term of order $O(r^{d-3})$. The coefficients appearing in the formula are some constant multiples of the intrinsic volumes V_0, V_1, V_2 of the convex hull of the centers. In Sect. 3, we show that if a Boolean expression $f(B_1, \ldots, B_n)$ of some balls is bounded, then its volume can be obtained as a linear combination of the volumes of the unions of some of the balls. The coefficients of this inclusion-exclusion type formula, given in Proposition 5, depend purely on the Boolean expression f. These coefficients are used to define the Boolean analogues of the intrinsic volumes of the convex hull of a point set in Sect. 4. Theorem 1 gives a generalization of Eq. (2) for Boolean expressions of large balls using Boolean intrinsic volumes. In Sect. 5, some classical facts on intrinsic volumes are generalized for Boolean intrinsic volumes. For example, it is known that the kth intrinsic volume of a polytope can be expressed in terms of the volumes of the k-dimensional faces and the angular measures of the normal cones of these faces. This formula is generalized for Boolean intrinsic volumes in Theorem 2. As an application of Theorem 2, we prove that the kth Boolean intrinsic volumes corresponding to dual Boolean expressions differ only in a sign $(-1)^k$. This explains why the coefficients of r^{d-1} in the Eqs. (2) and (3) are opposite to one another. Theorem 3 provides a Boolean extension of the fact that the first intrinsic volume of a convex set is a constant multiple of the integral of its support function. Section 6 is devoted to the proof of Theorem 4 on the monotonicity of the Boolean analogue of the first intrinsic volume.

2 Comparison of the Volume of a Union of Balls and the Volume of Its Convex Hull

Every convex polytope $K \subset \mathbb{R}^d$ defines a decomposition of the space as follows. Denote by $\mathcal{F}(K)$ the set of all faces of K, including K, and by $\mathcal{F}_k(K)$ the set of its k-dimensional faces. Let $\pi : \mathbb{R}^d \to K$ be the map assigning to a point $\mathbf{x} \in \mathbb{R}^d$ the unique point of K that is closest to $\mathbf{x}$. For a face $L \in \mathcal{F}(K)$, denote by $V(L, K)$ the preimage $\pi^{-1}(\operatorname{relint} L)$ of the relative interior of L. As K is the disjoint union of the relative interiors of its faces, $\mathbb{R}^d$ is the disjoint union of the sets $V(L, K)$, where L is running over $\mathcal{F}(K)$. If $L \in \mathcal{F}_k(K)$, then $V(L, K)$ is the Minkowski sum of the relative interior of L and the normal cone

$$N(L, K) = \{\mathbf{u} \in \mathbb{R}^d \mid \mathbf{u} \perp [L] \text{ and } \max_{\mathbf{x} \in K} \langle \mathbf{u}, \mathbf{x} \rangle \text{ is attained at a point } \mathbf{x} \in L\} \quad (4)$$

of K at L, where $[L]$ denotes the affine subspace spanned by L. Set $n(L, K) = N(L, K) \cap B^d(\mathbf{0}, 1)$ and $\nu(L, K) = \operatorname{vol}_{d-k}(n(L, K))/\kappa_{d-k}$. Division by κ_{d-k} in the definition of $\nu(L, K)$ is advantageous because it makes the angle measure $\nu(L, K)$ of the normal cone $N(L, K)$ independent of the dimension d of the ambient space $\mathbb{R}^d$, though the normal cone itself changes if we embed K into a higher dimensional space.

Denote by $K_r = K + B^d(\mathbf{0}, r)$ the distance r parallel body of K. The decomposition

$$\mathbb{R}^d = \bigcup_{L \in \mathcal{F}(K)} N(L, K) \quad (5)$$

induces a decomposition of the parallel body K_r, which enables us to write the volume of K_r as a polynomial of r

$$\begin{aligned} \mathrm{vol}_d(K_r) &= \sum_{L\in\mathcal{F}(K)} \mathrm{vol}_d(K_r \cap N(L,K)) = \sum_{L\in\mathcal{F}(K)} \mathrm{vol}_d(L + r\,(n(L,K)) \\ &= \sum_{k=0}^{d} \kappa_{d-k} \left(\sum_{L\in\mathcal{F}_k(K)} \mathrm{vol}_k(L)\nu(L,K) \right) r^{d-k}. \end{aligned} \tag{6}$$

Equation (6) is a special case of Steiner's classical formula (see, e.g., [12, Eq. (4.2.27)])

$$\mathrm{vol}_d(K + B(\mathbf{0}, r)) = \sum_{k=0}^{d} \binom{d}{k} W_k^d(K) r^k = \sum_{k=0}^{d} \kappa_{d-k} V_k(K) r^{d-k}, \tag{7}$$

expressing the volume of the distance r parallel body of an arbitrary compact convex set K as a polynomial of r, in which the normalized coefficients $W_k^d(K)$ and $V_k(K)$ are the quermassintegrals and intrinsic volumes of K respectively. It is known that the intrinsic volumes are continuous functions on the space of compact convex sets endowed with the Hausdorff metric (see [12, Sect. 4.2]), and $V_0(K) \equiv 1$. Comparing (6) and (7) we obtain the formula

$$V_k(K) = \sum_{L\in\mathcal{F}_k(K)} \mathrm{vol}_k(L)\nu(L,K) \tag{8}$$

expressing the intrinsic volumes of a polytope K.

Proposition 1 *Let* $\mathbf{p}_1, \ldots, \mathbf{p}_N$ *be a fixed set of points in* $\mathbb{R}^d$, $K = \mathrm{conv}(\{\mathbf{p}_1, \ldots, \mathbf{p}_N\})$ *be the convex hull of the points. Denote by* $B_i = B^d(\mathbf{p}_i, r)$ *the ball of radius* r *centered at* $\mathbf{p}_i$. *Then we have*

$$\left| \mathrm{vol}_d(K_r) - \mathrm{vol}_d\left(\bigcup_{i=1}^{N} B_i \right) \right| = O(r^{d-3}) \tag{9}$$

for large values of r.

Proof Denote by Δ the diameter of K, and set $r' = r - \Delta^2/r$. It is easy to see that if $r \geq \Delta$, then $K_{r'} \subseteq \bigcup_{i=1}^{N} B_i \subseteq K_r$, (see [3]). Intersecting the decomposition (5) with the union of the balls, we get

$$\bigcup_{i=1}^{N} B_i = \bigcup_{L\in\mathcal{F}(K)} \left(N(L,K) \cap \left(\bigcup_{i=1}^{N} B_i \right) \right).$$

When $L \in \mathcal{F}_0(K)$ is a vertex, we have $N(L, K) \cap \left(\bigcup_{i=1}^N B_i\right) = N(L, K) \cap K_r$. Thus,

$$K_r \setminus \left(\bigcup_{i=1}^{N} B_i\right) \subseteq \bigcup_{k=1}^{d} \bigcup_{L \in \mathcal{F}_k(K)} N(L, K) \cap (K_r \setminus K_{r'}),$$

and

$$\left|\mathrm{vol}_d(K_r) - \mathrm{vol}_d\left(\bigcup_{i=1}^{N} B_i\right)\right| \leq \sum_{k=1}^{d} \mathrm{vol}_k(L)\kappa_{d-k}\nu(L, K)\left(r^{d-k} - \left(r - \frac{\Delta^2}{r}\right)^{d-k}\right) = O(r^{d-3}), \quad (10)$$

as claimed.

Corollary 1 *Using the notations of Proposition 1, we have*

$$\mathrm{vol}_d\left(\bigcup_{i=1}^{N} B_i\right) = \kappa_d r^d + \kappa_{d-1} V_1(K) r^{d-1} + \kappa_{d-2} V_2(K) r^{d-2} + O(r^{d-3}). \quad (11)$$

3 Combinatorics of Boolean Expressions

For a subset I of the set $[N] = \{1, \ldots, N\}$, define $a_I \in \mathcal{B}_N$ by $a_I = (\bigcap_{j \notin I} x_j) \setminus (\bigcup_{i \in I} x_i)$. The elements a_I, $(I \subseteq [N])$ are the *atomic elements* of $\mathcal{B}_N$. Any $f \in \mathcal{B}_N$ can be decomposed uniquely as $f = \bigcup_{a_I \subseteq f} a_I$. In particular, $\mathcal{B}_N$ has 2^{2^N} elements.

Definition 1 The *reduced Euler characteristic* $\tilde{\chi}_N(f)$ *of* $f \in \mathcal{B}_N$ is the integer $\tilde{\chi}_N(f) = \sum_{a_I \subseteq f} (-1)^{|I|+1}$.

Obviously, the reduced Euler characteristic of a Boolean expression is an integer number in the interval $[-2^{N-1}, 2^{N-1}]$.

Proposition 2 *If* $f \in \mathcal{B}_N$ *can be represented by a formal expression which does not contain all the variables* $x_1, \ldots, x_N$, *then* $\tilde{\chi}_N(f) = 0$.

Proof We may assume without loss of generality that f can be written as an expression not using the variable x_N. This means that if $\iota \colon \mathcal{B}_{N-1} \to \mathcal{B}_N$ is the natural embedding, then $f = \iota(g)$ for some $g \in \mathcal{B}_{N-1}$. If $I \subseteq [N-1]$, and $a_I \in \mathcal{B}_{N-1}$ is the corresponding atomic expression in $\mathcal{B}_{N-1}$, then $\iota(a_I) \cap x_N$ and $\iota(a_I) \cap \bar{x}_N$ are atomic expressions in $\mathcal{B}_N$ corresponding to the index sets $I \subseteq [N]$ and $I \cup \{N\} \subseteq [N]$ respectively, furthermore,

$$a_I \subseteq g \iff \iota(a_I) \cap x_N \subseteq f \iff \iota(a_I) \cap \bar{x}_N \subseteq f.$$

Thus,

$$\tilde{\chi}_N(f) = \sum_{a_I \subseteq g} ((-1)^{|I|} + (-1)^{|I \cup \{N\}|}) = 0.$$

□

Proposition 3 *If $\bar{f}$ is the complement of $f \in \mathcal{B}_N$, then $\tilde{\chi}_N(\bar{f}) = -\tilde{\chi}_N(f)$.*

Proof It is clear from the definition of the reduced Euler characteristic that if $f \cap g = \emptyset$, then $\tilde{\chi}_N(f \cup g) = \tilde{\chi}_N(f) + \tilde{\chi}_N(g)$. We also have

$$\tilde{\chi}_N(X) = \sum_{i=0}^{N} (-1)^{i+1} \binom{N}{i} = 0,$$

so $\tilde{\chi}_N(f) + \tilde{\chi}_N(\bar{f}) = \tilde{\chi}_N(f \cup \bar{f}) = \tilde{\chi}_N(X) = 0$.

Recall that the contradual $f^{\bar{*}}$ of $f \in \mathcal{B}_N$ is formed by replacing each variable x_i by its complement $\bar{x}_i$, while the dual $f^* = \overline{f^{\bar{*}}}$ of f is the complement of the contradual of f.

Proposition 4 *For any $f \in \mathcal{B}_N$, we have*

$$-\tilde{\chi}_N(f^*) = \tilde{\chi}_N(f^{\bar{*}}) = (-1)^N \tilde{\chi}_N(f).$$

Proof The first equation is a corollary of Proposition 3, so it is enough to show the second one. The contradual operation preserves the ordering and maps the atom a_I to $a_{[N]\setminus I}$. Consequently,

$$\tilde{\chi}_N(f^{\bar{*}}) = \sum_{a_I \subseteq f^{\bar{*}}} (-1)^{|I|+1} = (-1)^N \sum_{a_{[N]\setminus I} \subseteq f} (-1)^{|[N]\setminus I|+1} = (-1)^N \tilde{\chi}_N(f).$$

□

Let $\mathcal{L}_N$ be the sublattice of $\mathcal{B}_N$ generated by the elements $x_1, \ldots, x_N$ and the operations $\cup$ and $\cap$. An element $f \in \mathcal{B}_n$ belongs to $\mathcal{L}_n$ if and only if $f \neq \emptyset$ and whenever $a_I \subseteq f$ and $J \subseteq I$ we also have $a_J \subseteq f$. This means that we can associate to any element $f \in \mathcal{L}_N$ an abstract simplicial complex $P_f = \{I \subset [N] \mid a_I \subseteq f\}$. This assignment gives a bijection between $\mathcal{L}_n$ and abstract simplicial complexes on the vertex set $[N]$ different from the abstract $(N-1)$-dimensional simplex. In this special case, the reduced Euler characteristic of f is one less than the ordinary Euler characteristic of P_f. The difference is due to the fact that $\emptyset$ is not counted as a -1-dimensional face when we compute the Euler characteristic, but it is taken into account in the computation of $\tilde{\chi}_N(f)$. The number of elements of $\mathcal{L}_N$ is $M_N - 2$, where M_N is the Nth Dedekind number.

There is a sublattice $\mathcal{C}_N \supset \mathcal{L}_N$ of $\mathcal{B}_N$ consisting of expressions that can be built from the variables $x_1, \ldots, x_N$ using only the operations $\cup$, $\cap$, and $\setminus$. The lattice $\mathcal{C}_N$

contains exactly those elements of $\mathcal{B}_N$ that do not contain the atomic expression $a_{[N]}$. This way, $\mathcal{C}_N$ has 2^{2^N-1} elements.

Denote by $\mathcal{M}_N$ the linear space of real valued functions $\mu\colon \mathcal{C}_N \to \mathbb{R}$ such that $\mu(f \cup g) = \mu(f) + \mu(g)$ if $f \cap g = \emptyset$. As $\mu \in \mathcal{C}_N$ is uniquely determined by its values on the atomic expressions a_I, $(I \subsetneq [N])$, $\dim \mathcal{M}_N = 2^N - 1$.

For $\emptyset \neq I \subseteq [N]$, let $u_I \in \mathcal{L}_N$ be the union $u_I = \bigcup_{i\in I} x_i$.

Proposition 5 *For any $f \in \mathcal{C}_N$, there is a unique collection of integers $m_{f,I} \in \mathbb{Z}$ for $(\emptyset \neq I \subseteq [N])$ such that for any $\mu \in \mathcal{M}_N$, we have*

$$\mu(f) = \sum_{\emptyset \neq I \subseteq [N]} m_{f,I}\,\mu(u_I). \tag{12}$$

Proof There is a natural embedding $\mathrm{ev}\colon \mathcal{C}_N \to \mathcal{M}_N^*$ of $\mathcal{C}_N$ into the dual space of $\mathcal{M}_N$ given by the evaluation map $\mathrm{ev}\colon f \mapsto \mathrm{ev}_f$, where $\mathrm{ev}_f(\mu) = \mu(f)$ for any $\mu \in \mathcal{M}_N$. The proposition claims that for any $f \in \mathcal{C}_N$, ev_f can be decomposed uniquely as an integer coefficient linear combination of the evaluations ev_{u_I}, $(\emptyset \neq I \subseteq [N])$.

Any $f \in \mathcal{C}_N$ has an atomic decomposition $f = \bigcup_{a_I \subseteq f} a_I$, showing that

$$\mathrm{ev}_f = \sum_{a_I \subseteq f} \mathrm{ev}_{a_I}. \tag{13}$$

Applying the inclusion–exclusion formula

$$\mu\left(\bigcap_{k\in K} A_k\right) = \sum_{\emptyset \neq J \subseteq K} (-1)^{|J|+1}\mu\left(\bigcup_{j\in J} A_j\right)$$

for the Boolean expressions $A_k = x_k \setminus u_I$, $k \in K = [N] \setminus I$, we obtain

$$\begin{aligned}\mu(a_I) &= \sum_{\emptyset \neq J \subseteq ([N]\setminus I)} (-1)^{|J|+1}\mu(u_J \setminus u_I) = \sum_{\emptyset \neq J \subseteq ([N]\setminus I)} (-1)^{|J|+1}(\mu(u_{I\cup J}) - \mu(u_I))\\ &= \sum_{I \subseteq K \subseteq [N]} (-1)^{|K\setminus I|+1}\mu(u_K),\end{aligned} \tag{14}$$

for any $\mu \in \mathcal{M}_N$ and $I \neq [N]$.

Equations (13) and (14) show that ev_f can be written as a linear combination of the evaluations ev_{u_I}, $(\emptyset \neq I \subseteq [N])$) with integer coefficients.

To show uniqueness of the coefficients $m_{f,I}$, observe that the evaluations ev_{a_I}, $(\emptyset \neq I \subseteq [N])$ form a basis of $\mathcal{M}_N^*$, and as the linear space spanned by the $2^N - 1$ evaluations ev_{u_I}, $(\emptyset \neq I \subseteq [N])$ contains this basis, it is the whole space $\mathcal{M}_N^*$. Since $\dim \mathcal{M}_N^* = 2^N - 1$, the evaluations ev_{u_I}, $(\emptyset \neq I \subseteq [N])$ are linearly independent.

Proposition 6 *The sum* $\sum_{\emptyset \neq I \subseteq [N]} m_{f,I}$ *of the coefficients is* 1 *if* $a_\emptyset \subseteq f$ *and* 0 *otherwise.*

Proof Let $\mu \in \mathcal{M}_N$ be the additive function the values of which on atoms are given by

$$\mu(a_I) = \begin{cases} 0 & \text{if } I \neq \emptyset \\ 1 & \text{if } I = \emptyset. \end{cases}$$

Then $\mu(u_I) = 1$, for all $\emptyset \neq I \subseteq [N]$. Applying (12) to μ, we obtain

$$\sum_{\emptyset \neq I \subseteq [N]} m_{f,I} = \sum_{\emptyset \neq I \subseteq [N]} m_{f,I}\mu(u_I) = \mu(f) = \sum_{a_I \subseteq f} \mu(a_I) = \begin{cases} 0 & \text{if } a_\emptyset \not\subseteq f \\ 1 & \text{if } a_\emptyset \subseteq f. \end{cases}$$

□

Proposition 7 *For any* $\emptyset \neq I \subseteq [N]$, $m_{u_I,J} = \delta_{I,J}$ *holds, where* $\delta_{I,J}$ *is the Kronecker delta symbol.*

Proof It is clear that $\mu(u_I) = \sum_{\emptyset \neq J \subseteq [N]} \delta_{I,J}\mu(u_J)$. By the uniqueness of the coefficients $m_{u_I,J}$, this equation implies $m_{u_I,J} = \delta_{I,J}$.

Proposition 8 *If* $f, g \in \mathcal{C}_N$ *and* $f \cap g = \emptyset$, *then* $m_{f\cup g,I} = m_{f,I} + m_{g,I}$ *for every* $\emptyset \neq I \subseteq [N]$.

Proof Since for any $\mu \in \mathcal{M}_N$, equation

$$\sum_{\emptyset \neq I \subseteq [N]} m_{f\cup g,I}\mu(u_I) = \mu(f \cup g) = \mu(f) + \mu(g) = \sum_{\emptyset \neq I \subseteq [N]} (m_{f,I} + m_{g,I})\mu(u_I)$$

holds, uniqueness of the coefficients $m_{f\cup g,I}$ implies the statement.

If $\mu \in \mathcal{M}_N$, then there are infinitely many ways to extend μ to a map $\mu\colon \mathcal{B}_N \to \mathbb{R}$ preserving the additivity property $\mu(f \cup g) = \mu(f) + \mu(g) - \mu(f \cap g)$. Since such a map is uniquely defined by its values on the atomic expressions a_I, and $\mu(a_I)$ is already given for $I \neq [N]$, the extension of μ is uniquely given if we prescribe the value $\mu(a_{[N]}) \in \mathbb{R}$. This value is uniquely determined if we require that $\mu(X) = 0$, since this equation holds if and only $\mu(a_{[N]}) = -\sum_{I \subsetneq [N]} \mu(a_I)$.

Definition 2 The unique extension of $\mu \in \mathcal{M}_N$ to a map $\mu\colon \mathcal{B}_N$ satisfying the conditions $\mu(f \cup g) = \mu(f) + \mu(g) - \mu(f \cap g)$ and $\mu(X) = 0$ will be called the *0-weight extension of* μ.

4 Asymptotics for the Volume of Boolean Expressions of Large Balls

Let $f \in \mathcal{C}_N$ be a Boolean expression built from the variables $x_1, \ldots, x_N$ and the operations $\cup$, $\cap$ and $\setminus$. For a system of N points $\mathbf{p} = (\mathbf{p}_1, \ldots, \mathbf{p}_N) \in (\mathbb{R}^d)^N$ and a given radius $r > 0$, consider the body

$$B_f^d(\mathbf{p}, r) = f(B^d(\mathbf{p}_1, r), \ldots, B^d(\mathbf{p}_N, r))$$

obtained by evaluating f on the balls $x_i = B^d(\mathbf{p}_i, r)$. We are interested in the asymptotic behaviour of the volume $\mathcal{V}_f^d(\mathbf{p}, r) = \mathrm{vol}_d(B_f^d(\mathbf{p}, r))$ of this body.

For a system of points $\mathbf{p} = (\mathbf{p}_1, \ldots, \mathbf{p}_N) \in (\mathbb{R}^d)^N$ and a set $I \subseteq [N]$, denote by $K_I(\mathbf{p})$ the convex hull of the points $\{\mathbf{p}_i \mid i \in I\}$.

Definition 3 For $f \in \mathcal{C}_N$ and a system of points $\mathbf{p} \in (\mathbb{R}^d)^N$, define the *Boolean quermassintegrals* $W_{f,k}^d(\mathbf{p})$ and *Boolean intrinsic volumes* $V_{f,k}(\mathbf{p})$ by the equations

$$W_{f,k}^d(\mathbf{p}) = \sum_{\emptyset \neq I \subseteq [N]} m_{f,I} W_k^d(K_I(\mathbf{p})) \quad \text{and} \quad V_{f,k}(\mathbf{p}) = \sum_{\emptyset \neq I \subseteq [N]} m_{f,I} V_k(K_I(\mathbf{p})).$$

By Proposition 8, for any k and $\mathbf{p} \in (\mathbb{R}^d)^N$, the maps $\mathcal{C}_N \ni f \mapsto W_{f,k}^d(\mathbf{p})$ and $\mathcal{C}_N \ni f \mapsto V_{f,k}(\mathbf{p})$ are in $\mathcal{M}_N$. We define $W_{f,k}^d(\mathbf{p})$ and $V_{f,k}(\mathbf{p})$ *for arbitrary* $f \in \mathcal{B}_N$ as the 0-weight extension of these maps, respectively.

Theorem 1 *For any Boolean expression $f \in \mathcal{C}_N$, and any fixed system of centers $\mathbf{p} \in (\mathbb{R}^d)^N$ we have*

$$\mathcal{V}_f^d(\mathbf{p}, r) = \sum_{k=d-2}^{d} \binom{d}{k} W_{f,k}^d(\mathbf{p}) r^k + O(r^{d-3}) = \sum_{k=0}^{2} \kappa_{d-k} V_{f,k}(\mathbf{p}) r^{d-k} + O(r^{d-3}).$$

Proof Let $\mu \colon \mathcal{C}_N \to \mathbb{R}$ be the additive function defined by $\mu(g) = \mathcal{V}_g^d(\mathbf{p}, r)$. Applying Eq. (12) for μ, we obtain

$$\mathcal{V}_f^d(\mathbf{p}, r) = \mu(f) = \sum_{\emptyset \neq I \subseteq [N]} m_{f,I} \mu(u_I) = \sum_{\emptyset \neq I \subseteq [N]} m_{f,I} \mathcal{V}_{u_I}^d(\mathbf{p}, r).$$

For each I, $\mathcal{V}_{u_I}^d$ is the volume of the union of some balls, to which we can apply Corollary 1. This gives

$$\mathcal{V}_{u_I}^d = \kappa_d r^d + \kappa_{d-1} V_1(K_I) r^{d-1} + \kappa_{d-2} V_2(K_I) r^{d-2} + O(r^{d-3}). \tag{15}$$

The last two equations together with the definition of the Boolean quermassintegrals and Boolean intrisic volumes imply the theorem.

Remark One of the main goals set in the introduction was to extend Eqs. (2), (3), and (11) for the volumes of Boolean expressions of large congruent balls, finding suitable generalizations of the intrinsic volumes V_0, V_1, V_2, appearing in (11). Theorem 1 gives the desired extension and justifies our definition of the Boolean intrinsic volumes.

5 Properties of Boolean Intrinsic Volumes

The following properties are straightforward corollaries of the analogous properties of intrinsic volumes of convex bodies and the definitions.

Proposition 9

(a) $V_{f,0}(\mathbf{p})$ *does not depend on* $\mathbf{p}$. *Its value* $V_{f,0} \equiv \sum_{\emptyset \neq I \subset [N]} m_{f,I}$ *is* 1 *if* $a_\emptyset \subseteq f$, *and* 0 *otherwise.*

(b) The Boolean intrinsic volume $V_{f,k}(\mathbf{p})$ *does not depend on the dimension* d. *In particular,*

$$W^d_{f,k} = \frac{\kappa_k}{\binom{d}{k}} V_{f,d-k} = \frac{\kappa_k}{\binom{d}{k}} V_{f,(d+s)-(k+s)} = \frac{\binom{d+s}{k+s}\kappa_k}{\binom{d}{k}\kappa_{k+s}} W^{d+s}_{f,k+s} = \frac{(d+1)\cdots(d+s)\kappa_k}{(k+1)\cdots(k+s)\kappa_{k+s}} W^{d+s}_{f,k+s}$$

for any $s \in \mathbb{N}$.

(c) $V_{f,k}$ *is a continuous function on* $(\mathbb{R}^d)^N$ *for every* $d > 0$.

(d) If $f, g \in \mathcal{B}_N$ *and* $f \cap g = \emptyset$, *then* $W^d_{f\cup g,k} = W^d_{f,k} + W^d_{g,k}$ *and* $V_{f\cup g,k} = V_{f,k} + V_{g,k}$.

(e) $W^d_{\bar{f},k} = -W^d_{f,k}$ *and* $V_{\bar{f},k} = -V_{f,k}$ *for any* $f \in \mathcal{B}_N$.

We are going to find a formula for the Boolean intrinsic volumes that generalizes Eq. (8). Assume that any $k+2$ points of the system $\mathbf{p} = (\mathbf{p}_1, \dots, \mathbf{p}_N) \in (\mathbb{R}^d)^N$ are affinely independent. This can always be achieved by a small perturbation of the points if $d \geq k+1$. Choose a $k+1$ element index set $S = \{i_1, \dots, i_{k+1}\} \subset [N]$ and denote by σ_S the convex hull of the points $\mathbf{p}_{i_1}, \dots, \mathbf{p}_{i_{k+1}}$. By the general position assumption on $\mathbf{p}$, σ_S is a k-dimensional simplex and the affine subspace $[\sigma_S]$ spanned by it does not contain any of the points $\mathbf{p}_j$ for $j \notin S$.

Define an integer valued function $n_{f,S,\mathbf{p}} : \mathbb{S}^{d-k-1}_S \to \mathbb{Z}$ on the unit sphere $\mathbb{S}^{d-k-1}_S = \{\mathbf{u} \in \mathbb{S}^{d-1} \mid \mathbf{u} \perp [\sigma_S]\}$ as follows. Choose a vector $\mathbf{u} \in \mathbb{S}^{d-k-1}_S$. Split the index set $[N]$ into three parts depending on the position of the point $\mathbf{p}_i$ relative to the hyperplane orthogonal to $\mathbf{u}$, containing the simplex σ_S by setting

$$\Pi_+ = \{j \in [N] \mid \langle \mathbf{p}_j - \mathbf{p}_{i_1}, \mathbf{u}\rangle > 0\},$$
$$\Pi_0 = \{j \in [N] \mid \langle \mathbf{p}_j - \mathbf{p}_{i_1}, \mathbf{u}\rangle = 0\},$$
$$\Pi_- = \{j \in [N] \mid \langle \mathbf{p}_j - \mathbf{p}_{i_1}, \mathbf{u}\rangle < 0\}.$$

It is clear that $S \subseteq \Pi_0$ and $S = \Pi_0$ for almost all $\mathbf{u}$. Define the elements $y_1, \dots, y_N \in \mathcal{B}_N$ by the rule

$$y_j = \begin{cases} X & \text{if } j \in \Pi_+ \cup (\Pi_0 \setminus S), \\ x_j & \text{if } j \in S, \\ \emptyset & \text{if } j \in \Pi_-. \end{cases}$$

Evaluating the Boolean expression f on the y_j's we obtain an element $f(y_1, \ldots, y_N) \in \mathcal{B}_{k+1}(x_{i_1}, \ldots, x_{i_{k+1}})$ in the free Boolean algebra generated by the elements $x_{i_1}, \ldots, x_{i_{k+1}}$. Set $n_{f,S,\mathbf{p}}(\mathbf{u}) = (-1)^{k+1}\tilde{\chi}_{k+1}(f(y_1, \ldots, y_N))$.

The values of $n_{f,S,\mathbf{p}}$ are integers in the interval $[-2^k, 2^k]$. Let

$$\nu_{f,S,\mathbf{p}} = \frac{1}{(d-k)\kappa_{d-k}} \int_{\mathbb{S}_S^{d-k-1}} n_{f,S,\mathbf{p}}(\mathbf{u})\mathrm{d}\mathbf{u}$$

be the average value of $n_{f,S,\mathbf{p}}$.

Theorem 2 *If $f \in \mathcal{B}_N$ and $\mathbf{p} \in (\mathbb{R}^d)^N$ satisfies that any $k+2$ points of $\mathbf{p}$ are affinely independent, then we have*

$$V_{f,k}(\mathbf{p}) = \sum_{\substack{S \subseteq [N] \\ |S|=k+1}} \nu_{f,S,\mathbf{p}} \mathrm{vol}_k(\sigma_S). \tag{16}$$

Proof If $f, g \in \mathcal{B}_N$ are disjoint, that is $f \cap g = \emptyset$, then $V_{f \cup g,k} = V_{f,k} + V_{g,k}$, furthermore, $f(y_1, \ldots, y_N) \cap g(y_1, \ldots, y_N) = \emptyset$ for any choice of the variables y_i, and since the reduced Euler characteristic is an additive function, $\nu_{f \cup g,k} = \nu_{f,k} + \nu_{g,k}$. Thus, both sides of Eq. (16) are additive functions of the Boolean expression f. Since both sides vanish for $f = X$, the two sides are equal for any $f \in \mathcal{B}_N$ if they are equal for any $f \in \mathcal{C}_N$. As it was shown in the proof of Proposition 5, the evaluations ev_{u_I}, for $\emptyset \neq I \subseteq [N]$, form a basis of $\mathcal{M}_N^*$, so it is enough to check the proposition for the unions u_I.

Assume $f = u_I$. Then $V_{f,k}(\mathbf{p}) = V_k(K_I(\mathbf{p}))$ by Proposition 7. Let $S = \{i_1, \ldots, i_{k+1}\} \subseteq [N]$ be a set of $k+1$ indices. To understand the geometrical meaning of $n_{f,S,\mathbf{p}}(\mathbf{u})$, consider first the value of $f(y_1, \ldots, y_N) = \bigcup_{j \in I} y_j$.

If $y_j = X$ for an index $j \in I$, then $f(y_1, \ldots, y_N) = X$ and $\nu_{f,S}(\mathbf{u}) = \tilde{\chi}_{k+1}(X) = 0$. Hence $n_{f,S,\mathbf{p}}(\mathbf{u})$ vanishes if $I \not\subseteq \Pi_- \cup \Pi_0$. By Proposition 2, $n_{f,S,\mathbf{p}}(\mathbf{u})$ vanishes also in the case when one of the variables $x_{i_1}, \ldots, x_{i_{k+1}}$ does not appear in $f(y_1, \ldots, y_N)$. These variables appear in $f(y_1, \ldots, y_N)$ if and only if $S \subseteq I \cap \Pi_0$. This means that if $n_{f,S,\mathbf{p}}(\mathbf{u}) \neq 0$, then $K_I(\mathbf{p})$ is contained in the halfspace $\{\mathbf{x} \in \mathbb{R}^d \mid \langle \mathbf{u}, \mathbf{x} - \mathbf{p}_{i_1} \rangle \leq 0\}$ and the boundary hyperplane of this halfspace intersects the polytope $K_I(\mathbf{p})$ in a face that contains the k-dimensional simplex σ_S. What is the value of $n_{f,S,\mathbf{p}}(\mathbf{u})$ in this case? If $I \subseteq \Pi_- \cup \Pi_0$ and $S \subseteq I \cap \Pi_0$, then

$$n_{f,S,\mathbf{p}}(\mathbf{u}) = (-1)^{k+1}\tilde{\chi}_{k+1}(x_{i_1} \cup \cdots \cup x_{i_{k+1}}) = -\tilde{\chi}_{k+1}(x_{i_1} \cap \cdots \cap x_{i_{k+1}}) = 1.$$

If the simplex σ_S is not a face of $K_I(\mathbf{p})$, then the smallest face of $K_I(\mathbf{p})$ that contains σ_S has dimension bigger than k because of the general position assumption on $\mathbf{p}$.

In this case, the support of the function $n_{f,S,\mathbf{p}}$ is contained in a great subsphere of $\mathbb{S}_S^{d-k-1}$, and $\nu_{f,S,\mathbf{p}} = 0$.

If σ_S is a face of $K_I(\mathbf{p})$, then $n_{f,S,\mathbf{p}}$ is the indicator function of the intersection of the cone $N(\sigma_S, K_I(\mathbf{p}))$ and the sphere $\mathbb{S}_S^{d-k-1}$, therefore

$$\nu_{f,S,\mathbf{p}} = \frac{1}{(d-k)\kappa_{d-k}} \int_{\mathbb{S}_S^{d-k-1}} n_{f,S,\mathbf{p}}(\mathbf{u})d\mathbf{u} = \frac{\mathrm{vol}_{d-k}(n(\sigma_S, K_I(\mathbf{p})))}{\kappa_{d-k}} = \nu(\sigma_S, K_I(\mathbf{p})).$$

As all the k-dimensional faces of $K_I(\mathbf{p})$ are simplicies, we conclude that for $f = u_I$, we have

$$\sum_{\substack{S\subseteq[N] \\ |S|=k+1}} \nu_{f,S,\mathbf{p}} \mathrm{vol}_k(\sigma_S) = \sum_{\sigma\in\mathcal{F}_k(K_I(\mathbf{p}))} \nu(\sigma, K_I(\mathbf{p}))\mathrm{vol}_k(L) = V_k(K_I(\mathbf{p})) = V_{f,k}(\mathbf{p}),$$

as desired.

Proposition 10 *If $f \in \mathcal{B}_N$, f^* and $f^{\bar{*}}$ are the dual and contradual of f respectively, then*

$$V_{f^*,k} = -V_{f^{\bar{*}},k} = (-1)^k V_{f,k} \text{ and } W^d_{f^*,k} = -W^d_{f^{\bar{*}},k} = (-1)^{d-k} W^d_{f,k}.$$

Proof Due to Proposition 9 (e) and (b), it is enough to show the equality $V_{f^*,k} = (-1)^k V_{f,k}$. As $V_{f,k}(\mathbf{p})$ does not depend on the dimension of the ambient space $\mathbb{R}^d$, we may assume that $d > k$. Then the set of configurations $\mathbf{p} = (\mathbf{p}_1, \dots, \mathbf{p}_N) \in (\mathbb{R}^d)^N$ satisfying that any $k+2$ of the points $\mathbf{p}_1, \dots, \mathbf{p}_N$ are affinely independent is dense in $(\mathbb{R}^d)^N$. Since $V_{f,k}$ is continuous on $(\mathbb{R}^d)^N$ for all $f \in \mathcal{B}_N$, it suffices to prove the equation $V_{f^*,k}(\mathbf{p}) = (-1)^k V_{f,k}(\mathbf{p})$ for configurations satisfying this general position condition. Under this assumption, Theorem 2 implies the statement if we show the equations $\nu_{f^*,S,\mathbf{p}} = (-1)^k \nu_{f,S,\mathbf{p}}$.

Consider the function $n_{f,S,\mathbf{p}}(\mathbf{u}) = (-1)^{k+1}\tilde{\chi}_{k+1}(f(y_1, \dots, y_N))$ in the definition of $\nu_{f,S,\mathbf{p}}$. It is not difficult to see that $f^{\bar{*}}(y_1, \dots, y_N)$ is the contradual of $f(y_1, \dots, y_N)$ and $f^*(y_1, \dots, y_N)$ is the dual of it, so applying Proposition 4, we obtain $n_{f^*,S,\mathbf{p}} = (-1)^k n_{f,S,\mathbf{p}}$. Taking the mean value of both sides over the unit sphere $\mathbb{S}_S^{d-k-1}$ we get the desired equation $\nu_{f^*,S,\mathbf{p}} = (-1)^k \nu_{f,S,\mathbf{p}}$.

Denote by $l_{\mathbf{u}}\colon \mathbb{R}^d \to \mathbb{R}$ the linear function $l_{\mathbf{u}}\colon \mathbf{x} \mapsto \langle \mathbf{u}, \mathbf{x}\rangle$. If $\mathbf{u}$ is a unit vector, and K is a bounded convex set, then the length of the interval $l_{\mathbf{u}}(K)$ is the width $w_K(\mathbf{u})$ of K in the direction of $\mathbf{u}$. It is known that $V_1(K)$ is proportional to the mean width of K, namely,

$$V_1(K) = \frac{1}{2\kappa_{d-1}} \int_{\mathbb{S}^{d-1}} w_K(\mathbf{u})d\mathbf{u} = \frac{d\kappa_d}{2\kappa_{d-1}} \omega_d(K).$$

The width and the mean width can be expressed with the help of the support function of K. Recall that the support function of a bounded set $X \subset \mathbb{R}^d$ is defined as the function $h_X\colon \mathbb{S}^{d-1} \to \mathbb{R}$, $h_X(\mathbf{u}) = \sup_{\mathbf{x}\in X}\langle \mathbf{x}, \mathbf{u}\rangle$. It is clear that $w_K(\mathbf{u}) = h_K(\mathbf{u}) + h_K(-\mathbf{u})$, and

$$V_1(K) = \frac{1}{2\kappa_{d-1}} \int_{\mathbb{S}^{d-1}} (h_K(\mathbf{u}) + h_K(-\mathbf{u}))\mathrm{d}\mathbf{u} = \frac{1}{\kappa_{d-1}} \int_{\mathbb{S}^{d-1}} h_K(\mathbf{u})\mathrm{d}\mathbf{u}.$$

We can extend this formula for the case when $f \in \mathcal{L}_N$. Then f can be evaluated on real numbers by setting $a \cup b = \max\{a, b\}$ and $a \cap b = \min\{a, b\}$ for $a, b \in \mathbb{R}$.

Theorem 3 *If $f \in \mathcal{L}_N$, then for any $\mathbf{p} \in (\mathbb{R}^d)^N$, we have*

$$V_{f,1}(\mathbf{p}) = \frac{1}{\kappa_{d-1}} \int_{\mathbb{S}^{d-1}} f(\langle \mathbf{u}, \mathbf{p}_1\rangle, \dots, \langle \mathbf{u}, \mathbf{p}_N\rangle)\mathrm{d}\mathbf{u}.$$

Proof Suppose that the points $\mathbf{p}_i$ are all contained in the interior of the ball $B_R = B^d(\mathbf{0}, R)$. Since $f \in \mathcal{L}_N$, $a_\emptyset \subseteq f$, therefore $\sum_{\emptyset\neq I\subseteq[N]} m_{f,I} = 1$, and

$$V_{f,1}(\mathbf{p}) = \sum_{\emptyset\neq I\subseteq[N]} m_{f,I} V_1(K_I(\mathbf{p})) = \left(\sum_{\emptyset\neq I\subseteq[N]} m_{f,I} V_1(K_I(\mathbf{p}) + B_R)\right) - V_1(B_R).$$

Denote by $S_i(\mathbf{u})$ the interval $l_\mathbf{u}(\{\mathbf{p}_i\} + B_R) = [\langle \mathbf{u}, \mathbf{p}_i\rangle - R, \langle \mathbf{u}, \mathbf{p}_i\rangle + R]$. Then

$$V_1(K_I(\mathbf{p}) + B_R) = \frac{1}{2\kappa_{d-1}} \int_{\mathbb{S}^{d-1}} \mathrm{vol}_1(l_\mathbf{u}(K_I + B_R)\mathrm{d}\mathbf{u} = \frac{1}{2\kappa_{d-1}} \int_{\mathbb{S}^{d-1}} \mathrm{vol}_1\left(\bigcup_{i\in I} S_i(\mathbf{u})\right)\mathrm{d}\mathbf{u},$$

and

$$V_{f,1}(\mathbf{p}) = \frac{1}{2\kappa_{d-1}} \int_{\mathbb{S}^{d-1}} \Big(\sum_{\emptyset\neq I\subseteq[N]} m_{f,I}\mathrm{vol}_1\Big(\bigcup_{i\in I} S_i(\mathbf{u})\Big)\Big)\mathrm{d}\mathbf{u} - \frac{d\kappa_d R}{\kappa_{d-1}}.$$

For any fixed $\mathbf{u} \in \mathbb{S}^{d-1}$, the function $\mu\colon \mathcal{C}_N \to \mathbb{R}$ defined by $\mu(f) = \mathrm{vol}_1(f(S_1(\mathbf{u}), \dots, S_N(\mathbf{u})))$ is in $\mathcal{M}_N$, therefore Proposition 5 yields

$$\sum_{\emptyset\neq I\subseteq[N]} m_{f,I}\mathrm{vol}_1\Big(\bigcup_{i\in I} S_i(\mathbf{u})\Big) = \sum_{\emptyset\neq I\subseteq[N]} m_{f,I}\mu(u_I) = \mu(f) = \mathrm{vol}_1(f(S_1(\mathbf{u}), \dots, S_N(\mathbf{u}))).$$

By the choice of R, 0 is a common interior point of all the intervals $S_i(\mathbf{u})$. For this reason, all the sets that can be obtained from these intervals using the operations $\cup$ and $\cap$ are also intervals. In particular,

$$f(S_1(\mathbf{u}), \dots, S_N(\mathbf{u})) = [-f(-\langle \mathbf{u}, \mathbf{p}_1\rangle, \dots, -\langle \mathbf{u}, \mathbf{p}_N\rangle) - R, f(\langle \mathbf{u}, \mathbf{p}_1\rangle, \dots, \langle \mathbf{u}, \mathbf{p}_N\rangle) + R],$$

and

$$\mathrm{vol}_1(f(S_1(\mathbf{u}), \dots, S_N(\mathbf{u})) = f(\langle \mathbf{u}, \mathbf{p}_1\rangle, \dots, \langle \mathbf{u}, \mathbf{p}_N\rangle) + f(\langle -\mathbf{u}, \mathbf{p}_1\rangle, \dots, \langle -\mathbf{u}, \mathbf{p}_N\rangle) + 2R.$$

Using the fact that for any integrable function $h\colon \mathbb{S}^{d-1} \to \mathbb{R}$, we have $\int_{\mathbb{S}^{d-1}} h(\mathbf{u})\mathrm{d}\mathbf{u} = \int_{\mathbb{S}^{d-1}} h(-\mathbf{u})\mathrm{d}\mathbf{u}$, these equations give

$$V_{f,1}(\mathbf{p}) = \frac{1}{2\kappa_{d-1}} \int_{\mathbb{S}^{d-1}} (f(\langle \mathbf{u}, \mathbf{p}_1\rangle, \dots, \langle \mathbf{u}, \mathbf{p}_N\rangle) + f(\langle -\mathbf{u}, \mathbf{p}_1\rangle, \dots, \langle -\mathbf{u}, \mathbf{p}_N\rangle) + 2R)\mathrm{d}\mathbf{u} - \frac{d\kappa_d R}{\kappa_{d-1}}$$
$$= \frac{1}{\kappa_{d-1}} \int_{\mathbb{S}^{d-1}} f(\langle \mathbf{u}, \mathbf{p}_1\rangle, \dots, \langle \mathbf{u}, \mathbf{p}_N\rangle)\mathrm{d}\mathbf{u},$$

as we wanted to show.

6 Monotonocity of the Boolean Intrinsic Volume $V_{f,1}$

In this section, we prove the following result.

Theorem 4 *Assume that the Boolean expression $f \in C_N$ can be represented by a formula in which each of the variables occurs exactly once. Define the signs ϵ_{ij}^f, for $1 \le i < j \le N$, as in the introduction. If the configurations $\mathbf{p} = (\mathbf{p}_1, \dots, \mathbf{p}_N)$ and $\mathbf{q} = (\mathbf{q}_1, \dots, \mathbf{q}_N) \in (\mathbb{R}^d)^N$ satisfy the inequalities $\epsilon_{ij}^f(\mathrm{d}(\mathbf{p}_i, \mathbf{p}_j) - \mathrm{d}(\mathbf{q}_i, \mathbf{q}_j)) \ge 0$ for all $0 \le i < j \le N$, then we have*

$$V_{f,1}(\mathbf{p}) \ge V_{f,1}(\mathbf{q}). \tag{17}$$

Proof It is proved in [5], that if there exist piecewise analytic continuous maps $\mathbf{z}_i\colon [0,1] \to \mathbb{R}^d$ for $1 \le i \le N$, such that $\mathbf{z}_i(0) = \mathbf{p}_i$, $\mathbf{z}_i(1) = \mathbf{q}_i$, and the distances $\mathrm{d}(\mathbf{z}_i(t), \mathbf{z}_j(t))$ are weakly monotonous functions of t for all i and j, then ineqality (1) is true for any choice of the radii. It is not difficult to see that the analytic curves $\mathbf{z}_i\colon [0,1] \to \mathbb{R}^d \times \mathbb{R}^d$ defined by $\mathbf{z}_i(t) = (\cos(t\pi/2)\mathbf{p}_i, \sin(t\pi/2)\mathbf{q}_i)$ connect the points $(\mathbf{p}_i, \mathbf{0})$ to the points $(\mathbf{0}, \mathbf{q}_i)$ in the required way, but jumping into $\mathbb{R}^{2d}$. Thus, embedding the centers into $\mathbb{R}^{2d}$, our assumptions imply the inequality

$$\mathcal{V}_f^{2d}(\mathbf{p}, r) = \mathrm{vol}_{2d}\left(B_f^{2d}(\mathbf{p}, r)\right) \ge \mathrm{vol}_{2d}\left(f(B_f^{2d}(\mathbf{q}, r)\right) = \mathcal{V}_f^{2d}(\mathbf{q}, r) \tag{18}$$

for any choice of the radius r. By Proposition 9 (a), $V_{f,0}(\mathbf{p}) = V_{f,0}(\mathbf{q})$, therefore Theorem 1 gives

$$0 \le \mathcal{V}_f^{2d}(\mathbf{p}, r) - \mathcal{V}_f^{2d}(\mathbf{q}, r) = \kappa_{2d-1}(V_{f,1}(\mathbf{p}) - V_{f,1}(\mathbf{q}))r^{2d-1} + O(r^{2d-2}).$$

This inequality can hold for large r only if the coefficient of the dominant term is nonnegative, i.e., $V_{f,1}(\mathbf{p}) \ge V_{f,1}(\mathbf{q})$.

It seems to be an interesting question whether we can write strict inequality in (17) if, in addition to the assumptions of Theorem 4, we know that the configurations $\mathbf{p}$ and $\mathbf{q}$ are not congruent. An affirmative answer would imply that the generalized Kneser–Poulsen conjecture holds for Boolean expression of congruent balls if the radius of the balls is greater than a certain number depending on the system of the centers.

Acknowledgements This research was supported by the Hungarian National Science and Research Foundation OTKA K 112703. Part of the research was done in the academic year 2014/15, while the author enjoyed the hospitality of the MTA Alfréd Rényi Institute of Mathematics as a guest researcher.

References

1. K. Bezdek, Private communication. Retrospective Workshop on Discrete Geometry, Optimization, and Symmetry, The Fields Institute, Toronto, November 25–29, 2013
2. K. Bezdek, R. Connelly, Pushing disks apart–the Kneser-Poulsen conjecture in the plane. J. Reine Angew. Math. **553**, 221–236 (2002). https://doi.org/10.1515/crll.2002.101
3. V. Capoyleas, J. Pach, On the perimeter of a point set in the plane, in *Discrete and computational geometry (New Brunswick, NJ, 1989/1990)*, DIMACS Ser. Discrete Math. Theoret. Comput. Sci., vol. 6 (Amer. Math. Soc., Providence, RI, 1991), pp. 67–76
4. B. Csikós, On the volume of flowers in space forms. Geom. Dedicata **86**(1–3), 59–79 (2001). https://doi.org/10.1023/A:1011983123985
5. B. Csikós, A Schläfli-type formula for polytopes with curved faces and its application to the Kneser-Poulsen conjecture. Monatsh. Math. **147**(4), 273–292 (2006). https://doi.org/10.1007/s00605-005-0363-7
6. S. Givant, P. Halmos, Introduction to Boolean algebras, in *Undergraduate Texts in Mathematics* (Springer, New York, 2009). https://doi.org/10.1007/978-0-387-68436-9
7. I. Gorbovickis, Strict Kneser-Poulsen conjecture for large radii. Geom. Dedicata **162**, 95–107 (2013). https://doi.org/10.1007/s10711-012-9718-0
8. Y. Gordon, M. Meyer, On the volume of unions and intersections of balls in Euclidean space, in *Geometric aspects of functional analysis* (Israel, 1992–1994), Oper. Theory Adv. Appl., vol. 77 (Birkhäuser, Basel, 1995), pp. 91–101
9. M. Gromov, Monotonicity of the volume of intersection of balls, in *Geometrical aspects of functional analysis* (1985/86), Lecture Notes in Math., vol. 1267 (Springer, Berlin, 1987), pp. 1–4. https://doi.org/10.1007/BFb0078131
10. M. Kneser, Einige Bemerkungen über das Minkowskische Flächenmass. Arch. Math. (Basel) **6**, 382–390 (1955)
11. E.T. Poulsen, Problem 10. Math. Scandinavica **2**, 346 (1954)
12. R. Schneider, Convex bodies: the Brunn-Minkowski theory, in *Encyclopedia of Mathematics and its Applications*, vol. 151, expanded edn. (Cambridge University Press, Cambridge, 2014)

Small Primitive Zonotopes

Antoine Deza, George Manoussakis and Shmuel Onn

Abstract We study a family of lattice polytopes, called *primitive zonotopes*, describe instances with small parameters, and discuss connections to the largest diameter of lattice polytopes and to the computational complexity of multicriteria matroid optimization. Complexity results and open questions are also presented.

Keywords Lattice polytopes · Primitive integer vectors · Matroid optimization · Diameter

1 Introduction

Recent results dealing with the combinatorial, geometric, and algorithmic aspects of linear optimization include Santos' counterexample [27] to the Hirsch conjecture, and Allamigeon, Benchimol, Gaubert, and Joswig's counterexample [2] to a continuous analogue of the polynomial Hirsch conjecture. Borgwardt, De Loera, and Finhold [4] showed that the Hirsch bound holds for transportation polytopes. Kalai and Kleitman's upper bound [18] for the diameter of polytopes was strengthened by Todd [32] and by Sukegawa [30].

A. Deza (✉)
Advanced Optimization Laboratory, Faculty of Engineering,
McMaster University, Hamilton, ON, Canada
e-mail: deza@mcmaster.ca; deza@lri.fr

A. Deza · G. Manoussakis
Laboratoire de Recherche en Informatique,
CNRS – Université de Paris Sud, Orsay, France

G. Manoussakis
e-mail: george@lri.fr

S. Onn
Operations Research, Davidson faculty of IE & M,
Technion - Israel Institute of Technology, Haifa, Israel
e-mail: onn@ie.technion.ac.il

M. D. E. Conder et al. (eds.), *Discrete Geometry and Symmetry*,
Springer Proceedings in Mathematics & Statistics 234,
https://doi.org/10.1007/978-3-319-78434-2_5

Focusing on lattice polytopes; that is bounded polytopes whose vertices are integer-valued, Del Pia and Michini [7] strengthened Kleinschmidt and Onn's upper bound [19] for the diameter of lattice polytopes. Multicriteria matroid optimization is a generalization of standard linear matroid optimization introduced by Onn and Rothblum [26] where each basis is evaluated according to several, rather than one, criteria, and these values are traded-in by a convex function.

The article pursues the study of the *primitive zonotopes* initiated in [10] and is organized as follows. After recalling their definition and providing some of their combinatorial properties, we highlight in Sect. 2 connections to convex multicriteria matroid optimization, and to the diameter of lattice polytopes. In particular, we strengthen the bounds on the maximum number $m(d, 1)$ of greedily solvable linear single criterion counterparts needed to solve any d-criteria 1-bounded instance. Section 3 focuses on primitive zonotopes of small dimension d, norm q, and order p. The diameter, grid embedding size, and number of vertices are given for values of (d, q, p) yielding computationally tractable primitive zonotopes. Complexity results and open questions are discussed in Sect. 4. In particular, we show that linear optimization and separation over primitive zonotopes can be done in polynomial time, as well as deciding whether a given point, respectively a pair of points, is a vertex, respectively an edge. Proofs for Sects. 2.2 and 3 are given in Sect. 5.

2 Primitive Zonotopes

2.1 *Zonotopes Generated by Short Primitive Vectors*

The convex hull of integer-valued points is called a *lattice polytope* and, if all the vertices are drawn from $\{0, 1, \ldots, k\}^d$, is refereed to as a *lattice* (d, k)*-polytope*. For simplicity, we only consider full dimensional lattice (d, k)-polytopes. Given a finite set G of vectors, also called the generators, the *zonotope* generated by G is the convex hull of all signed sums of the elements of G. We consider zonotopes generated by short integer vectors in order to keep the grid embedding size relatively small. In addition, we restrict to integer vectors which are pairwise linearly independent in order to maximize the diameter. Thus, for $q = \infty$ or a positive integer, and d, p positive integers, we consider the *primitive zonotope* $Z_q(d, p)$ defined as the zonotope generated by the primitive integer vectors of q-norm at most p:

$$\begin{aligned} Z_q(d, p) &= \sum [-1, 1]\{v \in \mathbb{Z}^d \ : \ \|v\|_q \leq p \,,\ \gcd(v) = 1 \,,\ v \succ 0\} \\ &= \operatorname{conv}\Big(\sum \{\lambda_v v \ : \ v \in \mathbb{Z}^d \,,\ \|v\|_q \leq p \,,\ \gcd(v) = 1 \,,\ v \succ 0\} \ : \ \lambda_v = \pm 1\Big) \end{aligned}$$

where $\gcd(v)$ is the largest integer dividing all entries of v, and $\succ$ the lexicographic order on $\mathbb{R}^d$, i.e. $v \succ 0$ if the first nonzero coordinate of v is positive. Similarly, we consider the Minkowski sum $H_q(d, p)$ of the generators of $Z_q(d, p)$:

$$H_q(d, p) = \sum [0, 1]\{v \in \mathbb{Z}^d : \|v\|_q \le p, \ \gcd(v) = 1, \ v \succ 0\}.$$

In other words, $H_q(d, p)$ is, up to translation, the image of $Z_q(d, p)$ by a homothety of factor $1/2$. We also consider the *positive primitive zonotope* $Z_q^+(d, p)$ defined as the zonotope generated by the primitive integer vectors of q-norm at most p with nonnegative coordinates:

$$Z_q^+(d, p) = \sum [-1, 1]\{v \in \mathbb{Z}_+^d : \|v\|_q \le p, \ \gcd(v) = 1\}$$

where $\mathbb{Z}_+ = \{0, 1, \dots\}$. Similarly, we consider the Minkowski sum of the generators of $Z_q^+(d, p)$:

$$H_q^+(d, p) = \sum [0, 1]\{v \in \mathbb{Z}_+^d : \|v\|_q \le p, \ \gcd(v) = 1\}.$$

We illustrate the primitive zonotopes with a few examples:

(i) For finite q, $Z_q(d, 1)$ is generated by the d unit vectors and forms the $\{-1, 1\}^d$-cube. $H_q(d, 1)$ is the $\{0, 1\}^d$-cube.

(ii) $Z_1(d, 2)$ is the permutahedron of type B_d and thus, $H_1(d, 2)$ is, up to translation, a lattice $(d, 2d-1)$-polytope with $2^d d!$ vertices and diameter d^2. For example, $Z_1(2, 2)$ is generated by $\{(0, 1), (1, 0), (1, 1), (1, -1)\}$ and forms the octagon whose vertices are $\{(-3, -1), (-3, 1), (-1, 3), (1, 3), (3, 1), (3, -1), (1, -3), (-1, -3)\}$. $H_1(2, 2)$ is, up to translation, a lattice $(2, 3)$-polygon, see Fig. 1. $Z_1(3, 2)$ is congruent to the truncated cuboctahedron, see Fig. 2 for an illustration, which is also called the great rhombicuboctahedron and is the Minkowski sum of an octahedron and a cuboctahedron, see for instance Eppstein [12]. $H_1(3, 2)$ is, up to translation, a lattice $(3, 5)$-polytope with diameter 9 and 48 vertices.

(iii) $H_1^+(d, 2)$ is the Minkowski sum of the permutahedron with the $\{0, 1\}^d$-cube. Thus, $H_1^+(d, 2)$ is a lattice (d, d)-polytope with diameter $\binom{d+1}{2}$.

(iv) $Z_\infty(3, 1)$ is congruent to the truncated small rhombicuboctahedron, see Fig. 3 for an illustration, which is the Minkowski sum of a cube, a truncated octahedron,

Fig. 1 $H_1(2, 2)$

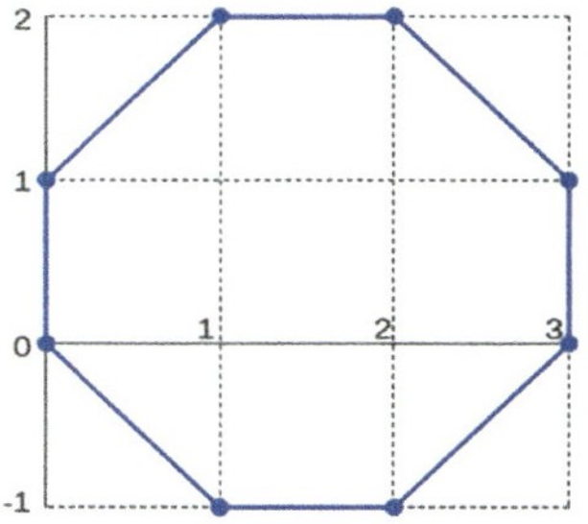

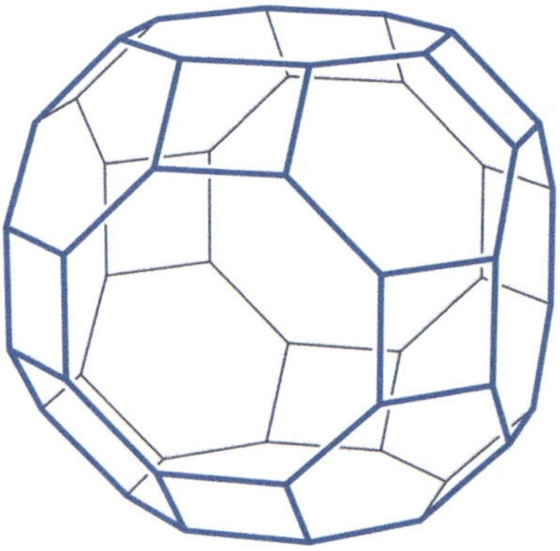

Fig. 2 $Z_1(3,2)$ is congruent to the truncated cuboctahedron

Fig. 3 $Z_\infty(3,1)$ is congruent to the truncated small rhombicuboctahedron

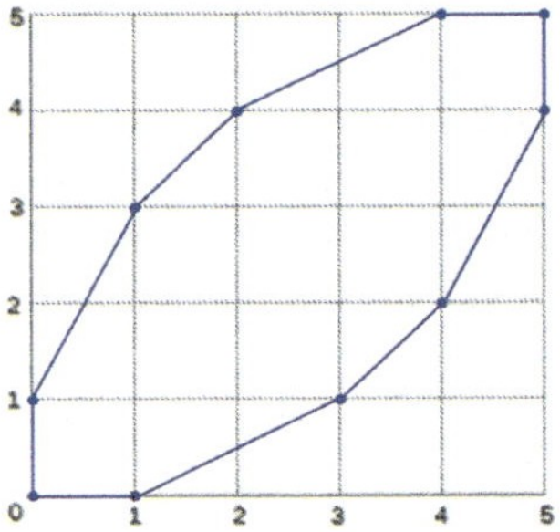

Fig. 4 $H_\infty^+(2,2)$

and a rhombic dodecahedron, see for instance Eppstein [12]. $H_\infty(3,1)$ is, up to translation, a lattice $(3,9)$-polytope with diameter 13 and 96 vertices.

(v) $Z_\infty^+(2,2)$ is generated by $\{(0,1),(1,0),(1,1),(1,2),(2,1)\}$ and forms the decagon whose vertices are $\{(-5,-5),(-5,-3),(-3,-5),(-3,1),(-1,3),(1,-3),(3,-1),(3,5),(5,3),(5,5)\}$. $H_\infty^+(2,2)$ is a lattice $(2,5)$-polygon, see Fig. 4.

2.2 Combinatorial Properties of the Primitive Zonotopes

We recall properties concerning $Z_q(d, p)$ and $Z_q^+(d, p)$, and in particular their symmetry group, diameter, and vertices. $Z_1(d, 2)$ is the permutahedron of type B_d as its generators form the root system of type B_d, see [17]. Thus, $Z_1(d, 2)$ has $2^d d!$ vertices and its symmetry group is B_d. The properties listed in this section are extensions to $Z_q(d, p)$ of known properties of $Z_1(d, 2)$ whose proofs are given in Sect. 5.1. We refer to Fukuda [14], Grünbaum [16], and Ziegler [33] for polytopes and, in particular, zonotopes.

Property 2.1

(i) *$Z_q(d, p)$ is invariant under the symmetries induced by coordinate permutations and the reflections induced by sign flips.*
(ii) *The sum $\sigma_q(d, p)$ of all the generators of $Z_q(d, p)$ is a vertex of both $Z_q(d, p)$ and $H_q(d, p)$. The origin is a vertex of $H_q(d, p)$, and $-\sigma_q(d, p)$ is a vertex of $Z_q(d, p)$.*
(iii) *The coordinates of the vertices of $Z_q(d, p)$ are odd. Thus, the number of vertices of $Z_q(d, p)$ is a multiple of 2^d.*
(iv) *$H_q(d, p)$ is, up to translation, a lattice (d, k)-polytope where k is the sum of the first coordinates of all generators of $Z_q(d, p)$*
(v) *The diameter of $Z_q(d, p)$, respectively $Z_q^+(d, p)$, is equal to the number of its generators.*

Property 2.2

(i) *$Z_q^+(d, p)$ is centrally symmetric and invariant under the symmetries induced by coordinate permutations.*
(ii) *The sum $\sigma_q^+(d, p)$ of all the generators of $Z_q^+(d, p)$ is a vertex of both $Z_q^+(d, p)$ and $H_q^+(d, p)$. The origin is a vertex of $H_q^+(d, p)$, and $-\sigma_q^+(d, p)$ is a vertex of $Z_q^+(d, p)$.*

A vertex v of $Z_q(d, p)$ is called *canonical* if $v_1 \geq \cdots \geq v_d > 0$. Property 2.1 item (i) implies that the vertices of $Z_q(d, p)$ are all the coordinate permutations and sign flips of its canonical vertices.

Property 2.3

(i) *A canonical vertex v of $Z_q(d, p)$ is the unique maximizer of $\{\max c^T x : x \in Z_q(d, p)\}$ for some vector c satisfying $c_1 > c_2 > \cdots > c_d > 0$.*
(ii) *$Z_1(d, 2)$ has $2^d d!$ vertices corresponding to all coordinate permutations and sign flips of the unique canonical vertex $\sigma_1(d, 2) = (2d-1, 2d-3, \ldots, 1)$.*
(iii) *For $q = \infty$ or $p \neq 1$, $Z_q(d, p)$ has at least $2^d d!$ vertices including all coordinate permutations and sign flips of the canonical vertex $\sigma_q(d, p)$.*
(iv) *$Z_\infty^+(d, 1)$ has at least $2 + 2d!$ vertices including the $2d!$ permutations of $\pm\sigma(d)$ where $\sigma(d)$ is a vertex with pairwise distinct coordinates, and the 2 vertices $\pm\sigma_\infty^+(d, 1)$.*

2.3 Primitive Zonotopes as Lattice Polytopes with Large Diameter

Let $\delta(d,k)$ be the maximum possible edge-diameter over all lattice (d,k)-polytopes. Finding lattice polygons with the largest diameter; that is, to determine $\delta(2,k)$, was investigated independently in the early nineties by Thiele [31], Balog and Bárány [3], and Acketa and Žunić [1]. This question can be found in Ziegler's book [33] as Exercise 4.15. The answer is summarized in Proposition 2.4, with the role of primitive zonotopes highlighted.

Proposition 2.4 *$\delta(2,k)$ is achieved, up to translation, by the Minkowski sum of a subset of the generators of $H_1(2,p)$ for a proper p. In particular, for $k = \sum_{1\le j\le p} j\phi(j)$ for some p, $\delta(2,k)$ is uniquely achieved, up to translation, by $H_1(2,p)$.*

In general dimension, Naddef [24] showed in 1989 that $\delta(d,1) = d$, Kleinschmidt and Onn [19] generalized this result in 1992 showing that $\delta(d,k) \le kd$, before Del Pia and Michini [7] strengthened the upper bound to $\delta(d,k) \le kd - \lceil d/2 \rceil$ for $k \ge 2$, and showed that $\delta(d,2) = \lfloor 3d/2 \rfloor$. Deza and Pournin [11] further strengthened the upper bound to $kd - \lceil 2d/3 \rceil - (k-3)$ for $k \ge 3$ and showed that $\delta(4,3) = 8$. The quantities $\delta(3,4) = 7$ and $\delta(3,5) = 9$, respectively $\delta(3,6) = 10$ and $\delta(5,3) = 10$, were computationally determined in [5], respectively [8]. Concerning the lower bound, Deza, Manoussakis, and Onn [10] showed that $\delta(d,k) \ge \lfloor (k+1)d/2 \rfloor$ for $k < 2d$. These bounds are summarized in Proposition 2.5, and Conjecture 2.6 given in [10] is recalled.

Proposition 2.5

(i) $\delta(d,k) = \lfloor (k+1)d/2 \rfloor$ *for* $(d,k) = (d,1), (d,2), (2,3), (3,3), (4,3), (5,3), (3,4), (3,5)$, *and* $(3,6)$.
(ii) $2d \le \delta(d,3) \le \lfloor 7d/3 \rfloor - 1$ *for* $d \not\equiv 2 \mod 3$, *and* $\delta(d,3) \le \lfloor 7d/3 \rfloor$ *otherwise,*
(iii) $\delta(d,k) \ge \lfloor (k+1)d/2 \rfloor$ *for* $k < 2d$,
(iv) $\delta(d,k) \le kd - \lceil 2d/3 \rceil - (k-2)$ *for* $k \ge 4$

Conjecture 2.6 $\delta(d,k)$ is achieved, up to translation, by a Minkowski sum of lattice vectors. In particular, $\delta(d,k) \le \lfloor (k+1)d/2 \rfloor$ for any d and k, and $\delta(d,k) = \lfloor (k+1)d/2 \rfloor$ when $k < 2d$.

Note that Conjecture 2.6 holds for all known values of $\delta(d,k)$ given in Table 1, and hypothesizes, in particular, that $\delta(d,3) = 2d$.

Soprunov and Soprunova [29] considered the Minkowski length of a lattice polytope P; that is, the largest number of lattice segments whose Minkowski sum is contained in P. For example, the Minkowski length of the $\{0,k\}^d$-cube is kd. We consider a variant of the Minkowski length and the special case when P is the $\{0,k\}^d$-cube. Let $L(d,k)$ denote the largest number of pairwise linearly independent lattice segments

Table 1 Largest diameter $\delta(d, k)$ over all lattice (d, k)-polytopes

		k									
	$\delta(d, k)$	1	2	3	4	5	6	7	8	9	10
d	1	1	1	1	1	1	1	1	1	1	1
	2	2	3	4	4	5	6	6	7	8	8
	3	3	4	6	7	9	10				
	4	4	6	8							
	5	5	7	10							
	$\vdots$	$\vdots$	$\vdots$								
	d	d	$\lfloor 3d/2 \rfloor$								

whose Minkowski sum is contained in the $\{0, k\}^d$-cube. One can check that the generators of $H_1(d, 2)$ form the largest, and unique, set of primitive lattice vectors which Minkowski sum fits within the $\{0, k\}^d$-cube for $k = 2d - 1$; that is, for k being the sum of the first coordinates of the d^2 generators of $H_1(d, 2)$. Thus, $L(d, 2d - 1) = \delta(H_1(d, 2)) = d^2$. Similarly, $L(2, k) = \delta(2, k)$ for all k, and $L(d, k) = \lfloor (k + 1)d/2 \rfloor$ for $k \leq 2d - 1$.

2.4 *Primitive Zonotopes and Convex Matroid Optimization*

We consider the convex multicriteria matroid optimization framework of Melamed, Onn and Rothblum, see [22, 25, 26]. Call $S \subset \{0, 1\}^n$ a *matroid* if it is the set of the indicators of bases of a matroid over $\{1, \dots, n\}$. For instance, S can be the set of indicators of spanning trees in a connected graph with n edges. For a $d \times n$ matrix W, let $WS = \{Wx : x \in S\}$, and let $\mathrm{conv}(WS) = W\mathrm{conv}(S)$ be the projection to $\mathbb{R}^d$ of $\mathrm{conv}(S)$ by W. Given a convex function $f : \mathbb{R}^d \to \mathbb{R}$, *convex matroid optimization* deals with maximizing the composite function $f(Wx)$ over S; that is, $\max \{f(Wx) : x \in S\}$, and is concerned with $\mathrm{conv}(WS)$; that is, the projection of the set of the feasible points. The maximization problem can be interpreted as a problem of multicriteria optimization, where each row of W gives a linear criterion $W_i x$ and f compromises these criteria. Thus, W is called the *criteria* matrix or *weight* matrix. The projection polytope $\mathrm{conv}(WS)$ and its vertices play a key role in solving the maximization problem as, for any convex function f, there is an optimal solution x whose projection $y = Wx$ is a vertex of $\mathrm{conv}(WS)$. In particular, the enumeration of all vertices of $\mathrm{conv}(WS)$ enables to compute the optimal objective value by picking a vertex attaining the optimal value $f(y) = f(Wx)$. Thus, it suffices that f is presented by a *comparison oracle* that, queried on vectors $y, z \in \mathbb{R}^d$, asserts whether or not $f(y) < f(z)$. *Coarse* criteria matrices; that is, W whose entries are small or in $\{0, 1, \dots, p\}$, are of particular interest. In multicriteria combinatorial optimization, this case corresponds to the weight $W_{i,j}$ attributed to element j of the ground set $\{1, \dots, n\}$ under criterion i being small or in $\{0, 1, \dots, p\}$ for all i, j. In the remainder, we only consider $\{0, 1, \dots, p\}$-valued W.

Let $m(d, p)$ denote the number of vertices of $H_\infty(d, p)$. Theorem 2.7, given in [10], settles the computational complexity of the multicriteria optimization problem by showing that the maximum number of vertices of the projection polytope $\mathrm{conv}(WS)$ of any matroid S on n elements and any d-criteria p-bounded utility matrix; that is, $W \in \{0, 1, \ldots, p\}^{d\times n}$, is equal to $m(d, p)$, and hence is in particular *independent* of n, S, and W.

Theorem 2.7 *Let d, p be any positive integers. Then, for any positive integer n, any matroid $S \subset \{0, 1\}^n$, and any d-criteria p-bounded utility matrix W, the primitive zonotope $H_\infty(d, p)$ refines* $\mathrm{conv}(WS)$. *Moreover, $H_\infty(d, p)$ is a translation of* $\mathrm{conv}(WS)$ *for some matroid S and d-criteria p-bounded utility matrix W. Thus, the maximum number of vertices of* $\mathrm{conv}(WS)$ *for any n, any matroid $S \subset \{0, 1\}^n$, and any d-criteria p-bounded utility matrix W, equals the number $m(d, p)$ of vertices of $H_\infty(d, p)$, and hence is independent of n, S, and W. Also, for any fixed d and convex $f : \mathbb{R}^d \to \mathbb{R}$, the multicriteria matroid optimization problem can be solved using a number of arithmetic operations and queries to the oracles of S and f which is polynomial in n and p using $m(d, p)$ greedily solvable linear matroid optimization counterparts.*

Theorem 2.8 *The number $m(d, 1)$ of vertices of $H_\infty(d, 1)$ satisfies*

$$2^d d! \le m(d, 1) \le 2 \sum_{0\le i\le d-1} \binom{(3^d-3)/2}{i} - 2\binom{(3^{d-1}-3)/2}{d-1}.$$

Proof The first inequality restates item (iii) of Property 2.3 where (q, d, p) is set to $(\infty, d, 1)$. The second inequality is obtained by exploiting the structure of the generators of $H_\infty(d, 1)$. One can check that $H_\infty(d, 1)$ has $(3^d - 1)/2$ generators and that removing the first zero of the generators of $H_\infty(d, 1)$ starting with zero yields exactly the $(3^{d-1} - 1)/2$ generators of $H_\infty(d - 1, 1)$. We recall that the number of vertices $f_0(Z)$ of a d-dimensional zonotope Z generated by m generators is bounded by $\bar{f}(d, m) = 2 \sum_{0\le i\le d-1} \binom{m-1}{i}$. By duality, the number $f_0(Z)$ of vertices of a zonotope Z is equal to the number $f_{d-1}(\mathcal{A})$ of cells of the associate hyperplane arrangement $\mathcal{A}$ where each generator m^j of Z corresponds to an hyperplane h^j of $\mathcal{A}$. The inequality $f_0(Z) \le \bar{f}(d, m)$ is based on the inequality $f_{d-1}(\mathcal{A}) \le f_{d-1}(\mathcal{A} \setminus h^j) + f_{d-1}(\mathcal{A} \cap h^j)$ for any hyperplane h^j of $\mathcal{A}$ where $\mathcal{A} \setminus h^j$ denotes the arrangement obtained by removing h^j from $\mathcal{A}$, and $\mathcal{A} \cap h^j$ denotes the arrangement obtained by intersecting $\mathcal{A}$ with h^j. This last inequality and the duality between zonotopes and hyperplane arrangements are detailed, for example, in [14]. Recursively applying this inequality to the arrangement $\mathcal{A}_\infty(d, 1)$ associated to $H_\infty(d, 1)$ till the remaining $(3^{d-1} - 1)/2$ hyperplanes form a $(d - 1)$-dimensional arrangement equivalent to $\mathcal{A}_\infty(d - 1, 1)$ yields: $f_{d-1}(\mathcal{A}_\infty(d, 1)) \le \bar{f}(d, (3^d - 1)/2) - \big(\bar{f}(d, (3^{d-1} - 1)/2) - \bar{f}(d - 1, (3^{d-1} - 1)/2)\big)$ which completes the proof since $f_{d-1}(\mathcal{A}_\infty(d, 1)) = f_0(H_\infty(d, 1))$ and $\bar{f}(d, m) - \bar{f}(d - 1, m) = 2\binom{m-1}{d}$. In other words, the inequality is based on the inductive build-up of $H_\infty(d, 1)$ starting with the $(3^{d-1}-3)/2$ generators with zero as first coordinate, and noticing that these $(3^{d-1}-3)/2$ generators belong to a lower dimensional space. □

3 Small Primitive Zonotopes $H_q(d, p)$ and $H_q^+(d, p)$

In this section we provide the number of vertices, the diameter; that is, the number of generators, and the grid embedding size for $H_q(d, p)$ and $H_q^+(d, p)$ for small d and p, and $q = 1, 2$, and ∞. We recall that, up to translation, $Z_q(d, p)$, respectively $Z_q^+(d, p)$, is the image of $H_q(d, p)$, respectively $H_q^+(d, p)$, by a homothety of factor 2. Thus $Z_q(d, p)$ and $H_q(d, p)$, respectively $Z_q^+(d, p)$ and $H_q^+(d, p)$, have the same number of vertices and the same diameter, while the grid embedding size of the $Z_q(d, p)$, respectively $Z_q^+(d, p)$, is twice the one of $H_q(d, p)$, respectively $H_q^+(d, p)$. Since both $H_q(d, 1)$ and $H_q^+(d, 1)$ are equal to the $\{0, 1\}^d$-cube for finite q, both are omitted from the tables provided in this section. The vertex enumeration was performed using standard algorithms described, for instance, in [14]. The Euler totient function counting positive integers less than or equal to j and relatively prime with j is denoted by $\phi(j)$. Note that $\phi(1)$ is set to 1.

Enumerative questions concerning $H_q(d, p)$ and $H_q^+(d, p)$ have been studied in various settings. We list a few instances, and the associated OEI sequences, see [28] for details and references therein.

(i) $f_0(H_\infty^+(d, 1))$ corresponds to the OEI sequence A034997 giving the number of generalized retarded functions in quantum field theory. The value of $f_0(H_\infty^+(d, 1))$ was determined till $d = 8$.
(ii) $f_0(H_\infty(d, 1))$, which is the number of regions of hyperplane arrangements with $\{-1, 0.1\}$-valued normals in dimension d, corresponds to the OEI sequence A009997 giving $f_0(H_\infty(d, 1))/(2^d d!)$. The value of $f_0(H_\infty(d, 1))$ was determined till $d = 7$.
(iii) $\delta(H_\infty^+(d, p))$ corresponds to the OEI sequence A090030 with further cross-referenced sequences for $d \leq 7$ and $p \leq 8$.
(iv) $\delta(H_1^+(3, p))$, respectively $\delta(H_2^+(2, p))$, $\delta(H_\infty(d, 2))$, $\delta(H_\infty(2, p))/4$, $\delta(H_2(2, p))/2$, $\delta(H_1^+(d, 3))$, and $\delta(H_2^+(d, 2))$, corresponds to the OEI sequence A048134, respectively A049715, A005059, A002088, A175341, A008778, and A055795.
(v) the grid embedding size of $H_2(d, 2)$, respectively $H_\infty(d, 2)$ and $H_1^+(d, 3)$, corresponds to the OEI sequence A161712, respectively A080961 and A052905.

3.1 Small Primitive Zonotopes $H_q(d, p)$

In Tables 2, 3, and 4, the number of vertices $f_0(H_q(d, p))$ is divided by $2^d d!$ and followed by its diameter $\delta(H_q(d, p))$ and grid embedding size. For instance, the entry

Table 2 Small primitive zonotopes $H_1(d, p)$

	$H_1(d, p)$	p 2	3	4	5	6
d	2	1 (4, 3)	2 (8, 9)	3 (12, 17)	5 (20, 37)	6 (24, 49)
	3	1 (9, 5)	7 (25, 21)	26 (49, 53)	102 (97, 133)	227 (145, 229)
	4	1 (16, 7)	40 (56, 37)	531 (136, 117)	6741 (312, 337)	? (560, 709)
	5	1 (25, 9)	339 (105, 57)	? (305, 217)	? (801, 713)	? (1681, 1769)

Table 3 Small primitive zonotopes $H_2(d, p)$

	$H_2(d, p)$	p 2	3	4	5
d	2	1 (4, 3)	2 (8, 9)	4 (16, 27)	6 (24, 51)
	3	2 (13, 9)	26 (49, 57)	126 (109, 161)	443 (205, 377)
	4	14 (40, 27)	1427 (192, 193)	? (592, 795)	? (1424, 2411)
	5	273 (105, 65)	? (641, 577)		

Table 4 Small primitive zonotopes $H_\infty(d, p)$

	$H_\infty(d, p)$	p 1	2	3	4
d	2	1 (4, 3)	2 (8, 9)	4 (16, 27)	6 (24, 51)
	3	2 (13, 9)	26 (49, 57)	228 (145, 249)	910 (289, 633)
	4	14 (40, 27)	4333 (272, 321)	? (1120, 1923)	? (2928, 6459)
	5	516 (121, 81)			
	6	124,187 (364, 243)			
	7	214,580,603 (1093, 729)			

26(49, 53) for $(q, d, p) = (1, 3, 4)$ in Table 2 indicates that $H_1(3, 4)$ has $26 \times 2^3 3! = 1248$ vertices, diameter 49, and is, up to translation, a lattice $(3, 53)$-polytope. The rather straightforward proofs are given in Sect. 5.2.

3.1.1 Small Primitive Zonotopes $H_1(d, p)$

Property 3.1

(i) $H_1(d, 1)$ *is the* $\{0, 1\}^d$*-cube,*
(ii) $H_1(d, 2)$ *is, up to translation, a lattice* (d, k)*-polytope with* $k = 2d - 1$*, and diameter* d^2*, and* $2^d d!$ *vertices,*
(iii) $H_1(d, 3)$ *is, up to translation, a lattice* (d, k)*-polytope with* $k = 2d^2 + 2d - 3$*, and diameter* $d(d + 2)(2d - 1)/3$*,*

(iv) $H_1(d,4)$ *is, up to translation, a lattice* (d,k)*-polytope with* $k = \binom{d-1}{0} + 16\binom{d-1}{1} + 20\binom{d-1}{2} + 8\binom{d-1}{3}$, *and diameter* $d(d^3+2d^2+2d-2)/3$,
(v) $H_1(2,p)$ *is, up to translation, a lattice* $(2,k)$*-polygon with* $k = \sum_{1\le j\le p} j\phi(j)$, *and diameter* $2\sum_{1\le j\le p} \phi(j)$.

3.1.2 Small Primitive Zonotopes $H_2(d, p)$

Property 3.2

(i) $H_2(d,1)$ *is the* $\{0,1\}^d$*-cube,*
(ii) $H_2(d,2)$ *is, up to translation, a lattice* (d,k)*-polytope with* $k = \sum_{0\le j\le 3} 2^j\binom{d-1}{j}$, *and diameter* $\sum_{0\le j\le 3} 2^j\binom{d}{j+1}$.

3.1.3 Small Primitive Zonotopes $H_\infty(d, p)$

Property 3.3

(i) $H_\infty(d,1)$ *is, up to translation, a lattice* (d,k)*-polytope with* $k = 3^{d-1}$, *and diameter* $(3^d-1)/2$,
(ii) $H_\infty(d,2)$ *is, up to translation, a lattice* (d,k)*-polytope with* $k = 3\times 5^{d-1} - 2\times 3^{d-1}$, *and diameter* $(5^d-3^d)/2$,
(iii) $H_\infty(2,p)$ *is, up to translation, a lattice* $(2,k)$*-polygon with diameter* $4\sum_{1\le j\le p} \phi(j)$.

3.2 Small Positive Primitive Zonotopes $H_q^+(d, p)$

In Tables 5, 6, and 7, the number of vertices $f_0(H_q^+(d,p))$ is followed by its diameter $\delta(H_q^+(d,p))$ and grid embedding size. For instance, the entry 1082(15, 5) for $(q,d,p) = (1,5,2)$ in Table 5 indicates that $H_1^+(5,1)$ has 1082 vertices, diameter 15, and is a lattice (5, 5)-polytope.

3.2.1 Small Positive Primitive Zonotopes $H_1^+(d, p)$

Property 3.4

(i) $H_1^+(d,1)$ *is the* $\{0,1\}^d$*-cube,*
(ii) $H_1^+(d,2)$ *is a lattice* (d,k)*-polytope with* $k = d$, *and diameter* $\binom{d+1}{2}$,

Table 5 Small positive primitive zonotopes $H_1^+(d, p)$

	$H_1^+(d,p)$	p 2	3	4	5	6
d	2	6 (3, 2)	10 (5, 5)	14 (7, 9)	22 (11, 19)	26 (13, 25)
	3	26 (6, 3)	110 (13, 10)	314 (22, 22)	1022 (40, 52)	1970 (55, 82)
	4	150 (10, 4)	2194 (26, 16)	17,534 (51, 41)	145,198 (103, 106)	593,402 (161, 193)
	5	1082 (15, 5)	71,582 (45, 23)	2,062,682 (100, 67)	? (221, 188)	? (386, 386)
	6	9366 (21, 6)	? (71, 31)	?(176, 106)		

Table 6 Small positive primitive zonotopes $H_2^+(d, p)$

	$H_2^+(d,p)$	p 2	3	4	5
d	2	6 (3, 2)	10 (5, 5)	18 (9, 14)	26 (13, 26)
	3	32 (7, 4)	212 (19, 19)	1010 (40, 54)	3074 (70, 120)
	4	370 (15, 8)	19,438 (55, 49)	362,962 (141, 170)	3,497,862 (299, 462)
	5	10,922 (30, 15)	? (136, 108)	? (441, 487)	

Table 7 Small positive primitive zonotopes $H_\infty^+(d, p)$

	$H_\infty^+(d,p)$	p 1	2	3	4
d	2	6 (3, 2)	10 (5, 5)	18 (9, 14)	26 (13, 26)
	3	32 (7, 4)	212 (19, 19)	1418 (49, 76)	4916 (91, 184)
	4	370 (15, 8)	27,778 (65, 65)	1,275,842 (225, 344)	? (529, 1064)
	5	11,292 (31, 16)	? (211, 211)	? (961, 1456)	? (2851, 5716)
	6	1,066,044 (63, 32)			
	7	347,326,352 (127, 64)			
	8	419,172,756,930 (255, 128)			

(iii) $H_1^+(d, 3)$ *is a lattice* (d, k)*-polytope with* $k = (d^2 + 5d - 4)/2$ *and diameter* $d(d^2 + 6d - 1)/6$.

(iv) $H_1^+(2, p)$ *is a lattice* $(2, k)$*-polygon with* $k = 1 + \sum_{2\le j\le p} j\phi(j)/2$, *and diameter* $1 + \sum_{1\le j\le p} \phi(j)$.

3.2.2 Small Positive Primitive Zonotopes $H_2^+(d, p)$

Property 3.5

(i) $H_2^+(d, 1)$ *is the* $\{0, 1\}^d$*-cube,*

(ii) $H_2^+(d, 2)$ *is a* (d, k) *polytope with* $k = \binom{d}{1} + \binom{d}{3}$, *and diameter* $\binom{d+1}{2} + \binom{d+1}{4}$.

3.2.3 Small Positive Primitive Zonotopes $H_\infty^+(d, p)$

Property 3.6

(i) $H_\infty^+(d, 1)$ *is, a lattice* (d, k)*-polytope with* $k = 2^{d-1}$*, and diameter* $2^d - 1$,
(ii) $H_\infty^+(d, 2)$ *is a lattice* (d, k)*-polytope with* $k = 3^d - 2^d$*, and diameter* $3^d - 2^d$,
(iii) $H_\infty^+(2, p)$ *is a lattice* $(2, k)$*-polygon with diameter* $1 + 2 \sum_{1 \le j \le p} \phi(j)$.

4 Complexity Issues

We discuss a few complexity issues related to primitive zonotopes. While we mainly focus on $Z_q(d, p)$, the discussion and results, such as Propositions 4.1 and 4.2, can be adapted to $Z_q^+(d, p)$. As $H_q(d, p)$, respectively $H_q^+(d, p)$, is the translation of the image by a homothety of $Z_q(d, p)$, respectively $Z_q^+(d, p)$, the complexity results are the same.

4.1 Complexity Properties

Proposition 4.1 *For fixed positive integers* p *and* q*, linear optimization over* $Z_q(d, p)$ *is polynomial-time solvable, even in variable dimension* d.

Proof Since the q-norm of a generator of $Z_q(d, p)$ is bounded by p, it has at most p^q nonzero entries – which is attained for the vector of all ones and $d = p^q$. Thus, the number of generators of $Z_q(d, p)$ is bounded by $\binom{d}{p^q}(2p)^{p^q} = O\left(d^{p^q}\right)$. Hence, one can explicitly write all the generators of $Z_q(d, p)$ in polynomial time. Consequently, one can compute in polynomial time the following signed sum of generators of $Z_q(d, p)$ for any given rational $c \in \mathbb{R}^d$: $v^* = \sum_{v \in G_q(d,p)} \operatorname{sign}(c^T v) v$ where $G_q(d, p)$ denotes the set of generators of $Z_q(d, p)$. Note that $\operatorname{sign}(0)$ is set to 0. Then, one can show that v^* is a maximizer of $\{\max c^T x : x \in Z_q(d, p)\}$. □

The algorithmic theory developed by Grötschel, Lovász, and Schrijver [15] shows that polynomial-time solvability for linear optimization over a polytope implies polynomial-time solvability for other questions. In particular, Proposition 4.1 implies Proposition 4.2.

Proposition 4.2 *For fixed positive integers* p *and* q*, the following problems are polynomial-time solvable.*

(i) Extremality*: Given* $v \in \mathbb{Z}^d$*, decide if* v *is a vertex of* $Z_q(d, p)$*,*
(ii) Adjacency*: Given* $v^1, v^2 \in \mathbb{Z}^d$*, decide if* $[v^1, v^2]$ *is an edge of* $Z_q(d, p)$*;*
(iii) Separation*: Given rational* $y \in \mathbb{R}^d$*, either assert* $y \in Z_q(d, p)$*, or find* $h \in \mathbb{Z}^d$ *separating* y *from* $Z_q(d, p)$*; that is, satisfying* $h^T y > h^T x$ *for all* $x \in Z_q(d, p)$.

4.2 Open Problems

A natural open problem is to find direct algorithms to solve, over both $Z_q(d, p)$ and $Z_q^+(d, p)$, the extremality, adjacency, and separation questions given in Proposition 4.2.

Note that the case $q = \infty$, even for $p = 1$, seems to be significantly harder as the number of nonzero entries in a generator of $Z_\infty(d, p)$ can not bounded by a constant independent of d. Thus, the number of generators of $Z_\infty(d, p)$ is exponential in d. Hence, the complexity of linear optimization, extremality, adjacency, and separation over both $Z_\infty(d, p)$ and $Z_\infty^+(d, p)$, is open. In particular, it is not clear if deciding if a given point is a vertex of $Z_\infty(d, 1)$, or of $Z_\infty^+(d, p)$, is in NP or in coNP.

The remaining open questions deal with a reformulation in term of degree sequence of hypergraphs. The question is presented within the context of $H_q^+(d, p)$ but could be adapted to $H_q(d, p)$. Each subset $H \subseteq \{0, 1\}^d$ can be associated to a hypergraph with ground set $[d]$. The vector $\sum_{h \in H} h$ is called the *degree sequence* of H, and the convex hull of the degree sequences of all hypergraphs with ground set $[d]$ is called the *hypergraph polytope* D_d; and thus $D_d = H_\infty^+(d, 1)$. Considering only k-uniform hypergraphs; that is, subsets $H \subseteq \{0, 1\}^d$ where all vectors in H have k nonzero entries, one obtains the *k-uniform hypergraph polytope* $D_d(k)$ as the convex hull of the degree sequences of all k-uniform hypergraphs. The k-uniform hypergraph polytope, in particular $D_d(2)$ and $D_d(3)$, have been extensively studied, see [6, 13, 20, 23] and references therein. A natural question raised in the literature asks for suitable necessary and sufficient conditions to check whether a vector $h \in D_d(k) \cap \mathbb{Z}^d$ is the degree sequence of some k-uniform hypergraph. A trivial necessary condition is that the sum of the coordinates of h is a multiple of k. For $k = 2$; that is for graphs, the celebrated Erdős-Gallai Theorem [13] shows that the trivial necessary condition is also sufficient. For $k = 3$; that is for 3-uniform hypergraphs, the question was raised by Klivans and Reiner [20]. Liu [21] exhibited counterexamples by constructing *holes* for $d \geq 16$; that is, vectors h in $D_d(3) \cap \mathbb{Z}^d$ such that the sum of the coordinates of h is a multiple of 3, but h is not the degree sequence of a 3-uniform hypergraph. Deza et al. [9] answered a question raised in 1986 by Colbourn, Kocay, and Stinson [6] by showing that deciding whether a given sequence is the degree sequence of a 3-uniform hypergraph is NP-complete.

As there is no trivial congruence necessary condition, we call a vector in $H_q^+(d, p) \cap \mathbb{Z}^d$ a *hole* if it cannot be written as the sum of a subset of the generators of $H_q^+(d, p)$. While the answer to the question "*Does $H_q^+(d, p)$ have holes?*" is likely yes for most p, q, d, it would be interesting to explicitly find such holes and better understand them. A natural follow-up question, provided there are holes, is "*For given fixed positive integers p and q, what is the complexity of deciding if a given vector $h \in H_q^+(d, p) \cap \mathbb{Z}^d$ is a hole, and if not, of writing h as the sum of a subset of generators of $H_q^+(d, p)$?*".

As noted in the proof of Proposition 4.1, there are polynomially many generators for fixed integer p and q. Thus, the above follow-up question is in coNP as, if h is not a hole, it is possible to write h as a sum of a subset of generators $H_q^+(d, p)$. The last question is thus "*Is this problem coNP-complete?*".

As for the linear optimization related questions, the hole related questions seem to be significantly harder for $q = \infty$. In particular, for $(q, d, p) = (\infty, d, 1)$, the questions investigate the holes of D_d.

5 Proofs for Sections 2.2 and 3

Let $G_q(d, p)$, respectively $G_q^+(d, p)$, denote the generators of $Z_q(d, p)$, respectively $Z_q^+(d, p)$. Recall that $\sigma_q(d, p)$, respectively $\sigma_q^+(d, p)$, denotes the sum of the generators of $Z_q(d, p)$, respectively $Z_q^+(d, p)$.

5.1 Proof for Section 2.2

5.1.1 Proof of Item (i) of Property 2.1

Proof Note that if the set G of generators of a zonotope Z is invariant under coordinate permutation or sign flip, then the same holds for Z. Let π denote a permutation or a sign flip, and consider a signed sum $\sum_{g \in G} \epsilon_g g$. Then, $\pi(\sum_{g \in G} \epsilon_g g) = \sum_{g \in G} \epsilon_g \pi(g)$ is also a signed sum of generators since G is permutation and sign flip invariant. In other words, the set of all signed sums is invariant under permutations and sign flips, and thus the same holds for the convex hull Z of all signed sums. Let $J_q(d, p)$ be the set of all $-g$ for $g \in G_q(d, p)$. The zonotope $\tilde{Z}_q(d, p)$ generated by $G_q(d, p) \cup J_q(d, p)$ is the image of $Z_q(d, p)$ by a homothety of factor 2, and thus shares the same symmetry group. One can check that the set of generators of $\tilde{Z}_q(d, p)$ is invariant under coordinate permutation or sign flip, thus the same holds for $\tilde{Z}_q(d, p)$, and consequently holds for $Z_q(d, p)$. □

5.1.2 Proof of Item (ii) of Property 2.1

Proof Consider the minimization problem $\{\min c^T x : x \in H_q(d, p)\}$ or, equivalently, $\min c^T x$ over all integer valued points of $H_q(d, p)$. Set $c = (d!\bar{x}^d, (d-1)!\bar{x}^{d-1}, \dots, \bar{x})$ where $\bar{x} = (2p+1)^{d+1}$. Assuming that x is not the origin, let x_{i_0} denotes the first nonzero coordinate of x. Note that $x_{i_0} \geq 1$ by definition of $G_q(d, p)$, and $|x_i| \leq \bar{x}$. Thus, $c^T x \geq (d+1-i_0)!\bar{x}^{d+1-i_0} - \bar{x} \sum_{i_0 < i \leq d} (d+1-i)!\bar{x}^{d+1-i} > 0$.

In other words, the origin is the unique minimizer of a linear optimization instance over $H_q(d, p)$; that is, the origin is a vertex of $H_q(d, p)$. As $Z_q(d, p) = 2H_q(d, p) - \sigma_q(d, p)$, the point $-\sigma_q(d, p)$ is a vertex of $Z_q(d, p)$. By item (i) of Proposition 2.1, the point $\sigma_q(d, p)$ is a vertex of $Z_q(d, p)$, and thus $(\sigma_q(d, p) + \sigma_q(d, p))/2$ is a vertex of $H_q(d, p)$. □

5.1.3 Proof of Item (iii) of Property 2.1

Proof We first show that the coordinates of the vertex $\sigma_q(d, p)$ are odd. As noted in the proof of item (iii) of Property 2.3, the i-th coordinate of $\sigma_q(d, p)$ is equal to the first coordinate of $\sigma_q(d - i + 1, p)$. Thus, it is enough to show that the first coordinate of $\sigma_q(d, p)$ is odd. Except for the first unit vector $(1, 0, \ldots, 0)$, any generator g of $Z_q(d, p)$ with nonzero first coordinate can be paired with the generator $\bar{g}$ where $\bar{g}_1 = g_1$ and $\bar{g}_i = -g_i$ for $i \neq 1$. Thus, the sum of the first coordinates of the generators of $Z_q(d, p)$, excluding the first unit vector, is even. Hence, the first coordinate of $\sigma_q(d, p)$ is odd, and thus all the coordinates of $\sigma_q(d, p)$ are odd. Consider a vertex $v = \sum_{g \in G_q(d,p)} \epsilon(g) g$ of $Z_q(d, p)$. Since flipping the sign of an $\epsilon(g)$ does not change the parity of a coordinate of v, the coordinates of v have the same parity as the ones of $\sigma_q(d, p)$; i.e. are odd. In particular, the coordinates of a vertex of $Z_q(d, p)$ are nonzero and item (i) of Proposition 2.1 implies that the number of vertices of $Z_q(d, p)$ is a multiple of 2^d. □

5.1.4 Proof of Items (iv) and (v) of Property 2.1

Proof Let Z be a zonotope generated by integer-valued generators $m^j : j = 1, \ldots, m(Z)$. Then, Z is, up to translation, a lattice (d, k)-polytope with $k \leq \max_{i=1,\ldots,d} \sum_{1 \leq j \leq m(Z)} |m_i^j|$. Item (i) of Property 2.1 implies that the integer range of its coordinates is independent of the chosen coordinate. The same holds for $H_q(d, p)$, and, thus to determine the integer range of $H_q(d, p)$, it is enough to consider the first coordinates of its generators. Since the origin is a vertex of $H_q(d, p)$ and the first coordinate of its generator is nonnegative, the integer range of $H_q(d, p)$ is the sum of the first coordinates of its generators. For item (v), recall that the diameter of a zonotope is at most the number of its generators, and this inequality is satisfied with equality if no pair of generators are linearly dependent – which is the case for $Z_q(d, p)$ and $Z_q^+(d, p)$. □

5.1.5 Proof of Property 2.2

Proof Consider a generator $g \in G_q^+(d, p)$ and a coordinate permutation π. Since $\pi(g) \in G_q^+(d, p)$, $\pi(Z_q^+(d, p)) = \pi(\sum[-1, 1]G_q^+(d, p)) = \sum[-1, 1]$

$\pi(G_q^+(d,p)) = \sum[-1,1]G_q^+(d,p) = Z_q^+(d,p)$. As in the proof of item (ii) of Property 2.1, one can check that the origin is the unique minimizer of $\{\min c^T x : x \in H_q(d,p)\}$ with $c = (1,1,\ldots,1)$. Thus, the origin is a vertex of $H_q^+(d,p)$. As $Z_q^+(d,p) = 2H_q^+(d,p) - \sigma_q(d,p)$, the point $-\sigma_q(d,p)$ is a vertex of $Z_q^+(d,p)$. Since $Z_q^+(d,p)$ is invariant under the symmetries induced by coordinate permutations, $\sigma_q(d,p)$ is a vertex of $Z_q^+(d,p)$, and thus $(\sigma_q(d,p)+\sigma_q(d,p))/2$ is a vertex of $H_q^+(d,p)$. □

5.1.6 Proof of Items (i) and (ii) of Property 2.3

Proof Given a canonical vertex v of $Z_q(d,p)$, let c be a vector such that v is the unique maximizer of $\{\max c^T x : x \in Z_q(d,p)\}$. Up to infinitesimal perturbations, we can assume that the coordinates of c are pairwise distinct and nonzero. Note that each coordinate c_i of c is positive as otherwise flipping the sign of $v_i > 0$ would yield a point in $Z_q(d,p)$ with higher objective value than v. Assume that $c_i < c_j$ for some $i < j$. Then, $v_i = v_j$ as otherwise permuting v_i and v_j would yield a point in $Z_q(d,p)$ with higher objective value than v. Let $\pi_{ij}(c)$ be obtained by permuting c_i and c_j. Then, one can check that v is the unique maximizer of $\{\max \pi_{ij}(c)^T x : x \in Z_q(d,p)\}$. Assume, by contradiction, that $v' \in Z_q(d,p)$ satisfies $\pi_{ij}(c)^T v' \geq \pi_{ij}(c)^T v$. Then, $c^T \pi_{ij}(v') = \pi_{ij}(c)^T v' \geq \pi_{ij}(c)^T v = c^T v$ which implies $\pi_{ij}(v') = v$, and hence $v' = v$, since v is the unique maximizer of $\{\max c^T x : x \in Z_q(d,p)\}$. Thus, successive appropriate permutations yield a vector $\pi(c)$ with $\pi(c)_1 > \cdots > \pi(c)_d > 0$ such that v is the unique maximizer of $\{\max c^T x : x \in Z_q(d,p)\}$. For item (ii), one can check that $\sigma_1(d,2) = (2d-1, 2d-3, \ldots, 1)$ is the unique maximizer of $\{\max c^T x : x \in Z_1(2,p)\}$ for any c satisfying $c_1 > \cdots > c_d > 0$. Thus, by item (i) of Property 2.3, $\sigma_1(d,2)$ is the unique canonical vertex of $Z_1(d,2)$ and the vertices of $Z_1(d,2)$ are the $2^d d!$ coordinate permutations and sign flips of $\sigma_1(d,2)$. □

5.1.7 Proof of Item (iii) of Property 2.3

Proof We first note that the i-th coordinate of $\sigma_q(d,p)$ is equal to the first coordinate of $\sigma_q(d-i+1,p)$. The statement trivially holds for $i=1$. For $i>1$, consider a generator g of $Z_q(d,p)$ with $g_i \neq 0$ and $g_{i_0} > 0$ for some $i_0 < i$, then g can be paired with the generator $\bar{g}$ where $g_i = -\bar{g}_i$ and $g_{i_0} = \bar{g}_{i_0}$. Thus, the sum of all the i-th coordinates of the generators of $Z_q(d,p)$ is equal to the sum of the generators of $Z_q(d,p)$ such that the first $i-1$ coordinates are zero. In other words, the i-th coordinate of $\sigma_q(d,p)$ is equal to the first coordinate of $\sigma_q(d-i+1,p)$. For example, for finite q, $\sigma_q(d,1) = (1,\ldots,1)$ and $Z_q(d,1)$ is the $\{-1,1\}^d$-cube. Then, note that for $q=\infty$ or $p \neq 1$ the first coordinate of $\sigma_q(d-i+1,p)$, which is the grid embedding size of $H_q(d-i+1,p)$, is strictly decreasing with i increasing. Thus, the action of the symmetry group of $Z_q(d,p)$ on $\sigma_q(d,p)$ generates $2^d d!$ distinct vertices of $Z_q(d,p)$. For instance, one can check the i-th coordinate of $\sigma_\infty(d,1)$ is 3^{d-i}. □

5.1.8 Proof of Item (iv) of Property 2.3

Proof The statement trivially holds for $d = 1$. For $d \geq 2$, we show by induction that the vertices of $Z^+_\infty(d, 1)$ include $\sigma(d)$ satisfying $0 = \sigma_1(d) < \cdots < \sigma_d(d) = 2^{d-1}$. The base case holds for $d = 2$ as $\sigma(2) = (0, 2)$ is a vertex of $Z^+_\infty(2, 1)$. Assume such a vertex $\sigma(d)$ exists, and thus $\sigma(d) = \sum_{g \in G^+_\infty(d,1)} \epsilon(g)g$ for some $\epsilon(g)$ and $\sigma(d)$ is the unique maximizer of $\{\max c(d)^T x : x \in Z^+_\infty(d, 1)\}$ for some $c(d)$. The generators of $Z^+_\infty(d+1, 1)$ consist of the $2^d - 1$ vectors $(g, 0)$ obtained by appending 0 to a generator of $Z^+_\infty(d, 1)$, the $2^d - 1$ vectors $(g, 1)$ obtained by appending 1, and the unit vector e_{d+1}. Consider the point $s(d+1) = e_{d+1} + \sum_{g \in G^+_\infty(d,1)} (g, 1) - \sum_{g \in G^+_\infty(d,1)} \epsilon(g)(g, 0) = (2^{d-1}, \ldots, 2^{d-1}, 2^d) - (\sigma(d), 0)$; that is, $s(d+1) = (2^{d-1} - \sigma_1(d), \ldots, 2^{d-1} - \sigma_{d-1}(d), 0, 2^d)$. Thus, the coordinates of $s(d+1)$ are pairwise distinct and a suitable permutation of $s(d+1)$ yields a point $\sigma(d+1)$ satisfying $0 = \sigma_1(d+1) < \cdots < \sigma_{d+1}(d+1) = 2^d$. To show that $\sigma(d+1)$ is a vertex of $Z^+_\infty(d+1, 1)$, one can check that $\sigma(d+1)$ is the unique maximizer of $\{\max c(d+1)^T x : x \in Z^+_\infty(d+1, 1)\}$ where $c(d+1) = (-c(d), c_{d+1})$ for sufficiently large c_{d+1}. Thus, for $d \geq 2$, a point $\sigma(d)$ satisfying $0 = \sigma_1(d) < \cdots < \sigma_d(d) = 2^{d-1}$ is a vertex of $Z^+_q(d, p)$. Zonotopes being centrally symmetric, $-\sigma(d)$ is a vertex of $Z^+_q(d, p)$ and the same holds for the distinct $2d!$ permutations of $\pm\sigma(d)$. □

5.2 *Proof for Section 3*

5.2.1 Proof of Property 3.1

Proof One can check that the generators of $H_1(d, 2)$ consist of $\binom{d}{1}$ unity vectors and $2\binom{d}{2}$ vectors $\{\ldots, 1, \ldots, \pm 1, \ldots\}$. Thus, the diameter of $H_1(d, 2)$ is $\binom{d}{1} + 2\binom{d}{2} = d^2$. Similarly, one can check that the sum of the first coordinates of the generators of $H_1(d, 2)$ is $2d - 1$. Note that $H_1(d, 2)$ is the permutahedron of type B_d. Then, one can check that, in addition to the previously determined generators of $H_1(d, 2)$, the generators of $H_1(d, 3)$ consist of $2\binom{d}{2}$ vectors $\{\ldots, 1, \ldots, \pm 2, \ldots\}$, $2\binom{d}{2}$ vectors $\{\ldots, 2, \ldots, \pm 1, \ldots\}$, and $4\binom{d}{3}$ vectors $\{\ldots, 1, \ldots, \pm 1, \ldots, \pm 1, \ldots\}$. Thus, the diameter of $H_1(d, 3)$ is $\binom{d}{1} + 6\binom{d}{2} + 4\binom{d}{3} = d(d+2)(2d-1)/3$. Similarly, one can check that the sum of the first coordinates of the generators of $H_1(d, 3)$ is $\binom{d-1}{0} + 8\binom{d-1}{1} + 4\binom{d-1}{2} = 2d^2 + 2d - 3$. Furthermore, one can check that, in addition to the previously determined generators of $H_1(d, 3)$, the generators of $H_1(d, 4)$ consist of $2\binom{d}{2}$ vectors $\{\ldots, 1, \ldots, \pm 3, \ldots\}$, $2\binom{d}{2}$ vectors $\{\ldots, 3, \ldots, \pm 1, \ldots\}$, $4\binom{d}{3}$ vectors $\{\ldots, 1, \ldots, \pm 1, \ldots, \pm 2, \ldots\}$, $4\binom{d}{3}$ vectors $\{\ldots, 1, \ldots, \pm 2, \ldots, \pm 1, \ldots\}$, $4\binom{d}{3}$ vectors $\{\ldots, 2, \ldots, \pm 1, \ldots, \pm 1, \ldots\}$, and $8\binom{d}{4}$ vectors $\{\ldots, 1, \ldots, \pm 1, \ldots,$

$\pm 1, \ldots, \pm 1, \ldots\}$. Thus, the diameter of $H_1(d, 4)$ is $\binom{d}{1} + 10\binom{d}{2} + 16\binom{d}{3} + 8\binom{d}{4} = d(d^3 + 2d^2 + 2d - 2)/3$. Similarly, one can check that the sum of the first coordinates of the generators of $H_1(d, 4)$ is $\binom{d-1}{0} + 16\binom{d-1}{1} + 20\binom{d-1}{2} + 8\binom{d-1}{3}$. Finally, item (v) corresponds to Proposition 2.4. □

5.2.2 Proof of Property 3.2

Proof One can check that the generators of $H_2(d, 2)$ consist of $\binom{d}{1}$ unity vectors, $2\binom{d}{2}$ vectors $\{\ldots, 1, \ldots, \pm 1, \ldots\}$, $4\binom{d}{3}$ vectors $\{\ldots, 1, \ldots, \pm 1, \ldots, \pm 1, \ldots\}$, and $8\binom{d}{4}$ vectors $\{\ldots, 1, \ldots, \pm 1, \ldots, \pm 1, \ldots, \pm 1, \ldots\}$. Thus, the diameter of $H_2(d, 2)$ is $\sum_{0 \le j \le 3} 2^j \binom{d}{j+1}$. Similarly, one can check that the sum of the first coordinates of the generators of $H_2(d, 2)$ is $\sum_{0 \le j \le 3} 2^j \binom{d-1}{j}$. □

5.2.3 Proof of Property 3.3

Proof One can check that $H_\infty(d, 1)$ has $(3^d - 1)/2$ generators consisting of all $\{-1, 0, 1\}$-valued vectors which first nonzero coordinate is positive. Out of the 5^d $\{-2, -1, 0, 1, 2\}$-valued vectors, 3^d are $\{-2, 0, 2\}$-valued. Thus, keeping the ones which first nonzero coordinate is positive, $H_\infty(d, 2)$ has $(5^d - 3^d)/2$ generators. Similarly, one can check that the sum of the first coordinates of the generators of $H_\infty(d, 2)$ is $3 \times 5^d - 5 \times 3^d$. The generators (i, j) of $H_\infty(2, p)$ such that $||(i, j)||_\infty \le 1$ are $(1, 0)$, $(0, 1)$, $(1, 1)$ and $(1, -1)$. For a given $i > 1$, there are $2\phi(i)$ generators (i, j) such that $||(i, j)||_\infty > 1$ and $j < i$. Thus, there are $4 \sum_{2 \le j \le p} \phi(j)$ generators (i, j) such that $||(i, j)||_\infty > 1$. Thus, the diameter of $H_\infty(2, p)$ is $4 \sum_{1 \le j \le p} \phi(j)$. □

5.2.4 Proof of Property 3.4

Proof One can check that the generators of $H_1^+(d, 2)$ consist of $\binom{d}{1}$ unity vectors and $\binom{d}{2}$ vectors $\{\ldots, 1, \ldots, 1, \ldots\}$. Thus, the diameter of $H_1^+(d, 2)$ is $\binom{d}{1} + \binom{d}{2} = \binom{d+1}{2}$. Similarly, one can check that the sum of the first coordinates of the generators of $H_1^+(d, 2)$ is d. Note that $H_1^+(2, p)$ is the Minkowski sum of the permutahedron with the $\{0, 1\}^d$-cube. One can check that, in addition to the previously determined generators of $H_1^+(2, p)$, the generators of $H_1^+(d, 3)$ consist of $\binom{d}{3}$ vectors $\{\ldots, 1, \ldots, 1, \ldots, 1, \ldots,\}$, $\binom{d}{2}$ vectors $\{\ldots, 1, \ldots, 2, \ldots\}$, and $\binom{d}{2}$ vectors $\{\ldots, 2, \ldots, 1, \ldots\}$. Thus $H_1^+(d, 3)$ has $\binom{d}{3} + 3\binom{d}{2} + \binom{d}{1}$ generators. Similarly, one can check that the sum of the first coordinates of the generators of $H_1^+(d, 3)$ is

$\binom{d-1}{2} + 4\binom{d-1}{1} + \binom{d}{0}$. Out of the generators of $H_1(2, p)$, $\sum_{2 \le j \le p} \phi(j)$ have a negative coordinate. Thus, the diameter of $H_1^+(2, p)$ is $1 + \sum_{1 \le j \le p} \phi(j)$. Similarly, one can check that the sum of the first coordinates of the generators of $H_1^+(2, p)$ is $1 + \sum_{2 \le j \le p} j\phi(j)/2$. □

5.2.5 Proof of Property 3.5

Proof One can check that the generators of $H_2^+(d, 2)$ consist of $\binom{d}{i}$ vectors with exactly i ones for $i = 1, 2, 3$, and 4. Thus, the diameter of $H_2^+(d, 2)$ is $\binom{d+1}{2} + \binom{d+1}{4}$. Similarly, one can check that the sum of the first coordinates of the generators of $H_2^+(d, 2)$ is $\binom{d}{1} + \binom{d}{3}$. □

5.2.6 Proof of Property 3.6

Proof One can check that $H_\infty^+(d, 1)$ has $2^d - 1$ generators consisting of all $\{0, 1\}$-valued vectors except the origin. Thus, the diameter of $H_\infty^+(d, 1)$ is $2^d - 1$. Similarly, one can check that the sum of the first coordinates of the generators of $H_\infty^+(d, 1)$ is 2^{d-1}. Out of the 3^d $\{0, 1, 2\}$-valued vectors, 2^d are $\{0, 2\}$-valued. Thus, the diameter of $H_\infty^+(d, 2)$ is $3^d - 2^d$. Similarly, one can check that the sum of the first coordinates of the generators of $H_\infty^+(d, 2)$ is $3^d - 2^d$. The generators (i, j) of $H_\infty^+(2, p)$ such that $||(i, j)||_\infty \le 1$ are $(1, 0)$, $(0, 1)$, and $(1, 1)$. For a given $i > 1$, there are $\phi(i)$ generators (i, j) such that $||(i, j)||_\infty > 1$ and $j < i$. Thus, there are $2 \sum_{2 \le j \le p} \phi(j)$ generators (i, j) such that $||(i, j)||_\infty > 1$. Thus, the diameter of $H_\infty^+(2, p)$ is $1 + 2 \sum_{1 \le j \le p} \phi(j)$. □

Acknowledgements The authors thank the anonymous referees, Johanne Cohen, Nathann Cohen, Komei Fukuda, and Aladin Virmaux for valuable comments and for informing us of reference [31], Emo Welzl and Günter Ziegler for helping us access Thorsten Thiele's Diplomarbeit, Dmitrii Pasechnik for pointing out reference [29], and Vincent Pilaud for pointing out that $Z_1(d, 2)$ is the permutahedron of type B_d. This work was partially supported by the Natural Sciences and Engineering Research Council of Canada Discovery Grant program (RGPIN-2015-06163), by the Digiteo Chair C&O program at Université de Paris Sud, and by the Dresner Chair at the Technion.

References

1. D. Acketa, J. Žunić, On the maximal number of edges of convex digital polygons included into an $m \times m$-grid. J. Comb. Theory A **69**, 358–368 (1995)
2. X. Allamigeon, P. Benchimol, S. Gaubert, M. Joswig, Log-barrier interior point methods are not strongly polynomial. SIAM J. Appl. Algebra Geom. **2**, 140–178 (2018)

3. A. Balog, I. Bárány, On the convex hull of the integer points in a disc, in *Proceedings of the Seventh Annual Symposium on Computational Geometry* (1991), pp. 162–165
4. S. Borgwardt, J. De Loera, E. Finhold. The diameters of transportation polytopes satisfy the Hirsch conjecture. Mathematical Programming, (to appear)
5. N. Chadder, A. Deza, Computational determination of the largest lattice polytope diameter. Electron. Notes Discrete Math. **62**, 105–110 (2017)
6. C. Colbourn, W. Kocay, D. Stinson, Some NP-complete problems for hypergraph degree sequences. Discrete Appl. Math. **14**, 239–254 (1986)
7. A. Del Pia, C. Michini, On the diameter of lattice polytopes. Discrete Comput. Geom. **55**, 681–687 (2016)
8. A. Deza, A. Deza, Z. Guan, L. Pournin, Distance between vertices of lattice polytopes. AdvOL Report 2018/1, McMaster University, 2018
9. A. Deza, A. Levin, S.M. Meesum, S. Onn, Optimization over degree sequence. arXiv:1706.03951, 2017
10. A. Deza, G. Manoussakis, S. Onn, Primitive zonotopes. Discrete Comput. Geom. **60**, 27–39 (2018)
11. A. Deza, L. Pournin, Improved bounds on the diameter of lattice polytopes. Acta Mathematica Hungarica, **154**, 457–469 (2018)
12. D. Eppstein, Zonohedra and zonotopes. Math. Educ. Res. **5**, 15–21 (1996)
13. P. Erdős, T. Gallai, Graphs with prescribed degrees of vertices (in Hungarian). Matematikai Lopak **11**, 264–274 (1960)
14. K. Fukuda, Lecture notes: Polyhedral computation. http://www-oldurls.inf.ethz.ch/personal/fukudak/lect/pclect/notes2015/
15. M. Grötschel, L. Lovász, A. Schrijver, *Geometric Algorithms and Combinatorial Optimization* (Springer, Berlin, 1993)
16. B. Grünbaum, *Convex Polytopes* (Springer, Graduate Texts in Mathematics, Berlin, 2003)
17. J. Humphreys, *Reflection Groups and Coxeter Groups* (Cambridge University Press, Cambridge Studies in Advanced Mathematics, Cambridge, 1990)
18. G. Kalai, D. Kleitman, A quasi-polynomial bound for the diameter of graphs of polyhedra. Bull. Am. Math. Soc. **26**, 315–316 (1992)
19. P. Kleinschmidt, S. Onn, On the diameter of convex polytopes. Discrete Math. **102**, 75–77 (1992)
20. C. Klivans, V. Reiner, Shifted set families, degree sequences, and plethysm. Electron. J. Comb. **15** (2008)
21. R. Liu, Nonconvexity of the set of hypergraph degree sequences. Electron. J. Comb. **20**(1) (2013)
22. M. Melamed, S. Onn, Convex integer optimization by constantly many linear counterparts. Linear Algebra Its Appl. **447**, 88–109 (2014)
23. N.L.B. Murthy, M.K. Srinivasan, The polytope of degree sequences of hypergraphs. Linear Algebra Its Appl. **350**, 147–170 (2002)
24. D. Naddef, The Hirsch conjecture is true for (0, 1)-polytopes. Math. Program. **45**, 109–110 (1989)
25. S. Onn, *Nonlinear Discrete Optimization*. (European Mathematical Society, Zurich Lectures in Advanced Mathematics, 2010)
26. S. Onn, U.G. Rothblum, Convex combinatorial optimization. Discrete Comput. Geom. **32**, 549–566 (2004)
27. F. Santos, A counterexample to the Hirsch conjecture. Ann. Math. **176**, 383–412 (2012)
28. N. Sloane (ed.), The on-line encyclopedia of integer sequences. https://oeis.org
29. I. Soprunov, J. Soprunova, Eventual quasi-linearity of the Minkowski length. Eur. J. Comb. **58**, 110–117 (2016)
30. N. Sukegawa, Improving bounds on the diameter of a polyhedron in high dimensions. Discrete Math. **340**, 2134–2142 (2017)
31. T. Thiele, *Extremalprobleme für Punktmengen* (Diplomarbeit, Freie Universität Berlin, 1991)
32. M. Todd, An improved Kalai-Kleitman bound for the diameter of a polyhedron. SIAM J. Discrete Math. **28**, 1944–1947 (2014)
33. G. Ziegler, *Lectures on Polytopes* (Springer, Graduate Texts in Mathematics, Berlin, 1995)

Delone Sets: Local Identity and Global Symmetry

Nikolay Dolbilin

To Friends who love and do Geometry

Abstract In the paper we present a proof of the local criterion for crystalline structures which generalizes the local criterion for regular systems. A Delone set is called a crystal if it is invariant with respect to a crystallographic group. Locally antipodal Delone sets, i.e. those in which all $2R$-clusters are centrally symmetrical, are considered and we prove that they have crystalline structure. Moreover, if in a locally antipodal set all $2R$-clusters are the same, then the set is a regular system, i.e. a Delone set whose symmetry group operates transitively on the set.

Keywords Delone (Delaunay) set · Regular system · Crystal · Locally antipodal set · Crystallographic group · Symmetry group · Cluster · Local criterion for crystals · Cluster counting function

1 Introduction

This paper continues the investigative line started in the pioneering work [1] on local conditions for a Delone set X to be either a regular system, i.e. a crystallographic orbit of a single point, or a crystal, i.e. the orbit of a few points. In Fig. 1 one can see the set X_1 of points that are nodes of the square grid and the set X_2 of point quadruples. Each of these sets is a regular system because each of them is an orbit of a 2D-crystallographic group p4m (the full group of the standard square grid on the plane). The union $X = X_1 \cup X_2$ of the sets X_1 and X_2 is a crystallographic orbit of two points, i.e. it is an example of a crystal.

Work is supported by the RNF grant 14-11-00414.

N. Dolbilin (✉)
Steklov Mathematical Institute of the Russian Academy of Sciences, Moscow, Russia
e-mail: dolbilin@mi.ras.ru

M. D. E. Conder et al. (eds.), *Discrete Geometry and Symmetry*,
Springer Proceedings in Mathematics & Statistics 234,
https://doi.org/10.1007/978-3-319-78434-2_6

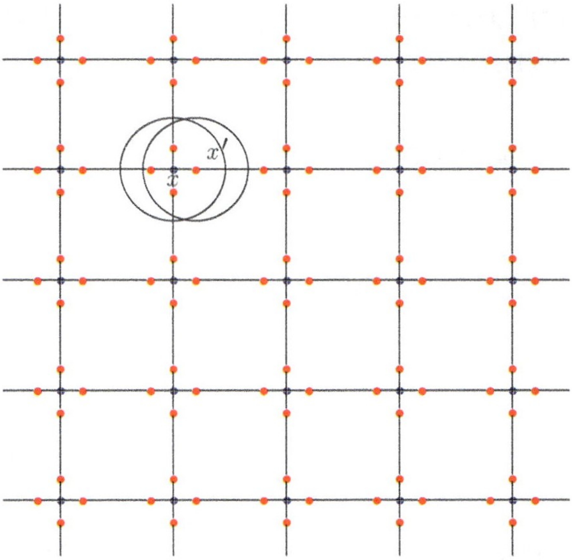

Fig. 1 Regular systems, crystals, and clusters

After Fedorov [2], a mathematical model of a mono-crystalline matter is defined as a discrete set which is invariant with respect to some crystallographic group. One should emphasize that under this definition the well-known periodicity of a crystal in all 3 dimensions is not an additional requirement. By the famous Schönflies-Bieberbach theorem [3, 4], any crystallographic group contains a translational subgroup with a finite index.

Since crystallization results from mutual interaction of nearby atoms, it is believed (R. Feynman, N.V. Belov, et al., see, for instance, [5]) that the long-range order of atomic structures of crystals (and quasicrystals too) results from local rules restricting the arrangement of nearby atoms.

Before 1970s there were no rigorous arguments explaining the link between properties of local patterns and the global order in the internal structure of crystals. However, in 1976 Delone and his students initiated the local theory of crystals [1]. One of two main aims of the local theory was (and is) the rigorous derivation of space group symmetry of a crystalline structure from the pairwise identity of local arrangements around each atom. At that time this appeared to be a purely abstract goal that would be of interest only to mathematicians.

However, the subsequent discovery of Penrose patterns (1977) and the discovery by D. Shechtman of real quasicrystals (1982, Nobel Prize in 2011) showed that there are also **non-periodic** Delone sets in which several different local patterns are repeated over and over again. On the other hand, **periodic**[1] crystals (which are either

[1]The concept of a 'periodic' crystal as the union of several crystallographic point orbits goes back to Fedorov. After Shechtman's discovery of 'aperiodic' quasicrystals, the International Union for Crystallography has extended the concept of crystal by including periodic (in Fedorov's sense) as

regular or multi-regular systems) are built also of a few species of local patterns. This circumstance suggests that the connection between the recurrence in a Delone set of a few different local patterns and the global order of the set is not so obvious. One of the goals of the local theory for periodic crystals was to look for right wordings of theorems and then prove them.

The local theory has been developed for Delone sets as well as for polyhedral tilings (see e.g. [6–8]).

The paper is organized as follows. In the next section we give definitions of all necessary concepts and a short survey of some of the earlier results. Then we give formulations of the local criterion for a crystal and of several new 'local' theorems on locally antipodal Delone sets (Theorems 1–5), which will be proved in concluding sections of the paper.

2 Basic Definitions and Results

Definition 2.1 A point set $X \subset \mathbb{R}^d$ is called a *Delone set* with parameters r and R, where r and $R > 0$, (or an (r, R)-*system*, see [9, 10]), if two conditions hold:

(1) an open d-ball $B^o_y(r) = \{z \in \mathbb{R}^d : |yz| < r\}$ of radius r centered at an *arbitrary* point $y \in \mathbb{R}^d$ contains at most one point from X:

$$|B^o_y(r) \cap X| \leq 1; \qquad (r)$$

(2) a closed d-ball $B_y(R) = \{z \in \mathbb{R}^d : |yz| \leq R\}$ of radius R centered at an arbitrary point y contains at least one point of X:

$$|B_y(R) \cap X| \geq 1. \qquad (R)$$

We note that by condition (r) the distance between any two points x and $x' \in X$ is not less than $2r$.

For $x \in X$ we call $C_x(\rho) := X \cap B_x(\rho)$ a ρ-*cluster* of point x. Thus, a ρ-cluster $C_x(\rho)$ consists of all points of X whose distance from x is at most ρ. It is easy to see that for $\rho < 2r$ $C_x(\rho) = \{x\}$. It is known that for $\rho \geq 2R$, the ρ-cluster $C_x(\rho)$ of any point $x \in X$ has the full rank: the dimension of $\mathrm{conv}(C_x(\rho)) = d$, e.g. [1].

We emphasize that we distinguish between ρ-clusters $C_x(\rho)$ and $C_{x'}(\rho)$ of different points x and x', even if the two sets coincide (see Fig. 1).

Definition 2.2 Two ρ-clusters $C_x(\rho)$ and $C_{x'}(\rho)$ are called *equivalent*, if there is an isometry $g \in O(d)$ such that

$$g(x) = x' \text{ and } g(C_x(\rho)) = C_{x'}(\rho).$$

---

well as aperiodic (in Shechtman's sense) crystals. Though a search for local conditions for aperiodic crystals remains extremely interesting and unsolved problem, throughout the paper we will mean under 'crystal' periodic crystals only.

We emphasize that the equivalence of two clusters is stronger than congruence of sets of points they contain. The two clusters depicted in Fig. 1 around two points x and x' coincide as subsets of X. However, since this subset of X surrounds the points x and x' in different ways it is natural to distinguish between the ρ-clusters $C_x(\rho)$ and $C_{x'}(\rho)$. Indeed, the clusters $C_x(\rho)$ and $C_{x'}(\rho)$ are non-equivalent because there is no isometry that moves both x and $C_x(\rho)$ onto x' and $C_{x'}(\rho)$, respectively.

In a Delone set for any $\rho > 0$ the set of all ρ-clusters is partitioned into classes of equivalent ρ-clusters. For any given $\rho < 2r$, the ρ-cluster at any point of X consists of a single point: $C_x(\rho) = \{x\}$, i.e. all "small" ρ-clusters in X are equivalent. We denote by $N(\rho)$ the cardinality of a set of equivalence classes of ρ-clusters in X.

For any Delone set X $N(\rho) = 1$ for $\rho < 2r$. However, for $\rho > 2r$, $N(\rho)$ is (in the general case) infinite.

Definition 2.3 A Delone set X is said to be *of finite type* if for each $\rho > 0$ the number $N(\rho)$ of classes of ρ-clusters is finite.

From now on, we will consider Delone sets of finite type only. We note that in this case the number $N(\rho)$ of ρ-clusters is a positive, integer-valued, non-decreasing, piece-wise constant function.

Very important examples of Delone sets include the *regular systems* and *crystals*. Here is an equivalent definition in terms of a Delone set.

Definition 2.4 A *regular system* is a Delone set $X \subset \mathbb{R}^d$ whose symmetry group acts transitively, i.e. for any two points x and $x' \in X$ there is an isometry $g \in Iso(d)$ such that

$$g(x) = x' \text{ и } g(X) = X.$$

Recall that a group $G \subset Iso(d)$ is called a *crystallographic group* if

(1) G operates discontinuously at each point $y \in \mathbb{R}^d$, i.e. if for any point $y \in \mathbb{R}^d$ the orbit $G \cdot y$ is a discrete set;
(2) G has a compact fundamental domain.

Now we are going to formulate two known statements on point sets with a crystallographic symmetry group.

Statement 2.1 *A point set $X \subset \mathbb{R}^d$ is a regular system if and only if the set X is an orbit of a point $x \in \mathbb{R}^d$ with respect to a crystallographic group $G \subset Iso(d)$.*

A regular set is an important particular case of the more general concept of a crystal.

Definition 2.5 A (periodic) *crystal* is a Delone set X such that X is a finite collection of orbits with respect to its symmetry group Sym (X): $X = \mathrm{Sym}(X) \cdot X_0$, where X_0 is a finite point set.

It is not hard to prove that the symmetry group of a periodic crystal is a crystallographic group. Thus we have the following statement.

Statement 2.2 *A set $X \subset \mathbb{R}^d$ is a crystal if and only if it is an orbit of a finite point set X_0 with respect to a crystallographic group G, i.e. $X = G \cdot X_0$.*

Thus, crystals can be described as Delone sets of finite type in terms of the cluster counting function $N(\rho)$ as follows. A Delone set of finite type is a regular system if and only if $N(\rho) \equiv 1$ on R_+. A Delone set is a crystal if and only if its cluster counting function is bounded:

$$N(\rho) \leq m < \infty, \quad \text{where } m \leq |X_0|.$$

Earlier, before Fedorov's work, a crystal had been considered as the finite union of pairwise congruent and parallel lattices. The definition of a crystal in terms of regular systems seemed to generalize the Haüi-Bravais concept of crystal as a periodic set. But due to the Schönflies-Bieberbach theorem, the more general structure of a regular system in fact is also the union of lattices.

Indeed, let h be the index of the translational subgroup T of a crystallographic group $G \subset Iso(d)$, and $X_0 \subset \mathbb{R}^d$ a finite point set. Then a crystal $G \cdot X_0$ splits into m pairwise congruent and parallel lattices of rank d, where $m \leq h \cdot |X_0|$. In fact, we have:

$$G \cdot X_0 = \cup_i^m (T \cdot x_i \cup T \cdot g_2(x_i) \cup \ldots \cup T \cdot g_h(x_i)), \quad x_i \in X_0.$$

We note that m is strictly smaller than $h \cdot |X_0|$ if, for instance, some $x_i \in X_0$ is a fixed point for $g \in G$ or points x_i and x_j in X_0 belong to the same G-orbit.

Now we define the *group* $S_x(\rho)$ of the ρ-cluster $C_x(\rho)$ as a subgroup of Iso(d) to consist of isometries s such that

$$s(x) = x \text{ and } s(C_x(\rho)) = C_x(\rho).$$

Let us denote by $M_x(\rho)$ the order of the group $S_x(\rho)$. Since the rank of $C_x(2R)$ equals d, the order $1 \leq M_x(\rho) < \infty$ for all $\rho \geq 2R$.

The function $M_x(\rho)$ for all $\rho \geq 2R$ takes positive integer values and is non-increasing. Moreover, the ratio $M_x(\rho) : M_x(\rho')$ is integer for $\rho' > \rho$. In fact the group $S_x(\rho')$ of a bigger cluster $C_x(\rho')$ either coincides with $S_x(\rho)$, or it is a proper subgroup of $S_x(\rho)$.

Let X be a Delone set of finite type. Then for a given positive ρ the set X splits into a finite number $N(\rho)$ of subsets $X_1, X_2, \ldots, X_{N(\rho)}$, such that points x and x' from every subset X_i have equivalent ρ-clusters $C_x(\rho)$ and $C_{x'}(\rho)$. But if the points x and x' are from different subsets X_i and X_j the ρ-clusters $C_x(\rho)$ and $C_{x'}(\rho)$ are not equivalent. The groups of equivalent ρ-clusters are conjugate in $Iso(d)$ and consequently have the same order $M_i(\rho)$, where $i \in [1, N(\rho)]$.

One of main goals of the local theory for regular systems is to determine a radius $\hat{\rho}$ such that for a Delone set X (with given parameters r and R) the condition $N(\hat{\rho}) = 1$ implies X to be a regular system. Certainly, the answer may depend on the dimension. It is easy to see that a Delone set on a line is a regular system if $N(2R) = 1$. The value $2R$ cannot be improved: in fact, for any $\varepsilon > 0$ there are Delone sets with $N(2R - \varepsilon) = 1$ that are not regular systems. The first important result in the local theory of regular systems was obtained in [1].

Theorem 2.1 (Local criterion for regular systems) *A Delone set $X \subset \mathbb{R}^d$ is a regular system if and only if for some $\rho_0 > 0$ the following conditions hold:*

(I) $N(\rho_0 + 2R) = 1$;
(II) $M(\rho_0) = M(\rho_0 + 2R)$.

Condition (I) says that $(\rho_0 + 2R)$-clusters at all points $x \in X$ are equivalent. Therefore the groups $S_x(\rho_0 + 2R)$ of the clusters are pairwise conjugate. Condition (II) ensures that for each point $x \in X$ the groups $S_x(\rho_0)$ and $S_x(\rho_0 + 2R)$ coincide.

Let us select among Delone sets with $N(2R) = 1$ the *locally asymmetric* sets, i.e. those for which the group $S_x(2R)$ is trivial. Then Theorem 2.1 immediately implies

Theorem 2.2 (Locally asymmetric sets) *Let a Delone set $X \subset \mathbb{R}^d$ be a locally asymmetric set and $N(4R) = 1$. Then X is a regular system, i.e. $N(\rho) \equiv 1, \forall \rho > 2R$.*

We now show that the condition $N(4R) = 1$ can not be reduced.

Theorem 2.3 ($(4R - \varepsilon)$-theorem) *For any given $\varepsilon > 0$ there is a Delone set $X \subset \mathbb{R}^2$ such that $N(4R - \varepsilon) = 1$, but X is not a regular system.*

Below we present an explicit construction.[2]

We begin with a rectangular lattice Λ (Fig. 2, on the left) whose fundamental rectangle has side lengths a and b where, by assumption, $a << b$. It is clear that the parameter $R = \frac{\sqrt{b^2+a^2}}{2}$.

Since $a << b$ we have

$$2R \sim b + \frac{a}{2}.$$

The horizontal rows of Λ form a bi-infinite sequence with indices $i \in \mathbb{Z}$. The set of the rows splits into pairs $P_j = (i, i + 1)$, $j = 2i$ of rows with sequel indices $(i, i + 1)$ where i is even. We choose c so that $0 < c < a/2$ and shift **each** pair P_{j+1} relatively to P_j by c to the left or to the right.

The sequence of shifted rows can be encoded by a bi-infinite sequence $l = \ldots RLLRL \ldots$. There are uncountably many different bi-infinite binary sequences $\{l\}$ and corresponding pairwise non-congruent Delone sets $\{X_l\}$. The Delone sets X_l have the same parameters r and R. Among the sequences $\{l\}$ there are exactly 3 whose corresponding Delone sets are regular systems. Two sequences $\ldots LLLL \ldots$ and $\ldots RRRR \ldots$ generate congruent regular systems. The third bi-infinite sequence $\ldots RLRLRL \ldots$ encodes the third regular system which is mirror symmetrical. No other Delone sets from the family are regular systems, though they all have the same b-clusters $C_x(b)$. Since $b \sim 2R - a/2$ and $a > 0$ can be chosen arbitrarily small, we have the theorem.

Estimates for ρ_0 in Theorem 2.1 have been determined in dimensions $d = 2$ and 3.

---

[2]In September 2016 the author presented this construction in his talk at the American Institute of Mathematics on a workshop "Soft Packings, Nested Clusters, and Condensed Matter". The construction raised in frames of the workshop a fruitful discussion on possible extending this example for any dimension. The discussion led a group of the workshop's participants to the following result: *For any dimension d and $\varepsilon > 0$ there is a non-regular Delone set with $N(d2R - \varepsilon) = 1$.*

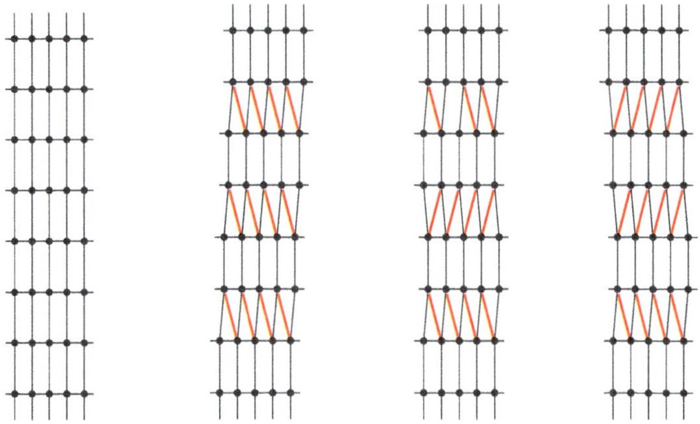

Fig. 2 Regular and non-regular systems with $N(4R - \varepsilon) = 1$

Theorem 2.5 (Regular systems, $d = 2, 3$)

(1) *Let* $X \subset \mathbb{R}^2$ *be a Delone set in the plane, if* $N(4R) = 1$, *then* X *is a regular system.*
(2) *Let* $X \subset \mathbb{R}^3$ *be a Delone set. If* $N(10R) = 1$, *then* X *is a regular system.*

This result was obtained by M. Stogrin and by N. Dolbilin independently many years ago. However, a detailed proof was published recently [11]. As for point (1) of Theorem 2.5, the case $d = 2$ can be derived from the following theorem:

Theorem 2.6 ([12]) *A tiling of Euclidean plane by convex polygons is regular, i.e. a tiling with a transitive symmetry group, if all first coronas are equivalent.*

Emphasize that, due to the $(4R - \varepsilon)$-theorem, the estimate $4R$ for plane is the best estimate. As for the estimate $10R$ for 3D-space, it seems to be bigger than the actual one. The difficulty lies in the fact that we can not deal effectively with the $2R$-cluster group.

In contrast, if the $2R$-cluster group contains the central symmetry, an extremely simple condition holds in every dimension.

Definition 2.6 A Delone set X is said to be a *locally antipodal* if the $2R$-cluster $C_x(2R)$ for each point $x \in X$ is centrally symmetric about the cluster center x. In the next sections we will prove:

Theorem 1 *If* X *is a locally antipodal set and* $N(2R) = 1$, *then* X *is a regular system.*

Theorem 2 *A locally antipodal Delone set* X *is centrally symmetrical about each point* $x \in X$ *globally.*

Notice that no condition is imposed on the cluster counting function $N(\rho)$. We do not even require X to be of finite type.

Theorem 3 *A locally antipodal Delone set $X \subset \mathbb{R}^d$ is a crystal. Moreover, X is the disjoint union of at most $2^d - 1$ congruent and parallel lattices:*

$$X = \bigsqcup_{i=1}^{n} (x + \lambda_i/2 + \Lambda),$$

where $x \in \mathbb{R}^d$, Λ is a lattice of the rank d, and $\lambda_1, \ldots \lambda_n$ are representatives of some n of 2^d cosets of the factor $\Lambda/(2\Lambda)$, $1 \le n \le 2^d - 1$.

Theorems 1–3 have been published in part in [13, 14]. In the present paper Theorems 1 and 2 are easily derived from the following theorem.

Theorem 4 (Uniqueness theorem) *Let X and Y be Delone locally antipodal sets. Suppose that they have a point x in common and the $2R$-clusters of X and Y centered at this point x coincide, i.e. $C_x(2R) = C'_x(2R)$, where $C_x(\rho)$ stands for a cluster in X and $C'_y(\rho)$ for a cluster in Y. Then $X = Y$.*

To conclude this section we present a local criterion for a crystal that generalizes the local criterion for regular systems. It was announced [15] and proved a while ago, a full proof was published recently [13] (in Russian). The proof in this paper is a slight improvement of that proof.

Theorem 5 (Local criterion for a crystal) *A Delone set X of finite type is a crystal which consists of m regular systems if and only if there is some $\rho_0 > 0$ such that two conditions hold:*

(1) $N(\rho_0) = N(\rho_0 + 2R) = m$;
(2) $S_x(\rho_0) = S_x(\rho_0 + 2R), \forall x \in X$.

It is obvious that the local criterion for regular systems (Theorem 2.1) is a particular case of Theorem 5.

3 Proof of the Local Criterion for Crystal

We note that crystals are meant here to be only 'periodic' crystals (see a comment in Sect. 2).

First of all we comment on Conditions (1) and (2) of Theorem 5. Condition (1) means that when radius ρ increases from ρ_0 to $\rho_0 + 2R$, the number of cluster classes on segment $[\rho_0, \rho_0 + 2R]$ remains unchanged $N(\rho_0) = N(\rho_0 + 2R)$.

Due to Condition (2), the cluster group $S_x(\rho_0), \forall x \in X$, does not get smaller on the segment $[\rho_0, \rho_0 + 2R]$: $S_x(\rho_0) = S_x(\rho_0 + 2R)$. The key point of Theorem 5 is that stabilization of the cluster counting function and the cluster groups on the segment $[\rho_0, \rho_0 + 2R]$ implies the stabilization of these parameters on the ray $[\rho_0, \infty)$.

Lemma 3.1 (on $2R$-chain) *For any pair of points x and $x' \in X$, where X is a Delone set, in X there is a finite sequence $x_1 = x, x_2, \dots, x_k = x'$, such that $|x_i x_{i+1}| < 2R$ for $i \in [1, k-1]$.*

A proof of Lemma 3.1 can be found in [1]. Now recall that $X_i \subset X$ is a subset of all points of X whose ρ_0-clusters are equivalent and belong to the i-th class.

Lemma 3.2 (on $2R$-extension) *Let a Delone set X fulfil Conditions* (1) *and* (2) *of Theorem 5 and $x, x' \in X_i$. Let $f \in Iso(d)$ be an isometry such that*

$$f(x) = x' \text{ and } f(C_x(\rho_0)) = C_{x'}(\rho_0). \tag{1}$$

Then the same isometry f superposes the bigger cluster $C_x(\rho_0 + 2R)$ onto cluster $C_{x'}(\rho_0 + 2R)$:

$$f(C_x(\rho_0 + 2R)) = C_{x'}(\rho_0 + 2R).$$

Proof If the ρ_0-clusters $C_x(\rho_0)$ and $C_{x'}(\rho_0)$ are equivalent, then, by Condition (1) of Theorem 5, the corresponding $(\rho_0 + 2R)$-clusters are equivalent too. Therefore there is an isometry g such that

$$g(C_x(\rho_0 + 2R)) = C_{x'}(\rho_0 + 2R).$$

If $g = f$ there is nothing to prove. Assume $f \neq g$ and consider the superposition of isometries $f^{-1} \circ g$. The order here is from the right to the left:

$$(f^{-1} \circ)(C_x(\rho_0)) = f^{-1}(g(C_x(\rho_0))) = f^{-1}(C_{x'}(\rho_0)) = C_x(\rho_0).$$

So, we have:

$$(f^{-1} \circ g)(x) = x \text{ and } (f^{-1} \circ g)(C_x(\rho_0)) = C_x(\rho_0). \tag{2}$$

By (2) $f^{-1} \circ g = s$, where $s \in S_x(\rho_0)$. By Condition (2) of Theorem 5 we have $s \in S_x(\rho_0 + 2R)$. Since $f = g \circ s^{-1}$, we have

$$f(C_x(\rho_0 + 2R)) = (g \circ s^{-1})(C_x(\rho_0 + 2R)) =$$

$$= g(s^{-1}(C_x(\rho_0 + 2R))) = g(C_x(\rho_0 + 2R)) = C_{x'}(\rho_0 + 2R).$$

□

Lemma 3.3 *Let X fulfil Conditions* (1) *and* (2) *of Theorem 5 and X_i a subset of X of all the points whose ρ_0-clusters belong to the i-th class, $i \in [1, m]$. Let a group G_i be generated by all isometries f that superpose ρ_0-clusters from the i-th class:*

$$G_i = < \{f \mid f(C_x(\rho_0)) = C_{x'}(\rho_0), \text{ where } x, x' \in X_i\} > . \tag{3}$$

Then G_i is a group of symmetries of X and operates transitively on each subset X_j, $\forall j \in [1, m]$. Moreover, the group G_i does not depend on i and $G_i = Sym(X)$ for $\forall i \in [1, m]$.

Proof First of all we will prove that any isometry f from (3) is a symmetry of X. Since for x and $x' \in X_i$ the ρ_0-clusters $C_x(\rho_0)$ and $C_{x'}(\rho_0)$ are equivalent and their groups $S_x(\rho_0)$ and $S_{x'}(\rho_0)$ are conjugate. Therefore, there are as many isometries superposing these clusters as the order $|S_x(\rho_0)|$ of the cluster group.

We will prove that if f is any of those isometries, then f is a symmetry of the whole X. Take an arbitrary point $y \in X$ and prove that its image $f(y)$ belongs to X. Let us connect points x and y with a $2R$-chain $\mathcal{L}$ (see Fig. 3):

$$\mathcal{L} = \{x_1 = x, x_2, \ldots, x_n = y \,:\, |x_i x_{i+1}| < 2R, \forall i \in [1, n-1]|\}.$$

Since $f(C_{x_1}(\rho_0)) = C_{x'_1}(\rho_0)$, by Lemma 3.2

$$f(C_{x_1}(\rho_0 + 2R)) = C_{x'_1}(\rho_0 + 2R). \tag{4}$$

Since $|x_1 x_2| < 2R$, we have that $C_{x_2}(\rho_0)$ as a point set belongs to $C_{x_1}(\rho_0 + 2R)$. Therefore relation (4) implies:

$$f(C_{x_2}(\rho_0)) = C_{x'_2}(\rho_0).$$

By Lemma 3.2 we have

$$f(C_{x_2}(\rho_0 + 2R)) = C_{x'_2}(\rho_0 + 2R).$$

Therefore, since inequality $|x_2 x_3| < 2R$ implies that

$$C_{x_3}(\rho_0) \subset C_{x_2}(\rho_0 + 2R),$$

we have:

$$f(C_{x_3}(\rho_0)) = C_{x'_3}(\rho_0).$$

By Lemma 3.2 we have again:

$$f(C_{x_3}(\rho_0 + 2R)) = C_{x'_3}(\rho_0 + 2R).$$

Moving along the $2R$-chain $\mathcal{L}$ and repeating the same argument as many times as the length of $\mathcal{L}$, we get that the isometry f moves $\mathcal{L}$ into a $2R$-chain $\mathcal{L}' \subset X$. The endpoint y of the chain $\mathcal{L}$ moves into the endpoint y' of L'. Thus, we have proved that the isometry f maps X **into** X: $f(X) \subseteq X$.

To prove that the isometry f maps X **onto** itself, we notice that the inverse isometry f^{-1} also maps X onto itself. It is the case because f^{-1} maps x' onto x and $f^{-1}(C_{x'}(\rho_0)) = C_x(\rho_0)$. Therefore f^{-1} also maps X into itself and, consequently,

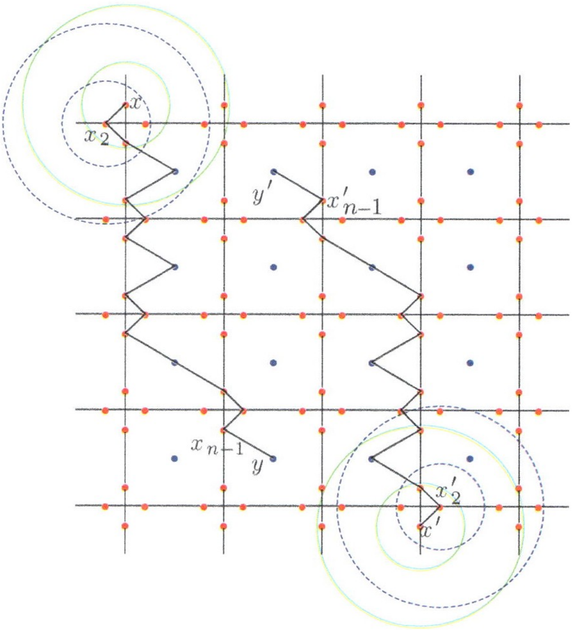

Fig. 3 Extention of a mapping along a 2R-chain

for f and any point $y \in X$ $f^{-1}(y) \in X$, i.e. f maps X onto itself. So, any isometry f in (3) is a symmetry of X.

Take the group G_i described in (3). We proved that $G_i \subseteq \mathrm{Sym}\,(X)$ and G_i operates on X_i transitively. Now we will see that $G_i \supseteq \mathrm{Sym}\,(X)$. Indeed, if $g \in \mathrm{Sym}\,(X)$, then, in particular, g moves any point $x \in X_i$ and its ρ_0-cluster $C_x(\rho_0)$ into the point $g(x) \in X_i$ and the cluster $C_{g(y)}(\rho_0)$, respectively. In other words, the symmetry $g \in G_i$ and therefore $G_i \supseteq G$. So, we proved that $G_i = \mathrm{Sym}(X)$ for any $i \in [1, m]$. □

So, as we proved above, a Delone set X with conditions (1) and (2) of Theorem 5 is the disjoint union of m subsets sets X_i which are $Sym\,(X)$-orbits of points x_i, $i \in [1, m]$:

$$X = \bigsqcup_{i=1}^{m} \mathrm{Sym}\,(X) \cdot x_i \tag{5}$$

To prove Theorem 5 we will see that $Sym\,(X)$ is a crystallographic group. This fact will follow from Lemmas 3.4 and 3.5.

Lemma 3.4 *Assume for a group $G \subset Iso(d)$ and for some point $x \in \mathbb{R}^d$ the orbit $G \cdot x$ is a Delone set, then G is a crystallographic group.*

Lemma 3.5 *In partition $X = \bigsqcup_{i=1}^{m} X_i$ each X_i is a Delone set.*

In fact, from these two lemmas it follows that X is a crystallographic orbit of finite set of m points. Let us take a Delone subset X_i (by Lemma 3.5) and a group $G := G_i = \mathrm{Sym}(X)$ generated by isometries f in (3). Since $G \cdot x_i = X_i$, where $x_i \in X_i$, by Lemma 2.4 G is a crystallographic group. Let a finite point set $X_0 = \{x_1, \dots, x_m\}$ consist of m representatives of subsets X_i, then $X = G \cdot X_0$, We get that that X is a G-orbit of X_0, i.e. X is a crystal. This completes the proof of the local criterion.

It remains to prove the last two lemmas.

Proof of Lemma 3.4 Assume that a set $Y := G \cdot x$ is a Delone set. Note that we do not assume that G is the full symmetry group Sym (Y) of Y, i.e. $G \subseteq Sym\ (Y)$. Let $\mathcal{V}$ denote the Voronoi tiling of space $\mathbb{R}^d$ for Y. A Voronoi domain V_y for the point $y \in Y$ is a convex d-polytope with a finite number of facets. The number of facets of V_y as well as the order of the symmetry group Sym (V_y) are bounded from above depending on the parameters r and R of the Delone set Y.

Any symmetry of Y leaves the Voronoi tiling $\mathcal{V}$ invariant. Therefore, since the group G operates on Y transitively, G also operates transitively on the set of all cells of $\mathcal{V}$. The following inclusions are also true: $G \subseteq \mathrm{Sym}\,(Y) \subseteq \mathrm{Sym}\,(\mathcal{V})$.

The orbit Sym $(\mathcal{V}) \cdot z$ for any point z of space $\mathbb{R}^d$ is a discrete set because the intersection of Sym $(\mathcal{V}) \cdot z$ and V_y is a finite point set:

$$|\mathrm{Sym}(\mathcal{V}) \cdot z \cap V_y| \le |\mathrm{Sym}(V_y)| < c = c(r, R, d).$$

Therefore Sym $\mathcal{V}$ and its subgroups Sym (Y) and G are discrete groups.

Now we will see that the fundamental domain $F(G)$ is compact. Indeed, the domain can be chosen as $V_y/\mathrm{stab}(y)$ where $\mathrm{stab}(y)$ is the stabilizer of y in G. In particular, if $\mathrm{stab}(y)$ is a trivial group then the fundamental domain $F(G)$ is V_y. Thus, the fundamental domain $F(G)$ is compact. So, G is a crystallographic group. □

Proof of Lemma 3.5 First of all, we note that since X is a Delone set any its subset X_i fulfils Condition (1) from Definition 2.1 with some parameter r', where $r' \ge r$.

As for Condition (2) from Definition 2.1, assume that it does not hold for at least one subset X_i, i.e. we suppose that X_i does not satisfy Condition (2) for an arbitrary finite value R'. In this case there is an infinite sequence of balls $B_1, B_2, \dots, B_k, \dots$ empty of points of X_i with infinitely increasing radii: $R_1 < R_2 < \dots < R_k < \dots$, where $R_k \to \infty$ as $k \to \infty$.

Since X_i is discrete, without loss of generality, one can suppose that each ball B_k, free from points of X_i, contains on its boundary a point $x_k \in X_i$. Since the set X_i is a G-orbit, each point x_k along with the ball B_k can be moved by a suitable symmetry $g_k \in G$ to some chosen point $y \in X_i$. The symmetry g_k also takes an empty ball B_k to an empty ball B'_k with a point y on its boundary. Thus, the point y is on the boundary of an empty ball B'_k of radius R_k for all $k = 1, 2, \dots$ centered at o_k. Let $\mathbf{n}_k$ denote a unit vector

$$\mathbf{n}_k := \frac{1}{|\overrightarrow{yo_k}|}\overrightarrow{yo_k}.$$

Select from a sequence $\{\mathbf{n}_k\}$ a convergent subsequence $\mathbf{n}_{k_j} \to \mathbf{n}$, where $\mathbf{n} = \overrightarrow{yo}$ is a limit unit vector.

Let Π be a hyperplane through the point y orthogonal to $\mathbf{n}$ and Π^+ denotes the half-space which the normal vector $\mathbf{n}$ looks in. The open half-space Π^+ contains no points of X_i. In fact, given a point $z \in \Pi^+$, in the subsequence of balls B_{k_j} with infinitely increasing radii one can find a ball which contains z in the interior. Since all balls B_{k_j} are free from points of X_i, $z \notin X_i$.

Thus, all points of X_i are in the opposite closed half-space Π^-. In particular, we do not exclude the case when all points of X_i lay on the hyperplane Π itself.

Now we prove that the open half-space Π^+ cannot be free from points of X_i. In fact, since X is a Delone set, the open half-space Π^+ contains points of X. So, we have $y \in X_i \cap \Pi$. For $j \in [1, m]$, $j \neq i$, we choose in X_j a point z closest to y. Since X_j is a discrete set, the closest point z does exist. Generally, there are finitely many points closest to y. Let us denote $\delta(y, X_j) := \min_{z' \in Y_j} |yz'|$. Since G operates transitively on both sets X_i and X_j, the minimum $\delta(y, X_j)$ does not depend on the choice of point $y \in Y_i$, i.e. for any point $y' \in X_i$ there is a point $z' \in X_j$ with condition $|y', z'| = |y, z|$. Therefore, $\delta(y, X_j) = \delta(y', X_j)$. One can denote $\delta(y, X_j) := \delta_{ij}$. It is obvious that $\delta_{ij} = \delta_{ji}$. Now one can denote

$$\delta_i := \max_{j \in [1,m]} \delta_{ij}.$$

It is clear that for every $j \in [1, m]$, $j \neq i$, and $\forall y \in X_j$ there is a point x of X_i at distance from y of no bigger than δ_i.

Therefore, since X_i is supposed to be located in the closed half-space Π^- the whole set X is located in the half-space $(\Pi + \delta_i \mathbf{n})^-$ determined by hyperplane $\Pi + \delta_i \mathbf{n}$. The obtained contradiction to the R-condition of the Delone set X completes a Proof of Lemma 3.5. □

4 Proof of Theorem 4

We note that in Theorem 4 the locally antipodal Delone sets X and Y are not required to be sets of finite type a priori.

In the beginning let X be an arbitrary Delone set and $x \in X$. We will call the set of distances between x and all the other points of X

$$\mathfrak{R}_x := \{\rho > 0 \mid \exists x' \in X,\ x' \neq x,\ |xx'| = \rho\}$$

the *distance spectrum* of X at the point x.

By Condition (r) of Definition 2.1 for X the spectrum $\mathfrak{R}_x$ is discrete and has no limit points (with the exception of ∞) for any given $x \in X$. Now we consider the union $\cup_{x \in X} \mathfrak{R}_x$ over all $x \in X$. It is easy to see that the union $\cup_{x \in X} \mathfrak{R}_x$ is a discrete set with no proper limit point if and only if X is of finite type.

Recall conditions of Theorem 4 for the Delone locally antipodal sets X and Y: $x \in X \cap Y$ and $C_x(2R) = C'_x(2R)$. $C'_y(\rho)$ stands for the ρ-cluster in the set Y for $y \in Y$. We take the point x and the two distance spectra $\mathfrak{R}_x = \{\rho_1 < \rho_2 < \ldots\}$ in X and $\mathfrak{R}'_x = \{\rho'_1, \rho'_2, \ldots\}$ in Y and prove the coincidence of the spectra $\mathfrak{R}_x$ and $\mathfrak{R}'_x$ and the sets X and Y.

Due to Condition $C_x(2R) = C'_x(2R)$, some initial segments of the sequences $\mathfrak{R}_x$ and $\mathfrak{R}'_x$ at the given point x coincide . Assume that we have already established the equality of the first k distances $\rho_1 = \rho'_1, \ldots, \rho_k = \rho'_k$ in the spectra and the coincidence of the corresponding clusters $C_x(\rho_k) = C'_x(\rho_k)$. This is our inductive assumption. We emphasize that we assume these clusters to coincide as point sets.

Now we prove that $\rho_{k+1} = \rho'_{k+1}$ and $C_x(\rho_{k+1}) = C'_x(\rho_{k+1})$. Without loss of generality we can assume that $\rho_{k+1} \le \rho'_{k+1}$. The ball $B_x(\rho_{k+1})$ has on its boundary at least one point $x_1 \in X$, $|xx_1| = \rho_{k+1}$ (see Fig. 4). We will show that $x_1 \in Y$, and, consequently, $\rho'_{k+1} = \rho_{k+1}$.

Let $z \in \mathbb{R}^d$ be such that z is on the segment $[xx_1]$ at distance R from x_1, i.e. $|zx_1| = R$. We note that z, generally, does not belong to X. The ball $B_z(R)$ centered at z of radius R touches the sphere $\partial B_x(\rho_{k+1})$ at point x_1. Now we apply a homothety with the center x_1 and a coefficient 2 to the ball $B_z(R)$. $B_z(R)$ is mapped onto the ball $B_{z'}(2R)$ centered at z', where $|z'x_1| = 2R$ (Fig. 4). It is obvious that we have

$$B_z(R) \subset B_{z'}(2R) \subset B^o_x(\rho_{k+1}) \cup \{x_1\}.$$

Here $B^o_x(\rho_{k+1})$ means an open ball.

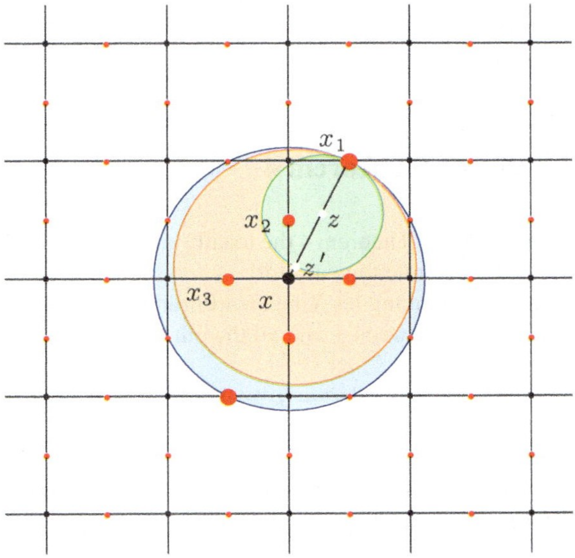

Fig. 4 Proof of Theorem 4

By Condition (R) in Definition 2.1 of a Delone set, in $B_z(R)$ there is at least one point $x_2 \in X$, $x_2 \neq x_1$. Since x_1 is the only point of the ball $B_z(R)$ which is located on the boundary $\partial B_x(\rho_{k+1})$, all other points of $B_z(R)$, including the point x_2, lay in the interior of $B_x(\rho_{k+1})$. Therefore $|xx_2| < |xx_1| = \rho_{k+1}$, i.e. $|xx_2| \leq \rho_k$. By the induction assumption, $C_x(\rho_k) = C'_x(\rho_k)$, hence $x_2 \in X \cap Y$.

Since $|x_1x_2| \leq 2R$ the point x_1 belongs to the cluster $C_{x_2}(2R)$. $C_{x_2}(2R)$ is antipodal about x_2. Therefore in $C_{x_2}(2R)$ there is a point $x_3 \in X$ which is antipodal to x_1. Recall that the coefficient of the homothety equals 2, hence we have

$$x_3 \in B_{z'}(2R) \subset B^o_x(\rho_{k+1}) \cup \{x_1\}.$$

Therefore we have two inequalities $|xx_3| \leq \rho_k$ and $|xx_2| \leq \rho_k$. By the inductive assumption, $x_2, x_3 \in Y$. Since $|x_2x_3| \leq 2R$, we have that $x_3 \in C'_{x_2}(2R)$. Now, the cluster $C'_{x_2}(2R)$ in Y is also antipodal about x_2, therefore, the point x_1 as antipodal to x_3 about x_2, also belongs to $C'_{x_2}(2R)$. Hence we have also $x_1 \in Y$. This inclusion is true for any $x'_1 \in X$ with $|xx'_1| = \rho_{k+1}$. Thus, it has been proved that if $\rho_{k+1} \leq \rho'_{k+1}$ we actually have $\rho_{k+1} = \rho'_{k+1}$ and $C_x(\rho_{k+1}) \subseteq C'_x(\rho_{k+1})$. However, in the case $\rho_{k+1} = \rho'_{k+1}$ one can also take any point $y_1 \in Y$ with $|xy_1| = \rho_{k+1}$ and, by the same argument, prove that $y_1 \in X$. Thus, the inductive step is established: one has proved that $C_x(\rho_{k+1}) = C'_x(\rho_{k+1})$. □

5 Proofs of Theorems 1, 2 and 3

Theorems 1 and 2 immediately follow from Theorem 4.

Proof of Theorem 1 Since $N(2R) = 1$, for any x' and $x \in X$ there is an isometry g such that $g(x') = x$ and $g(C_{x'}(2R)) = C_x(2R)$. Let us denote $Y := g(X)$. We have two local antipodal sets X and Y such that $X \cap Y \supseteq C_x(2R)$. By Theorem 4, the relationship $C_x(2R) = C'_x(2R)$ implies $X = Y$, i.e. g is a symmetry of X. Thus Sym (X) possesses a transitive symmetry group, i.e. X is a regular system. □

Proof of Theorem 2 Let σ_x be the inversion about a point $x \in X$. Since X is assumed to be a locally antipodal Delone set. $\sigma_x(C_x(2R)) = C_x(2R)$. Let us denote $Y := \sigma(X)$. Then we have again two sets X and Y with the $2R$-cluster $C_x(2R)$ in common. By Theorem 4 we have $X = Y$, i.e. the inversion σ_x maps the set X onto itself.

Proof of Theorem 3 Given a locally antipodal set $X \subset \mathbb{R}^d$, let Λ be a set of vectors λ such that $X + \lambda = X$. Since X is a discrete set, the vector set Λ is a lattice.

Now we show that the lattice Λ has rank d. Indeed, let σ_x and $\sigma_{x'}$ be inversions of clusters $C_x(2R)$ and $C_{x'}(2R)$ at points x and x', respectively. By Theorem 2, the inversions are both symmetries of X. On the other hand, the superposition $\sigma_x \circ \sigma_{x'}$ is a translation by the vector $2(x - x')$. Since the set X is a Delone set in $\mathbb{R}^d$, the translational group Λ containing all possible $2(x - x')$, $x, x' \in X$, has rank d. Λ is the maximum lattice such that $X + \Lambda = X$. Therefore the Delone set X is the union of finitely many lattices which are congruent and parallel to Λ, i.e.

$$X = \bigsqcup_{i=1}^{n} (x_i + \Lambda).$$

As said above, for $i = 1, 2, \dots, n$ we have $x_i - x_1 \in \Lambda/2$. By putting $x := x_1$, $\lambda_i/2 = x_i - x_1$ $(i = 1, 2, \dots, n)$ we come to:

$$X = \bigsqcup_{i=1}^{n} (x + \lambda_i/2 + \Lambda), \text{ where } \lambda_i \in \Lambda. \tag{6}$$

Now, if $\lambda_i \equiv \lambda_j \mod 2\Lambda$, i.e. if $\lambda_i - \lambda_j = 2\Lambda$, then subsets $x + \lambda_i/2 + \Lambda$ and $x + \lambda_j/2 + \Lambda$ obviously coincide. Therefore in (6) $n \leq 2^d$. Moreover, the value n of different subsets $x + \lambda_i/2 + X$ in (6) cannot be equal to 2^d because in this case $X = x + \Lambda/2$ and hence $X + \Lambda/2 = X$. This contradicts the assumption that Λ is the maximum lattice with the condition $X + \Lambda = X$. So, $n \leq 2^d - 1$. □

Acknowledgements The author thanks Nikolay Andreev (Moscow) for making drawings and Andrey Ordine (Toronto) for his help in editing the English text. The author is very grateful to the anonymous reviewer for having made numerous significant comments that helped to improve this paper.

References

1. B.N. Delone, N.P. Dolbilin, M.I. Stogrin, R.V. Galiuilin, A local criterion for regularity of a system of points. Soviet Math. Dokl. **17**, 319–322 (1976)
2. E.S. Fedorov, Elements of the study of figures. Zap. Mineral. Imper. S.Peterburgskogo Obschestva **21**(2), 1–279 (1985)
3. A. Schönflies, *Kristallsysteme und Kristallstruktur* (Druck und verlag von BG Teubner, Leipzig, 1891)
4. L. Bieberbach, Über die Bewegungsgruppen des n-dimensionalen Euklidischen Räumes I. Math. Ann. **70**, 207–336; II. Math. Ann. **72**(1912), 400–412 (1911)
5. R. Feynman, R. Leighton, M. Sands, *Feynman Lectures on Physics*, vol. II (Addison-Wesley, Reading, MA, 1964)
6. N.P. Dolbilin, J.C. Lagarias, M. Senechal, Multiregular point systems. Discr. Comput. Geom. **20**, 477–498 (1998)
7. N. Dolbilin, E. Schulte, The local theorem for monotypic tilings. Electron. J. Combin. **11**, 2 (2004). (Research Paper 7, 19pp)
8. N. Dolbilin, E. Schulte, A local characterization of combinatorial multihedrality in tilings. Contrib. Discrete Math. **4**(1), 1–11 (2009)
9. B. Delaunay, Sur la sphère vide. A la mémoire de Georges Voronoï. Bull. de l'Académie des Sci. de l'URSS **6**, 793–800 (1934)
10. B.N. Delone, Geometry of positive quadratic forms, Uspekhi Matem. Nauk **3**, 16–62 (1937). (in Russian)
11. N. Dolbilin, delone sets in $\mathbb{R}^3$: regularity conditions. Fundam. Appl. Math. **21**, (6) (2016) (in Russian, English translation will appear in Journal of Mathematical Sciences in 2018)

12. D. Schattschneider, N. Dolbilin, One corona is enough for the Euclidean plane, in *Quasicrystals and Discrete Geometry*, Fields Inst. Monogr., vol 10 (American Mathematical Society, Providence RI, 1998), pp. 207–246
13. N.P. Dolbilin, A criterion for crystal and locally antipodal Delone sets. Vestnik Chelyabinskogo Gos. Universiteta **3**(358), 6–17 (2015). (in Russian)
14. N.P. Dolbilin, A.N. Magazinov, Locally antipodal Delauney sets. Russian Math. Surveys. **70**(5), 958–960 (2015)
15. N.P. Dolbilin, M.I. Shtogrin, A local criterion for a crystal structure, in Abstracts of the IXth All-Union Geometrical Conference (Kishinev, 1988) p. 99 (in Russian)

The Twist Operator on Maniplexes

Ian Douglas, Isabel Hubard, Daniel Pellicer and Steve Wilson

Abstract Maniplexes are combinatorial objects that generalize, simultaneously, maps on surfaces and abstract polytopes. We are interested on studying highly symmetric maniplexes, particularly those having maximal 'rotational' symmetry. This paper introduces an operation on polytopes and maniplexes which, in its simplest form, can be interpreted as twisting the connection between facets. This is first described in detail in dimension 4 and then generalized to higher dimensions. Since the twist on a maniplex preserves all the orientation preserving symmetries of the original maniplex, we apply the operation to reflexible maniplexes, to attack the problem of finding chiral polytopes in higher dimensions.

Keywords Graph · Automorphism group · Symmetry · Polytope · Maniplex
Map · Flag · Transitivity · Rotary · Reflexible · Chiral

I. Douglas
17 E Pennington St, Tucson, AZ 85701, USA
e-mail: ifd2@nau.edu

I. Hubard
Instituto de Matemáticas, Universidad Nacional Autónoma de México
Circuito Exterior, C.U., Coyoacán, 04510 Mexico D.F., Mexico
e-mail: isahubard@im.unam.mx

D. Pellicer
Centro de Ciencias Matemáticas, Universidad Nacional Autónoma de México,
Antigua Carretera a Pátzcuaro 8701, Ex Hacienda de San José de la Huerta,
58089 Morelia, Michoacán, Mexico
e-mail: pellicer@matmor.unam.mx

S. Wilson (✉)
Department of Mathematics and Statistics, Northern Arizona University,
Box 5717, Flagstaff, AZ 86011, USA
e-mail: stephen.wilson@nau.edu

M. D. E. Conder et al. (eds.), *Discrete Geometry and Symmetry*,
Springer Proceedings in Mathematics & Statistics 234,
https://doi.org/10.1007/978-3-319-78434-2_7

1 Introduction

We have been struck by the beauty of the Platonic solids for thousands of years.

We saw them first when we asked this question: How can we make a polyhedron in such a way that the faces are identical regular polygons and there are the same number of them meeting at every vertex? In answer, a simple argument that the sum of the angles around a vertex must be less than 360° shows that there are exactly 5 possibilities:

(1) triangles meeting three around a vertex (the tetrahedron);
(2) triangles meeting four around a vertex (the octahedron);
(3) triangles meeting five around a vertex (the icosahedron);
(4) squares meeting three around a vertex (the cube);
(5) pentagons meeting three around a vertex (the dodecahedron).

These are often given the name of their Schläfli symbol: $\{3, 3\}$, $\{3, 4\}$, $\{3, 5\}$, $\{4, 3\}$, and $\{5, 3\}$, respectively.

It is worth noting that these five objects, besides having the requested local niceness, also have a global niceness, *symmetry*. They have *rotational* symmetry; we can rotate any of these objects about any of their faces and about any of their vertices. Moreover, they have *reflectional* symmetry; we can reflect about planes through face-centers, through vertices, across edges and along edges.

The discovery and proof that there are five and only five regular convex polyhedra was an interesting bit of reasoning. But it was over so soon, we hardly had a chance to enjoy it. How can we work in a more general but similar field?

There are at least three viable generalizations:

1. by regarding the cube, for instance, not as a solid hewn from stone, but as an assemblage of squares connected by hinges;
2. by regarding the cube as the convex hull of a finite set of points in 3-space;
3. by regarding the cube as having *faces* of many kinds: 2-faces (the squares), 1-faces (the edges) and 0-faces(the vertices);

The first of these viewpoints generalizes to *maps* on a surface and the second generalizes to *convex polytopes* in higher dimensions. The third generalizes to the idea of an *abstract polytope*. The first and third have *maniplexes* as their common generalization. All of these we define in the next section.

Then in Sect. 3 we discuss symmetry of maniplexes, with emphasis on chirality, meaning the property of having maximal rotational symmetry but no reflections. The aim we pursue is to devise a technique to construct higher rank chiral maniplexes, a task that has proved very difficult (see [15]).

2 Polyhedra, Maps, Maniplexes and Polytopes

A map is often defined as an embedding of a graph on a (compact, connected) surface so that components of the complement of the embedding (called faces) are topologically open discs. In some contexts graphs are allowed to have multiple edges, loops and semi-edges. We can, for example, regard the cube as an embedding of the graph Q_3 on the sphere.

To look more closely at the structure of a map, we find the following subdivision useful: choose a point in the interior of each face to be its *center* and a point in the relative interior of each edge to be its *midpoint*. Draw dotted lines to connect each face-center with each incidence of the surrounding vertices and edge-midpoints. The original edges and these dotted lines divide the surface into triangles called *flags*. Figure 1 shows the subdivision of the cube into flags.

Each flag corresponds to a mutual incidence of face, edge, and vertex, though several different flags may correspond to the same triple. For instance, consider the map shown in Fig. 2.

The map has one face A, an octagon, with opposite edges identified orientably. As a result, it has exactly one vertex v as well. The dotted lines divide it into its 16 flags. Each of the four flags marked with a dot correspond to the same triple (vertex, edge, face), namely, their vertex is v, their edge is 1 and their face is A.

Let Ω be the set of flags of a map $\mathcal{M}$. Then let r_0, r_1, r_2 be the permutations on Ω which match each flag f with its three immediate neighbors, as in Fig. 3. Define C to be the *connection group*, i.e., the group $\langle r_0, r_1, r_2\rangle$.

In this paper, we will write elements of the connection group on the left: the image of the flag f under the connection r_2 is written $r_2 f$.

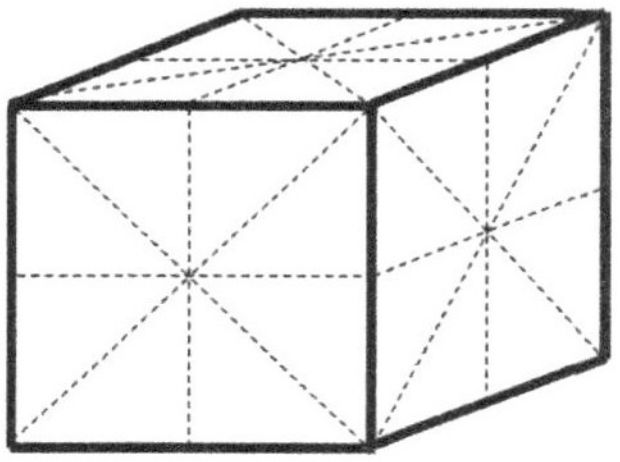

Fig. 1 The cube divided into flags

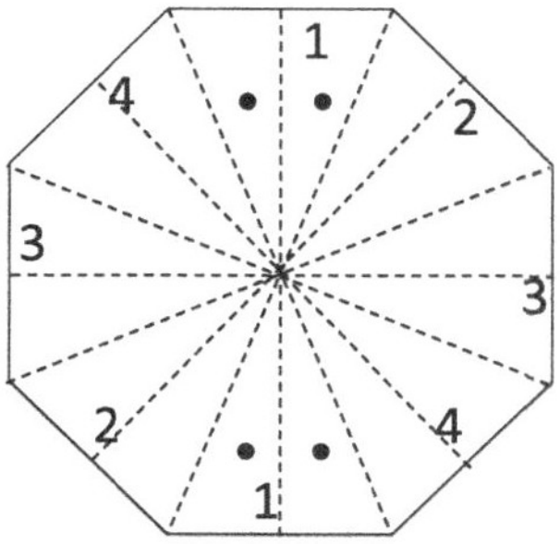

Fig. 2 A map with one face

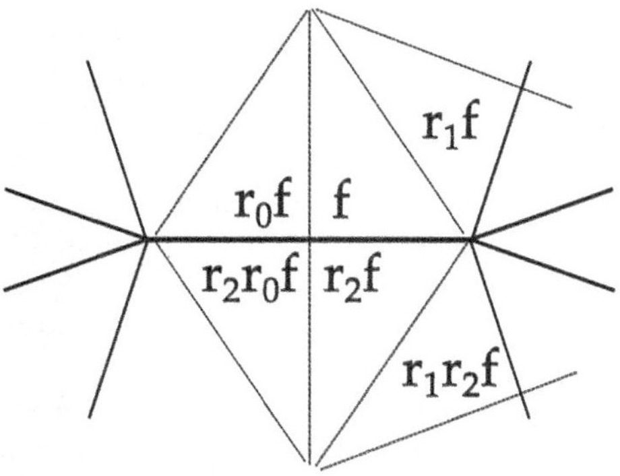

Fig. 3 Flags in a map

In Fig. 3, we see that f and $r_0 f$ are adjacent along a face-center-to-edge-midpoint line. Thus f and $r_0 f$ are incident to the same face and edge; they differ, if at all, in their incidences to a vertex, a 0-dimensional face of $\mathcal{M}$. We say these two flags are r_0-*adjacent*, or just *0-adjacent*. Similarly, f and $r_1 f$ meet the same 2-face and 0-face, while f and $r_2 f$ meet the same 0-face and 1-face. Notice from Fig. 3 that the flag r_2-adjacent to $r_0 f$ is also r_0-adjacent to $r_2 f$. In other words, as permutations on Ω, r_0 and r_2 commute.

We next take a slightly more abstract point of view by defining a map $\mathcal{M}$ to be a pair $(\Omega, [r_0, r_1, r_2])$ where Ω is a set of things called *flags*, the r_i's are permutations of order 2 on Ω, the *connection group* $C(\mathcal{M}) = \langle r_0, r_1, r_2 \rangle$ is transitive on Ω, and r_0 and r_2 commute. This $C(\mathcal{M})$ is often called the *monodromy group* of the map (see for example [9], and for higher ranks see also [13]). We can then think of vertices in $\mathcal{M}$ as orbits of $\langle r_1, r_2 \rangle$ in Ω. Similarly, edges correspond to orbits of $\langle r_0, r_2 \rangle$ and faces to orbits of $\langle r_0, r_1 \rangle$.

2.1 Maniplexes

This leads to the notion, introduced in [19], of a *maniplex*. An *(n+1)-dimensional maniplex* $\mathcal{M}$ is a pair $(\Omega, [r_0, r_1, \ldots, r_n])$, where Ω is a set of things called *flags* and each r_i is an involutory permutation on Ω such that (1) the *connection group* $C = \langle r_0, r_1, \ldots, r_n \rangle$ acts transitively on Ω, and (2) for all $0 \leq i < j - 1 < n - 1$, we have that $(r_i r_j)^2 = I$, where I is the identity in C. One can easily verify that every map on a surface is a 3-maniplex with Ω being its set of (triangular) flags. Furthermore, every 3-maniplex can be realised as a map on a surface. When we desire to avoid degeneracies, such as semi-edges or maps on a surface with boundary, we often also require that (3) each r_i and $r_i r_j$ are fixed-point-free, whenever $i \neq j$.

The *type* of a maniplex is the sequence $\{p_1, p_2, \ldots, p_n\}$, where each p_i is the order of $r_{i-1} r_i$ in C. The cube, then, is of type $\{4, 3\}$, the simplex is of type $\{3, 3, \ldots, 3\}$, and the 600-cell is of type $\{3, 3, 5\}$ (see [1, Chap. VII]). Type is well-defined even if not all faces have the same size. For example, the cuboctahedron, which is also the medial map of the cube, has triangles and squares, two of each meeting at each vertex. We say, then, that this map is of type $\{12, 4\}$.

Let C_i be the subgroup of C generated by all of the r_j's except r_i. Then an orbit of flags under C_i is called an *i-face*. A 0-face is a *vertex*, a 1-face is an *edge*, a 2-face is a *face*, an n-face is a *facet*. A facet of a facet is a *subfacet*; this is an orbit under $\langle r_0, r_1, \dots, r_{n-2}\rangle$. The restriction to a subfacet of the permutation r_n acts as an isomorphism from that subfacet to some subfacet.

We wish to assign colors, red and white, to flags so that for any given two i-adjacent flags, either one is colored red (and not white) and one is colored white (and not red), or both flags are colored both red and white. Choose a *root* flag (sometimes called also *base flag*) and call it I. Let $\mathcal{R}_0 = \{I\}$. Recursively let $\mathcal{W}_i$ be the set of all flags adjacent to flags in $\mathcal{R}_i$, and let $\mathcal{R}_{i+1}$ be the set of all flags adjacent to flags in $\mathcal{W}_i$. Finally, let $\mathcal{R}$ be the union of all $\mathcal{R}_i$'s and similarly let $\mathcal{W}$ be the union of all $\mathcal{W}_i$'s. We often say this another way: let C^+ be the subgroup of C generated by all products of the form $r_i r_j$. Then $\mathcal{R}$ is the orbit of I under C^+ and $\mathcal{W}$ is the orbit of $r_0 I$ under C^+. Consider these as assignments of the colors red and white, respectively to the flags. There are two possibilities for the result:

1. it could happen that $\mathcal{R}$ and $\mathcal{W}$ are disjoint; in this case we say that $\mathcal{M}$ is *orientable*;
2. otherwise it must happen that $\mathcal{R} = \mathcal{W} = \Omega$, and in this case we say that $\mathcal{M}$ is *non-orientable*.

See [10] for more information about bi-colorings of flags.

The idea of having one flag designated as a 'root' flag helps us in several constructions and theorems. Henceforward, we will assume that any maniplex does have a root flag chosen, and that isomorphisms and projections are required to send root flag to root flag. Notice that the choice of root affects the colors of flags. In particular, let $\mathcal{M}'$ be the maniplex identical to $\mathcal{M}$ except that $I' = r_0 I$ is chosen as its root. We will refer to $\mathcal{M}'$ as the *mirror-image* of $\mathcal{M}$. The red flags of $\mathcal{M}$ are the white flags of $\mathcal{M}'$ and vice versa.

If $\mathcal{M} = (\Omega, [r_0, r_1, \dots, r_{n-1}])$ is any n-maniplex, we can make an $(n+1)$-maniplex, called the *trivial* maniplex over $\mathcal{M}$, by using $\Omega \times \mathbb{Z}_2$ as a flag set (though we will write f_i instead of (f, i)), and connections $[s_0, s_1, \dots, s_{n-1}, s_n]$, where $s_j f_i = (r_j f)_i$ for all $j = 0, 1, 2, \dots, n-1$, $f \in \Omega$, $i \in \mathbb{Z}_2$, and $s_n f_i = f_{1-i}$. For example, the trivial maniplex over an n-gon has only two n-gonal faces over the same vertices and edges, and can be realised as a map on the sphere in such a way that the n vertices and the n edges lie on the equator. Note that if a maniplex $\mathcal{M}$ has type $\{p_1, p_2, \dots, p_{n-1}\}$, then the trivial maniplex over $\mathcal{M}$ has type $\{p_1, p_2, \dots, p_{n-1}, 2\}$. In particular, the trivial maniplex over an n-gon has type $\{n, 2\}$.

2.2 Polytopes

A convex polytope $\mathcal{P}$ is the convex hull of a finite set S of points in some Euclidean space. A *face* in $\mathcal{P}$ is the intersection of $\mathcal{P}$ with some hyperplane which does not separate S; it is an i-face if its affine hull has dimension i. The set of faces of $\mathcal{P}$ (of all dimensions) is partially ordered by inclusion. This partial ordering has certain properties, and these form the axiomatics for abstract polytopes. An *abstract polytope* is a partially ordered set $(\mathcal{P}, \leq)$ (whose elements are called *faces*) satisfying the following axioms:

(1) $\mathcal{P}$ contains a unique maximal and a unique minimal element.

(2) All maximal chains (these are called *flags*) have the same length. This allows us to assign a "rank" or "dimension" to each face. The unique minimal face (usually called "∅") is given rank -1.

(3) If $f < g < h$ are consecutive in some flag, then there exists exactly one $g' \neq g$ such that $f < g' < h$. This axiom is usually called the *diamond condition*.

(4) For any $f \leq h$, the *section* $[f, h]$ is the sub-poset consisting of all faces g such that $f \leq g \leq h$. We require it to be true in any section that if Φ_1 and Φ_2 are any two flags of the section, then there is a sequence of flags of the section, beginning at Φ_1 and ending at Φ_2, such that any two consecutive flags differ in exactly one rank. This condition is called *strong flag-connectivity*.

See [2, 12, 16, 17] for illuminating work on polytopes and their symmetry.

In particular, if the rank of the maximal element is n, we call $\mathcal{P}$ an n-polytope. If f is any flag, let f_i be its face of rank i, let f_i' be the unique face of rank i other than f_i such that $f_{i-1} \leq f_i' \leq f_{i+1}$, and let f^i be the flag identical to f except that the face of rank i is f_i'. From a given n-polytope, we can form its *flag graph* in the following way: the vertex set is Ω, the set of all flags (maximal chains) in $\mathcal{P}$. It has edges of colors $0, 1, 2, \ldots, n-1$. The edges of color i are all $\{f, f^i\}$ for $f \in \Omega$. Thus, two vertices of the flag graph are joined (by an edge colored i) if they are flags which are identical except at rank i. Let r_i be the set of all edges colored i. Because all flags have the same entry at rank -1 and at rank n, r_i will be defined only for $i = 0, 1, \ldots, n-1$.

Thus, for every abstract polytope $\mathcal{P}$, the flag graph of $\mathcal{P}$ is a maniplex. The converse does not hold. Briefly, and very loosely, the flag graphs of polytopes are those maniplexes in which no contact between a facet and itself is permitted. We refer the reader to [7] for examples of non-polytopal maniplexes, as well as for necessary and sufficient conditions on a maniplex to be polytopal.

3 Symmetry

We define a *symmetry* of a maniplex $\mathcal{M}$ as a permutation of the flags which preserves the connections. We write symmetries on the right, so that the image of the flag f under the symmetry α is $f\alpha$. We denote the group of symmetries of $\mathcal{M}$ by $\mathrm{Aut}(\mathcal{M})$, and the notation gives the nice statement that for all $i \in \{0, 1, 2, \ldots, n\}$ and all $\alpha \in \mathrm{Aut}(\mathcal{M})$, we have that

$$(r_i f)\alpha = r_i(f\alpha).$$

There are two levels of symmetry that are particularly interesting in maps and maniplexes. First, we say that $\mathcal{M}$ is *rotary* provided that $\mathrm{Aut}(\mathcal{M})$ acts transitively on $\mathcal{R}$, the set of red flags. Also, $\mathcal{M}$ is *reflexible* provided that $\mathrm{Aut}(\mathcal{M})$ acts transitively on Ω. It follows trivially, then, that if $\mathcal{M}$ is rotary and non-orientable, then it is reflexible. If $\mathcal{M}$ is rotary but not reflexible, we say it is *chiral*. If $\mathcal{M}$ is orientable, it is often useful to consider $\mathrm{Aut}^+(\mathcal{M})$; this is the group of all symmetries which send $\mathcal{R}$ (the set of red flags) to itself (and so send $\mathcal{W}$ to itself).

A reflexible maniplex is nice in several ways. First, $C = C(\mathcal{M})$ is isomorphic to $\mathrm{Aut}(\mathcal{M})$ [19]. Further, each of these groups acts regularly on Ω and so has the same cardinality as Ω. These correspondences allow us to label each flag with the element g of C which sends the root flag I there. In short, "g" is the label of the flag gI. Then, we claim, elements of C can act on the right as symmetries. For each $h \in C$, and for each pair of i-adjacent flags g and $r_i g$, the element h sends them to gh and $r_i gh$, respectively, and these two are also i-adjacent. Thus h, acting on the right, acts as a symmetry of $\mathcal{M}$.

Consider, for instance, Fig. 4, and observe how multiplication on the right by r_2 acts as a reflection about the horizontal edge, while multiplication on the right by r_0 acts as a reflection about the vertical axis.

Thus when the maniplex is reflexible, we can use the same names for the elements of $C(\mathcal{M})$, Ω and $\mathrm{Aut}(\mathcal{M})$. We must, though, be aware that multiplication by, say, r_0 on the left is a different permutation of the group elements than multiplication on the right.

When we have a reflexible maniplex $\mathcal{M}$ with $C(\mathcal{M})$ and $\mathrm{Aut}(\mathcal{M})$ expressed as permutations of some neutral set Ω, we can still talk about the symmetry corresponding to an element of the connection set by referring to the root flag; specifically, we

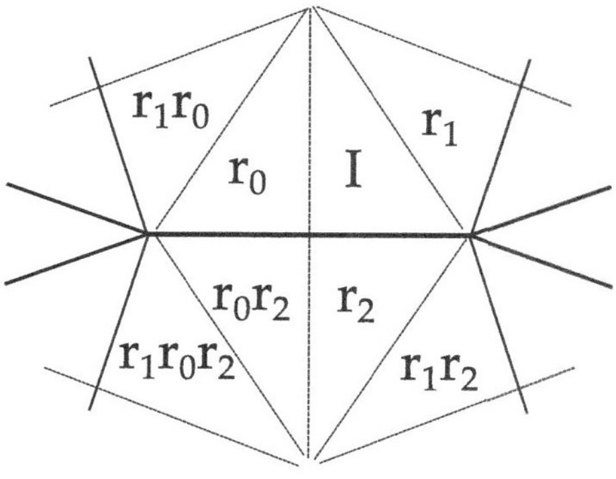

Fig. 4 Connections as symmetries

say that α is the symmetry *corresponding to* the connection x when xI is the same as $I\alpha$. This is a one-to-one correspondance and is operation preserving, and this is an isomorphism from $C(\mathcal{M})$ to $\mathrm{Aut}(\mathcal{M})$.

A note on language: Map theorists, starting with Brahana, have used the word *regular* to describe maps with rotational symmetries. Polytope theorists, though, use the word 'regular' to describe polytopes that we would call reflexible. In this paper, we remove the perhaps overused word 'regular' and instead use words that mean what they say. Also, we recognize that the English word 'chiral' simply means 'without reflections'. In our context, where generally only rotary maniplexes and polytopes are of interest, we will permit ourselves to use it to mean 'rotary but not reflexible'. And yes, we do recognize the contradictory flavors of these two preferences.

4 The Twist

We begin this section by presenting a very interesting maniplex which has only two facets, but is chiral.

4.1 The Krughoff Cubes

Consider the cube shown on the left in Fig. 5. The edges have been colored in such a way that each of the six possible circular orderings of the four colors appears exactly once clockwise about some face. Notice that this coloring is chiral; i.e., every rotation of the cube permutes the colors, while any reflection sends edges of any one fixed color to edges of different colors.

It is not obvious but a careful examination of the cube on the right shows that, ignoring the letter face-labels, the arrangement of colors is the same as on the left. Then each face of the left cube matches a face of the right with orientation reversed; these matching faces have the same letter. For instance face A-left has colors blue-yellow-red-green in order clockwise, while face A-right has colors blue-yellow-red-green in order counterclockwise. Thus, when we identify matching faces and colors, the result is a chiral 4-maniplex $\mathcal{K}$ with two cubical facets and four edges, one for each color. It was first discovered by Krughoff [11].

We introduce $\mathcal{K}$ in this paper because it can be formed from the trivial maniplex over the cube in a simple but very interesting way: separate each pair of attached squares and re-attach them after making a twist (locally) clockwise. The purpose of this paper is to generalize and re-generalize this operation, investigating the resulting chiral maniplexes.

4.2 The Twist in 4 dimensions

Let $\mathcal{M}$ be any *orientable* 4-maniplex. Recall that this means that $C = \langle r_0, r_1, r_2, r_3 \rangle$ is its connection group, that its facets are maps, that r_3 connects a face of each facet to some face (of the same size) in some facet, and that $\mathcal{R}$ and $\mathcal{W}$ are disjoint.

We construct the maniplex $T_j(\mathcal{M})$ to be $(\Omega, [s_0, s_1, s_2, s_3])$, where $s_0 = r_0, s_1 = r_1, s_2 = r_2$ and

$$s_3 f = \begin{cases} (r_0 r_1)^j r_3 f & \text{if } f \in \mathcal{R} \\ (r_1 r_0)^j r_3 f & \text{if } f \in \mathcal{W} \end{cases}$$

for all $f \in \Omega$. The index j indicates how much twist is performed to the faces of $\mathcal{M}$, after being separated, before gluing them back. This construction first appeared in [4].

Theorem 4.1 *For any orientable 4-maniplex $\mathcal{M}$ and any integer j, $T_j(\mathcal{M})$ is a maniplex.*

Proof We need to show that s_3 is an involution and that it commutes with r_0 and r_1. Suppose that $f \in \mathcal{R}$. Then $s_3^2 f = s_3(s_3 f) = s_3((r_0 r_1)^j r_3 f)$. Since $(r_0 r_1)^j r_3 f \in \mathcal{W}$, this is equal to $(r_1 r_0)^j r_3 (r_0 r_1)^j r_3 f$. Since r_3 commutes with r_0 and r_1, and $r_0 r_1$ is the inverse of $r_1 r_0$, this evaluates to f. Also, since $f \in \mathcal{R}$, $r_0 f \in \mathcal{W}$ and $s_3 r_0 f = (r_1 r_0)^j r_3 r_0 f = (r_1 r_0)^{j-1} r_1 r_3 f = r_0 (r_0 r_1)^j r_3 f = r_0 s_3 f$. Similar computations for r_1 and for $f \in \mathcal{W}$ show that the result holds. ∎

Theorem 4.2 *If $\mathcal{M}$ is an orientable 4-maniplex and j is any integer, then every α in* $\mathrm{Aut}^+(\mathcal{M})$ *is also in* $\mathrm{Aut}^+(T_j(\mathcal{M}))$.

Fig. 5 Krughoff's two-cube maniplex

Proof Since the 0, 1, and 2-connections are the same in both maps, we only need to show that α preserves the 3-connections. Consider any red flag f and its neighbor $s_3 f = (r_0 r_1)^j r_3 f$. Then $f\alpha$ is also red, and its s_3-neighbor is $(r_0 r_1)^j r_3(f\alpha) = ((r_0 r_1)^j r_3 f)\alpha = (s_3 f)\alpha$. Thus α preserves all connections, and so is a symmetry of $T_j(\mathcal{M})$. ■

Because $r_0 r_1$ is color-preserving, the bicoloring of flags which results from the orientability of $\mathcal{M}$ shows that each $T_j(\mathcal{M})$ is orientable as well. Thus we have:

Corollary 4.3 *If a 4-maniplex $\mathcal{M}$ is orientable and rotary, then so is every $T_j(\mathcal{M})$.*

Corollary 4.4 *If a 4-maniplex $\mathcal{M}$ is reflexible, then $T_{-j}(\mathcal{M})$ is the mirror image of $T_j(\mathcal{M})$.*

There are examples of reflexible maniplexes for which some, all or none of the T_j's result in chiral maniplexes. It is most common, though, that the result of the Twist operation on a reflexible maniplex is chiral. For example, T_1 of the 4-dimensional cube is a chiral maniplex of type $\{4, 3, 8\}$, while the 4-cube itself has type $\{4, 3, 3\}$. There are also examples of chiral maniplexes for which every $T_j(\mathcal{M})$ is chiral.

We will address the question of the chirality or reflexibility of a Twist of a reflexible maniplex after we have introduced a more general form of the definition.

4.3 The General Twist

Let $\mathcal{M} = (\Omega, [r_0, r_1, \ldots, r_n])$ be an orientable $(n+1)$-maniplex of dimension at least 4. Let $B = \langle r_0, r_1, \ldots, r_{n-2}\rangle$; this is the connection group of the root sub-facet. Further, let $B^+ = \langle r_0 r_1, r_1 r_2, \ldots, r_{n-3} r_{n-2}\rangle$; this is the subgroup of B which preserves $\mathcal{R}$. Let w be an element of B^+, such that for $i = 0, 1, 2, \ldots, n-2$ we have $(w r_i)^2 = I$; i.e., that the conjugate of w by r_i is w^{-1}. Call such a w *sub-invertible* because it is invertible within the sub-facet group B. Define the twist $T_w(\mathcal{M})$ of $\mathcal{M}$ to be $(\Omega, [s_0, s_1, \ldots, s_n])$, where $s_i = r_i$ for $i < n$ and

$$s_n f = \begin{cases} w r_n f & \text{if } f \in \mathcal{R}, \\ w^{-1} r_n f & \text{if } f \in \mathcal{W}, \end{cases}$$

for all $f \in \Omega$. Note that, since r_n commutes with all r_i with $i \leq n-2$, $r_n w = w r_n$. Imitating the proofs of Theorems 4.1 and 4.2 yields these results:

Theorem 4.5 *For any orientable maniplex $\mathcal{M}$ and any sub-invertible w, $T_w(\mathcal{M})$ is a maniplex.*

Note that if $w \in B^+$ is not sub-invertible, then $T_w(\mathcal{M})$ is not a maniplex, but *is* a complex in the sense of [19] (or a combinatorial map in the sense of [18]). Moreover, in this case, some $T_w(\mathcal{M})$ could be a chiral hypertope, in the sense of [6].

Theorem 4.6 *If $\mathcal{M}$ is an orientable maniplex and w is sub-invertible, then every α in* $\mathrm{Aut}^+(\mathcal{M})$ *is also in* $\mathrm{Aut}^+(T_w(\mathcal{M}))$.

Corollary 4.7 *If $\mathcal{M}$ is orientable and rotary, then so is every $T_w(\mathcal{M}))$.*

For a 4-maniplex, the subfacets are polygons, and so the only candidates for sub-invertible elements are the powers of r_0r_1, and these are sub-invertible. For higher dimensions, there are no obvious non-trivial candidates for w, and indeed, some sub-facets have no such elements. We claim that the simplex is one such maniplex. To see that, first notice that if w is sub-invertible, then w is central in B^+. Thus if the sub-facet of some maniplex $\mathcal{M}$ is a simplex of dimension $n-2$ for n greater than 4, then B^+ is A_{n-1}. This has a trivial center, and hence no viable w.

Contrast this with the cube of dimension $n-2$. Here, when n is even, the central inversion is orientation-preserving and is central and thus sub-invertible.

In a maniplex $\mathcal{M}$ of any rank, if its symmetry group has k orbits on flags, then:

1. If $\mathrm{Aut}(\mathcal{M})$ contains an orientation-reversing element then $T_w(\mathcal{M})$ has either k or $2k$ orbits on flags.
2. If $\mathrm{Aut}(\mathcal{M})$ does not contain an orientation-reversing element then $T_w(\mathcal{M})$ has either k or $\frac{k}{2}$ orbits on flags.

5 Chirality

In the paper [14], the third author demonstrated the existence of a series of chiral polytopes of all dimensions. By using the twist operator, we hope to produce such maniples in a simpler way. In this section we address the following question: What are the conditions on an orientable reflexible maniplex $\mathcal{M}$ and a sub-invertible element w that would force $T_w(\mathcal{M})$ to be reflexible?

So, suppose that $\mathcal{M}$ is an orientable and reflexible $(n+1)$-maniplex; suppose that w is sub-invertible in $\mathcal{M}$; finally suppose that $T_w(\mathcal{M})$ is reflexible. Since $\mathcal{M}$ is reflexible, its set of flags is (or can be considered as) the group $C(\mathcal{M})$, which we will call G for convenience. Remember how nice this is: elements of G are the flags, elements of G are the connections (acting by multiplication on the left), and elements of G are symmetries (acting by multiplication on the right). Hence "r_0r_1" is the name of a flag. It is 0-adjacent to the flag r_1. It is the image of r_0 under the symmetry sending each flag g to the flag gr_1.

Because $\mathcal{M}$ is orientable, its flags come in two colors, red and white, and the identification gives us that $\mathcal{R} = C^+$, the subgroup consisting of products of even lengths in the generators.

In $T_w(\mathcal{M}) = (G, [s_0, s_1, \dots, s_n])$, all of the connections are in G except perhaps s_n. Since $T_w(\mathcal{M})$ is reflexible, it must have a symmetry α_0 which sends the flag I to the flag s_0. This α_0 is probably not in G. Let $H = \langle r_0, r_1, \dots, r_{n-1}\rangle$. This group is the stabilizer (in $\mathcal{M}$ and in $T_w(\mathcal{M})$) of the 'central' facet; i.e., the facet containing the root flag I. Hence, on one hand, we can regard the elements of H as the flags of the central facet. On the other hand, we can think of the elements in H as paths of the graph with colours in $\{0, 1, \ldots, n-1\}$. This means that if we have $h \in H$, for every flag f, the flags f and hf are in the same facet (of both $\mathcal{M}$ and $T_w(\mathcal{M})$).

We will deduce the action of α_0 first in the central facet and then in facets farther and farther away.

Remembering that the identity I of G is assigned to the root flag I of Ω, we have that for $h \in H$, i.e. for flags in the central facet, the action of α_0 must be the same as in $\mathcal{M}$: thus, $h\alpha_0 = (hI)\alpha_0 = h(I\alpha_0) = h(r_0 I) = hr_0$.

Given $h \in H$, $r_n h$ is a flag in one of the facets adjacent to the central facet in $\mathcal{M}$. Since the twist operator preserves facets, and facet-adjacency, each flag $r_n h$ is in a facet adjacent to the central facet in $T_w(\mathcal{M})$ as well.

Let $g = r_n h$ be such a flag, for some $h \in H$, and suppose that g is red. Then it is n-adjacent in $T_w(\mathcal{M})$ to $s_n r_n h = w r_n r_n h = wh$, which is a white flag in H since $T_w(\mathcal{M})$ is orientable, and so its image under α_0 is whr_0, a red flag. Thus the flag $g\alpha_0$ must be n-adjacent to that red flag and so must be $wr_n whr_0 = w^2 r_n hr_0$. Similarly, if g is white then $g\alpha_0 = w^{-2} r_n hr_0$.

Now, every flag in a facet adjacent to the central facet is of the form $g = h_1 r_n h_0$, where the h_i's are from H. Then $g\alpha_0 = (h_1 r_n h_0)\alpha_0 = h_1(r_n h_0 \alpha_0) = h_1(w^{\pm 2} r_n h_0 r_0)$, where the exponent is $+2$ if $r_n h_0$ is red (i.e., if a product of generators equalling $r_n h_0$ has even length) and -2 otherwise.

Thus, we know the effect of α_0 on the central facet and on each facet adjacent to it. Next, consider a flag g in the layer of facets two steps away from the central facet, but n-adjacent to a flag in a facet adjacent to the central facet. Then $g = r_n h_1 r_n h_0$, where the h_i's are from H. If g is red, then g is n-adjacent to $wr_n r_n h_1 r_n h_0 = wh_1 r_n h_0$, a white flag. Then the image of this white flag under α_0 is the red flag $wh_1 w^{\pm 2} r_n h_0 r_0$; again, the exponent depends on the color of $r_n h_0$. Then $g\alpha_0$ is the flag n-adjacent to this one, which is $wr_n wh_1 w^{\pm 2} r_n h_0 r_0 = w^2 r_n h_1 w^{\pm 2} r_n h_0 r_0$; similarly, if g is white, then $g\alpha_0 = w^{-2} r_n h_1 w^{\pm 2} r_n h_0 r_0$.

In general, then, if $g = h_{k+1}(r_n h_k)(r_n h_{k-1})(r_n h_{k-2}) \ldots (r_n h_0)$, define

$$P(g) = h_{k+1}(t_k r_n h_k)(t_{k-1} r_n h_{k-1})(t_{k-2} r_n h_{k-2}) \ldots (t_0 r_n h_0),$$

where each t_j is w^2 if $(r_n h_j) \ldots (r_n h_0)$ is red and w^{-2} if it is white. Then it must be that $g\alpha_0 = P(g)r_0$, and a similar argument shows that, if α_i is the symmetry which sends I to s_i, then

$$g\alpha_i = \begin{cases} P(g)r_i & \text{if } i = 0, 1, \ldots, n-1; \\ P(g)wr_i & \text{if } i = n. \end{cases} \tag{1}$$

To recap: if $\mathcal{M}$ is reflexible and orientable and *if* we know that $T_w(\mathcal{M})$ is reflexible, then we have that for $i = 0, 1, \ldots, n-1$, $g\alpha_i = P(g)r_i$, and $g\alpha_n = P(g)wr_n$. This implies that P is well-defined. Well-definedness *is* an issue, for we see that $P(g)$ is defined in terms of a product p of generators which evaluates to g. The well-definedness of P means that if p_1, p_2 are two products of generators which both equal g then $P(p_1) = P(p_2)$. Moreover, this must hold even if the generator r_n appears in the words p_1 and p_2 with different multiplicity.

We claim that this is equivalent to saying that if some product p evaluates to I, then $P(p)$ must also evaluate to I. To see that, one uses the following Lemma:

Lemma 5.1 *For any word p let $Q(p)$ be obtained from $P(p)$ by replacing each t_i by t_i^{-1}. Then:*

1. *If p_2 is white, then $P(p_1p_2) = Q(p_1)P(p_2)$, while if p_2 is red, $P(p_1p_2) = P(p_1)P(p_2)$.*
2. *$(P(p^{-1}))^{-1} = P(p)$ if p is red and $(P(p^{-1}))^{-1} = Q(p)$ if p is white.*

On the other hand, if P is well-defined, it is clear that the equations in (1) serve as definitions for reflective symmetries, making $T_w(\mathcal{M})$ reflexible.

At first glance, the process of checking, for a given $\mathcal{M}$ and w, that the set of words which evaluate to I is closed under P may seem to be a daunting task. Our hearts need not seize up in fear, however. When we consider a reflexible maniplex, we are quite often given the generator-and-relator form of G. In this case, the only products we need to check are the relators, since every other word evaluating to I is a consequence of those.

Theorem 5.2 *Suppose that $\mathcal{M}$ is an orientable reflexible n-maniplex for n at least 4, and $C(\mathcal{M})$ has presentation $C = \langle r_0, r_2, \ldots, r_{n-1} | I = W_1, W_2, W_3, \ldots, W_k \rangle$, where each W_j is a word in the r_i's. If w is a sub-invertible element then $T_w(\mathcal{M})$ is reflexible if and only if $P(W_j)$ evaluates to the identity in C for all j.*

For example, consider the trivial maniplex over the cube. Its generator-relator form (abbreviating 'r_i' by just 'i') is $G = \langle 0, 1, 2, 3 | I = 0^2 = 1^2 = 2^2 = 3^2 = (02)^2 = (30)^2 = (31)^2 = (01)^4 = (12)^3 = (32)^2 \rangle$. We will use $w = 01$. Because w^2 commutes with 0 and 1, it is clear that $P(3030) = I$ and $P(3131) = I$, and any g which includes no 3 has $P(g) = g$. Thus the only word we need to check is $(32)^2$.

Consider $P(3232) = 0101\ 32\ 0101\ 32 = 0101\ 323\ 0101\ 2 = 0101\ 2\ 0101\ 2$. This is a motion of the cube $\mathcal{Q}$, and it can be seen as a 2-step rotation about a face. It is certainly not the identity and so $T_{01}(\mathcal{Q})$ is not reflexible. Therefore the Krughoff maniplex $\mathcal{K} = T_{01}(\mathcal{Q})$ is chiral, as claimed.

Remark 5.3 If w is its own inverse, so that $w^2 = I$, then for all g we have $P(g) = g$, and so P is well-defined and $T_w(\mathcal{M})$ is reflexible.

6 The Maniplex $\hat{2}^{\mathcal{M}}$

In this section, we describe a construction of an $(n+1)$-maniplex whose facets are all isomorphic to a given n-maniplex. Our motivation is this: we will show that if $\mathcal{M}$ is an orientable n-maniplex such that $C(\mathcal{M})$ contains an element w such that $(wr_i)^2 = I$ for $i = 0, 1, \ldots, n-1$, then, applying this construction twice, the constructed maniplex is one on which we can perform a twist and get a chiral maniplex as a result.

Definition 6.1 Let $\mathcal{M} = (\Omega, [r_0, r_1, \ldots, r_{n-1}])$ be an n-maniplex which has $m \geq 2$ facets named $F_1, \ldots, F_m$. Define $\hat{2}^{\mathcal{M}}$ to be the $(n+1)$-maniplex $(\Omega \times \mathbb{Z}_2^m, [s_0, s_1, \ldots, s_{n-1}, s_n])$, where, for $f \in F_j, x \in \mathbb{Z}_2^m$, we have

$$s_i(f, x) = \begin{cases} (r_i f, x) \text{ if } i = 0, 1, \ldots, n-1; \\ (f, x^j) \text{ if } i = n. \end{cases}$$

Here, x^j stands for the bitstring which differs from x in the j-th place and there only. If I is the root flag for $\mathcal{M}$, let $\hat{I} = (I, 000\ldots0)$ be the root flag for $\hat{2}^{\mathcal{M}}$.

Notice that if $\mathcal{M}$ has only one facet, then the above construction only yields the trivial maniplex over $\mathcal{M}$. In general $\hat{2}^{\mathcal{M}}$ is a 2^{m-1} fold cover of the trivial maniplex over $\mathcal{M}$. In what follows we are mainly interested in maniplexes with more than one facet.

This construction very slightly generalizes one of Danzer (see [5]) and sets it in maniplex form. Here $\hat{2}^{\mathcal{M}}$ is the same as Danzer's $D(2^{D(\mathcal{M})})$, where D stands for the usual dual of a polytope or maniplex.

Proposition 6.2 *Let $\mathcal{M}$ be any n-maniplex with at least two facets. Then*

1. *$\hat{2}^{\mathcal{M}}$ is an $(n+1)$-maniplex,*
2. *all facets of $\hat{2}^{\mathcal{M}}$ are isomorphic to $\mathcal{M}$;*
3. *if $\mathcal{M}$ has type $\{p_1, \ldots, p_{n-1}\}$ then $\hat{2}^{\mathcal{M}}$ has type $\{p_1, \ldots, p_{n-1}, 4\}$.*

Proof For $i = 0, 1, \ldots, n-1$, s_i is an involution because r_i is, and s_n is an involution because $(x^j)^j = x$. For $i = 0, 1, \ldots, n-2$, each f and $r_i f$ are in the same F_j, and so $s_i s_n s_i s_n(f, x) = s_i s_n s_i(f, x^j) = s_i s_n(r_i f, x^j) = s_i(r_i f, (x^j)^j) = s_i(r_i f, x) = (r_i r_i f, x) = (f, x)$. Thus, s_i and s_n commute, and so $\hat{2}^{\mathcal{M}}$ is a maniplex.

For a fixed x, the set of flags of the form (f, x) for $f \in \Omega$ is a facet of $\hat{2}^{\mathcal{M}}$, and every facet of $\hat{2}^{\mathcal{M}}$ is of this kind. Then the function sending f to (f, x) is an isomorphism of $\mathcal{M}$ to that facet of $\hat{2}^{\mathcal{M}}$.

Finally, suppose that some flag f of $\mathcal{M}$ is in F_j and that $r_{n-1}f$ is in F_k. Then repeatedly applying s_n and then s_{n-1} to (f, x) yields:

$$(f, x) \to (f, x^j) \to (r_{n-1}f, x^j) \to (r_{n-1}f, (x^j)^k) \to (f, (x^j)^k)$$

$$= (f, (x^k)^j) \to (f, x^k) \to (r_{n-1}f, x^k) \to (r_{n-1}f, x) \to (f, x).$$

This shows that $(s_{n-1}s_n)$ has order 4, as claimed. ■

Let us now consider symmetries of $\hat{2}^{\mathcal{M}}$. First, suppose that σ is a symmetry of $\mathcal{M}$ and that it acts on the facets of $\mathcal{M}$ as a permutation also called σ; i.e., that for all i, we define $\sigma(i)$ to be the index of $F_i\sigma$. Thus $F_i\sigma = F_{\sigma(i)}$. And denote by $\sigma(x)$ the vector $(x_{\sigma^{-1}(1)}, x_{\sigma^{-1}(2)}, \ldots, x_{\sigma^{-1}(m)})$. Then define $\hat{\sigma}$ acting on $\hat{2}^{\mathcal{M}}$ by

$$(f, x)\hat{\sigma} = (f\sigma, \sigma(x)).$$

Noting that $\sigma(x^j) = [\sigma(x)]^{\sigma(j)}$, it is easy to check that this is a symmetry of $\hat{2}^{\mathcal{M}}$ fixing the facet consisting of all flags of the form $(f, (0, 0, \ldots, 0))$. Thus, all of $\mathrm{Aut}(\mathcal{M})$ appears in $\mathrm{Aut}(\hat{2}^{\mathcal{M}})$, with $\hat{\alpha}_i$ playing the role of α_i for $i = 0, 1, 2, \ldots, n-1$.

For any $y \in \mathbb{Z}_2^m$, the function τ_y defined by $(f, x)\tau_y = (f, x + y)$ is clearly a symmetry of $\hat{2}^{\mathcal{M}}$. Assuming the root flag is in F_1, then the symmetry $\tau_{(1,0,0,\ldots,0)}$ sends the root flag $\hat{I}$ to $s_n\hat{I}$, its n-adjacent flag.

This shows that

Proposition 6.3 *Let $\mathcal{M}$ be a reflexible n-maniplex with at least two facets. Then $\hat{2}^{\mathcal{M}}$ is reflexible. The stabilizer of the central facet is isomorphic to* $\mathrm{Aut}(\mathcal{M})$*, by an isomorphism sending α_i to $\hat{\alpha}_i$.*

Notice that even if $\mathcal{M}$ has no particular symmetry, the maniplex $\hat{2}^{\mathcal{M}}$ has the symmetry $\tau_{(1,0,0,\ldots,0)} = \hat{\alpha}_n$, which is a reflection. Hence, $\hat{2}^{\mathcal{M}}$ can never be a chiral maniplex. Moreover, since $\mathrm{Aut}(\mathcal{M})$ can be regarded as a subgroup of $\mathrm{Aut}(\hat{2}^{\mathcal{M}})$ and for all $y \in \mathbb{Z}_2^m$, $\tau_y \in \mathrm{Aut}(\hat{2}^{\mathcal{M}})$, if $\mathcal{M}$ is a k-orbit maniplex, then so is $\hat{2}^{\mathcal{M}}$. In particular, if $\mathcal{M}$ is a 2-orbit maniplex in class 2_J, $J \subset \{0, 1, 2, \ldots, n-1\}$ (in the sense of [3]), then $\hat{2}^{\mathcal{M}}$ is a 2-orbit maniplex in class $2_{J\cup\{n\}}$.

6.1 Color-Coded Extensions

Suppose that $\mathcal{M} = (\Omega, [r_0, r_1, \ldots, r_{n-1}])$ is an n-maniplex, and let $\mathcal{C}$ be a partition of its facets into k classes called 'colors'. Define a new $(n+1)$-maniplex called $2^{(\mathcal{M},\mathcal{C})} = (\Omega \times \mathbb{Z}_2^k, [s_0, s_1, \ldots, s_n])$, where for $i = 0, 1, \ldots, n-1, s_i(f, x) = (r_i f, x)$, and $s_n(f, x) = (f, x^j)$, where f is in a facet of color j and x^j is the bitstring which differs from x in place j and there only. This generalizes previous notions:

1. If each class in $\mathcal{C}$ contains just one facet, then $2^{(\mathcal{M},\mathcal{C})}$ is exactly $\hat{2}^{\mathcal{M}}$.
2. If $\mathcal{C}$ consists of just one class containing all facets, then $2^{(\mathcal{M},\mathcal{C})}$ is the trivial maniplex over $\mathcal{M}$.

Moreover, proofs similar to those about $\hat{2}^{\mathcal{M}}$ show that if $\mathcal{M}$ is an n-maniplex with at least two facets then:

1. $2^{(\mathcal{M},\mathcal{C})}$ is an $(n+1)$-maniplex,
2. all facets of $2^{(\mathcal{M},\mathcal{C})}$ are isomorphic to $\mathcal{M}$;
3. if $\mathcal{M}$ has type $\{p_1, \ldots, p_{n-1}\}$ then $2^{(\mathcal{M},\mathcal{C})}$ has type $\{p_1, \ldots, p_{n-1}, 4\}$.
4. If $\mathcal{M}$ is reflexible and $\mathcal{C}$ is $\mathrm{Aut}(\mathcal{M})-$invariant, then $2^{(\mathcal{M},\mathcal{C})}$ is also reflexible.

7 Example of Twist on Rank 5

7.1 The Map $n\mathcal{M}$

We first give a general construction for covering of maps:

Definition 7.1 Let $\mathcal{M} = (\Omega, [r_0, r_1, r_2])$ be a map, a 3-maniplex, which is orientable, and let n be any integer greater than 2. Define $n\mathcal{M}$ to be the map $(\Omega \times \mathbb{Z}_n, [t_0, t_1, t_2])$, where for each flag (f, i) of $n\mathcal{M}$,

$$t_0(f, i) = (r_0 f, i),$$
$$t_2(f, i) = (r_2 f, i),$$

and

$$t_1(f, i) = \begin{cases} (r_1 f, i + 1) & \text{if } f \text{ is red,} \\ (r_1 f, i - 1) & \text{if } f \text{ is white.} \end{cases}$$

It is easy to see that each t_i is an involution and that t_0 commutes with t_2, so $n\mathcal{M}$ is a map, whenever it is connected. Observe that $n\mathcal{M}$ is an n-fold cover of $\mathcal{M}$, and if $\mathcal{M}$ is of type $\{p, q\}$, then $n\mathcal{M}$ is of type $\{LCM(p, n), LCM(q, n)\}$ whenever it is a map.

The second entry i of a flag (f, i) of $n\mathcal{M}$ is preserved by r_0 and r_2, but changed by r_1, according of the color of the flag f. We next define a function h that counts, for a given word W on the elements r_0, r_1, r_2, the difference between the number of appearances of a factor r_1 in odd and even places of W. If W is any word in r_0, r_1, r_2, and $\hat{W}$ is the corresponding word in t_0, t_1, t_2, define $h(W)$ recursively by using $h(I) = 0$ and

$$h(r_i W) = \begin{cases} h(W) + 1 & \text{if } i = 1 \text{ and } W \text{ has even length} \\ h(W) - 1 & \text{if } i = 1 \text{ and } W \text{ has odd length} \\ h(W) & \text{otherwise.} \end{cases}$$

Then, for all f, $\hat{W}(f, i) = (Wf, j)$, where $j = \begin{cases} i + h(W) & \text{if } f \text{ is red,} \\ i - h(W) & \text{if } f \text{ is white} \end{cases}$.

Note that this fact holds because W and $\hat{W}$ are considered as *words* in their respective sets of generators.

If D is the greatest common divisor of n and all values of $h(W)$ for which W evaluates to the identity in $\mathcal{M}$, then $n\mathcal{M}$ has exactly D connected components.

If $\mathcal{M}$ is reflexible, and, as before, we denote by α_i the symmetry exchanging the root flag I with $r_i I$, and by $\hat{\alpha}_i$ the symmetry exchanging $\hat{I}$ with $t_i \hat{I}$, then these functions are the corresponding symmetries in $n\mathcal{M}$:

$$(f, i)\hat{\alpha}_0 = (f\alpha_0, -i)$$

$$(f, i)\hat{\alpha}_1 = (f\alpha_1, 1 - i)$$

$$(f, i)\hat{\alpha}_2 = (f\alpha_2, -i).$$

Thus, $n\mathcal{M}$ is reflexible as well. Among the symmetries of $n\mathcal{M}$ is the function β, whichs sends each (f, i) to $(f, i + 1)$; direct computation verifies that β is a symmetry. Further, a simple computation will show that for each $i = 0, 1, 2$, the relation $(\beta\hat{\alpha}_i)^2 = I$ holds.

7.2 A Series of 5-Maniplex Examples

We use this construction to produce a series of examples. Each starts with an orientable map $\mathcal{M}$ of type $\{p, p\}$, with p odd, having a word W such that $h(W)$ is relatively prime to p. We study the effect of the twist operation on the 5-maniplex $\mathcal{M}' = \hat{2}^{\hat{2}^{p\mathcal{M}}}$.

Consider an orientable 3-maniplex $\mathcal{M}$ with type $\{p, p\}$ and some $n \geq 5$. We then construct the n-fold cover $n\mathcal{M}$ of $\mathcal{M}$.

As long as the greatest common divisor D of p and all $h(W)$ for words evaluating to the identity, is 1, the map is connected.

As an example, consider the great dodecahedron $\mathcal{M} = \mathcal{P}_0$, a polyhedron and orientable map of type $\{5, 5\}$ with 12 vertices and 12 pentagonal faces, where every vertex is surrounded by 5 faces. It can be constructed from the icosahedron by disregarding the triangles and considering as faces the 2-*holes*, that is, the convex polygons (pentagons in this case) determined by the neighbours of some vertex. The triangles of the icosahedron can be recovered as the 2-holes of the great dodecahedron (see [1, Chap. VI]). The great dodecahedron is reflexible, and as before, we consider its symmetry group, its connection group and its flag set to all be the same group G. Its connection group satisfies the relation $(r_0r_1r_2r_1)^3 = I$, since this indicates that the 2-holes are triangles. Then $h((r_0r_1r_2r_1)^3) = 6 \equiv 1$ modulo 5, and hence $h((r_0r_1r_2r_1)^3)$ is relatively prime to 5. The polyhedral map $5\mathcal{P}_0 = (\Omega \times \mathbb{Z}_5, [t_0, t_1, t_2])$ is then connected, has type $\{5, 5\}$, 60 vertices and 60 facets. This polyhedron is denoted by $\{5, 5\} * 600$ in the atlas of Hartley [8].

Naturally, the element w of $C(5\mathcal{P}_0)$ corresponding to $\beta \in \text{Aut}(5\mathcal{P}_0)$ has order 5. Furthermore, because $(\beta\hat{\alpha}_i)^2 = I$ we have that $(wt_i)^2 = I$ for $i \in \{0, 1, 2\}$.

Now, the element $(r_0r_1r_2r_1)^3$ acts trivially on Ω. Therefore $(t_0t_1t_2t_1)^3$ sends $(I, 0)$ to $(I, 1)$, and so it must be equal to w.

Let $\mathcal{M}'$ be the 5-maniplex (5-polytope) $\hat{2}^{\hat{2}^{5\mathcal{P}_0}}$ of type $\{5, 5, 4, 4\}$. The subfacets of $\mathcal{M}'$ are isomorphic to $5\mathcal{P}_0$ and w satisfies the desired properties in Sect. 4, so we can construct $T_w(\mathcal{M}')$. In what follows we prove that $T_w(\mathcal{M}')$ is chiral.

In $C(\mathcal{M}') = \langle s_0, s_1, s_2, s_3, s_4\rangle$, the relation $(s_3s_4)^4 = I$ holds. Assuming that $T_w(\mathcal{M}')$ is reflexible we have that $P((s_3s_4)^4)$ also equals I. But $P((s_3s_4)^4) = (s_3w^2s_4)^4$, since all flags $(s_3s_4)^k$ are red. Conjugating by s_4 we get

$$I = s_4(s_3 w^2 s_4)^4 s_4 = (s_4 s_3 w^2)^4.$$

Now, every flag in $\mathcal{M}'$ is of the form $(((f, i), x), y)$, where f is a flag of $\mathcal{P}_0$, $i \in \mathbb{Z}_5$, and x and y are bitstrings of the appropriate lengths. The connection s_4 changes only y, and s_3 changes only x.

Thus the image of $(((I, 0), x), y)$ under $s_4 s_3 w^2$ is $(((I, 2), x'), y')$ for some x', y'. Thus $(((I, 0), x), y)(s_4 s_3 w^2)^4 = (((I, 8), x''), y'')$ for some x'', y''. Since 5 is odd, $P((s_3 s_4)^4)$ is *not* the identity and so $T_w(\mathcal{M}')$ is chiral.

This example generalizes: If $\mathcal{M}$ is any map of type $\{p, p\}$ for some $p \geq 5$ and has a defining word W such that $h(W)$ is relatively prime to p, then it must have some defining word w such that $h(w) = 1$ and corresponds to β in $p\mathcal{M}$. If, in addition, p is odd, then in $\mathcal{M}' = \hat{2}^{\hat{2}^{p\mathcal{M}}}$, the word $(s_3 s_4)^4$ evaluates to the identity, and a computation similar to the previous paragraph shows that $P((s_3 s_4)^4) = (s_3 w^2 s_4)^4$ sends any flag $(((f, i), x), y)$ to $(((f, i + 8), x''), y'')$ for some x'', y'' and since p is odd, this is not the identity and so $T_w(\mathcal{M}')$ is chiral.

Remark Now, the maniplex $\hat{2}^{\hat{2}^{5\mathcal{P}_0}}$ is huge: $5\mathcal{P}_0$ is 5 times as large as the Great Dodecahedron and so has $5 * 120 = 600$ flags in 60 facets of 10 flags each. The 4-maniplex $\hat{2}^{5\mathcal{P}_0}$ has 2^{60} such facets. Then $\hat{2}^{\hat{2}^{5\mathcal{P}_0}}$ has $2^{2^{60}}$ facets. Then 2^{60} is an 19-digit number and thus $2^{2^{60}}$ is simply too large to contemplate.

We can reduce the scale of our constructions by using the color-coded extensions: $\mathcal{P}_0$, the great dodecahedron, has a coloring in 6 colors, opposite faces having the same colors. Then $5\mathcal{P}_0$, as a covering of $\mathcal{P}_0$, inherits this same 6-coloring. Call that coloring S. Then the 4-maniplex $\mathcal{P}_1 = 2^{(5\mathcal{P}_0, S)}$ has only $2^6 = 64$ facets and has a 2-coloring B, and so the 5-maniplex $\mathcal{P}_2 = 2^{(\mathcal{P}_1, B)}$ has only 4 facets, for a total of $600 * 64 * 4 = 153{,}600$ flags.

8 Open Questions

There are many interesting open questions regarding the twist operator on maniplexes. Here are just a few of them:

1. In general, if $\mathcal{M}$ is polytopal, what are the conditions on w for the maniplex $T_w(\mathcal{M})$ to be polytopal? Are there even any special cases in which this question can be answered?
2. Our original intent was to use the twist construction to produce chiral maniplexes and hopefully polytopes of all possible dimensions in an elegant way. Our main difficulty is that we seem to have no examples of maniplexes of any rank above five with sub-invertible elements that are not involutions, and thus we can display no k-maniplexes for $k \geq 6$ for which the twist operator is defined. We are trying not to conjecture that there are none.

3. Is there a way to generalize the construction of the map $n\mathcal{M}$ from the map $\mathcal{M}$ to apply to any maniplex $\mathcal{P}$? If there were, then we could have examples of maniplexes $\hat{2}^{\hat{2}^{\mathcal{P}}}$ of all dimensions to which we could apply the Twist operation.
4. Given a chiral maniplex $\mathcal{M}$, what are the criteria for $\mathcal{M}$ to be a twist of some reflexible maniplex?
5. If $T_w(\mathcal{M})$ is isomorphic to the mirror image of $\mathcal{M}$, is some $T_{w'}(\mathcal{M})$ reflexible?

Acknowledgements We gratefully acknowledge financial support of the PAPIIT-DGAPA, under grant IN107015, and of CONACyT, under grant 166951. The completion of this work was done while the second author was on sabbatical at the Laboratoire d'Informatique de l'École Polytechnique. She thanks LIX and Vincent Pilaud for their hospitality, as well as the program PASPA-DGAPA and the UNAM for the support for this sabbatical stay.

References

1. H.S.M. Coxeter, *Regular Polytopes*, 3rd edn. (Dover, New York, 1973)
2. H.S.M. Coxeter, The edges and faces of a 4-dimensional polytope, in *Proceedings of the Eleventh Southeastern Conference on Combinatorics, Graph Theory and Computing* (Florida Atlantic University, Boca Raton, FL, 1980), Vol. I. *Congr. Numer.* **28** (1980), pp. 309–334
3. G. Cunningham, M. Del Rio-Francos, I. Hubard, M. Toledo, Symmetry type graphs of polytopes and maniplexes. Ann. Comb. **19**, 243–268 (2015)
4. I. Douglas, Operators on Maniplexes, NAU Thesis series (2012)
5. L. Danzer, Regular incidence-complexes and dimensionally unbounded sequences of such, I, In Convexity and Graph Theory (Jerusalem 1981), North-Holland Math. Stud. vol. 87 (North-Holland, Amsterdam, 1984), pp. 115–127
6. M.E. Fernandes, D. Leemans, A.I. Weiss, Highly symmetric hypertopes. Aequationes Math. **90**, 1045–1067 (2016)
7. J. Garza-Vargas, I. Hubard, Politopality of Maniplexes. Discrete mathematics (2018)
8. M.I. Hartley, An atlas of small regular polytopes. Period. Math. Hung.**53**, pp. 149–156 (2006) (Full atlas online at http://www.abstract-polytopes.com/atlas/)
9. I. Hubard, A. Orbanić, A.I. Weiss, Monodromy groups and self-invariance. Canad. J. Math. **61**, 1300–1324 (2009)
10. H. Koike, D. Pellicer, M. Raggi, S. Wilson, Flag bicolorings, pseudo-orientations, and double covers of maps (submitted)
11. S. Krughoff, Rotary maniplexes with one and two facets, NAU Thesis series (2012)
12. P. McMullen, E. Schulte, *Abstract Regular Polytopes*, 1st edn. (Cambridge University Press, Cambridge, 2002)
13. B. Monson, D. Pellicer, G. Williams, Mixing and monodromy of abstract polytopes. Trans. Am. Math. Soc. **366**, 2651–2681 (2014)
14. D. Pellicer, A construction of higher rank chiral polytopes. Discrete Math. **310**, 1222–1237 (2010)
15. D. Pellicer, Developments and open problems on chiral polytopes. Ars Math. Contemp. **5**, 333–354 (2012)
16. E. Schulte, *Reguläre Inzidenzkomplexe*, Universität Dortmund Dissertation (1980)
17. E. Schulte, Symmetry of polytopes and polyhedra, in *Handbook of Discrete and Computational Geometry* (CRC Press Ser. Discrete Math. Appl., CRC, Boca Raton, FL, 1997)
18. A. Vince, Combinatorial maps. J. Combin. Theory, Ser. B **34** (1983) 1-21
19. S. Wilson, Maniplexes: part 1: maps, polytopes, symmetry and operators. Symmetry **4**(2), 265–275 (2012)

Hexagonal Extensions of Toroidal Maps and Hypermaps

Maria Elisa Fernandes, Dimitri Leemans and Asia Ivić Weiss

Dedicated to Egon Schulte on the occasion of his 60th birthday

Abstract The rank 3 concept of a hypermap has recently been generalized to a higher rank structure in which hypermaps can be seen as "hyperfaces" but very few examples can be found in literature. We study finite rank 4 structures obtained by hexagonal extensions of toroidal hypermaps. Many new examples are produced that are regular or chiral, even when the extensions are polytopal. We also construct a new infinite family of finite nonlinear hexagonal extensions of the tetrahedron.

Keywords Regularity · Chirality · Thin geometries · Hypermaps · Abstract polytopes

2000 Math Subj. Class: 52B11 · 20D06

M. E. Fernandes
Center for Research and Development in Mathematics and Applications,
Department of Mathematics, University of Aveiro, Aveiro, Portugal
e-mail: maria.elisa@ua.pt

D. Leemans (✉)
Department of Mathematics, University of Auckland, Private Bag 92019,
Auckland 1142, New Zealand
e-mail: dleemans@ulb.ac.be

D. Leemans
Département de Mathématique, C.P. 216 - Algèbre et Combinatoire,
Université Libre de Bruxelles, Boulevard du Triomphe, 1050 Brussels, Belgium

A. I. Weiss
Department of Mathematics and Statistics, York University,
Toronto, ON M3J 1P3, Canada
e-mail: weiss@mathstat.yorku.ca

M. D. E. Conder et al. (eds.), *Discrete Geometry and Symmetry*,
Springer Proceedings in Mathematics & Statistics 234,
https://doi.org/10.1007/978-3-319-78434-2_8

1 Introduction

A concept of a homogeneous honeycomb in Euclidean space was introduced by Sommerville in 1929 in his book "An Introduction to the Geometry of n Dimensions" [33] as an object consisting of polyhedral cells, all alike, such that each rotation that is a symmetry operation of a cell is also a symmetry operation of the whole configuration. This definition inspired Coxeter to name a map regular whenever its group of automorphisms contains, for each of its faces, elements that cyclically permute edges of that face, and also contains automorphisms that, for each of its vertices, cyclically permute the edges meeting at the vertex. Coxeter distinguished two kinds of regular maps: reflexible and irreflexible [9]; nowadays commonly referred to respectively as regular and chiral maps (he called a regular map reflexible when the group of automorphisms of the map contains an element that fixes an edge and the two faces that contain that edge, but interchanges the two vertices of the edge, otherwise he called the map irreflexible).

Earliest known examples of chiral maps were produced by Heffter in 1898 [18] as a family of maps of Schläfli type $\{2^k - 1, 2^k - 1\}$ for $k > 2$ (the first number describes the size of the faces and the second the degrees of the vertices of the map). In the 1940s Coxeter classified regular and chiral maps on the torus [7]. In 1966, Sherk [32], a Ph.D. student of Coxeter, looked for chiral maps of small genus and constructed an infinite family of chiral maps of type $\{6, 6\}$ (with the smallest member of that family embeddable on a surface of genus 7). About the same time Edmonds – a well-known and controversial Canadian combinatorist – re-discovered, but never published, Heffter's map of type $\{7, 7\}$ (also on a surface of genus 7). In 1969, Garbe [15] enumerated all regular maps on orientable surfaces of genus 2, 3, 4, 5 and 6, and proved that there are no chiral maps among them. A number of papers appeared thereafter dealing with chiral maps. A first systematic search for regular and chiral maps of higher genus was conducted by Conder and Dobcsányi [3] and resulted in the complete list of regular and chiral maps on surfaces of genus 2 to 15 in the orientable case and regular maps on surfaces of genus 3 to 30 in the non-orientable case. Subsequently, Conder expanded this list several times to include maps of increasingly higher genus. It now contains maps up to genus 301 in the orientable case and up to genus 602 in the non-orientable case [5].

In 1970 [8], Coxeter extended the notion of a chiral map by introducing the concept of a twisted honeycomb, a finite abstract object or rank 4 derived from a honeycomb, which is chiral in a sense that it inherits all the rotations of its cells but not its reflections. Two similar examples of such structures, both with only one polyhedral cell, were described earlier by Weber and Seifert in 1933 [36]. Coxeter produced a number of non-trivial examples, which he constructed from 3-dimensional euclidean, spherical and hyperbolic tessellations with spherical facets and vertex-figures, by looking at their Petrie polygons which naturally occur, in left- and right-handed varieties (each such polygon has three, but not four, consecutive edges belonging to a cell). Identifying vertices of each such, say left-handed Petrie polygon, that are separated by a fixed number of edges he observed that the resulting object may be regular (or as he called it reflexible) or may have the vertices on its right-handed Petrie polygons separated by a different number of edges in which case he called such a rank 4 object twisted. Twisted honeycomb is not symmetrical by a "reflection" in a sense that its automorphism group contains no involution that fixes, for example, a rank 2 face of the object but interchanges the two cells sharing this rank 2 face. The concept of twisted honeycomb inspired the modern definition of a chiral polytope, an abstract object of any rank that is maximally symmetric by abstract rotations but never by an abstract reflection (see Sect. 2).

The twisted honeycombs are finite structures resembling classical polytopes combinatorially in the sense that their facets and vertex-figures are spherical. In 1977 Coxeter suggested to the last author to derive finite twisted honeycombs from 3-dimensional hyperbolic tessellations with horospherical facets and/or vertex-figures producing therefore rank 4 objects with toroidal facets and/or vertex-figures. About the same time Grünbaum [16] suggested to study abstract objects, which he called polystroma, whose faces and vertex-figures are not necessarily spherical. Inspired by the ideas of Coxeter and Grünbaum, in 1982 Schulte and Danzer [11] formalised and began developing the theory of regular abstract polytopes (which they named incidence polytopes). In 1984, Colbourn and Weiss [2] unaware of the work of Schulte and Danzer, published a census of regular and chiral finite rank 4 polystroma derived from hyperbolic tessellations by applying the "twisted honeycomb" method of Coxeter. Not all such objects satisfied a more restrictive condition of abstract polytopality of [11].

By late 1980s a number of sporadic examples of chiral abstract polytopes in rank 3 and 4 were found. In 1991 Schulte and the last author of this paper developed the basic structure theory of abstract chiral polytopes of any rank [29] and in particular characterised their automorphism groups. These objects are now quite well understood and have been studied extensively over the past 30 years. Schulte, Monson and Weiss developed various methods of constructing such polytopes in rank 4. However, the classical approach to constructing higher rank polytopes inductively from the lower rank ones proved to be impossible for chiral polytopes. Although in 1995 [31] there was a universal extension method found leading to rank 5 chiral polytopes with regular facets, no chiral finite polytopes were known to exist in rank 5 or higher. It was only in 2008 that Conder, Hubard and Pisanski produced the first examples of finite higher rank chiral polytopes [4] and in 2010 Pellicer [28] gave a construction for arbitrary rank.

In [17], Hartley, Hubard and Leemans constructed two atlases of chiral polytopes. Firstly they sought them as quotients of regular polytopes arising from the Atlas of Small Regular Polytopes (http://www.abstract-polytopes.com/atlas/); secondly, for each almost simple group Γ such that $S \leq \Gamma \leq Aut(S)$ where S is a simple group of order less than 900,000 listed in the Atlas of Finite Groups, they gave, up to isomorphism, the abstract chiral polytopes on which Γ acts regularly. Such an atlas existed already in the regular case [21]. These atlases turned out to be very inspiring to find patterns and get classification results (see [13, 19, 20] for instance).

An abstract regular or chiral polytope is an incidence geometry with a string diagram. Recently, the authors have defined the notion of hypertope in [14] with the idea to allow more general diagrams than string diagrams. The present paper can be viewed as the beginning of an ambitious project to construct rank 4 hypertopes from their rank 3 residues. Among all hypertopes having some prescribed residues of rank 3, that are either spherical or toroidal maps or hypermaps, we consider whenever possible, the universal one (that is the one covering all hypertopes of this kind). The possible (non-string) diagrams for such hypertopes are listed in Fig. 1. In this paper we only consider the first three diagrams (hence the "hexagonal extensions" in the title) leaving other diagrams for future work.

The paper is organised as follows. In Sect. 2, we give the definitions and notation needed to understand this paper. In Sect. 3, we explain what are rank four universal locally toroidal hypertopes. In Sect. 4, we study locally toroidal regular and chiral polytopes of type $\{6, 3, 6\}$. In Sect. 5, we study locally toroidal regular and chiral polytopes of type $\{3, 6, 3\}$. In Sect. 6, we give examples of hexagonal extensions of toroidal hypermaps of type $(3, 3, 3)$. In Sect. 7, we give examples of nonlinear hexagonal extensions of the tetrahedron and, among these examples, a new infinite family of finite regular hypertopes arises. In Sect. 8, we give examples of 4-circuits with hexagonal residues. Finally, we conclude the paper in Sect. 9 by stating some conjectures and open problems.

As to notation for groups, we follow the Atlas of Finite Groups [6].

2 Preliminaries

2.1 Hypertopes

As in [1], an *incidence system* $\Gamma := (X, *, t, I)$ is a 4-tuple such that

- X is a set whose elements are called the *elements* of Γ;
- I is a set whose elements are called the *types* of Γ;
- $t : X \to I$ is a *type function*, associating to each element $x \in X$ of Γ a type $t(x) \in I$;
- $*$ is a binary relation on X called *incidence*, that is reflexive, symmetric and such that for all $x, y \in X$, if $x * y$ and $t(x) = t(y)$ then $x = y$.

The *incidence graph* of Γ is the graph whose vertex set is X and where two vertices are joined provided the corresponding elements of Γ are incident. A *flag* is a set of pairwise incident elements of Γ, i.e. a clique of its incidence graph. The *type* of a flag F is $\{t(x) : x \in F\}$. A *chamber* is a flag of type I. An element x is *incident* to a flag F, and we write $x * F$ for that, provided x is incident to all elements of F. An incidence system Γ is a *geometry* or *incidence geometry* provided that every flag of Γ is contained in a chamber (or in other words, every maximal clique of the incidence graph is a chamber). The *rank* of Γ is the number of types of Γ, namely the cardinality of I.

Let $\Gamma := (X, *, t, I)$ be an incidence geometry and F a flag of Γ. The *residue* of F in Γ is the incidence system $\Gamma_F := (X_F, *_F, t_F, I_F)$ where

- $X_F := \{x \in X : x * F, x \notin F\}$;
- $I_F := I \setminus t(F)$;
- t_F and $*_F$ are the restrictions of t and $*$ to X_F and I_F.

An incidence system Γ is *connected* if its incidence graph is connected. It is *residually connected* when each residue of rank at least two of Γ has a connected incidence graph. It is called *thin* (resp. *firm*) when every residue of rank one of Γ contains exactly two (resp. at least two) elements.

An incidence system $\Gamma := (X, *, t, I)$ is *chamber-connected* when for each pair of chambers C and C', there exists a sequence of successive chambers $C =: C_0, C_1, \ldots, C_n := C'$ such that $|C_i \cap C_{i+1}| = |I| - 1$. An incidence system $\Gamma := (X, *, t, I)$ is *strongly chamber-connected* when all its residues of rank at least 2 of Γ (including Γ itself) are chamber-connected.

Proposition 2.1 ([14, Proposition 2.1]) *Let Γ be a firm incidence geometry. Then Γ is residually connected if and only if Γ is strongly chamber-connected.*

A *hypertope* is a thin incidence geometry which is strongly chamber connected or equivalently residually connected.

2.2 Regular and Chiral Hypertopes as C^+-Groups

Let $\Gamma := (X, *, t, I)$ be an incidence system. An *automorphism* of Γ is a mapping $\alpha : (X, I) \to (X, I)$ where

- α is a bijection on X;
- for each $x, y \in X$, $x * y$ if and only if $\alpha(x) * \alpha(y)$;
- for each $x, y \in X$, $t(x) = t(y)$ if and only if $t(\alpha(x)) = t(\alpha(y))$.

An automorphism α of Γ is called *type preserving* when for each $x \in X$, $t(\alpha(x)) = t(x)$ (i.e. α maps each element on an element of the same type).

The set of type-preserving automorphisms of Γ is a group denoted by $Aut_I(\Gamma)$. The set of automorphisms of Γ is a group denoted by $Aut(\Gamma)$. Elements of $Aut(\Gamma) \setminus Aut_I(\Gamma)$ are called *correlations*.

An incidence system Γ is *flag-transitive* if $Aut_I(\Gamma)$ is transitive on all flags of a given type J for each type $J \subseteq I$. An incidence system Γ is *chamber-transitive* if $Aut_I(\Gamma)$ is transitive on all chambers of Γ. Observe that if Γ is a firm incidence geometry, flag-transivity and chamber-transitivity are equivalent. Finally, an incidence system Γ is *regular* if $Aut_I(\Gamma)$ acts regularly on the chambers (i.e. the action is semi-regular and transitive). A *regular hypertope* is a flag-transitive hypertope (note that thinness implies that the action of $Aut_I(\Gamma)$ is free).

Given an incidence system Γ and a chamber C of Γ, we may associate to the pair (Γ, C) the pair consisting of the automorphism group $G := Aut_I(\Gamma)$ and the set

$\{G_i : i \in I\}$ of subgroups of G where G_i is the stabiliser in G of the element of type i in C.

In the case of a regular hypertope Γ, the subgroups $\cap_{j\in I\setminus\{i\}} G_j$ are cyclic groups of order 2 and we denote their generators ρ_i's. The set $\{\rho_i : i \in I\}$ generates $Aut_I(\Gamma)$ (see [12]) and for that reason is called the set of *distinguished generators* of $Aut_I(\Gamma)$.

The following proposition shows how to start from a group and some of its subgroups and construct an incidence system.

Proposition 2.2 ([34]) *Let n be a positive integer and $I := \{1, \ldots, n\}$. Let G be a group together with a family of subgroups $(G_i)_{i\in I}$, X the set consisting of all cosets $G_i g$, $g \in G$, $i \in I$ and $t : X \to I$ defined by $t(G_i g) = i$. Define an incidence relation $*$ on $X \times X$ by :*

$$G_i g_1 * G_j g_2 \textit{ iff } G_i g_1 \cap G_j g_2 \textit{ is non-empty in } G.$$

Then the 4-tuple $\Gamma := (X, *, t, I)$ is an incidence system having a chamber. Moreover, the group G acts by right multiplication on Γ as a group of type preserving automorphisms. Finally, the group G is transitive on the flags of rank less than 3.

When a geometry Γ is constructed using the proposition above, we denote it by $\Gamma(G; (G_i)_{i\in I})$.

Consider a pair (G^+, R) with G^+ being a group and $R := \{\alpha_1, \ldots, \alpha_{r-1}\}$ a set of generators of G^+. Define $\alpha_0 := 1_{G^+}$ and $\alpha_{ij} := \alpha_i^{-1}\alpha_j$ for all $0 \le i, j \le r-1$. Observe that $\alpha_{ji} = \alpha_{ij}^{-1}$. Let $G_J^+ := \langle \alpha_{ij} \mid i, j \in J\rangle$ for $J \subseteq \{0, \ldots, r-1\}$. If the pair (G^+, R) satisfies condition (2.1) below called the *intersection condition* IC$^+$, we say that (G^+, R) is a *C^+-group*.

$$\forall J, K \subseteq \{0, \ldots, r-1\}, \textit{with } |J|, |K| \ge 2, G_J^+ \cap G_K^+ = G_{J\cap K}^+. \tag{2.1}$$

Two chambers C and C' of an incidence geometry of rank r are called *i-adjacent* if C and C' differ only in their i-elements. When the geometry is thin we denote C' by C^i. Let $\Gamma(X, *, t, I)$ be a thin incidence geometry. We say that Γ is *chiral* if $Aut_I(\Gamma)$ has two orbits on the chambers of Γ such that any two adjacent chambers lie in distinct orbits.

Given a chiral hypertope $\Gamma(X, *, t, I)$ (with $I := \{0, \ldots, r-1\}$) and its automorphism group $G^+ := Aut_I(\Gamma)$, pick a chamber C. For any pair $i \ne j \in I$, there exists a unique automorphism $\alpha_{ij} \in G^+$ that maps C to $(C^i)^j$. Also, $C\alpha_{ii} = (C^i)^i = C$ and $\alpha_{ij}^{-1} = \alpha_{ji}$. We define the *distinguished generators* of G^+ with respect to a base chamber C as follows:

$$\alpha_0 := 1_{G^+}, \alpha_i := \alpha_{0i} \ (i = 1 \ldots, r-1). \tag{2.2}$$

Define $\alpha_{ij} := \alpha_i^{-1}\alpha_j$ for all $0 \le i, j \le r-1$. Let $G_J^+ := \langle \alpha_{ij} \mid i, j \in J\rangle$ for $J \subseteq \{0, \ldots, r-1\}$.

Theorem 2.3 [14, Theorem 7.1] *Let $I := \{0, \ldots, r-1\}$ and let Γ be a chiral hypertope of rank r. Let C be a chamber of Γ. The pair (G^+, R), where $G^+ = Aut_I(\Gamma)$ and R is the set of distinguished generators of Γ with respect to C, is a C^+-group.*

Regular and chiral hypertopes can be constructed from some C$^+$-groups. We recall how to construct a coset geometry from a group and an independent generating set of this group.

Construction 2.1 [14, Construction 8.1] *Let G^+ be a group and $R := \{\alpha_1, \dots, \alpha_{r-1}\}$ be an independent generating set of G^+. Define $G_i := \langle \alpha_j \mid j \neq i \rangle$ for $i = 1, \dots, r-1$ and $G_0 := \langle \alpha_1^{-1}\alpha_j | j \geq 2 \rangle$. The coset geometry $\Gamma(G^+, R) := \Gamma(G^+; (G_i)_{i \in \{0,\dots,r-1\}})$ constructed using Proposition 2.2 is the geometry associated to the pair (G^+, R).*

The coset geometry $\Gamma(G^+, R)$ gives an incidence system using Proposition 2.2. This incidence system is not necessarily a thin geometry, nor is it necessarily residually connected. But if it is, then it is a hypertope and if its automorphism group has at most two orbits on its flags, the following theorem gives a way to check whether this geometry is chiral or regular.

Theorem 2.4 [14, Theorem 8.2] *Let (G^+, R) be a C^+-group. Let $\Gamma := \Gamma(G^+, R)$ be the coset geometry associated to (G^+, R) using Construction 2.1. If Γ is a hypertope and G^+ has two orbits on the set of chambers of Γ, then Γ is chiral if and only if there is no automorphism of G^+ that inverts all the elements of R. Otherwise, there exists an automorphism $\sigma \in Aut(G^+)$ that inverts all the elements of R and the group G^+ extended by σ is regular on Γ.*

Later in the paper, when we will build hypertopes from their C^+-groups given as finitely presented groups, we will indeed check that the corresponding incidence system is thin and residually connected. This check is most of the time easily performed with MAGMA. We will list the hypertopes obtained in tables, not mentioning those presentations giving a C^+-group that does not yield a hypertope.

2.3 B-Diagrams

Let $R := \{\alpha_1, \dots, \alpha_{r-1}\}$ and $G^+ = \langle R \rangle$ be such that (G^+, R) is a C$^+$-group. It is convenient to represent (G^+, R) by the following complete graph with r vertices which we will call the *B-diagram* (short for Buekenhout) of (G^+, R) and denote by $\mathcal{B}(G^+, R)$. The vertex set of $\mathcal{B}$ is the set $\{\alpha_0, \dots, \alpha_{r-1}\}$. The edges $\{\alpha_i, \alpha_j\}$ of this graph are labelled by $o(\alpha_i^{-1}\alpha_j) = o(\alpha_j^{-1}\alpha_i) = o(\alpha_i\alpha_j^{-1})$. We take the convention of dropping an edge if its label is 2 and of not writing the label if it is 3. Vertices of $\mathcal{B}$ are represented by small circles in order to distinguish from the vertices of a Coxeter diagram, which represent involutions. A *regular or chiral polytope* can be defined as a regular or chiral hypertope with linear Coxeter diagram, or equivalently, with linear B-diagram.

Rank four extensions of rank three toroidal polytopes of type $\{6, 3\}_{(a,b)}$ have been studied by Schulte and Weiss [30], Nostrand and Schulte [26] and Monson and Weiss [25]. The rotation subgroup of the automorphism group of a rank three toroidal polytope $\mathcal{P} := \{6, 3\}_{(a,b)}$ is the group $G^+ := Aut(\mathcal{P})^+$ defined as follows.

$$G^+ := \langle x, y | x^6, y^3, (x^{-1}y)^2, (y^{-1}x^{-2})^a(yx^2)^b \rangle. \tag{2.3}$$

The B-diagram of $(G^+, \{x, y\})$ is the following.

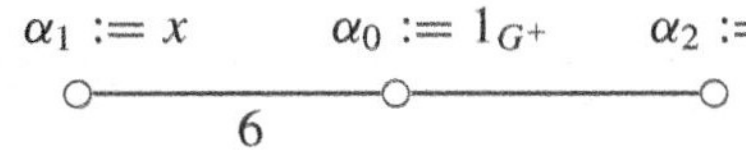

Recall that the polytope above is obtained by identifying opposite sides of a parallelogram in the tessellation of the Euclidean plane by hexagons to obtain the map $\{6, 3\}_{(a,b)}$.

The dual of $\mathcal{P}$ will be the polytope $\mathcal{P}^* := \{3, 6\}_{(a,b)}$ with rotation group $H^+ := Aut(\mathcal{P}^*)^+$ defined as follows.

$$H^+ := \langle x, y | x^3, y^6, (x^{-1}y)^2, (x^{-1}y^{-2})^a(xy^2)^b \rangle. \tag{2.4}$$

Observe that presentation (2.4) is obtained by interchanging x and y in presentation (2.3). The B-diagram of $(H^+, \{x, y\})$ is the following.

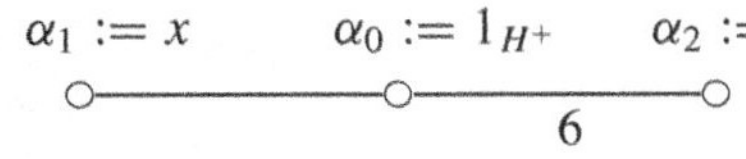

Indeed there is no distinction between the C$^+$-groups of a polytope and its dual. We make the distinction when we write the B-diagram (ranking the generators).

The rotation subgroup of the automorphism group of a rank three toroidal polytope $\mathcal{P} := \{4, 4\}_{(a,b)}$ is the group $G^+ := Aut(\mathcal{P})^+$ defined as follows.

$$G^+ := \langle x, y | x^4, y^4, (x^{-1}y)^2, (xy)^a(x^{-1}y^{-1})^b \rangle. \tag{2.5}$$

The B-diagram of $(G^+, \{x, y\})$ is the following.

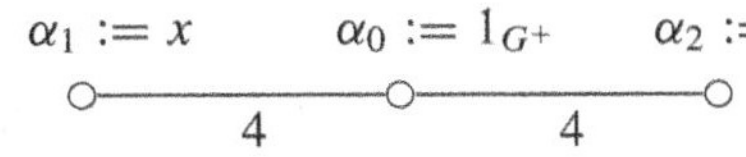

Observe that the dual $\mathcal{P}^*$ of $\mathcal{P}$ is $\{4, 4\}_{(a,-b)} = \{4, 4\}_{(b,a)}$ as the vectors $(a, -b)$ and (b, a) are orthogonal (the characterisation of dual polytopes it terms of the rotational groups can be found for instance in [29]). The rotational group for $\mathcal{P}^*$ is obtained by interchanging x with y.

The rotation subgroup of the automorphism group of a rank three toroidal hypermap $\mathcal{P} := (3, 3, 3)_{(a,b)}$ is the group $G^+ := Aut(\mathcal{P})^+$ defined as follows.

$$G^+ := \langle x, y | x^3, y^3, (x^{-1}y)^3, (xy^{-1}x)^a(xy)^b \rangle. \tag{2.6}$$

The B-diagram of $(G^+, \{x, y\})$ is the following.

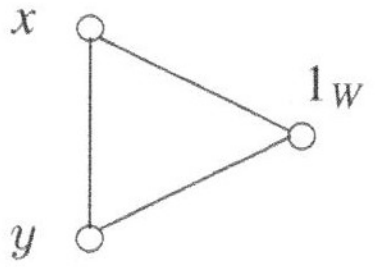

The elements corresponding to the three possible types in a hypermap are called hypervertices, hyperedges and hyperfaces. The dual of a hypermap is obtained interchanging hypervertices with hyperfaces. The dual $\mathcal{P}^*$ of $\mathcal{P}$ is $(3, 3, 3)_{(b,a)}$.

3 Rank 4 Universal Locally Toroidal Hypertopes

Spherical hypertopes (in the sense of Coxeter) of rank 3 are maps (polyhedra) on the sphere while toroidal hypertopes of rank 3 are either maps or hypermaps on the torus. The toroidal (regular or chiral) hypertopes of rank 3 are divided into the following families: the toroidal maps $\{3, 6\}_{(b,c)}$, $\{6, 3\}_{(b,c)}$, $\{4, 4\}_{(b,c)}$, and the hypermaps $(3, 3, 3)_{(b,c)}$ with $(b, c) \neq (1, 1)$. Note that the hypermap $(3, 3, 3)_{(b,c)}$ is obtained from the toroidal map $\{6, 3\}_{(b,c)}$ by doubling the fundamental region. Indeed as $\{6, 3\}_{(b,c)}$ is bipartite it is possible to take one monochromatic set of vertices to be the hyperedges of the hypermap $(3, 3, 3)_{(b,c)}$ (see [35]). But in the case $(b, c) = (1, 1)$ the corresponding incidence graph is a complete tripartite graph and, therefore, the geometry is not thin (see [14]). Indeed that is the unique highly symmetric (regular or chiral) toroidal hypermap that is not an hypertope.

Similarly to the theory of abstract regular polytopes it is possible to construct hypertopes inductively from hypertopes of lower rank. In the case of polytopes $\{P_1, P_2\}$ denotes a polytope having facets isomorphic to P_1 and vertex-figures isomorphic to P_2 (see [23]). More precisely, if the set of regular polytopes having facets P_1 and vertex-figures P_2, denoted by $\langle P_1, P_2\rangle$, is nonempty, there exists a regular polytope that covers every other element of the set $\langle P_1, P_2\rangle$, that is the universal regular polytope $\{P_1, P_2\}$. In addition if the automorphism group of the universal polytope $\{P_1, P_2\}$ is the group $\langle \rho_0, \dots, \rho_n\rangle$, the automorphism groups of P_1 and P_2 are $\langle \rho_0, \dots, \rho_{n-1}\rangle$ and $\langle \rho_1, \dots, \rho_n\rangle$, respectively.

In a similar way here, we construct rank 4 regular and chiral hypertopes that we call *universal* when the relations corresponding to each rank 3 residue of the resulting hypertope together with the relations implicit in the B-diagram of the hypertope determine the group.

Here we consider universal locally toroidal hypertopes of rank 4, meaning that all residues of rank 3 are either spherical or toroidal, with at least one being toroidal. These hypertopes are finite whenever their automorphism group is finite.

The existence of regular universal locally toroidal polytopes of rank 4 is investigated in [23], (see also [22] and [24]), moreover the authors give an enumeration of finite locally toroidal universal polytopes. For the universal polytopes $\{\{4, 4\}_{(b,c)}, \{4, 4\}_{(e,f)}\}$ a nearly complete finiteness characterisation is given, for

$\{\{4,4\}_{(b,c)},\{4,3\}\}$, $\{\{6,3\}_{(b,c)},\{3,p\}\}$ with $p \in \{3,4,5\}$ and $\{\{6,3\}_{(b,c)},\{3,6\}_{(e,f)}\}$ the enumeration is complete, and for the polytopes $\{\{3,6\}_{(b,c)},\{6,3\}_{(e,f)}\}$ partial results are known.

In [14] we also list the known finite universal chiral polytopes $\{\{6,3\}_{(b,c)},\{3,p\}\}$ for $p \in \{3,4,5\}$ and we conjecture that the list is complete. We give this list in Table 1.

In Fig. 1, we list the diagrams of all possible finite universal locally toroidal hypertopes of rank 4 having nonlinear diagram (where $p \in \{3,4,5,6\}$).

The finite universal locally toroidal hypertopes with diagram (1), when $p \neq 6$, have only one toroidal residue that is the hypermap $(3,3,3)_{(b,c)}$, all the remaining residues are spherical. We denote this hypertope by $(3,3,3;p)_{(b,c)}$.

In [14] we proved that when $p \in \{3,4,5\}$ and $(b,c) \neq (1,1)$, the regular hypertope $(3,3,3;p)_{(b,c)}$ exists (is finite) if and only if the universal regular polytope $\{\{6,3\}_{(b,c)},\{3,p\}\}$ exists (is finite).

In Sect. 6 we consider the diagram (1) with $p = 6$, that we call *hexagonal extension of the toroidal hypermaps* $(3,3,3)_{(b,c)}$. In this case there are three toroidal residues, that explains why the case $p = 6$ is substantially more complex than the cases $p \in \{3,4,5\}$ studied in [14].

Table 1 Known finite polytopes of type $\{\{6,3\}_{\mathbf{s}},\{3,p\}\}$ with $p \in \{3,4,5\}$ (having g flags)

p	s	g	Group	Chiral/Regular
3	(2, 0)	240	$S_5 \times C_2$	Regular
	(3, 0)	1296	$[1\ 1\ 2]^3 \rtimes C_2$	Regular
	(4, 0)	15,360	$[1\ 1\ 2]^4 \rtimes C_2$	Regular
	(1, 2)	336	$PGL_2(7)$	Chiral
	(1, 3)	2184	$L_2(13) \times C_2$	Chiral
	(1, 4)	8064	$SL_2(7) \rtimes A_4 \rtimes C_2$	Chiral
	(2, 2)	2880	$S_5 \times S_4$	Regular
	(2, 3)	6840	$PGL_2(19)$	Chiral
4	(1, 1)	288	$S_3 \rtimes [3,4]$	Regular
	(2, 0)	768	$[3,3,4] \times C_2$	Regular
	(1, 2)	2016	$PGL_2(7) \times S_3$	Chiral
5	(2, 0)	28,800	$[3,3,5] \times C_2$	Regular

Fig. 1 Possible nonlinear diagrams of rank 4 universal locally toroidal hypertopes, where $p \in \{3,4,5,6\}$

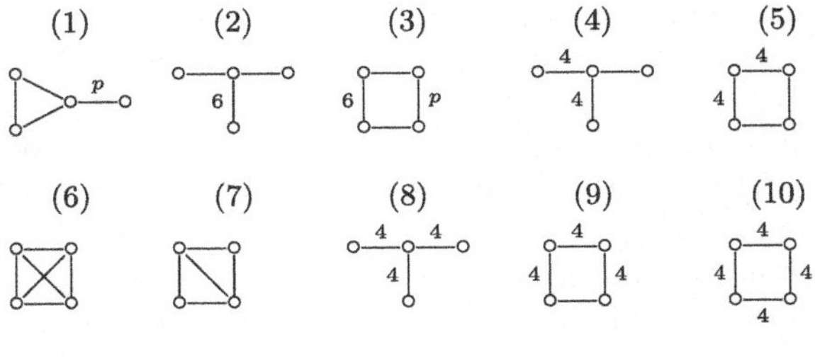

In Sect. 7 we deal with the finite universal polytopes with diagram (2) and in Sect. 8 with the universal hypertopes of diagram (3). The universal locally toroidal regular hypertopes with diagram (4) can be obtaining from the regular universal locally toroidal polytope of type $\{4, 4, 3\}$. Indeed each of them admits a correlation τ as shown in the following figure.

In the case that the toroidal residue is the map $\{4, 4\}_{(s,0)}$ satisfying a relation $(\rho_0\rho_1\rho_2\rho_1)^s = 1$ (the central vertex of the B-diagram being ρ_1), adding the automorphism τ, such that $\rho_2^\tau = \rho_0$ (and fixing the remaining generators), we get the automorphism group of $\{\{4, 4\}_{(s,s)}, \{4, 3\}\}$ (generated by ρ_0, ρ_1 and τ). Indeed apart from the relations corresponding to the type $\{4, 4, 3\}$ we get $(\rho_0\rho_1\rho_0^\tau\rho_1)^s = 1$. If the toroidal residue is $\{4, 4\}_{(s,s)}$ satisfying a relation $(\rho_0\rho_1\rho_2)^{2s} = 1$ then adding the automorphism τ we obtain the group of $\{\{4, 4\}_{(2s,0)}, \{4, 3\}\}$. Thus all regular universal locally toroidal hypertopes of this type are determined by the correspondent universal locally toroidal regular polytope of type $\{4, 4, 3\}$.

The universal locally toroidal regular hypertopes with diagram (5) also admit a correlation.

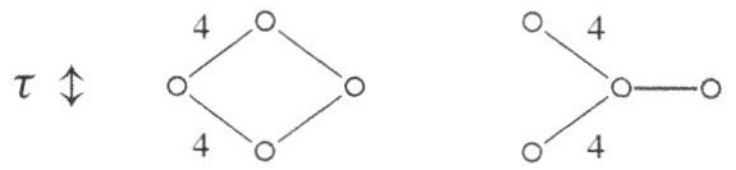

Computer experiments suggest that using this correlation, we get hypertopes. The hypertopes with this diagram and toroidal residue $\{4, 4\}_{(s,0)}$ can be derived from the universal locally toroidal hypertopes with diagram (4) and toroidal residue $\{4, 4\}_{(s,s)}$ (indeed if ρ_1 corresponds to the middle vertex of the toroidal residue, $(\rho_0\rho_1\rho_2\rho_1)^s = 1$, $\rho_0^\tau = \rho_2$ and $\rho_1^\tau = \rho_1$, conjugating $(\rho_0\rho_1\rho_2\rho_1)^s = 1$ by τ we get $(\rho_2\rho_1\tau)^{2s} = 1$). The hypertopes with diagram (5) and toroidal residue $\{4, 4\}_{(s,s)}$ can be derived from the universal locally toroidal hypertopes with diagram (4) and toroidal residue $\{4, 4\}_{(2s,0)}$. Again the regular universal locally toroidal hypertopes of these types are determined.

The universal locally toroidal regular hypertopes with diagram being a tetrahedron, as in (6), have four toroidal residues (hypermaps of type $(3, 3, 3)$) corresponding to the four faces of the tetrahedron. The case with all toroidal residues being regular hypermaps of type $(3, 3, 3)$ is completely studied in [23], where this diagrams are denoted by $\mathcal{T}_4(q_1, q_2, q_3, q_4)$ with reflexion group $G(q_1, q_2, q_3, q_4)$. The results are summarised in the following theorems.

Theorem 3.1 [23, Theorem 9E14] *$G(s, s, q, q)$ is finite if and only if $s = 2$ and $q \in \{2, 3, 4\}$ (up to an interchange of s and q).*

Theorem 3.2 [23, Theorem 9E15] $G(q_1, q_2, q_3, q_4)$ *is infinite in at least the following cases:*

(1) $p|q_1, \ldots, q_4$ *for some* $p \geq 3$;
(2) $p|q_1$ *and* $3p|q_2, q_3, q_4$ *for some* $p \geq 2$.

Theorem 3.3 [23, Theorem 9E17] *The group* $G(q_1, q_2, q_3, q_4)$ *with* $(q_1, q_2) \in \{(q_1, 3), (3, q_2), (4, 4), (4, 5), (5, 4)\}$ *and* $(q_3, q_4) \in \{(q_3, 3), (3, q_4), (4, 4), (4, 5), (5, 4)\}$ *is infinite except when* $(q_1, q_2, q_3, q_4) = (q_1, 3, 3, 2)$ *(up to an interchange of the pairs* (q_1, q_2) *and* (q_3, q_4)*).*

The remaining possibilities for a universal locally toroidal hypertope with diagram being a tetrahedron will be studied in future work, as well as the locally toroidal polytopes with diagrams (7), (8), (9) and (10).

4 Locally Toroidal Regular and Chiral Polytopes of Type {6, 3, 6}

Branko Grünbaum first posed the question of classifying regular locally toroidal polystromas in 1977 (see [16]). About the same time Coxeter and Shephard independently constructed in [10] such an object. Several attempts have been made at the classification, including Colbourn and Weiss who produced a computer-generated list of regular and chiral examples of all possible types (see [2]). In 2002 McMullen and Schulte in [24] succeeded in classifying all finite regular locally toroidal universal polytopes with Schläfli symbol $\{6, 3, p\}$ for $p = 3, 4, 5$ and 6 (see [22, 24], and also Chap. 11 of [23]). In [14], we present their classification and in addition provide a list of chiral polytopes with $p \leq 5$, for which we conjecture to be complete. Each such regular polytope is associated with a honeycomb of the hyperbolic 3-space. Since the horosphere is isomorphic to the euclidean plane, one can tesselate it by regular hexagons, three meeting at a vertex, to obtain a regular polytope embedded in a horosphere. The vertex figures of $\{6, 3, p\}$ for $p = 3, 4, 5$ are spherical polyhedra isomorphic to the tetrahedron, octahedron and icosahedron respectively. The size of each of these polyhedra is determined by the dihedral angle which has to be $\frac{2\pi}{p}$ in order that p facets fit around an edge without overlap. The facets of the hyperbolic honeycombs are therefore horospherical honeycombs $\{6, 3\}$ centered at the absolute. All the vertices of the honeycomb are (finite) points of the hyperbolic space. When $p = 6$, the vertices of the honeycomb $\{6, 3, 6\}$ also belong to the absolute. Hence there are no vertices of this honeycomb that belong to the hyperbolic space (but all the edges are there).

McMullen and Schulte used twisting operations on quotients of certain Coxeter groups that are associated with complex hermitian forms. Their results [23, Chap. 11] on existence of finite universal polytopes of type $\{6, 3, 6\}$ are summarised in Table 2.

Table 2 Finite regular polytopes of type $\{\{6,3\}_{\mathbf{s}}, \{3,6\}_{\mathbf{t}}\}$

s	t	g	Group
(2, 0)	(2, 0)	480	$S_5 \times 2^2$
(2, 0)	(3, 0)	2592	$D_{12} \times S_3 \times S_3 : S_3$
(2, 0)	(4, 0)	30,720	$2^{1+6} : (S_5 \times 2)$
(2, 0)	(2, 2)	5760	$S_5 \times 2^3 : S_3$
$(s, 0)$ with $s \equiv 0 \mod 3$	(1, 1)	$72s^2$	$S_3 : [3,6]_{(s,0)}$
(s, s) with $s \geq 1$	(1, 1)	$216s^2$	$S_3 : [3,6]_{(s,s)}$

Using MAGMA, we construct several new finite universal chiral polytopes of this type and we present our results in Table 3. In fact, only the first two polytopes appearing in this table were previously known. We conjecture this list to be complete.

Consider the Coxeter diagram with usual generators ρ_i $(i = 0, \ldots, 3)$ as follows.

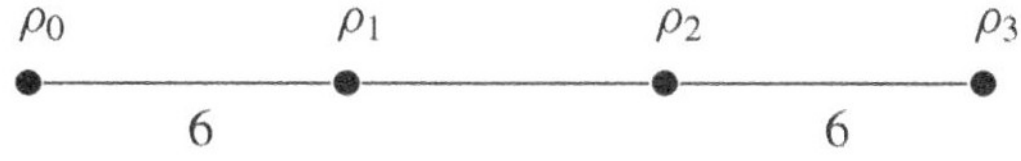

It gives the Coxeter group $W := [3, 6, 6]$. Its rotational subgroup $W^+ = \langle \alpha_1, \alpha_2, \alpha_3 \rangle$ with distinguished generators

$$\alpha_0 = 1_W;\ \alpha_1 = \rho_1\rho_0;\ \alpha_2 = \rho_1\rho_2;\ \alpha_3 = \rho_1\rho_3.$$

is a C$^+$-group. We now write the B-diagram associated to $(W^+, \{\alpha_1, \alpha_2, \alpha_3\})$ as follows.

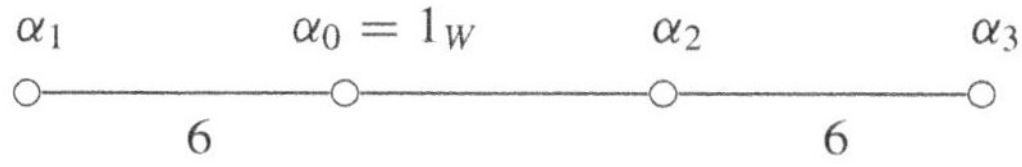

Given the base flag C of the universal polytope $\{6, 3, 6\}$ such that $\rho_i(C) = C^i$, $(i = 0, \ldots, 3)$, it follows that $\alpha_1(C) = (C^1)^0$, $\alpha_2(C) = (C^1)^2$ and $\alpha_3(C) = (C^1)^3$.

The automorphism groups of the polytopes of Tables 2 and 3 are obtained using the following presentation where $x := \alpha_1$, $y := \alpha_2$, $z := \alpha_3$, $\mathbf{s} = (a, b)$ and $\mathbf{t} = (c, d)$.

$$G^+ := \langle x, y, z | x^6, y^3, z^2, (x^{-1}y)^2, (y^{-1}z)^6, (zx)^2, (y^{-1}x^{-2})^a(yx^2)^b, (y(zy)^2)^c(yzy^{-1}z)^d \rangle$$

Note that, when the polytope is regular, the presentation above gives the rotational subgroup of the full automorphism group.

Extending the methods that were developed in [14], in Sect. 6 we will look at hypertopes with a nonlinear diagram, arising from these groups.

Table 3 Known finite chiral polytopes of type $\{\{6,3\}_{\mathbf{s}}, \{3,6\}_{\mathbf{t}}\}$

s	**t**	g	Group
(1, 2)	(1, 2)	1344	$2 \cdot L_2(7) : 2 : 2$
(1, 2)	(2, 1)	2060	$2 \cdot A_7 : 2 : 2$
(2, 0)	(1, 2)	672	$L_2(7) : 2 \times 2$
(2, 0)	(1, 3)	368	$L_2(13) \times 2 \times 2$
(2, 0)	(1, 4)	1628	$Q_8 \cdot (L_2(7) \times 2) : S_3$
(2, 0)	(2, 3)	1380	$L_2(19) : 2 \times 2$

5 Polytopes of Type {3, 6, 3}

As for the polytopes of the previous section, the polytopes of type {3, 6, 3} have also been studied is several articles and in [23]. We refer to [23, Sect. 11E] for more details. We now consider the polytopes of type $\{\{3,6\}_{(a,b)}, \{6,3\}_{(c,d)}\}$. We use the following B-diagram, where x, y and $y^{-1}z$ are rotations generating the infinite Coxeter group [3,6,3].

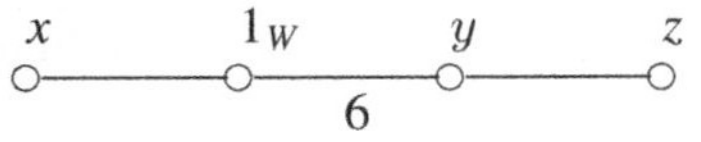

The type {3, 6} residues, given by the facets and vertex-figures, may be nonisomorphic. Thus we need four parameters a, b, c, d giving two additional relations in the following presentation for the rotation subgroup of the automorphism group G. The automorphism groups of the polytopes of Table 4 are obtained using the following presentation where $\mathbf{s} = (a, b)$ and $\mathbf{t} = (c, d)$.

$$G^+(a,b,c,d) := \langle x, y, z \mid x^3, y^6, z^2, (x^{-1}z)^2, (y^{-1}z)^3, (x^{-1}y)^2,$$
$$(x^{-1}y^{-2})^a(xy^2)^b, (zy^3)^c(y^{-1}zy^{-2})^d \rangle$$

We found several new universal hypertopes compared to Table 1 of [2]. In the regular case, lines 5 and 7 are not in [2], but they can be found in Table 11E1 of [23], line 8 of Table 4 is new. In the chiral case, all but the last two are new.

6 Hexagonal Extensions of Toroidal Hypermap (3, 3, 3)

In this section, as we did in [14], starting with the Coxeter group [3, 6, 6] with diagram

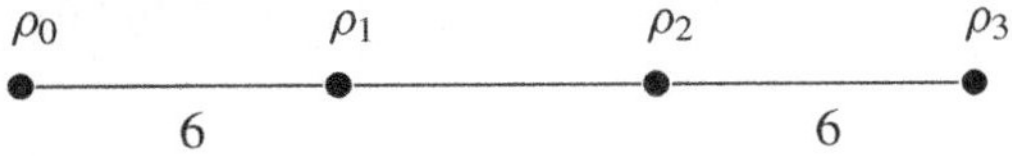

Table 4 Known finite regular and chiral universal polytopes of type $\{\{3, 6\}_{\mathbf{s}}, \{6, 3\}_{\mathbf{t}}\}$

s	**t**	g	Group	
(2, 0)	(2, 0)	240	$A_5 : 2 \times 2$	Regular
(2, 0)	(6, 0)	720	$S_5 \times S_3$	Regular
(3, 0)	(3, 0)	2916	$3^{1+2} \times 3 : S_3 : S_3$	Regular
(3, 0)	(4, 0)	241,920	$Aut(L_3(4))$	Regular
(3, 0)	(1, 1)	324	$3^{1+2} : 2 : S_3$	Regular
(3, 0)	(2, 2)	41,472	$2^{1+6} : 3^2 : S_3 : S_3$	Regular
(1, 1)	(1, 1)	108	$3^{1+2} : 2^2$	Regular
(2, 2)	(2, 2)	13,271,040	$2^{5+6} : A_5 \times 3 : S_3 : S_3$	Regular
(3, 0)	(1, 3)	33,696	$L_3(3) : S_3$	Chiral
(4, 0)	(1, 2)	12,096	$U_3(3) : 2$	Chiral
(6, 0)	(1, 2)	756,000	$U_3(5) : S_3$	Chiral
(1, 2)	(3, 6)	2016	$2 \cdot L_2(7) : S_3$	Chiral
(1, 2)	(4, 4)	36,288	$U_3(3) : S_3$	Chiral
(1, 2)	(6, 6)	2,268,000	$3 \cdot U_3(5) : S_3$	Chiral
(3, 5)	(2, 1)	672	$2 \cdot L_2(7) : 2$	Chiral
(1, 4)	(2, 1)	2016	$2 \cdot L_2(7) : S_3$	Chiral
(1, 2)	(1, 2)	672	$2 \cdot L_2(7) : 2$	Chiral

we double the fundamental region so that the resulting involutory generators give us the following Coxeter group that is of index two in [3, 6, 6].

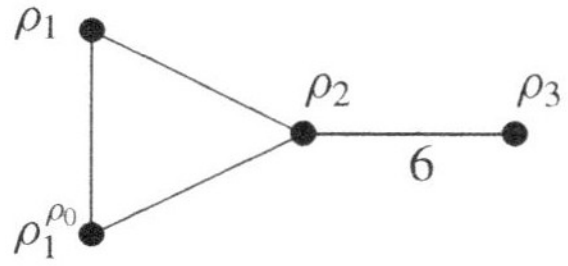

In the geometry constructed from this group, all rank three residues with connected diagrams are either Euclidean tessellations of horospheres of type {3, 6} (up to duality) or hypermaps of type (3, 3, 3). We denote by [(3, 3, 3), 6] the Coxeter group having this diagram.

In order to construct a finite hypertope $\mathcal{H}$ whose residues could be chiral, we consider the rotation subgroup $W^+ := [(3, 3, 3), 6]^+$ of the group $W := [(3, 3, 3), 6]$. The B-diagram of W^+ is as follows, where x, y and z are rotations generating this infinite group.

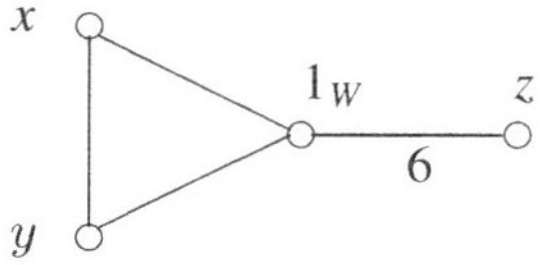

In $\mathcal{H}$ the two type {3, 6} residues may be nonisomorphic, and in that case they cannot be obtained from the polytopes of type $\{\{6,3\}_s, \{3,6\}_t\}$ as described before. To describe all regular and chiral hypertopes with the B-diagram above we need six parameters a, b, c, d, e, f giving three additional relations in the following presentation for the rotation subgroup of $G := \mathrm{Aut}(\mathcal{H})$.

$$G^+(a,b,c,d,e,f) := \langle x, y, z | x^3, y^3, z^6, (x^{-1}z)^2, (y^{-1}z)^2, (x^{-1}y)^3,$$

$$(yx^{-1}y)^a(yx)^b, (y^{-1}z^{-2})^c(yz^2)^d, (x^{-1}z^{-2})^e(xz^2)^f \rangle$$

where the subgroup $\langle x, z \rangle$ acts on a polytope of type $\{3,6\}_{(e,f)}$, $\langle y, z \rangle$ acts on a polytope of type $\{3,6\}_{(c,d)}$, and $\langle x, y \rangle$ acts of a hypermap of type $(3,3,3)_{(a,b)}$. We say that $\mathcal{H}$ has type $\{(3,3,3)_s, 6\}$ where $s = (a,b)$. If there exists a correlation δ of order two fixing z and interchanging x and y, the residues $\{3,6\}_{(e,f)}$ and $\{3,6\}_{(c,d)}$ are isomorphic and therefore $(c,d) = (e,f)$. The automorphism groups of the hypertopes of Table 5 are obtained using the presentation above.

Table 5 Known finite universal hypertopes of type $\{(3,3,3)_s, 6\}$

(a,b)	(c,d)	(e,f)	g	G	Regular/Chiral
(2, 0)	(2, 0)	(2, 0)	240	$S_5 \times 2$	Regular
(2, 0)	(3, 0)	(3, 0)	1296	$S_3 \times S_3 \times S_3 : S_3$	Regular
(2, 0)	(4, 0)	(4, 0)	15,360	$2^{1+6} : (A_5 : 2)$	Regular
(2, 0)	(6, 0)	(2, 2)	2880	$S_5 \times 2^2 : S_3$	Regular
(3, 0)	(2, 0)	(2, 0)	1296	$6 : S_3 : S_3 : S_3$	Regular
(3, 0)	(2, 0)	(3, 0)	13,824	$2^{1+6} : 3 : S_3 : S_3$	Regular
(3, 0)	(2, 0)	(4, 0)	165,888	$2^{1+6} : 6 : S_3 : S_3 : S_3$	Regular
(3, 0)	(2, 0)	(5, 0)	2,592,000	$2 \cdot (A_5 \times (A_5 \times A_5)) : 3 : 2$	Regular
(3, 0)	(2, 0)	(2, 2)	3888	$S_3 \times 3 : S_3 : S_3 : S_3$	Regular
(3, 0)	(2, 0)	(4, 4)	248,832	$2^{1+6} : (3 : S_3 : S_3 : 3) : S_3$	Regular
(3, 0)	(3, 0)	(1, 1)	972	$3^{1+2} : 6 : S_3$	Regular
(3, 0)	(1, 1)	(1, 1)	324	$3^2 \times S_3 : S_3$	Regular
(3, 0)	(1, 1)	(3, 3)	2916	$3^{1+2} \times 3 : S_3 : S_3$	Regular
(3, 0)	(2, 2)	(2, 0)	3888	$S_3 \times 3 : S_3 : S_3 : S_3$	Regular
(4, 0)	(2, 0)	(2, 0)	15,360	$4 \cdot (2^4 : A_5) : 2 \times 2$	Regular
(2, 0)	(1, 2)	(2, 1)	336	$L_2(7) : 2$	Chiral
(2, 0)	(1, 3)	(2, 5)	2184	$L_2(13) \times 2$	Chiral
(2, 0)	(1, 4)	(1, 2)	336	$L_2(7) : 2$	Chiral
(2, 0)	(1, 4)	(3, 6)	8064	$Q_8 \cdot L_2(7) : S_3$	Chiral
(2, 0)	(2, 3)	(3, 2)	6840	$L_2(19) : 2$	Chiral
(3, 0)	(1, 2)	(1, 2)	275,562	$3^{2+6} : 7 : 3 : 2$	Chiral

7 Nonlinear Hexagonal Extensions of the Tetrahedron

We now consider a hypertope $\mathcal{H}$ with the following B-diagram, where x, y and z are rotations generating the corresponding C$^+$-group.

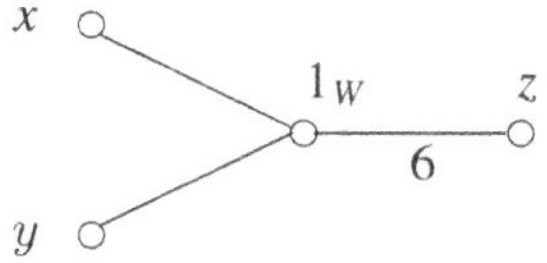

In $\mathcal{H}$ the two type $\{3, 6\}$ residues may be nonisomorphic. Thus we need four parameters a, b, c, d giving two additional relations in the following presentation for the the rotation subgroup of $G := \mathrm{Aut}(\mathcal{H})$.

$$G^+(a, b, c, d) := \langle x, y, z | x^3, y^3, z^6, (x^{-1}z)^2, (y^{-1}z)^2, (x^{-1}y)^2,$$
$$(y^{-1}z^{-2})^a(yz^2)^b, (x^{-1}z^{-2})^c(xz^2)^d\rangle.$$

The automorphism groups of the hypertopes of Table 6 are obtained using this presentation where $\mathbf{s} = (a, b)$ and $\mathbf{t} = (c, d)$.

There is an infinite family of regular locally toroidal hypertopes with the following Coxeter diagram having toroidal rank 3 residues $\{3, 6\}_{(2,0)}$ and $\{3, 6\}_{(s,0)}$ with $s \geq 3$.

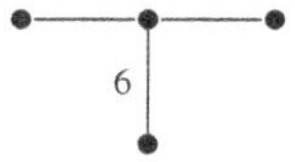

Table 6 Known finite universal hypertopes of type (2) in Fig. 1, with $\mathbf{s} = (a, b)$ and $\mathbf{t} = (c, d)$

s	t	g	G	
(2, 0)	(2, 0)	384	$2^{1+4} \times 2 : S_3$	Regular
(2, 0)	(3, 0)	1296	$S_3 \times S_3 \times S_3 : S_3$	Regular
(2, 0)	(4, 0)	3072	$2^{1+6} : 2^2 : S_3$	Regular
(2, 0)	(5, 0)	6000	$5^3 : 2 : 2 : 2 : S_3$	Regular
(2, 0)	(6, 0)	10,368	$2^3 : S_3 : S_3 : S_3 : S_3$	Regular
(3, 0)	(1, 1)	144	$S_3 \times 2^2 : S_3$	Regular
(6, 0)	(1, 1)	576	$2^4 : S_3 : S_3$	Regular
(3, 0)	(1, 3)	58,968	$L_2(27) : 3 : 2$	Chiral
(1, 2)	(1, 2)	2688	$2^6 : 7 : 3 : 2$	Chiral
(1, 2)	(2, 1)	1008	$L_2(7) \times 3 : 2$	Chiral
(1, 2)	(3, 1)	58,968	$L_2(27) : 3 \times 2$	Chiral

This family of hypertopes can be obtained from the cubic toroids $\{4, 3, 4\}_{(s,s,0)}$ whose automorphism group G is the Coxeter group $[4, 3, 4] = \langle \rho_0, \rho_1, \rho_2, \rho_3 \rangle$ factorized by the single extra relation $(\rho_0\rho_1\rho_2\rho_3\rho_2)^{2s}$ (see p.168 of [23]). First consider the hypertope that is obtain from $\{4, 3, 4\}_{(s,s,0)}$ using the Petrie operation defined by the correspondence $\alpha_0 \mapsto \rho_0$, $\alpha_1 \mapsto \rho_1$, $\alpha_2 \mapsto \rho_2$ and $\alpha_3 \mapsto \rho_1\rho_3$. We obtain a C-group with the following diagram and the extra relations $(\alpha_0\alpha_1\alpha_2\alpha_1\alpha_3\alpha_2)^{2s}$ and $(\alpha_1\alpha_2\alpha_3)^4 = 1$.

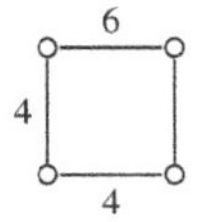

Now if we take the index 2 subgroup of G, $\langle \alpha_1^{\alpha_0}, \alpha_1, \alpha_2, \alpha_3 \rangle$, we obtain a nonlinear hexagonal extention of the tetrahedron, with residues $\{3, 6\}_{(2,0)}$ and $\{3, 6\}_{(s,0)}$ and order $48s^3$. In summary, these hypertopes are constructed from $\{4, 3, 4\}_{(s,s,0)}$ using a Petrie operation and then doubling the fundamental region of the Petrial.

It is interesting to see how we can obtain a permutation representation of the group of these locally toroidal hypertopes combining the permutation representation graphs of $\{3, 6\}_{(2,0)}$ and $\{3, 6\}_{(s,0)}$. For a better understanding about permutation representation graphs of polytopes, called CPR graphs see [27]. Let us first consider the case s even. We claim that the permutation representation graph of $\{3, 6\}_{(s,0)}$ is

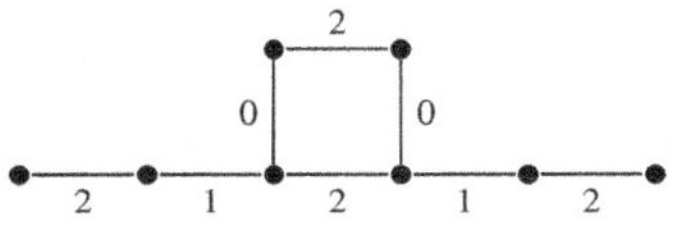

when $s = 2$,

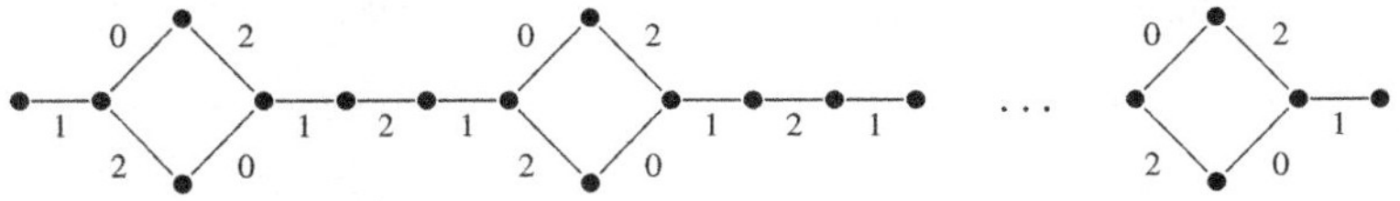

when s is even and $s \geq 4$, where the number of alternating squares of the permutation representation graph is $s/2$. Let $\alpha = \rho_0\rho_1\rho_2\rho_1\rho_2\rho_1$. To show that $\alpha^s = 1$ observe that α acts as a translation on the vertices of the permutation representation graph such that α^s fixes all vertices of the permutation representation graph (see the following figure where $s = 6$).

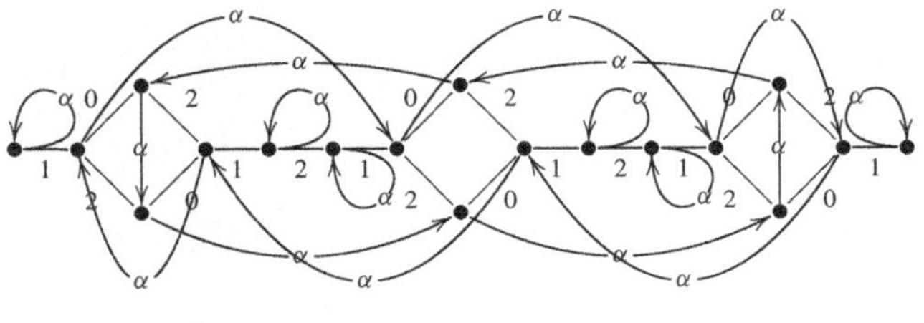

The size of the automorphism group of $\{3, 6\}_{(s,0)}$ is $12s^2$. Let us prove that $12s^2$ is also the size of $\langle \rho_0, \rho_1, \rho_2 \rangle$. Consider the first point x on the left of the permutation representation graph above. The group generated by α, ρ_0 and ρ_2 is in the stabiliser of x and has order $4s$ (as $\alpha^{\rho_0} = \alpha^{-1}$, $\alpha^{\rho_2} = \alpha$ and α_0 commute with α_2). The permutation representation graph has $3s$ vertices and $\langle \rho_0, \rho_1, \rho_2 \rangle$ acts transitively on it. Hence $|\langle \rho_0, \rho_1, \rho_2 \rangle| \geq 4s \cdot 3s = 12s^2$. Hence the graph above is a permutation representation graph of the automorphism group of $\{3, 6\}_{(s,0)}$.

When s is odd it can be shown that the permutation representation graph of $\{3, 6\}_{(s,0)}$ is as follows.

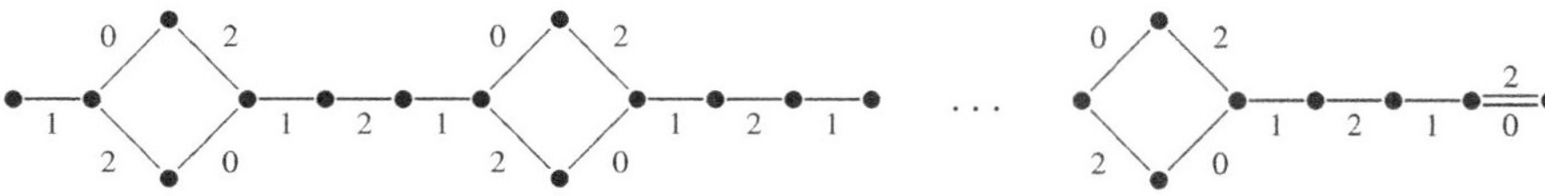

The number of alternating squares of the permutation representation graph is $\frac{s-1}{2}$. The proof that this graph is a permutation representation graph of the automorphism group of $\{3, 6\}_{(s,0)}$ when s is odd is similar to the proof in the case s even.

To obtain the permutation representation graph of the infinite family of locally toroidal hypertopes with residues $\{3, 6\}_{(2,0)}$ and $\{3, 6\}_{(s,0)}$ we combine the respective permutation representation graphs and we obtain the following graphs when $s \geq 3$ accordingly as if s is even or odd.

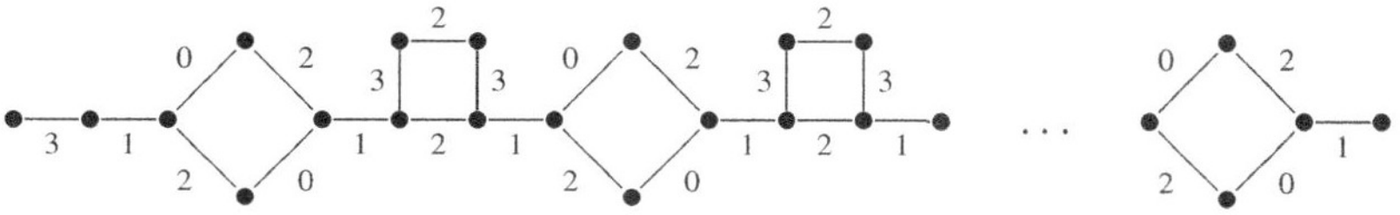

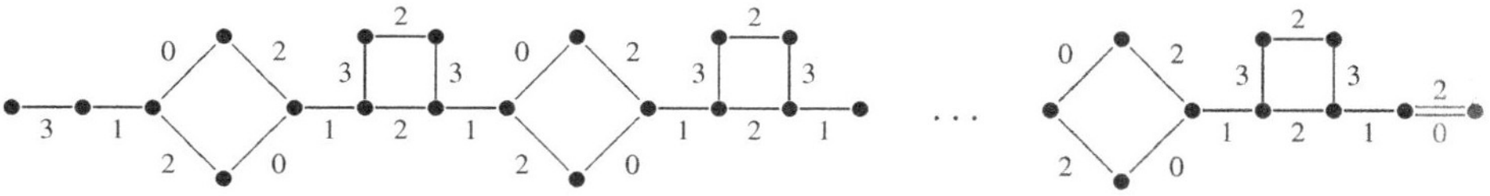

To prove that this is the group G generated by the permutations ρ_0, ρ_1, ρ_2 and ρ_3 and that this group has order $48s^3$, consider the vertex x on the left in each of the permutation representation graphs. The stabilizer of x contains the group generated by ρ_0, ρ_1 and ρ_2, of size $12s^2$, and G is transitive on the $4s$ vertices of the graph. Thus $|G| \geq 48s^3$.

Observe that lines 6 and 7 of Table 6 might suggest there could be a similar infinite family of regular locally toroidal hypertopes with diagram having toroidal rank 3 residues $\{3, 6\}_{(1,1)}$ and $\{3, 6\}_{(s,0)}$ with $s \equiv 0 \mod 3$. A check with MAGMA shows that the same groups appear in that case for $s \equiv 0 \mod 6$ and for $s \equiv 3 \mod 6$ with $s < 100$.

8 4-Circuits with Hexagonal Residues

We now consider a hypertope $\mathcal{H}$ with the following B-diagram, where x, y and z are rotations generating the corresponding C^+-group and $p = 3, 4, 5, 6$.

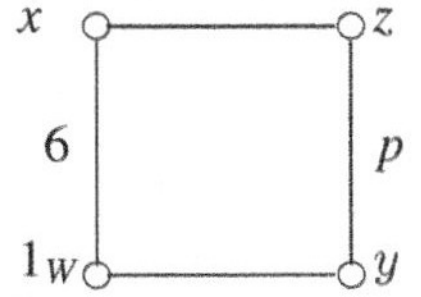

In $\mathcal{H}$ the two type $\{3, 6\}$ residues may be nonisomorphic. Thus when $p = 3, 4$ or 5, we need four parameters a, b, c, d giving two additional relations in the following presentation for the the rotation subgroup of $G := \mathrm{Aut}(\mathcal{H})$. The automorphism groups of the hypertopes of Table 7 are obtained using the following presentation where $\mathbf{s} = (a, b)$ and $\mathbf{t} = (c, d)$.

$$G^+(a,b,c,d) := \langle x, y, z | x^6, y^3, z^2, (zx)^3, (zy)^p, (x^{-1}y)^2,$$

$$(y^{-1}x^{-2})^a(yx^2)^b, (zx^3)^c(x^{-1}zx^{-2})^d\rangle.$$

Observe that no chiral universal hypertope was found for $p = 4$ with $\mathbf{s}, \mathbf{t} \in \{\{s, 0\}, \{s, s\}, \{s, t\} | s, t \in \{1, \ldots, 6\}\}$. Observe also that no universal hypertope (either regular or chiral) was found for $p = 5$ with $\mathbf{s}, \mathbf{t} \in \{\{s, 0\}, \{s, s\}, \{s, t\} | s, t \in \{1, \ldots, 6\}\}$.

When $p = 6$, we need eight parameters a, b, c, d, e, f, g, h giving four additional relations in the following presentation for the the rotation subgroup of $G := \mathrm{Aut}(\mathcal{H})$. The automorphism groups of the hypertopes of Table 8 are obtained using the following presentation where $\mathbf{s} = (a, b)$, $\mathbf{t} = (c, d)$, $\mathbf{u} = (e, f)$ and $\mathbf{v} = (g, h)$.

$$G^+(a,b,c,d,e,f,g,h) := \langle x, y, z | x^6, y^3, z^2, (zx)^3, (zy)^6, (x^{-1}y)^2,$$

$$(y^{-1}x^{-2})^a(yx^2)^b, (zx^3)^c(x^{-1}zx^{-2})^d, (x^{-1}z(y^{-1}z)^2)^e(zx(zy)^2)^f, (y(zy)^2)^g(y^{-1}(yz)^2)^h\rangle.$$

Table 7 Known finite universal hypertopes of type (3) in Fig. 1 with $p = 3, 4$, $\mathbf{s} = (a, b)$, $\mathbf{t} = (c, d)$

p	s	t	#G	G	
3	(3, 0)	(1, 1)	360	$A_5 : S_3$	Regular
	(6, 0)	(1, 1)	23040	$2^4 : A_5 : 2 : 2 : S_3$	Regular
	(1, 2)	(1, 1)	1512	$L_2(8) : 3$	Chiral
	(1, 4)	(1, 1)	90720	$L_2(8) \times A_5 : 3$	Chiral
	(2, 4)	(1, 1)	774144	$2^9 \cdot L_2(8) : 3$	Chiral
4	(2, 0)	(1, 1)	4320	$A_6 : 2 \times S_3$	Regular
	(1, 1)	(1, 1)	3456	$2^{1+4} : 3 : S_3 : S_3$	Regular

Table 8 Known finite universal hypertopes of type (3) in Fig. 1 with $p = 6$, $\mathbf{s} = (a, b)$, $\mathbf{t} = (c, d)$, $\mathbf{u} = (e, f)$, $\mathbf{v} = (g, h)$

s	t	u	v	#G	G	
(2, 0)	(1, 1)	(s, s)	(2, 0)	$1152s^4$	$s^4 : 2^{1+4} : S_3 : S_3$	Regular
(2, 0)	(2, 0)	(1, 1)	(3, 0)	720	$A_5 \times 2 : S_3$	Regular
(2, 0)	(2, 0)	(1, 1)	(6, 0)	46,080	$2^4 : A_5 : 2 \times 2 : 2 : S_3$	Regular
(2, 0)	(1, 1)	(4, 0)	(3, 0)	165,888	$2^{1+6} : S_3 : S_3 : S_3 : S_3$	Regular
(3, 0)	(2, 0)	(1, 1)	(2, 0)	1296	$S_3 \times S_3 \times S_3 : S_3$	Regular
(3, 0)	(2, 0)	(1, 1)	(3, 0)	13,824	$2^{1+6} : 3 : S_3 : S_3$	Regular
(3, 0)	(2, 0)	(1, 1)	(4, 0)	165,888	$2^{1+6} : S_3 : S_3 : S_3 : S_3$	Regular
(3, 0)	(2, 0)	(1, 1)	(5, 0)	5,184,000	$2 \cdot (A_5 \times (A_5 \times A_5)) : 2 : S_3$	Regular
(2, 0)	(1, 2)	(1, 1)	(6, 0)	15,120	$3 \times S_7$	Chiral
(2, 0)	(1, 2)	(1, 2)	(2, 0)	352,800	$L_2(49) \times 3 : 2$	Chiral
(2, 0)	(1, 2)	(1, 1)	(0, 3)	7,620,480	$S_7 \times (L_2(8) : 3)$	Chiral

We give several universal hypertopes with the help of MAGMA. These let us conjecture that two infinite families of finite locally toroidal hypertopes arise.

For each integer $s \geq 1$, the quotient T_s of the Coxeter group [3, 3, 4, 3] with diagram

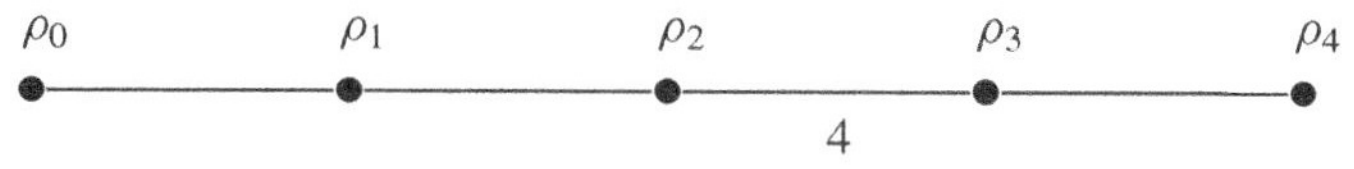

and additional relations

$$(\rho_0\rho_1(\rho_2\rho_3\rho_4)^3)^{2s} = 1_W$$

is the automorphism group of the rank 5 toroid $\{3, 3, 4, 3\}_{(s,0,0,0)}$ (see [23, Sect. 6E]). The subgroup of T_s generated by

$$\tau_0 := \rho_1,\ \tau_1 := \rho_2,\ \tau_2 := \rho_1\rho_3,\ \tau_3 := \rho_0\rho_4$$

is the automorphism group of the regular hypertope with diagram

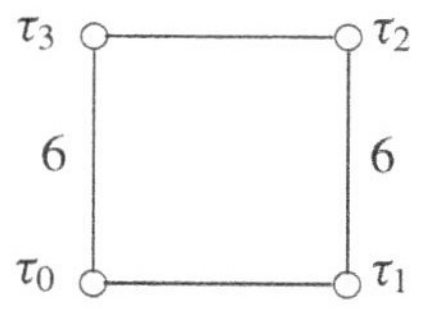

In fact, $\rho_4 = (\tau_0\tau_3)^3$ and therefore $\rho_0 = (\tau_0\tau_3)^2\tau_0$ yielding that $T_s = \langle \tau_0, \tau_1, \tau_2, \tau_3 \rangle$. This suggests a correspondence between $\{3, 3, 4, 3\}_{(s,0,0,0)}$ and the infinite family of finite hypertopes mentioned in the first line of Table 8.

Lines 5 to 8 of Table 8 also suggest the existence of an infinite family of finite hypertopes but we are unable to conjecture what will be the size of the automorphism group and what will be its structure. This is Problem 9.2 included in the next section.

9 Future Work and Open Problems

The basic theory of highly symmetric hypertopes was recently established in [14] but very few universal hypertopes were given. This paper, in a way, supplements it with numerous particularly interesting universal hypertopes. In each case, given a B-diagram and preassigned residues we establish the existence of the corresponding universal hypertope by checking the conditions established in [14]. Extensions of regular or chiral maps $\{4, 4\}_{(b,c)}$ will give a hypertope whose residues are not all either spherical or toroidal. For instance, $\{\{4, 4\}_{(4,0)}, \{4, 6\}_3\}$ gives a finite group of order 768. The $\{4, 6\}_3$ is non-orientable of genus 4. We decide not to study this case here but this case is definitely interesting for future research. Other similar hexagonal extensions include [3, 6, 6] and a star diagram with labels (4, 4, 6).

We conclude this paper with some open problems and conjectures.

Problem 9.1 Can Theorem 2E17 of [23] be generalised to regular and chiral hypertopes?

Problem 9.2 Determine whether or not lines 5 to 8 of Table 8 are part of an infinite family of hypertopes with $(\mathbf{s}, \mathbf{t}, \mathbf{u}, \mathbf{v}) = ((3, 0), (2, 0), (1, 1), (s, 0))$ with $s \geq 2$ an integer.

Conjecture 9.1 *Table 3 gives a complete list of finite universal chiral polytopes of type* $\{6, 3, 6\}$.

Conjecture 9.2 *There are no finite universal chiral hypertopes with the following diagram.*

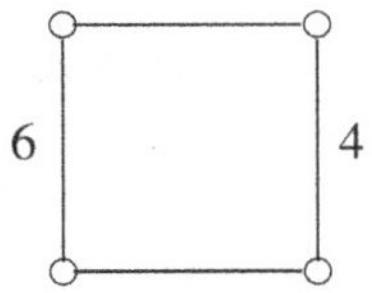

Conjecture 9.3 *There is no finite universal regular or chiral hypertope with the following diagram.*

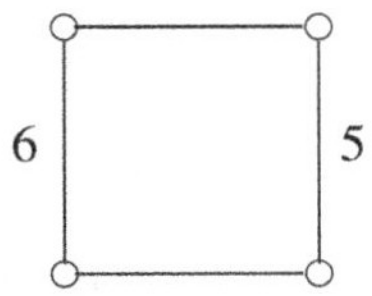

Acknowledgements The authors would like to thank two anonymous referees for their numerous helpful and insightful comments. This research was supported by a Marsden grant (UOA1218) of the Royal Society of New Zealand, by NSERC and by the Portuguese Foundation for Science and Technology (FCT-Fundação para a Ciência e a Tecnologia), through CIDMA - Center for Research and Development in Mathematics and Applications, within project UID/MAT/04106/2013.

References

1. F. Buekenhout, A.M. Cohen, *Diagram Geometry. Related to classical groups and buildings.* Ergebnisse der Mathematik und ihrer Grenzgebiete. 3. Folge. A Series of Modern Surveys in Mathematics [Results in Mathematics and Related Areas. 3rd Series. A Series of Modern Surveys in Mathematics], vol. 57 (Springer, Heidelberg, 2013), xiv+592 pp
2. J. Charles, Colbourn and Asia Ivić Weiss. A census of regular 3-polystroma arising from honeycombs. Discrete Math. **50**, 29–36 (1984)
3. M. Conder, P. Dobcsányi, Determination of all regular maps of small genus. J. Combin. Theory Ser. B **81**, 224–242 (2001)
4. M. Conder, I. Hubard, T. Pisanski, Constructions for chiral polytopes. J. Lond. Math. Soc. **2**(77), 115–129 (2008)
5. M. Conder's, website. http://www.math.auckland.ac.nz/~conder/. Last accessed 21 Sept 2014
6. J.H. Conway, R.T. Curtis, S.P. Norton, R.A. Parker, R.A. Wilson, *Atlas of Finite Groups* (Oxford U.P., Oxford, 1985)
7. H.S.M. Coxeter, Configurations and maps. Rep. Math. Colloq. **8**, 18–38 (1948)
8. H.S.M. Coxeter, *Twisted Honeycombs*, in Regional Conf. Ser. In Math. (American Mathematical Society, Providence, RI, 1970)
9. H.S.M. Coxeter, W.O.J. Moser, *Generators and Relations for Discrete Groups*, volume 14 of *Ergebnisse der Mathematik und ihrer Grenzgebiete [Results in Mathematics and Related Areas]*, 4th edn. (Springer, Berlin, 1980)
10. H.S.M. Coxeter, G.C. Shephard, Regular 3-complexes with toroidal cells. J. Comb. Theory Ser. B **22**(2), 131–138 (1977)
11. L. Danzer, E. Schulte, Regulare Inzidenzkomplexe I. Geom. Dedicata **13**, 295–308 (1982)
12. M. Dehon, Classifying geometries with Cayley. J. Symbolic Comput. **17**(3), 259–276 (1994)
13. D. Elliott, D. Leemans, E. O'Reilly-Regueiro, Chiral polyhedra and projective lines. Int. J. Math. Stat. **16**(1), 45–52 (2015)
14. M.E. Fernandes, D. Leemans, A.I. Weiss, Highly symmetric hypertopes. Aequationes Math. **90**, 1045–1067 (2016)
15. D. Garbe, Über die Regularen Zerlegungen geschlossener orientierbarer Flächen. J. Reine. Angew. Math. **237**, 39–55 (1969)
16. B. Grunbaum, *Regularity of graphs, complexes and designs*, in Problèmes Combinatoires et Théorie des Graphes, Coll. Internat. CNRS, No. 260, (Orsay, 1977), pp. 191–197
17. M.I. Hartley, I. Hubard, D. Leemans, Two atlases of abstract chiral polytopes for small groups. Ars Math. Contemp. **5**, 371–382 (2012)
18. L. Heffter, Ueber Metacyklische Gruppen und Nachbarcnfigurationem. Math. Ann. **50**, 261–268 (1898)
19. I. Hubard, D. Leemans, Chiral polytopes and Suzuki simple groups, in *Rigidity and Symmetry* vol 70, ed. by R. Connelly et al. (Fields Institute Communications, 2014), pp. 155–175
20. D. Leemans, J. Moerenhout, Chiral polyhedra arising from almost simple groups with socle $PSL(2,q)$. J. Algebraic Combin. **44**, 293–323 (2016)
21. D. Leemans, L. Vauthier, An atlas of abstract regular polytopes for small groups. Aequationes Math. **72**, 313–320 (2006)
22. P. McMullen, E. Schulte, Locally toroidal regular polytopes of rank 4. Comment. Math. Helv. **67**(1), 77–118 (1992)

23. P. McMullen, E. Schulte, in *Abstract Regular Polytopes*, vol. 92, Encyclopedia of Mathematics and its Applications (Cambridge University Press, Cambridge, 2002)
24. P. McMullen, E. Schulte, Locally unitary groups and regular polytopes. Adv. Appl. Math. **29**(1), 1–45 (2002)
25. B. Monson, A.I. Weiss, Eisenstein integers and related C-groups. Geom. Dedicata **66**, 99–117 (1997)
26. B. Nostrand, E. Schulte, Chiral polytopes from hyperbolic honeycombs. Discrete Comput. Geom. **13**, 17–39 (1995)
27. D. Pellicer, CPR graphs and regular polytopes. European J. Combin. **29**(1), 59–71 (2008)
28. D. Pellicer, A construction of higher rank chiral polytopes. Discrete Math. **310**, 1222–1237 (2010)
29. E. Schulte, A.I. Weiss, Chiral polytopes, in *Applied Geometry and Discrete Mathematics*, volume 4 of *DIMACS Ser. Discrete Math. Theoret. Comput. Sci.* (American Mathematical Society, Providence, RI, 1991), pp. 493–516
30. E. Schulte, A.I. Weiss, Chirality and projective linear groups. Discrete Math. **131**, 221–261 (1994)
31. E. Schulte, A.I. Weiss, Free extensions of chiral polytopes. Canad. J. Math. **47**, 641–654 (1995)
32. F.A. Sherk, A family of regular maps of type {6, 6}. Canad. Math. Bull. **5**, 13–20 (1962)
33. D.M.Y. Sommerville, *An Introduction to the Geometry of n Dimensions* (Methuen, London, 1929)
34. J. Tits, Sur les analogues algébriques des groupes semi-simples complexes. Colloque d'algébre supérieure, tenu á Bruxelles du 19 au 22 décembre, *Centre Belge de Recherches Mathématiques*, pp. 261–289, *Établissements Ceuterick* (Louvain, Librairie Gauthier-Villars, Paris, 1956), p. 1957
35. T.R.S. Walsh, Hypermaps versus bipartite maps. J. Combinatorial Theory Ser. B **18**(2), 155–163 (1975)
36. C. Weber, H. Seifert, Die beiden Dodekaederräume. Mat. Z. **37**, 237–253 (1933)

Noncongruent Equidissections of the Plane

D. Frettlöh

Abstract Nandakumar asked whether there is a tiling of the plane by pairwise non-congruent triangles of equal area and equal perimeter. Here a weaker result is obtained: there is a tiling of the plane by pairwise non-congruent triangles of equal area such that their perimeter is bounded by some common constant. Several variants of the problem are stated, some of them are answered.

Keywords Tilings · Equipartitions

1 Introduction

There are several problems in Discrete Geometry, old and new, that can be stated easily but are hard to solve. Tilings and dissections provide a large number of such problems, see for instance [1, Chapter C]. On his blog [3], R. Nandakumar asked in 2014:

Question 1 "Can the plane be filled by triangles of same area and perimeter with no two triangles congruent to each other?"

His webpage [3] contains several further interesting problems of this flavour. The main result of this paper, Theorem 2, answers a weaker form of the question above affirmatively. Section 4 is dedicated to the statement and the proof of this result, together with its analogues for quadrangles, pentagons and hexagons. Section 3 formulates several variants of this problem and gives a systematic overview. Section 2 contains some basic observations and a first result on a similar result for quadrangles.

D. Frettlöh (✉)
Bielefeld University, Postfach 100131, 33501 Bielefeld, Germany
e-mail: dfrettloeh@techfak.uni-bielefeld.de

M. D. E. Conder et al. (eds.), *Discrete Geometry and Symmetry*,
Springer Proceedings in Mathematics & Statistics 234,
https://doi.org/10.1007/978-3-319-78434-2_9

1.1 Notation

A *tiling* of $\mathbb{R}^2$ is a collection $\{T_1, T_2, \ldots\}$ of compact sets T_i (the *tiles*) that is a packing of $\mathbb{R}^2$ (i.e., the interiors of distinct tiles are disjoint) as well as a covering of $\mathbb{R}^2$ (i.e. the union of the tiles equals $\mathbb{R}^2$). In general, tile shapes may be pretty complicated, but for the purpose of this paper tiles are always simple convex polygons. A tiling is called *vertex-to-vertex*, if the intersection of any two tiles is either an entire edge of both tiles, or a point, or empty. A tiling $\mathcal{T}$ is *locally finite* if any compact set in $\mathbb{R}^2$ intersects only finitely many tiles of $\mathcal{T}$. A tiling $\mathcal{T}$ is *normal* if there are $R > r > 0$ such that (1) each tile in $\mathcal{T}$ contains some ball of radius r, and (2) each tile is contained in some ball of radius R. By [2, 3.2.1] we have that each normal tiling is locally finite.

2 Basic Observations

In Question 1 above, "filled" is to be understood in the sense that the plane is covered without overlaps. In other words: is there a tiling of the plane by pairwise noncongruent triangles all having the same area and the same perimeter? If one tries to find a solution one realises that the problem seems to be highly overdetermined. One possibility to relax the problem is to drop the requirement on the perimeter. So one may ask "Is there a tiling of the plane by pairwise noncongruent triangles all having the same area?" It is not too hard to find examples. One possibility is to partition the plane into half-strips and divide these half-strips into triangles, as shown in Fig. 1.

The image indicates how to fill the right half-plane by half-strips made of triangles. By choosing the widths of the half-strips appropriately all triangles can be made distinct. In particular, the first half-strip contains countably many distinct triangles, as well as the second one. For the width of the second half-strip one has uncountably many possibilities to choose from. Hence there is a width of the second half-strip that

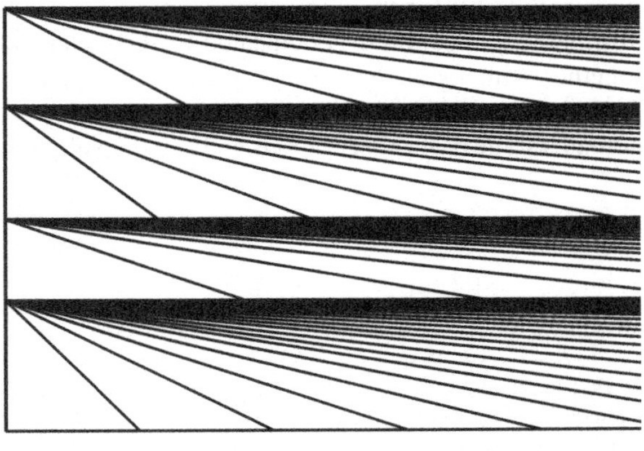

Fig. 1 Tiling of the plane by pairwise non-congruent triangles of unit area. The perimeters of the triangles are unbounded. Moreover, the tiling is not locally finite

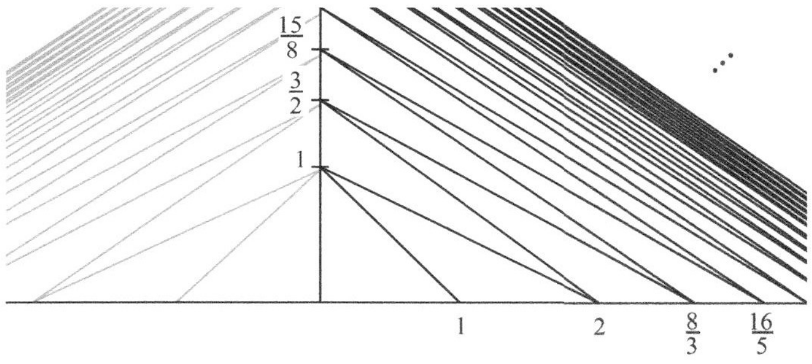

Fig. 2 Tiling of the plane by pairwise non-congruent triangles of area $\frac{1}{2}$

avoids that among the countably many triangles of the first and the second half-strip two triangles are congruent. In the same manner one can choose the width of the third, fourth, ...n-th ...half-strip avoiding that two congruent triangles occur.

The left half-plane can be filled in an analogous manner. Anyway, this tiling is not locally finite: the upper vertex of any half-strip is contained in infinitely many triangles. So one may ask "Is there a locally finite tiling of the plane by pairwise noncongruent triangles of unit area?" Even in this stronger form the question was already answered by R. Nandakumar. The image in Fig. 2 shows a solution, see also [4]. The idea is to dissect the upper right quadrant into triangles of area $\frac{1}{2}$ by zigzagging between the horizontal axis and the vertical axis. (Here we prefer to use area $\frac{1}{2}$ rather than one, because the particular coordinates given in Fig. 2 become nicer. For "area 1" just multiply everything with $\sqrt{2}$) The triangles become very long and thin soon. Nevertheless they are filling the quadrant. For the remaining three quadrants one uses an analogous construction, perturbing the coordinates slightly. (For instance, stretch a copy of the first quadrant by some irrational factor $q > 1$ in the horizontal direction and shrink it by $\frac{1}{q}$ in the vertical direction. See [4] for an alternative, more detailed explanation.) This tiling *is* locally finite. Nevertheless, this example is not really satisfying. More precisely, this tiling is not normal, since in this solution the perimeters of the triangles become arbitrary large. (Hence the inradii become arbitrary small). So it seems natural to ask:

Question 2 "Is there a normal tiling of the plane by pairwise noncongruent triangles of unit area?"

This question was already asked by Nandakumar, in the form whether there is a tiling of the plane by pairwise noncongruent triangles all having unit area such that the perimeter of the triangles is bounded by some common constant. Theorem 2 below answers this question affirmatively.

One possible approach to find a solution is the following. If one can partition a set $S \subset \mathbb{R}^2$ into triangles of unit area such that (1) S tiles the plane, and (2) all triangles in S can be distorted continuously, in a way such that any two triangles in S are distinct (but still having unit area), this solves the problem. We will illustrate this

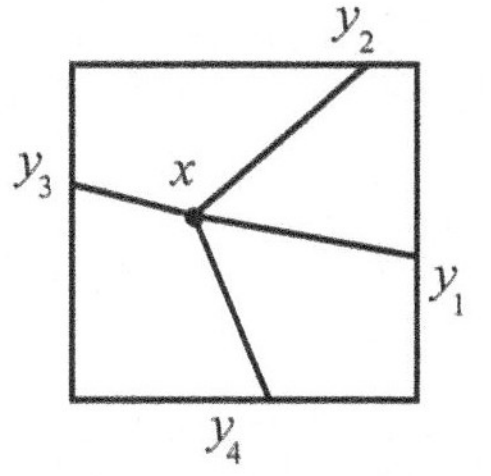

Fig. 3 Partitioning a square into 4 distinct quadrangles of equal area. There are uncountably many possibilities for such a dissection

concept (where "triangle" is replaced by "quadrangle") in the proof of the following result.

Theorem 1 *There is a normal tiling of the plane by pairwise non-congruent convex quadrangles of unit area.*

Proof Consider a square Q of edge length 2. Let $x \in Q$ be a point such that (1) for the distance d of x to the centre of Q holds $0 < d < \frac{1}{10}$, and (2) x is neither contained in the diagonals of Q nor in the line segments connecting mid-points of opposite edges of Q. Let y_1 be a point on the boundary of Q such that for the distance d' of y to the mid-point of the edge containing y_1 holds $0 < d' < \frac{1}{10}$.

The choice of x and y_1 determines three further unique points y_2, y_3, y_4 on the boundary of Q such that the line segments xy_i $(1 \leq i \leq 4)$ partition Q into four quadrangles of unit area. By the choice of x, avoiding the mirror axes of Q, it can always be achieved that all quadrangles in a single partition of Q are distinct. (In fact the author believes that no two congruent quadrangles can occur in a partition where x is not contained in the mirror axes of Q, but this might be tedious to prove. Here we prefer rather to use the free parameters to achieve that all quadrangles are distinct.) Fig. 3 indicates such a partition.

The two coordinates determining x can be changed continuously within a small range independently, yielding two free parameters. One coordinate of y_1 can be changed continuously, too, within some small range. Hence we obtain the desired tiling as follows: Tile the plane $\mathbb{R}^2$ with copies of the square Q. Dissect each copy of Q into four quadrangles of area 1, in some order. In each dissection, choose x and y_1 such that the resulting quadrangles have not shown up earlier in the construction. This is always possible since, in each step, there are only finitely many quadrangles constructed already, whereas there are uncountably many choices for x and y_1.

At this point it becomes obvious that Questions 1 and 2 lead to several variants. The example in the proof above yields a normal tiling, but in general not one that is vertex-to-vertex. One may ask the questions for triangles, for quadrangles, for pentagons, and in each case with or without requiring "equal perimeter", or "normal", or "vertex-to-vertex". The next section aims to give a systematic overview of the questions.

3 Variants of the Problem

The general property we will require throughout the paper is that a tiling consists of convex tiles of unit area such that all tiles are pairwise non-congruent. The tiles can be triangles (as in the original question), but also quadrangles, rectangles, pentagons or hexagons. We may or may not require additionally that all tiles have *equal perimeter*, or that the *perimeter is bounded* by some common constant, or that the tilings are *normal*, or just *locally finite*. Furthermore, it may be possible to construct a tiling analogous to the proof of Theorem 1, that is, by *tiling a tile* S in infinitely many ways, where S in turn can tile the plane. The connections between these properties is shown in the following diagram.

$$\left.\begin{array}{ll}\text{equal perimeter} & \Rightarrow \\ \text{tiling a tile} & \Rightarrow\end{array}\right\} \text{perimeter is bounded} \;\Leftrightarrow\; \text{normal} \;\Rightarrow\; \text{locally finite} \qquad (1)$$

For instance, Eq. (1) tells that if there is tiling obtained by tiling a tile S in infinitely many ways, then the perimeters of the tiles in this tiling are bounded by some common constant. In turn, the latter is equivalent to the tiling being normal (since all tiles are convex and have unit area), which in turn implies (by [2, 3.2.1]) that the tiling is locally finite.

These implications help to give an overview of the several variants of the questions. The following tables list, for each of the cases of triangles, convex quadrangles, convex pentagons, and convex hexagons, whether there is some tiling known fulfilling the properties in Eq. (1), and whether there is such a tiling that is even vertex-to-vertex. Because of the implications in Eq. (1), if there is "yes" in some column, then the entries above in the same column contain also a "yes".

Note that "not vtv" is usually a weaker condition than "vtv", but a tiling by convex hexagons that is *not* vtv is much harder to find than one that is vtv.

Table 1 Several variants of the problem of noncongruent equidissections. Entries with "yes" or "no" are proven in the text, entries with "?" are still open

Triangles	vtv	Not vtv	Quadrangles	vtv	Not vtv
Locally finite	?	Yes	Locally finite	Yes	Yes
Bounded perimeter	?	Yes	Bounded perimeter	Yes	Yes
Tiling a tile	?	?	Tiling a tile	Yes	Yes
Equal perimeter	No	?	Equal perimeter	?	?
Pentagons	vtv	Not vtv	Hexagons	vtv	Not vtv
Locally finite	?	Yes	Locally finite	Yes	?
Bounded perimeter	?	Yes	Bounded perimeter	Yes	?
Tiling a tile	?	Yes	Tiling a tile	Yes	?
Equal perimeter	?	?	Tiling a tile	Yes	?

Theorem 1 proves the case "quadrangles: tiling a tile, not vtv" (thus also the two cases above it in the same column in the corresponding table). Theorem 2 proves "triangles: bounded perimeter, not vtv", Theorem 3 proves "quadrangles: tiling a tile, vtv", Theorem 4 proves "hexagons: tiling a tile, vtv", and Theorem 5 proves "hexagons: tiling a tile, vtv".

The "no" in the table is due to the following observation: Given a fixed area and perimeter, then for each possible edge length there is at most one congruence class of triangle with that area, perimeter, and edge-length. In a vtv tiling two adjacent triangles share a common edge, hence have the same edge-length. Thus these two triangles are already congruent to each other. Even if we distinguish triangles if they are not directly congruent, but are mirror images of each other, two out of three triangles in the tiling need to be directly congruent.

4 Main Results

Theorem 2 *There is a normal tiling of the plane by pairwise non-congruent triangles of unit area.*

Proof The idea of the proof is a refinement of the construction in Fig. 2. Basically we add additional fault lines in each quadrant. Moreover, we make use of some free parameter in some range, allowing for uncountably many choices, where in each step of the construction only finitely many triangle shapes must be avoided.

Choose some constant c big enough. This serves as the upper bound on the perimeter of the triangles. For our purposes $c = 100$ will do. Consider the upper right quadrant Q_1. Pick a point x_0 on the positive horizontal axis with $|x_0| < \frac{c}{3}$. Let T_1 be the unique triangle in Q_1 with unit area, vertices 0, x_0 and the third vertex y_1 being on the y-axis. (For this and what follows compare Fig. 4.) Denote the third vertex of T_1 by y_1. Choose y_2 on the horizontal axis such that the triangle T_2 with vertices x_0, y_1 and y_2 has area 1. Continue zigzagging in this way between horizontal and vertical axis. i.e., choose y_{i+1} on the axis not containing y_i such that the triangle T_{i+1} with vertices y_{i-1}, y_i, y_{i+1} has area 1. Repeat this until the next triangle T_{i+2} would have

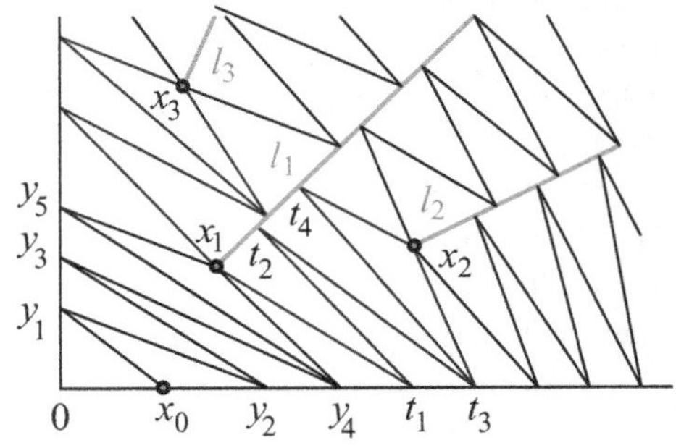

Fig. 4 Tiling the upper right quadrant Q_1 by pairwise non-congruent triangles of unit area and bounded perimeter

perimeter larger than c. Omit T_{i+2}. Pick x_1 such that the triangle y_i, y_{i+1}, x_1 has area 1.

There are uncountably many choices for x_1. For the sake of symmetry let x_1 be close to the bisector $\{(x, x) \mid x \in \mathbb{R}\}$ of Q_1. Choose a half-line ℓ_1 proceeding from x_1. There are uncountably many choices for ℓ_1. Again, for the sake of symmetry, let ℓ_1 be close to the bisector of Q_1. Continue by zigzagging in two regions, between the horizontal axis and ℓ_1, and between the vertical axis and ℓ_1. i.e., if y_{2k} is the last point on the horizontal axis, pick t_1 on the horizontal axis such that the triangle y_{2k}, x_1, t_1 has area 1. Continue by choosing t_2 on ℓ_1 such that the triangle x_1, t_1, t_2 has area 1 and so on, until the perimeter of the next triangle t_i, t_{i+1}, t_{i+2} would be larger than c. Omit this triangle. Choose x_2 such that the new triangle t_i, t_{i+1}, x_2 has area 1. Choose a half-line ℓ_2 proceeding from x_2. Again there are uncountably many choices for x_2 and ℓ_2.

Do the analogous construction in the upper region between ℓ_1 and the vertical axis. Continue in this manner. Whenever a triangle occurs with perimeter larger than c choose a new point x_k and a new line ℓ_k dividing the old region into two.

The uncountability of choices for x_k and ℓ_k ensures that we can always avoid adding a triangle that is congruent to some triangle added earlier. Indeed, whenever we are in the situation of choosing x_k and ℓ_k there are at most countably many triangles constructed already. Hence x_k can be chosen such that no triangle with vertex x_k is congruent to an already constructed one, and ℓ_k can be chosen such that no triangle occurring in the two new regions defined by ℓ_k is congruent to an already constructed one. Hence the quadrant Q_1 can be tiled by pairwise non-congruent triangles with area 1 and perimeter less than c.

The other quadrants can be tiled accordingly. Whenever a choice of a new point and a new half-line happens there are uncountably many possibilities, hence all (at most countably many) already constructed triangles can be avoided.

Theorem 3 *There is a normal vtv tiling of the plane by pairwise non-congruent quadrangles of unit area. The tiling consists of squares that are dissected into four distinct quadrangles of equal area.*

Proof The idea is to use the construction in the proof of Theorem 1, adding (dissected) squares consecutively, using the degrees of freedom to achieve vertex-to-vertex in neighbouring squares. Fig. 5 indicates the order in which squares are added, and the degrees of freedom in the dissection of each square.

Start with some square S, dissected as in the proof of Theorem 1. There are three degrees of freedom how to dissect S into four quadrangles, two for placing the centre of dissection, one for a point on the boundary. This square is indicated by a circled 1 in Fig. 5. Add four more dissected squares adjacent to S, such that the quadrangles are vertex-to-vertex. These squares are numbers 2–5 in the figure. In each of these squares there are still two degrees of freedom for placing the centre. The third parameter is determined uniquely by the vertex-to-vertex condition. Still one may use the two degrees of freedom to avoid adding a quadrangles that is congruent to one added already.

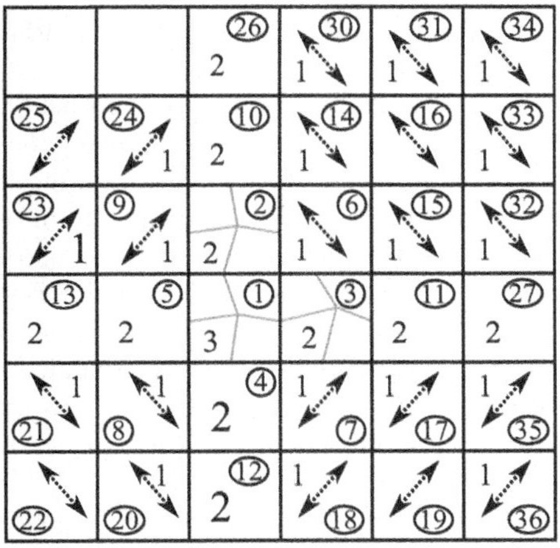

Fig. 5 Tiling a square with distinct quadrangles of unit area (compare Fig. 3) can be done in a way such that partitions of adjacent squares are vertex-to-vertex. Circled numbers indicate the order in which squares are added consecutively, ordinary numbers indicate the degrees of freedom in each square. A 2 means that the centre can be wiggled within a small ball. A 1 means that the centre can be shifted along some line by a small amount. The (approximate) directions of these lines are indicated by dashed line segments

Now add four more squares (6–9), each one adjacent to two edges of squares 2–5, respectively. Now the position of two points of the dissection are determined for each of the squares 6–9. Hence, by the area condition, the centre of the square is restricted to some line. Anyway, it can be shifted along a small segment of this line continuously. Hence there is still one free parameter that we can use to avoid adding a quadrangle that is congruent to some quadrangle added earlier.

In this way we continue filling the plane: add four squares along the horizontal and vertical axes (the next step would be adding squares 10–13 in the figure), add more squares to the pattern to complete a square pattern. Proceeding in this way ensures that in each step there is at least one free parameter that can be used to avoid adding a square congruent to one added earlier.

One last problem to solve is to ensure that the dependencies between the choices of points do not force us to destroy the desired properties, i.e. by pushing the vertices of the small quadrangles out of the squares they belong to. This can be achieved by limiting the choices of coordinates to small enough deviations from the centres of the squares respectively from the midpoints of the edges. The author suspects that it suffices to choose all deviations less than some common small constant. But to be on the safe side one may choose the deviations according to some rapidly decreasing series like 100^{-n}, such that all deviations can add up only to some number much less than one.

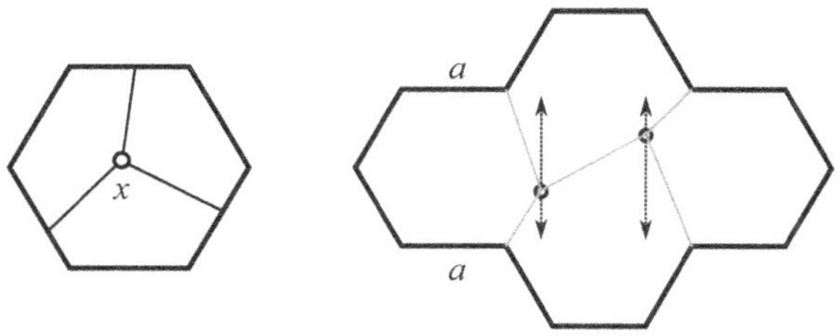

Fig. 6 A regular hexagon can be divided into three distinct pentagons of equal area in uncountably many ways (left). A non-convex 14-gon can be divided into four distinct hexagons of equal area in uncountably many ways

Theorem 4 *There is a normal tiling of the plane by pairwise non-congruent pentagons of unit area. The tiling consists of hexagons that are dissected into three distinct pentagons of equal area.*

Proof A regular hexagon of area three can be divided into three pentagons of unit area in uncountably many ways, compare the left part of Fig. 6.

Theorem 5 *There is a normal vtv tiling of the plane by pairwise non-congruent hexagons of unit area. The tiling consists of non-convex 14-gons that are dissected into four distinct hexagons of equal area.*

Proof Consider a non-convex 14-gon assembled from three regular hexagons and a fourth hexagon that is obtained from a regular hexagon by stretching it slightly in the direction of one of the edges, see Fig. 6 right. The longer edges are labelled with a in the figure. This 14-gon can be dissected into four hexagons of equal area. There is still one parameter of freedom: one vertex of the dissection can be shifted continuously along a line segment, the other interior vertex of the dissection is then determined uniquely by the area condition.

The 14-gons yield a tiling of the plane: gluing 14-gons together at their edges of length a yields biinfinite strips. These strips in turn can be assembled into a tiling.

During working on the problem the author tried several approaches. Based on this experience we want to highlight the following problems for further study.

(1) Is there a compact convex region in the plane that can be tiled by non-congruent triangles of unit area in infinitely many (uncountably many) ways?
(2) Is there a compact region in the plane that (a) can be tiled by non-congruent triangles of unit area in infinitely many (uncountably many) ways, and (b) tiles the plane?
(3) Is there a vertex-to-vertex tiling of the plane by pairwise non-congruent triangles of unit area?
(4) Is there a vertex-to-vertex tiling of the plane by pairwise non-congruent triangles of unit area such that the perimeter of the triangles is bounded by some common constant?

(5) Is there a tiling of the plane by pairwise non-congruent rectangles of unit area such that the perimeter of the rectangles is bounded by some common constant?

Acknowledgements The author expresses his gratitude to R. Nandakumar for providing several interesting problems. Particular thanks to one of the anonymous referees for providing the "no" in Table 1 and for other valuable remarks. Special thanks to Jens Schubert for helpful discussions on this topic during a pleasant summer weekend in Bochum.

References

1. H.T. Croft, K.J. Falconer, R.K. Guy, *Unsolved Problems in Geometry* (Springer, New York, 1991)
2. B. Grünbaum, G.C. Shephard, *Tilings and Patterns* (W.H. Freeman, New York, 1987)
3. R. Nandakumar, http://nandacumar.blogspot.in, post from Dec 2014 and Jan 2015
4. S. Wagon, http://mathforum.org/wagon/2015/p1199.html

Pascal's Triangle of Configurations

Gábor Gévay

Abstract We introduce an infinite class of configurations which we call Desargues–Cayley–Danzer configurations. The term is motivated by the fact that they generalize the classical (10_3) Desargues configuration and Danzer's (35_4) configuration; moreover, their construction goes back to Cayley. We show that these configurations can be arranged in a triangular array which resembles the classical Pascal triangle also in the sense that it can be recursively generated. As an interesting consequence, we show that all these configurations are connected to incidence theorems, like in the classical case of Desargues. We also show that these configurations can be represented not only by points and lines, but points and circles, too.

Keywords Combinatorial configuration · Desargues–Cayley–Danzer configuration · Geometric configuration · Incidence sum · Incidence theorem Point-circle configuration

1 Introduction

A *combinatorial* (or *abstract*) *configuration* of *type* (p_q, n_k) is an incidence structure with sets $\mathcal{P}$ and $\mathcal{B}$ of objects, called *points* and *blocks*, such that the following conditions hold:

(C1) $|\mathcal{P}| = p$;
(C2) $|\mathcal{B}| = n$;
(C3) each point is incident with q blocks;
(C4) each block is incident with k points;
(C5) two distinct points are incident with at most one block.

A *point-line configuration* is a geometric incidence structure consisting of points and (straight) lines, in the simplest case in Euclidean or real projective plane, such that the Conditions (C1)–(C4) hold. Note that in this case (C5) is fulfilled automatically.

G. Gévay (✉)
Bolyai Institute, University of Szeged, Szeged, Hungary
e-mail: gevay@math.u-szeged.hu

M. D. E. Conder et al. (eds.), *Discrete Geometry and Symmetry*,
Springer Proceedings in Mathematics & Statistics 234,
https://doi.org/10.1007/978-3-319-78434-2_10

Instead of plane, the ambient space can be some higher-dimensional (Euclidean, or projective) space as well [5, 9]. Also, the set of lines can be replaced by a set of circles; in this case we speak of a *point-circle configuration*.[1] In the simplest case this latter is also defined in Euclidean plane; however, a more natural context is inversive (or Möbius) geometry [10]). We note that for a point-circle configuration Condition (C5) does not necessarily hold; in case it still holds, we say that the configuration is *lineal* [10, 17].

Given two configurations, $\mathcal{C}_1$ and $\mathcal{C}_2$ with pairs $(\mathcal{P}_1, \mathcal{B}_1)$ and $(\mathcal{P}_2, \mathcal{B}_2)$, respectively, we say that they are *isomorphic* if there is a bijection which sends $\mathcal{P}_1$ to $\mathcal{P}_2$ and $\mathcal{B}_1$ to $\mathcal{B}_2$ such that incidences are preserved. For geometric configurations this is equivalent to saying that the underlying combinatorial configuration is essentially the same; we also say in this case that $\mathcal{C}_1$ and $\mathcal{C}_2$ are different *geometric realizations* of the same abstract configuration. Interesting examples are point-circle realizations of (in some cases, well-known) point-line configurations (see Sect. 6).

If in a configuration of type (p_q, n_k) we have $p = n$, then the equality $q = k$ also holds; in this case the more concise notation (n_k) is used. Formerly, such a configuration was called *symmetric*; however, it is appropriate to reserve this term for a configuration which has non-trivial automorphism, hence, following Grünbaum [12], we use the term *balanced configuration* instead.

Two configurations, $\mathcal{C}$ and $\mathcal{C}^*$ are said to be *dual* to each other if there is an incidence-preserving bijection which sends the set of points of $\mathcal{C}$ to the set of blocks of $\mathcal{C}^*$ and vice versa. A configuration dual to itself is called *self-dual*.

For further definitions and background material concerning configurations in general, the reader is referred to the recent monographs [12, 17].

Consider now seven 3-spaces in general position in the 4-dimensional real projective space. (General position means that no more than four of the 3-spaces meet in a point.) They meet by fours and by threes in 35 points and 35 lines, respectively; thus we obtain a (35_4) configuration in 4-space. By suitable projection, this configuration can be carried over to an isomorphic planar configuration. This construction has been given by Danzer, with the aim of answering a question of Grünbaum [11]; Danzer himself never published it. In fact, it appears in the literature first in the work of Grünbaum and Rigby [13], and later, in Grünbaum's "Musings" [11].

A recent paper by Boben, Gévay and Pisanski [2] is devoted again to Danzer's configuration, as well as to a class of configurations generalizing it. In that paper the authors point out that this (35_4) configuration occurs as early as more than 100 years ago, in the work of the Hungarian mathematician Klug, although in a different context [15]. On the other hand, the construction itself goes back to an earlier period, as it is related to some construction principles due to Cayley. The different context is rooted in Pascal's *Hexagrammum Mysticum*, which, just in the time of Cayley, was a subject of intense study (this topic experienced a revival of interest quite recently, see the papers by Conway and Ryba [3, 4]).

---

[1] We note that in a geometric configuration the blocks can also be other geometric elements, like planes, hyperplanes, spheres, etc. [5, 7, 8]; we shall not consider such examples here.

In the present paper (in Sect. 2) we further generalize the Desargues–Cayley–Danzer configurations introduced in [2]. In Sect. 3 we show that these more general configurations can be arranged in an array like Pascal's triangle. This scheme follows the pattern of Pascal's triangle also in the sense that each (non-terminal) entry is a certain "sum"of the two entries above it (Sect. 4). We also present some applications (Sects. 5 and 6).

2 The Configurations $DCD(n, d)$ and Their Geometric Realization

Denote the n-set $\{1, \ldots, n\}$ by $[n]$, and for any non-negative integer $k \leq n$, denote the set of all k-subsets of $[n]$ by $\binom{[n]}{k}$. Let $n \geq 1$ be an integer, and let d be an integer with $1 \leq d \leq n$. We form the incidence structure

$$\left(\binom{[n]}{d}, \binom{[n]}{d-1}, \supset\right), \tag{1}$$

whose set of points is $\binom{[n]}{d}$, the set of blocks is $\binom{[n]}{d-1}$, and incidences are defined by containment. It is easily seen that this structure is a combinatorial configuration of type

$$\left(\binom{n}{d}_{d}, \binom{n}{d-1}_{n-d+1}\right). \tag{2}$$

We denote this configuration by $DCD(n, d)$.

Proposition 2.1 *The configurations $DCD(n, d)$ and $DCD(n, n-d+1)$ are dual to each other.*

Proof The type of $DCD(n, n-d+1)$ is

$$\left(\binom{n}{n-d+1}_{n-d+1}, \binom{n}{n-d}_{n-(n-d+1)+1}\right) = \left(\binom{n}{d-1}_{n-d+1}, \binom{n}{d}_{d}\right),$$

as required for the dual of $DCD(n, d)$. Now we define the following correspondence δ. For each point P in $DCD(n, d)$, i.e. a d-subset of $[n]$, let $\delta(P)$ be precisely the line of $DCD(n, n-d+1)$ which is an $(n-d)$-subset complementary to P in $[n]$. Likewise, for each line l in $DCD(n, d)$, i.e. a $(d-1)$-subset of $[n]$, let $\delta(l)$ be precisely the point of $DCD(n, n-d+1)$ which is an $(n-d+1)$-subset complementary to l in $[n]$. Furthermore, clearly $P \supset l$ if and only if $\delta(l) \supset \delta(P)$. Thus we see that δ is an incidence-preserving bijection between $DCD(n, d)$ and $DCD(n, n-d+1)$ which interchanges the set of points and the set of blocks. ■

Now we construct a geometric realization of $DCD(n, d)$.

Construction 2.2

(a) For $d = 1$ and for all $n \geq 1$, $DCD(n, d)$ is realized in the real projective plane $\mathbb{P}^2$ by a configuration consisting of a single line and of n points incident with this line. Dually, for all $n \geq 1$, $DCD(n, n)$ is realized in $\mathbb{P}^2$ by a *pencil* with n lines, i.e. by a configuration consisting of a single point and of n lines passing through this point.

(b) Let $n \geq 3$ be an integer, and consider the d-dimensional (real) projective space $\mathbb{P}^d$ with $2 \leq d \leq n - 1$. Choose n hyperplanes in $\mathbb{P}^d$ in general position. (By general position we mean that no more than d hyperplanes meet in a common point.) Now every d hyperplanes of this arrangement meet in a point, and every $d - 1$ hyperplanes meet in a line. Thus we have altogether $\binom{n}{d}$ points and $\binom{n}{d-1}$ lines. These points and lines can be considered as labelled with the corresponding d-element and $(d - 1)$-element subsets, respectively, of the set with n elements we started from. Moreover, the incidence between the points and lines is determined by containment between the corresponding subsets. Hence it is clear that each point is incident with d of the $\binom{n}{d-1}$ lines, and each line is incident with $n - d + 1$ of the $\binom{n}{d}$ points. Thus, by a suitable projection, we obtain a planar point-line configuration of type

$$\left(\binom{n}{d}_{d}, \binom{n}{d-1}_{n-d+1}\right),$$

(cf. formula (2) above). Note that the same type is valid for $d = 1$ and for all $n \geq 1$. Hence we have constructed a geometric point-line realization of $DCD(n, d)$ for all integer $n \geq 1$ and for all integer d with $1 \leq d \leq n$. We denote this realization by $DCD[n, d]$

Definition 2.3 We call the configuration obtained by Construction 2.2(b) a *Desargues–Cayley–Danzer configuration.*

Remark 2.4 Clearly, $DCD[n, d]$ is balanced if and only if $n = 2d - 1$.

We note that in [2] the term Desargues–Cayley–Danzer configuration was applied to a less general notion; namely, for balanced $DCD[n, d]$. In that paper the aim was to find a common generalization of the Desargues' (10_3) and the Danzer's (35_4) configuration, as close as possible to them. Since both of these configurations are balanced, it was natural to choose the generalization balanced as well.

Observe that the only condition imposed on the arrangement of hyperplanes Construction 2.2(b) starts from is that they are in general position. A consequence is that for any arbitrary pair (n, d) with $n > 4$ and $d > 2$, there are infinitely many projectively inequivalent copies of configuration $DCD[n, d]$. Furthermore, note that the same statement is valid for all configurations $DCD[n, 1]$ (and dually, $DCD[n, n]$) with $n > 4$.

We formulate this property in the following definition (cf. Grünbaum [12], Sect. 5.7). We emphasize that this definition refers to point-line configurations realized in the (real) projective plane.

Definition 2.5 We say that a configuration is *rigid* if its geometric realizations form a single class under projective transformations. A configuration that is not rigid is called *movable*.

Hence our observation above can be formulated as follows.

Proposition 2.6 *For all $n > 4$, the configuration $DCD[n, d]$ is movable.* ■

Note that movability of a configuration implies movability of its dual. For an example, we recall that the Desargues configuration $DCD[5, 3]$ is well known to be movable.

We mention a further important property of $DCD[n, d]$. First, recall that an *automorphism* of a combinatorial configuration is an isomorphism to itself [5, 12]. Clearly, the set of all automorphisms of a configuration forms a group. By the automorphism group of a geometric configuration we mean the automorphism group of the underlying combinatorial configuration. We say that a configuration is *point-transitive* (resp. *line-transitive*) if its automorphism group is transitive on its points (resp. lines). A point-line pair of a configuration is called a *flag*. Similarly as before, we can speak of the *flag-transitivity* of a configuration.

Theorem 2.7 *The automorphism group of $DCD[n, d]$ is point-, line- and flag-transitive. Its automorphism group is isomorphic to $\mathbf{S}_n$ (the symmetric group of degree n).*

Proof Construction 2.2(b) starts from an arrangement of hyperplanes in general position. This implies that none of the hyperplanes is distinguished, i.e. all of them are equivalent under permutation. The same follows for the subsets of this arrangement defining the points (respectively, those defining either the lines or the flags) of the configuration. For $DCD[n, 1]$, and for its dual, the statements are trivial. ■

3 Pascal's Triangle of Configurations $DCD[n, d]$

Since the number of points and lines of configurations $DCD[n, d]$ are given by the binomial coefficients, it is obvious that these configurations can be arranged in a triangular array resembling Pascal's triangle. In this array, $DCD[n, d]$ is precisely the d-th entry in row number n. In Fig. 1 we show the first 7 lines of this scheme. In general, row number n is of the form

$$DCD[n, 1] \quad DCD[n, 2] \quad \ldots \quad DCD[n, n]; \tag{3}$$

the corresponding types in the same row are as follows:

$$\left(\binom{n}{1}_1, \binom{n}{0}_n\right) \quad \left(\binom{n}{2}_2, \binom{n}{1}_{n-1}\right) \quad \cdots \quad \left(\binom{n}{n}_n, \binom{n}{n-1}_1\right). \tag{4}$$

We call this array *Pascal's triangle of configurations*.

$$DCD[1,1]$$

$$DCD[2,1] \quad DCD[2,2]$$

$$DCD[3,1] \quad DCD[3,2] \quad DCD[3,3]$$

$$DCD[4,1] \quad DCD[4,2] \quad DCD[4,3] \quad DCD[4,4]$$

$$DCD[5,1] \quad DCD[5,2] \quad DCD[5,3] \quad DCD[5,4] \quad DCD[5,5]$$

$$DCD[6,1] \quad DCD[6,2] \quad DCD[6,3] \quad DCD[6,4] \quad DCD[6,5] \quad DCD[6,6]$$

$$DCD[7,1] \quad DCD[7,2] \quad DCD[7,3] \quad DCD[7,4] \quad DCD[7,5] \quad DCD[7,6] \quad DCD[7,7]$$

Fig. 1 The first 7 rows of Pascal's triangle of configurations $DCD[n, d]$

$$\left(\binom{3}{2}_2, \binom{3}{1}_2\right)$$

$$\left(\binom{4}{2}_2, \binom{4}{1}_3\right) \quad \left(\binom{4}{3}_3, \binom{4}{2}_2\right)$$

$$\left(\binom{5}{2}_2, \binom{5}{1}_4\right) \quad \left(\binom{5}{3}_3, \binom{5}{2}_3\right) \quad \left(\binom{5}{4}_4, \binom{5}{3}_2\right)$$

$$\left(\binom{6}{2}_2, \binom{6}{1}_5\right) \quad \left(\binom{6}{3}_3, \binom{6}{2}_4\right) \quad \left(\binom{6}{4}_4, \binom{6}{3}_3\right) \quad \left(\binom{6}{5}_5, \binom{6}{4}_2\right)$$

$$\left(\binom{7}{2}_2, \binom{7}{1}_6\right) \quad \left(\binom{7}{3}_3, \binom{7}{2}_5\right) \quad \left(\binom{7}{4}_4, \binom{7}{3}_4\right) \quad \left(\binom{7}{5}_5, \binom{7}{4}_3\right) \quad \left(\binom{7}{6}_6, \binom{7}{5}_2\right)$$

Fig. 2 Types in the part of Pascal's triangle of configurations $DCD[n, d]$ with $3 \le n \le 7$ and $2 \le d \le n-1$

For convenience, and for later reference, we also show in Fig. 2 the corresponding types in the part of the array with $3 \le n \le 7$ and $2 \le d \le n-1$. In Fig. 1 one easily recognizes some known examples of configurations (see also Fig. 2). $DCD[3, 2]$ is obviously the triangle. In the second row the second and the third entry is the complete quadrilateral and its dual, the complete quadrangle, respectively (the former is also known as the *Pasch configuration*).

More generally, the second and $(i-1)$th entry in the ith row is the complete i-lateral and the complete i-point, respectively. Here we recall that a *complete*

k-lateral ($k \geq 3$) is defined in the real projective plane as a configuration consisting, on the one hand, of k lines in general position, i.e. such that no more than two of them intersect in a point and, on the other hand, of all the lines pairwise connecting these points [9, 16]; a complete k-point is defined dually. (For the complete k-lateral, the term *complete k-line* is also used [21]).

The middle entry in the fifth row is nothing else than the Desargues configuration. The third and fourth entry in the 6th row are closely related to it: they are the *Steiner–Plücker configuration* and the *Cayley–Salmon configuration*, respectively. They arise e.g. from the following incidence theorems, which are extensions of Desargues' theorem (these theorems occur in Exercise II.16.6 in [21]; see also Exercise 2.3.2 in [6]).

Theorem 3.1 (Veblen, Young) *If three triangles are perspective from the same point, the three axes of perspectivity of the three pairs of triangles are concurrent.*

Theorem 3.2 (Veblen, Young) *If three triangles are perspective from the same line, the three centres of perspectivity of the three pairs of triangles are collinear.*

Observe that these two theorems are dual to each other; accordingly, so are the corresponding configurations (cf. Proposition 3.3 below). Both these theorems and the configurations go back to the 19th century; in two recent papers, Conway and Ryba discuss interesting new aspects of them [3, 4] (for further details, historical and other, see also [2]; in particular, a most recent result concerning the Steiner–Plücker configuration is that it can be interpreted as a *generalized Reye configuration* (see B. Servatius and H. Servatius [19]) (Figs. 3, 4).

The middle entry in the fifth row is Danzer's configuration $DCD[7, 4]$, of type (35_4). For drawings, as well as other details about it, see [2]. Recall that it is balanced (like Desargues' configuration). More generally, for each natural number k, the middle entry in the $(2k-1)$th row is a balanced configuration of type $\left(\binom{2k+1}{k+1}_{k+1}\right)$ (cf. Remark 2.4).

We conclude this section with a property of our triangular scheme, which is a direct consequence of Proposition 2.1.

Proposition 3.3 *The configurations $DCD[n, d]$ and $DCD[n, n-d+1]$ are dual to each other. Hence, reflection in the vertical mirror line of Pascal's triangle of configurations $DCD[n, d]$ takes its entries into their dual.* ■

In particular, the middle entries are self-dual. This particular case has been established in [2] (in fact, they have a stronger property, being *self-polar*; but we do not need this property here).

4 Generating the Entries as Incidence Sums

The following notion has been introduced recently [2, 9].

Definition 4.1 By the *incidence sum* of configurations $\mathcal{C}_1$ and $\mathcal{C}_2$ we mean the configuration $\mathcal{C}$ which is the disjoint union of $\mathcal{C}_1$ and $\mathcal{C}_2$, together with a specified set $I \subseteq \mathcal{P}_1 \times \mathcal{L}_2 \cup \mathcal{P}_2 \times \mathcal{L}_1$ of incident point-line pairs, where $\mathcal{P}_i$ denotes the point set and $\mathcal{L}_i$ denotes the line set of $\mathcal{C}_i$, for $i = 1, 2$. We denote it by $\mathcal{C}_1 \oplus_I \mathcal{C}_2$.

If the set I is clear from the context, it can be omitted from the operation symbol.

A simple example is provided by the (10_3) Desargues configuration, which is the incidence sum of a complete quadrilateral $(6_2, 4_3)$ and a complete quadrangle $(4_3, 6_2)$ (see [2], where this is also illustrated by a figure). We note that only the notion is new, the construction itself is not; for example, Grünbaum and Rigby point out in [13] that the $(12_4, 16_3)$ Reye configuration and its dual $(16_3, 12_4)$ form together a (28_4) configuration.

A much older example is due to Klug, who points out that, essentially, Danzer's (35_4) configuration is the incidence sum of the Steiner–Plücker $(20_3, 15_4)$ configuration and the Cayley–Salmon $(15_4, 20_3)$ configuration (in our notation, $DCD[6, 3]$ and $DCD[6, 4]$, respectively) [2, 15].

Theorem 4.2 (Klug) *The* $(20_3, 15_4)$ *Steiner-Plücker configuration and the* $(15_4, 20_3)$ *Cayley-Salmon configuration together form a* (35_4) *configuration. The 35 points and the 35 lines of this configuration are the 20 Steiner points and 15 Salmon points, and the 15 Plücker lines and 20 Cayley lines, respectively. On each Plücker line there are four Steiner points, and on each Cayley line there are three Salmon points and one Steiner point; moreover, three Plücker lines and one Cayley line pass through each Steiner point, and four Cayley lines pass through each Salmon point.*

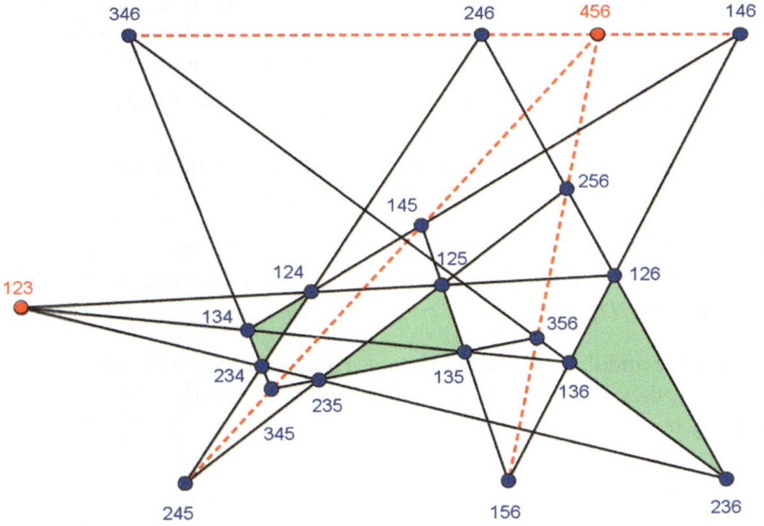

Fig. 3 The Steiner–Plücker configuration $(20_3, 15_4)$ (or $DCD[6, 3]$, in our notation). The shaded triangles are perspective from the same point (labelled by 123); the point of concurrency of the three axes of perspectivity is labelled by 456

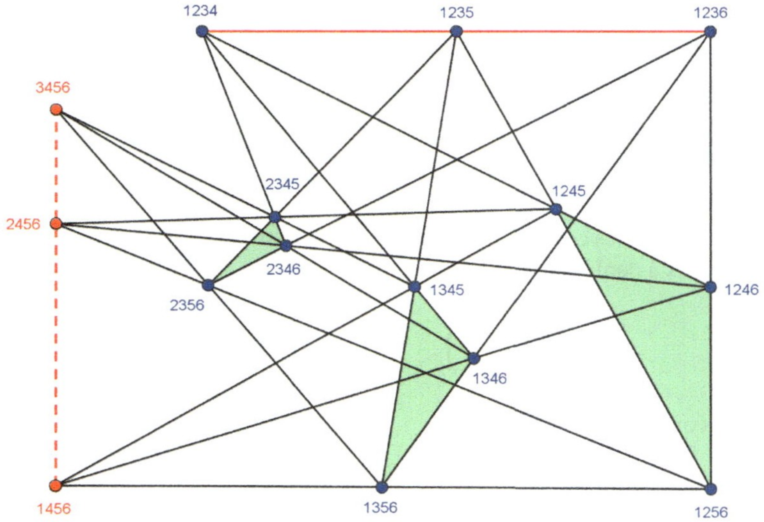

Fig. 4 The Cayley–Salmon configuration $(15_4, 20_3)$ (or $DCD[6, 4]$, in our notation). The points are labelled by quadruples of the hyperplanes which determine them in accordance with our Definition 2.3. The shaded triangles are perspective from the same line (this is the horizontal line on the top of the figure)

Remark 4.3 The following property holds in both of these examples. In the incidence sum $\mathcal{C}$ of configurations $\mathcal{C}_1$ and $\mathcal{C}_2$ the corresponding decomposition is determined by the partition of the set of points of $\mathcal{C}$ into two disjoint classes. In the one class the points are incident precisely with lines belonging to $\mathcal{C}_2$, while the other class consists of points which are incident with lines such that some of them belong to $\mathcal{C}_1$, and the rest belong to $\mathcal{C}_2$. (The lines of $\mathcal{C}$ obey a similar rule, with $\mathcal{C}_1$ and $\mathcal{C}_2$ interchanged.)

We emphasize that the notion of an incidence sum can be considered from two different aspects. On the one hand, we consider it as a configuration which is decomposed into two smaller configurations; on the other hand, it may mean that we have two distinct configurations (both existing in themselves), and we put them together in a suitable way to obtain a new configuration (in which I is the set of new incidences). In other words, in this latter sense, we consider it as a kind of operation (not *total operation*, but *partial operation*, in terms of abstract algebra; see [9] for some details of this particular context). The two aspects have some resemblance to the twofold meaning of the direct product of groups (recall that in group theory there is a distinction between *internal* and *external* direct product; see e.g. [18]).

One of the main results in [2] is that each balanced $DCD[n, d]$ can be decomposed into an incidence sum of two smaller (non-balanced) configurations (loc. cit., Theorem 4.1). Here we generalize this result showing that a similar decomposition is possible for each $DCD[n, d]$ (provided it is a non-terminal entry in the Pascal triangle introduced in the previous section).

Theorem 4.4 *For all integers $n \geq 3$ and d $(2 < d < n-1)$, the configuration $DCD[n, d]$ is the incidence sum of the form $C_1 \oplus_I C_2$ such that*

(1) $C_1 = DCD[n-1, d-1]$, $C_2 = DCD[n-1, d]$;
(2) the set I of new incidences can be given as follows: for each line in C_2, the number of points incident with it increases by one; the new points on these lines are precisely the points of C_1; hence $|I| = \binom{n-1}{d-1}$.

Proof The three cases corresponding to $n = 3, 4$ are straightforward. For $n \geq 5$, first recall that C_1 is a configuration of type

$$\left(\binom{n-1}{d-1}_{d-1}, \binom{n-1}{d-2}_{n-d+1}\right), \tag{5}$$

and C_2 is a configuration of type

$$\left(\binom{n-1}{d}_{d}, \binom{n-1}{d-1}_{n-d}\right). \tag{6}$$

Apply twice Pascal's rule:

$$\binom{n}{d} = \binom{n-1}{d} + \binom{n-1}{d-1}; \quad \binom{n}{d-1} = \binom{n-1}{d-1} + \binom{n-1}{d-2}. \tag{7}$$

The first sum corresponds to the decomposition of the set of the d-tuples of hyperplanes determining the points of $DCD[n, d]$, and the second one to the decomposition of the set of $(d-1)$-tuples determining the lines (cf. Construction 2.2). Let H an arbitrary but fixed hyperplane. We associate the first term in both sums to the d-tuples, respectively to the $d-1$-tuples, not containing H. Their number, and hence the number of the points and lines they determine is indeed $\binom{n-1}{d}$ and $\binom{n-1}{d-1}$, respectively. Moreover, one easily sees that each point is incident with precisely d lines, and each line is incident with precisely $n-d$ points. Hence they form the configuration C_2 (cf. 6).

Consider now the d-tuples and $(d-1)$-tuples of hyperplanes which contain H. Their number is clearly $\binom{n-1}{d-1}$, respectively $\binom{n-1}{d-2}$, in accordance with the second term in the sums (7) above. Thus the number of the points, respectively the lines determined by them, and required in (5), is obtained. On the other hand, for finding the incidence numbers, one observes that from a d-tuple of hyperplanes one can remove $d-1$ distinct hyperplanes different from H, to obtain a corresponding $(d-1)$-tuple; hence a point determined by such a d-tuple is incident with precisely $d-1$ lines. The incidence number $n-d+1$ in (5) is obtained similarly. Thus we have configuration C_1, indeed.

To see Condition (2), consider the d-tuples of hyperplanes containing H, which determine the points of C_1, as we have seen above. Recall that their number is $\binom{n-1}{d-1}$. We assign to each of these d-tuples a $(d-1)$-tuple obtained by removing H. The number of these $(d-1)$-tuples is still $\binom{n-1}{d-1}$, and, as established above, they determine

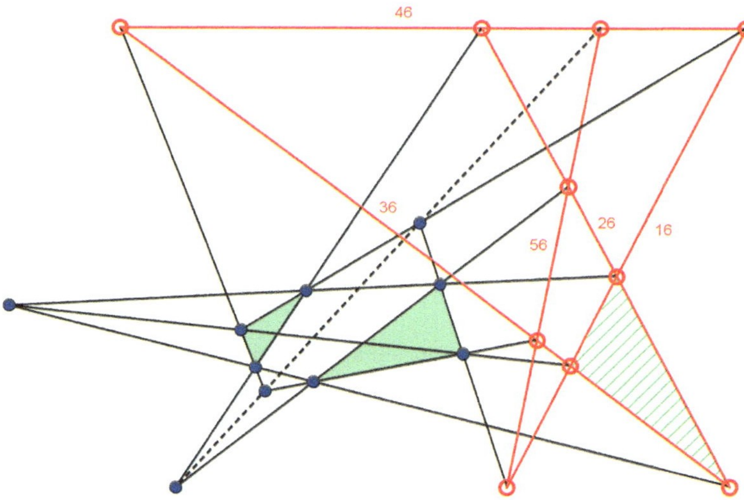

Fig. 5 Decomposition of the $(20_3, 15_4)$ Steiner–Plücker configuration into the incidence sum of a complete pentalateral $(10_2, 5_4)$ and a Desargues (10_3) configuration. The former is indicated by red; its lines are labelled with the pairs of hyperplanes determining them (cf. the labels of the corresponding points in Fig. 3)

precisely the lines of $\mathcal{C}_2$. Thus we have a bijective correspondence between the set of points of $\mathcal{C}_1$ and the set of lines of $\mathcal{C}_2$. The point-line pairs defined by this correspondence form precisely the set I. ∎

We note that when applying this decomposition theorem, the property considered in Remark 4.3 remains valid in each case. Figure 5 illustrates the example of the incidence sum in a non-balanced case.

With this result, the analogy with the classical Pascal triangle becomes complete: our triangular array is not merely an arrangement of objects, but it can also be recursively generated.

Corollary 4.5 *The Desargues–Cayley–Danzer configurations, i.e. the non-terminal entries of Pascal's triangle of configurations $DCD[n, d]$, whose general row is of the form (3) given at the begininning of Sect. 3, can be recursively generated: each of them is an incidence sum of the two entries directly above it.*

Proof Consider an arbitrarily chosen non-terminal entry $DCD[n, d]$), and the entries $DCD[n-1, d-1]$ and $DCD[n-1, d)$ directly above it. These configurations are realized independently of each other (see Construction 2.2), hence it my happen that neither $DCD[n-1, d-1]$, nor $DCD[n-1, d)$ is projectively equivalent with the summands occurring in the incidence sum of $DCD[n, d]$) given by Theorem 4.4. However, Proposition 2.6 guarantees that an isomorphism is still valid, and this is sufficient. ∎

We mention an additional consequence of Theorem 4.4. First, we define the following special subset of our triangle.

$$\mathcal{Q}(n, d) = \{DCD[m, e] \mid 1 \le m \le n,\ d - n + m \le e \le d\}. \tag{8}$$

It is easy to check that this subset forms a quadrangle such that its two sides consist of configurations lying on the left and the right boundary of Pascal's triangle, respectively, and the other two sides are parallel with these boundary lines. We call this set the *quadrangle of subconfigurations* belonging to $DCD[n, d]$. The name is justified by the following observation.

Corollary 4.6 *For a chosen entry* $DCD[n, d]$ *of Pascal's triangle of configurations, the set* $\mathcal{Q}(n, d)$ *consists precisely of the configurations which are subconfigurations of* $DCD[n, d]$.

Proof This is straightforward by repeated application of Theorem 4.4, taking into account Proposition 2.6. ■

Remark 4.7 In the particular case of a middle entry $DCD[2k + 1, k + 1]$, the set $\mathcal{Q}(2k + 1, k + 1)$ of subconfigurations belonging to it is contained in a quadrangle which is a rhombus such that its vertical symmetry axis coincides with the symmetry axis of the Pascal triangle. We shall need this property in Sect. 6.

Similarly to $\mathcal{Q}(n, d)$, we define the *triangle of superconfigurations* belonging to $DCD[n, d]$, as follows.

$$\mathcal{T}(n, d) = \{DCD[m, e] \mid m \ge n,\ d - n + m \ge e \ge d\}. \tag{9}$$

It is easy to see that this set consists of configurations which are *superconfigurations* of $DCD[n, d]$, in the sense that each of them contains $DCD[n, d]$ as a subconfiguration.

Remark 4.8 Having established the presence of subconfigurations and superconfigurations in our Pascal triangle, we see that the whole set of $DCD[n, d]$ configurations forms a partially ordered set, where the order is defined by the substructure relation. In fact, it is a *lattice*; the lattice operations can be defined by the sets $\mathcal{Q}$ and $\mathcal{T}$.

5 Incidence Theorems

Consider a (movable) configuration $\mathcal{C}$ which expresses an incidence theorem (like e.g. Desargues' (10_3) configuration [14]). Such a theorem essentially states that, when taking any particular realization, if all but one of the incidences in $\mathcal{C}$ are satisfied, then the "last" incidence is also satisfied[2] This intuitive description can be made more precise as follows.

[2]Not every incidence statement involves merely a single incidence in this sense. For example, the author presented a conjecture in [9] which is connected with a (100_4) configuration; the conjecture essentially states that if 350 (suitable) incidences in this configuration are satisfied, then the remaining 50 are also satisfied. (We note that the same (100_4) configuration also occurs in [17], in two different versions.).

Definition 5.1 Suppose we are given a geometric configuration or more generally, an incidence structure of points and lines $\mathcal{C} = (\mathcal{P}, \mathcal{B}, I)$ which is movable in the sense of Definition 2.5. We say that $\mathcal{C}$ *expresses an incidence theorem* if there exists a pair (PIb) such that any geometric realization of $(\mathcal{P}, \mathcal{B}, I) \setminus (P, b)$ is impossible unless the incidence (PIb) is present.

The following observation is straightforward.

Lemma 5.2 *Let* $\mathcal{C}$ *be a configuration of points and lines, and let* $\mathcal{C}_1$ *be one of its subconfigurations. If* $\mathcal{C}_1$ *expresses an incidence theorem then* $\mathcal{C}$ *is also expresses an incidence theorem.* ■

The next theorem is the main result of this section.

Theorem 5.3 *Every Desargues–Cayley–Danzer configuration* $DCD[n, d]$ *with* $n \geq 5$ *and* $2 < d < n - 2$ *expresses an incidence theorem.*

Proof Such a configuration falls into the triangle of superconfigurations belonging to $DCD[5, 3]$ (cf. Fig. 1 and the definition given by equality (9) in the previous section). $DCD[5, 3]$ is the Desargues configuration, and since it expresses an incidence theorem, Lemma 5.2 implies the statement. ■

It is worth emphasizing that one and the same configuration may express more than one distinct incidence theorems (in this respect, we do not consider a theorem and its converse as essentially distinct statements; cf. our Theorem 5.6 below). For example, in addition to Theorem 3.2, the following incidence statement (which occurs in the classical work of Veblen and Young, see [21], §17, Theorem 2; but was already known to Felix Klein, too [2]) also gives rise to the Cayley–Salmon configuration.

Theorem 5.4 (Veblen, Young) *If two tetrahedra are perspective from a point, the six pairs of lines of the corresponding edges intersect in coplanar points, and the planes of the four pairs of faces intersect in coplanar lines; i.e. the tetrahedra are perspective from a plane.*

Hence the dual of this theorem, like the double Desargues theorem as well, gives rise to the Steiner–Plücker configuration; moreover, a recently published incidence theorem, different from both, also yields the Steiner–Plücker configuration ([1], Theorem 4).

Note that the plane in the conclusion of Theorem 5.4 is the "axis" of perspectivity (being an analogue of the axis in the classical Desargues theorem). Conway and Ryba [3] use the term *perspector* for the center of perspectivity, and *perspectrix* for the (classical) axis of perspectivity. We adopt these terms and, in particular, use the latter in more general case as well (whatever is the dimension of the axis).

With this terminology, we present a further classical incidence theorem, which gives rise to the (35_4) Danzer configuration ([15], p. 34; for some further details, see [2]).

Theorem 5.5 (Klug) *If three tetrahedra have a common perspector, the three perspectrices belonging to the pairs of these tetrahedra have a point in common.*

Beyond the classical examples, what kind of incidence theorems can be associated to the Desargues–Cayley–Danzer configurations, in the light of our Theorem 5.3? We present some examples here, without proof (and with the hope that they have some novelty).

Consider the configuration $DCD[8, 4]$; its type is

$$\left(\binom{8}{4}_{4,} \binom{8}{3}_{5}\right) = (70_4, 56_5).$$

It can be shown that the first two of the following four incidence statements give rise alike to this configuration; the other two are associated to the dual configuration.

Theorem 5.6 *The following implications hold for four distinct tetrahedra in the real projective 3-space.*

(A) *If four tetrahedra have a common perspector, the six perspectrices of the $\binom{4}{2}$ pairs of tetrahedra have a point in common.*
(B) *Assume that each pair of the tetrahedra has a perspectrix. If these $\binom{4}{2}$ perspectrices have a point in common, the four tetrahedra have a common perspector.*
(C) *If the four tetrahedra have a common perspectrix, the six perspectors of the $\binom{4}{2}$ pairs of the tetrahedra lie in a common plane.*
(D) *Assume that each pair of the tetrahedra has a perspector. If these $\binom{4}{2}$ perspectors lie in a common plane, the four tetrahedra have a common perspectrix.*

Clearly, (A) and (B), respectively (C) and (D), are the converse of each other.

As a preparation for our last theorem in this section, we recall the following common generalization of Desargues' theorem and Theorem 5.4 (see Veblen and Young [21], p. 54, Exercise 26).

Theorem 5.7 (Veblen, Young) *Let $d \geq 2$ be an integer. If two $(d+1)$-points in a d-space are perspective from a point, their corresponding r-spaces meet in $(r-1)$-spaces which lie in the same $(d-1)$-space ($r = 1, 2, \ldots d-1$) and form a complete configuration of $(d+1)$ $(d-2)$-spaces in $(d-1)$-space.*

Here the term $(d+1)$-*point* refers to the projective analogue of a Euclidean simplex of dimension d. This figure consists of a set of $d+1$ points in general position as well as of all the $(2^{d+1}-2)$ proper projective subspaces which are spanned by the subsets of this set. In addition, we say that $d+1$ points in a projective d-space are in general position if they do not lie within the same (projective) hyperplane. (Here we adopted the term used by Veblen and Young [21].)

We note that we disregard the structure formed within the perspectrix and only use the fact that that the dimension of the perspectrix is $(d-1)$.

Theorem 5.8 *Let $d \geq 2$ be an integer. If $d+1$ $(d+1)$-points in a d-space have a common perspector, the $\binom{d+1}{2}$ perspectrices belonging to the pairs of these $(d+1)$-points have a point in common.*

This theorem is a common generalization of Theorems 3.1 and 5.6(A). It gives rise to a $DCD[2d+2, d+1]$ configuration, whose type is

$$\left(\binom{2d+2}{d+1}_{d+1}, \binom{2d+2}{d}_{d+2}\right).$$

6 Point-Circle Realizations

In this section we show that every configuration $DCD[n, d]$ can also be realized geometrically as a configuration of points and circles. Moreover, all the circles in such a configuration are of equal size; we call a point-circle configuration with this particular property *isometric*.

This is an extension of a former result (see Theorem 7.1 in [2]), by which the same holds in the restricted case of the *middle entries* of our Pascal triangle. Recall that such a configuration is of the form $DCD[2k+1, k+1]$ if it takes place in the $(2k+1)$th row $(k = 1, 2, \ldots)$.

Theorem 6.1 *Every configuration $DCD(n, d)$ admits an isometric point-circle realization.*

Proof Start from the fact that for a configuration $DCD[n, d]$, we have the following three distinct cases: (L) $n > 2d - 1$; (R) $n < 2d - 1$; (B) $n = 2d - 1$. Now observe that

- in case (L), $DCD[n, d]$ lies in a diagonal line of Pascal's triangle determined by a constant $c = n - d$ such that this line bounds from the left the quadrangle $\mathcal{Q}(2c+1, c+1)$;
- in case (R), $DCD[n, d]$ lies in a diagonal line determined by the condition $d =$ const.; this line bounds from the right the quadrangle $\mathcal{Q}(2d-1, d)$;
- in case (B), $DCD[n, d]$ lies in both lines considered above, which means that it coincides with the configuration defining the quadrangle $\mathcal{Q}(2d-1, d)$; in other words, it is a middle term of Pascal's triangle.

We see that in all the three cases $DCD[n, d]$ is within a quadrangle of subconfigurations which is in fact a rhombus (cf. Remark 4.7); in other words, the configuration defining this quadrangle is a middle term. On the other hand, we know (by Theorem 7.1 in [2]; see also Theorem 5.2 in [10]) that each of these middle terms admits an isometric point-circle realization. Clearly, the same is true for any of their subconfigurations, and hence, for $DCD[n, d]$, too. ■

We note that in [10] it is shown that the point-circle representation of the configuration $DCD[2k+1, k+1]$ occurs as a subconfiguration of the point-circle representation of a *Clifford configuration* of type (2^{2k}_{2k+1}). Now we see that in fact all the point-circle representations considered in Theorem 6.1 have this property. In [10] this is illustrated through the example of the Desargues configuration. In papers [2,

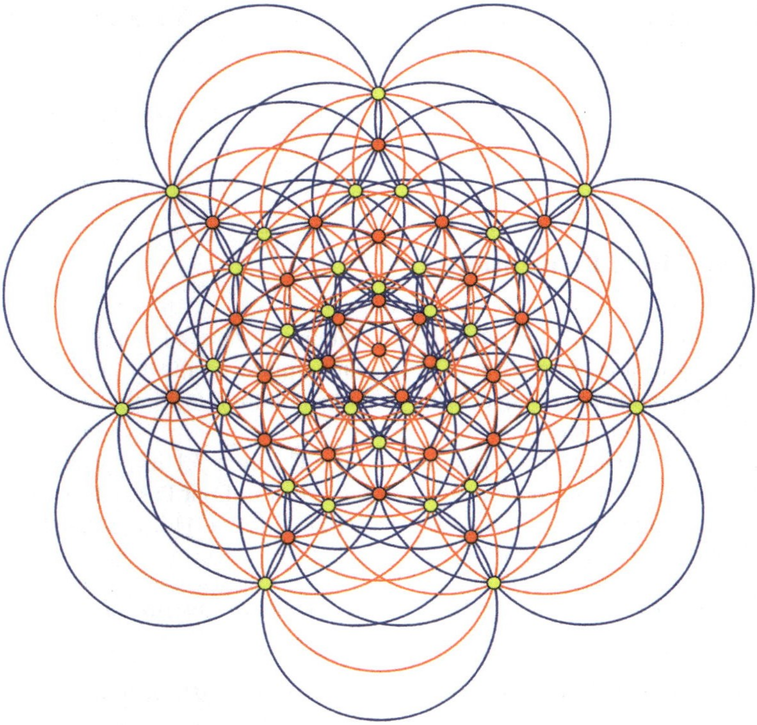

Fig. 6 A point-circle representation of the (35_4) Danzer configuration embedded in the (64_7) Clifford configuration. (The Danzer configuration is shown by yellow points and blue circles.)

10] various point-circle versions of the Danzer configuration are also presented. Here we show (in Fig. 6) an embedding of this latter in the (64_7) Clifford configuration.[3]

Figure 7 shows another example. It also illustrates the fact that a decomposition given in Theorem 4.4 is possible in the case of point-circle representations as well.

Symmetry is another interesting aspect of these representations (observe the D_7 symmetry of Fig. 6, and the C_5 symmetry of Fig. 7). Few is known about the possibility of symmetric representation of the Desargues–Cayley–Danzer configurations. In particular, we have the following conjecture ([2], Conjecture 5.2).

Conjecture 6.2 *There is no realization of Danzer's* (35_4) *point-line configuration with five- or seven-fold rotational symmetry.*

We note that even less is known about the symmetric realizability of the point-line $DCD[n, d]$ configurations in general.

[3]The drawing in Fig. 6 has been prepared by Ákos Varga (University of Szeged) [20].

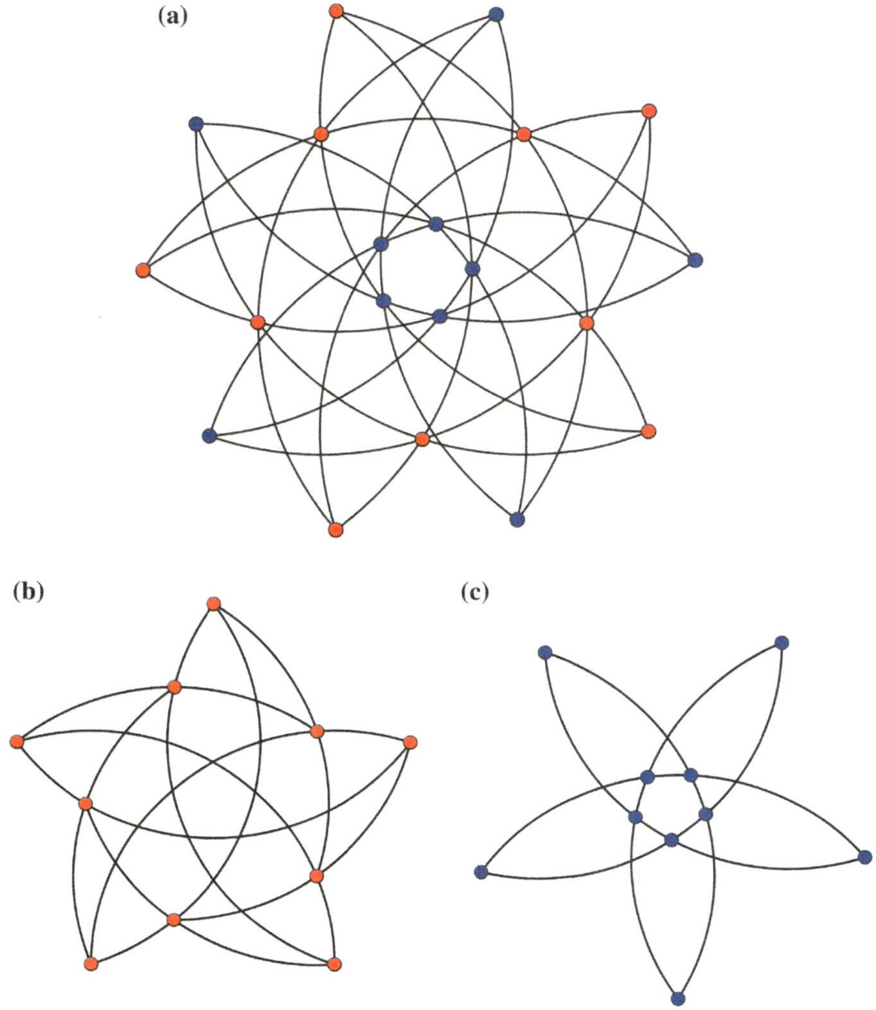

Fig. 7 A point-circle representation of the $(20_3, 15_4)$ Steiner–Plücker configuration (**a**) together with its decomposition into the incidence sum of a copy of the Desargues configuration (**b**) and of the complete pentalateral (**c**) (cf. Fig. 5). Observe the 5-fold rotational symmetry. (For better visibility, we used arcs instead of full circles.)

On the other hand, it seems that a symmetric realization can be more easily achieved in the case of point-circle representations. In particular, here we formulate the following conjecture.

Conjecture 6.3 *For each natural number k,* $DCD[2k+1, k+1]$ *has a point-circle realization with* $(2k+1)$*-fold rotational symmetry.*

The validity of this conjecture depends on the possibility of a symmetric representation of the so-called *odd graphs* (for the graph-theoretical background of constructing point-circle configurations, see [10]).

We expect that if there will be some progress on the graph-theoretical side of this problem, our Theorem 4.4 could be a suitable tool in applying the results to the $DCD[n, d]$ configurations.

Acknowledgements The author is grateful to the (anonymous) reviewers for their helpful comments. Special thanks go to Reviewer #1 for the insightful remarks and valuable suggestions that improved the exposition, results and proofs of this paper. This research is supported by the OTKA grant NN-114614.

References

1. V.E. Adler, Some incidence theorems and integrable discrete equations. Discrete Comput. Geom. **36**, 489–498 (2006)
2. M. Boben, G. Gévay, T. Pisanski, Danzer's configuration revisited. Adv. Geom. **15**, 393–408 (2015)
3. J. Conway, A. Ryba, The Pascal mysticum demystified. Math. Intelligencer **34**, 4–8 (2012)
4. J. Conway, A. Ryba, Extending the Pascal mysticum. Math. Intelligencer **35**, 44–51 (2013)
5. H.S.M. Coxeter, Self-dual configurations and regular graphs. Bull. Amer. Math. Soc. **56** (1950), 413–455. Reprinted in: H.S.M. Coxeter, *Twelve Geometric Essays*, Southern Illinois University Press, Carbondale, 1968, pp. 106–149
6. H.S.M. Coxeter, *Projective Geometry* (University of Toronto Press, Toronto, 1974)
7. I. Dolgachev, Abstract configurations in algebraic geometry, in *Proceedings Fano Conference, Torino, 2002* ed. by A. Collino, A. Conte, M. Marchisio (University of Torino, 2004), pp. 423–462
8. G. Gévay, Symmetric configurations and the different levels of their symmetry. Symmetry Cult. Sci. **20**, 309–329 (2009)
9. G. Gévay, Constructions for large spatial point-line (n_k) configurations. Ars Math. Contemp. **7**, 175–199 (2014)
10. G. Gévay, T. Pisanski, Kronecker covers, V-construction, unit-distance graphs and isometric point-circle configurations. Ars Math. Contemp. **7**, 317–336 (2014)
11. B. Grünbaum, Musings on an example of Danzer's. European J. Combin. **29**, 1910–1918 (2008)
12. B. Grünbaum, *Configurations of Points and Lines, Graduate Texts in Mathematics*, vol. 103 (American Mathematical Society, Providence, Rhode Island, 2009)
13. B. Grünbaum, J.F. Rigby, The real configuration (21_4). J. London Math. Soc. **41**, 336–346 (1990)
14. D. Hilbert, S. Cohn-Vossen, *Anschauliche Geometrie, Springer, Berlin, 1932* (Szemléletes geometria, Gondolat, Budapest, Hungarian translation, 1982)
15. L. Klug, *Az általános és négy különös Pascal-hatszög configuratiója* (*The Configuration of the General and Four Special Pascal Hexagons; in Hungarian*), Ajtai K. Albert Könyvnyomdája, Kolozsvár, 1898. Reprinted in: L. Klug, *Die Configuration Des Pascal'schen Sechseckes Im Allgemeinen Und in Vier Speciellen Fällen (German Edition)*, Nabu Press, 2010

16. F. Levi, *Geometrische Konfigurationen* (Hirzel, Leipzig, 1929)
17. T. Pisanski, B. Servatius, *Configurations from a Graphical Viewpoint, Birkhäuser Advanced Texts Basler Lehrbücher Series* (Birkhäuser Boston Inc., Boston, 2013)
18. D.J.S. Robinson, *A Course in the Theory of Groups* (Springer-Verlag, New York, Berlin, Heidelberg, 1995)
19. B. Servatius, H. Servatius, The generalized Reye configuration. Ars Math. Contemp. **3**, 21–27 (2010)
20. Á. Varga, personal communication (2015)
21. O. Veblen, J.W. Young, *Projective Geometry*, vol. 1 (Ginn and Company, Boston, 1910)

Volume of Convex Hull of Two Bodies and Related Problems

Ákos G. Horváth

Abstract In this paper we deal with problems concerning the volume of the convex hull of two "connecting" bodies. After a historical background we collect some results, methods and open problems, respectively.

Keywords Isoperimetric problem · Volume inequality · Polytope · Simplex

2010 Mathematics Subject Classification. 52B60 · 52A40 · 52A38

1 Introduction

To find the convex polyhedra in Euclidean 3-space $\mathbb{R}^3$, with a given number of faces and with minimal isoperimetric quotient, is a centuries old question of geometry: research in this direction perhaps started with the work of Lhuilier in the 18th century. A famous result of Lindelöf [1], published in the 19th century, yields a necessary condition for such a polyhedron: it states that any optimal polyhedron is circumscribed about a Euclidean ball, and this ball touches each face at its centroid. In particular, it follows from his result that, instead of fixing surface area while looking for minimal volume, we may fix the inradius of the polyhedron. Since the publication of this result, the same condition for polytopes in n-dimensional space $\mathbb{R}^n$ has been established (cf. [2]), and many variants of this problem have been investigated (cf., e.g. [3]). For references and open problems of this kind, the interested reader is referred to [4, 5] or [6]. For polytopes with $(n+2)$ vertices this question was answered by Kind and Kleinschmidt [7]. The solution for polytopes with $n+3$ vertices was published in [8], which later turned out to be incomplete (cf. [9]), and thus, this case is still open. We mention two problems in more detail:

Á. G. Horváth (✉)
Department of Geometry, Budapest University of Technology,
Egry József u. 1, Budapest 1111, Hungary
e-mail: ghorvath@math.bme.hu

M. D. E. Conder et al. (eds.), *Discrete Geometry and Symmetry*,
Springer Proceedings in Mathematics & Statistics 234,
https://doi.org/10.1007/978-3-319-78434-2_11

- The dual of the original problem: to find, among d-polytopes with a given number of vertices and inscribed in the unit sphere, the ones with maximal volume, and
- to find the extremity of the volume of the convex hull of two "connecting" bodies.

The first problem that to find the maximal volume polyhedra in $\mathbb{R}^3$ with a given number of vertices and inscribed in the unit sphere, was first mentioned in [10] in 1964. A systematic investigation of this question was started with the paper [11] of Berman and Hanes in 1970, who found a necessary condition for optimal polyhedra, and determined those with $n \leq 8$ vertices. The same problem was examined in [12], where the author presented the results of a computer-aided search for optimal polyhedra with $4 \leq n \leq 30$ vertices. Nevertheless, according to our knowledge, this question, which is listed in both research problem books [4, 6], is still open for polyhedra with $n > 8$ vertices.

The second problem connected with the first one on the following way: If the given points form the respective vertex sets of two polyhedra (inscribed in the unit sphere) then the volume of the convex hull of these points is the volume of the convex hull of two "connecting" bodies, too. It is interesting that the case of two regular simplices with common center gives another maximum as the global isodiametric problem on eight points inscribed in the unit sphere.

The examination of the volume of the convex hull of two congruent copies of a convex body in Euclidean d-space (for special subgroups) investigated systematically first by Rogers, Shepard and Macbeath in 1950s (see in [13–15]). Fifty years later a problem similar to that of the simplices arose that lead to new investigations by new methods which obtained fresh results (see in [16–18]). In particular, a related conjecture of Rogers and Shephard has been proved in [17].

Finally we review some important consequences of the icosahedron inequality of L. Fejes-Tóth. In particular, it is needed for the proof of the statement that the maximal volume polyhedron spanned by the vertices of two regular simplices with common centroid is the cube. It is also used in the proof of that the maximal volume polyhedron with eight vertices and inscribed in the unit sphere is a triangular one distinct from the cube.

2 Maximal Volume Polytopes Inscribed in the Unit Sphere

The aim of this section is to review the results on the first problem mentioned in the introduction.

Let for any $p, q \in \mathbb{R}^d$, $|p|$ and $[p, q]$ denote the standard Euclidean norm of p, and the closed segment with endpoints p and q, respectively. The origin of the standard coordinate system of $\mathbb{R}^d$ is denoted by o. If $v_1, v_2, \ldots, v_d \in \mathbb{R}^d$, then the $d \times d$ determinant with columns $v_1, v_2, \ldots, v_d$, in this order, is denoted by $|v_1, \ldots, v_d|$. The unit ball of $\mathbb{R}^d$, with o as its center, is denoted by B^d, and we set $\mathbb{S}^{d-1} = \operatorname{bd} \mathbf{B}^d$.

Throughout this section, by a polytope we mean a convex polytope. The vertex set of a polytope P is denoted by $V(P)$. We denote the family of d-dimensional polytopes, with n vertices and inscribed in the unit sphere $\mathbb{S}^{d-1}$, by $\mathcal{P}_d(n)$. The d-dimensional volume denotes by vol_d, and set $v_d(n) = \max\{\mathrm{vol}_d(P) : P \in \mathcal{P}_d(n)\}$. Note that by compactness, $v_d(n)$ exists for any value of d and n.

Let P be a d-polytope inscribed in the unit sphere $\mathbb{S}^{d-1}$, and let $V(P) = \{p_1, p_2, \dots, p_n\}$.

Let $\mathcal{C}(P)$ be a simplicial complex with the property that $|\mathcal{C}(P)| = \mathrm{bd}\, P$, and that the vertices of $\mathcal{C}(P)$ are exactly the points of $V(P)$. Observe that such a complex exist. Indeed, for any positive integer k, and for $i = 1, 2, \dots, n$, consider a point p_i^k such that $|p_i^k - p_i| < \frac{1}{k}$, and the polytope $P_k = \mathrm{conv}\{q_i^k : i = 1, 2, \dots, n\}$ is simplicial. Define $\mathcal{C}(P_k)$ as the family of the faces of P_k. We may choose a subsequence of the sequence $\{\mathcal{C}(P_k)\}$ with the property that the facets of the complexes belong to vertices with the same indices. Then the limit of this subsequence yields a complex with the required properties. Note that if P is simplicial, then $\mathcal{C}(P)$ is the family of the faces of P.

Now we orient $\mathcal{C}(P)$ in such a way that for each $(d-1)$-simplex $(p_{i_1}, p_{i_2}, \dots, p_{i_d})$ (where $i_1 \le i_2 \le \dots \le i_d$) in $C(P)$, the determinant $|p_{i_1}, \dots, p_{i_d}|$ is positive; and call the d-simplex $\mathrm{conv}\{o, p_{i_1}, \dots, p_{i_d}\}$ a *facial simplex* of P. We call the $(d-1)$-dimensional simplices of $\mathcal{C}(P)$ the *facets* of $\mathcal{C}(P)$.

2.1 3-Dimensional Results

The problem investigated in this section was raised by Fejes-Tóth in [10]. His famous inequality (called by icosahedron inequality) can be formulated as follows.

Theorem 2.1 ([10] on p.263) *If V denotes the volume, r the inradius and R the circumradius of a convex polyhedron having f faces, v vertices and e edges, then*

$$\frac{e}{3}\sin\frac{\pi f}{e}\left(\tan^2\frac{\pi f}{2e}\tan^2\frac{\pi v}{2e}\right)r^3 \le V \le \frac{2e}{3}\cos^2\frac{\pi f}{2e}\cot\frac{\pi v}{2e}\left(1-\cot^2\frac{\pi f}{2e}\cot^2\frac{\pi v}{2e}\right)R^3. \tag{1}$$

Equality holds in both inequalities only for regular polyhedra.

He noted that *"a polyhedron with a given number of faces f is always a limiting figure of a trihedral polyhedron with f faces. Similarly, a polyhedron with a given number v of vertices is always the limiting figure of a trigonal polyhedron with v vertices. Hence introducing the notation*

$$\omega_n = \frac{n}{n-2}\frac{\pi}{6}$$

we have the following inequalities

$$(f-2)\sin 2\omega_f\left(3\tan^2\omega_f - 1\right)r^3 \le V \le \frac{2\sqrt{3}}{9}(f-2)\cos^2\omega_f\left(3-\cot^2\omega_f\right)R^3, \tag{2}$$

$$\frac{\sqrt{3}}{2}(v-2)\left(3\tan^2\omega_v - 1\right) r^3 \leq V \leq \frac{1}{6}(v-2)\cot\omega_v\left(3-\cot^2\omega_v\right) R^3. \quad (3)$$

Equality holds in the first two inequalities only for regular tetrahedron, hexahedron and dodecahedron (f = 4, 6, 12) and in the last two inequalities only for the regular tetrahedron, octahedron and icosahedron (v = 4, 6, 12)."

The right hand side of inequality (1) immediately solves our first problem in the cases when the number of vertices is $v = 4, 6, 12$; the maximal volume polyhedra with $4, 6$ and 12 vertices inscribed in the unit sphere are the regular tetrahedron, octahedron and icosahedron, respectively.

The second milestone in the investigation of this problem is the paper of Berman and Hanes ([11]) written in 1970. They solved the problem for $v = 5, 7, 8$ vertices, respectively. Their methods are based on a combinatorial classification of the possible spherical tilings due to Bowen and Fisk ([19]) and a geometric result which gives a condition for the local optimal positions. They characterized these positions by a property called Property Z. We now give the definitions with respect to the d-dimensional space.

Definition 2.1 Let $P \in \mathcal{P}_d(n)$ be a d-polytope with $V(P) = \{p_1, p_2, \ldots, p_n\}$. If for each i, there is an open set $U_i \subset \mathbb{S}^{d-1}$ such that $p_i \in U_i$, and for any $q \in U_i$, we have

$$\mathrm{vol}_d\left(\mathrm{conv}\left((V(P) \setminus \{p_i\}) \cup \{q\}\right)\right) \leq \mathrm{vol}_d(P),$$

then we say that P satisfies *Property Z*.

Returning to the three-dimensional case if p_i and p_j are vertices of P, denote the line segment whose endpoints are p_i and p_j by s_{ij} and its length by $|s_{ij}|$. Also, let $n_{ij} = 1/6\left(p_i \times p_j\right)$ where $\times$ denotes the vector product in E^3.

Lemma 2.1 (Lemma 1 in [11]) *Let P with vertices $p_1, \ldots, p_n$ have property Z. Let $C(P)$ be any oriented complex associated with P such that* $\mathrm{vol}(C(P)) \geq 0$. *Suppose $s_{12}, \ldots, s_{1r}$ are all the edges of $C(P)$ incident with p_1 and that p_2, p_3, p_1; p_3, p_4, p_1; ...; p_r, p_2, p_1 are orders for faces consistent with the orientation of $C(P)$.*

i. *Then $p_1 = m/|m|$ where $m = n_{23} + n_{34} + \cdots + n_{r2}$.*
ii. *Furthermore, each face of P is triangular.*

Let the *valence* of a vertex of $C(P)$ be the number of edges of $C(P)$ incident with that vertex. By Euler's formula the average of the valences is $6 - 12/n$. If n is such that $6 - 12/n$ is an integer then $C(P)$ is *medial* if the valence of each vertex is $6 - 12/n$. If $6 - 12/n$ is not an integer then $C(P)$ is *medial* provided the valence of each vertex is either m or $m + 1$ where $m < 6 - 12/n < m + 1$. P is said to be *medial* provided all faces of P are triangular and $C(P)$ is medial. Goldberg in [20] made a conjecture whose dual was formulated by Grace in [21]: The polyhedron with n vertices in the unit sphere whose volume is a maximum is a medial polyhedron provided a medial polyhedron exists for that n. Connecting to this conjecture Berman and Hanes proved that if $n = 4, 5, 6, 7, 8$ then the polyhedra with maximal volume inscribed in the unit sphere are medial polyhedra with Property Z, respectively.

Note that in the proofs of the above results (on $n \geq 5$) is an important step to show that the valences of the vertices of a polyhedron with maximal volume are at least 4. This follows from inequality (2).

The maximal volume polyhedron for $n = 4$ is the regular simplex. For $n = 5, 6, 7$ they are the so-called double n-pyramids, with $n = 5, 6, 7$, respectively. (By a *double n-pyramid* (for $n \geq 5$), is meant a complex of n vertices with two vertices of valence $n - 2$ each of which is connected by an edge to each of the remaining $n - 2$ vertices, all of which have valence 4. The $2(n - 2)$ faces of a double n-pyramid are all triangular. A polyhedron P is a *double n-pyramid* provided each of its faces is triangular and some $C(P)$ is a double n-pyramid.) An interesting observation (see Lemma 2 in [11]) is that if P is a double n-pyramid with property Z then P is unique up to congruence and its volume is $[(n-2)/3] \sin 2\pi/(n-2)$.

For $n = 8$ there exists only two non-isomorphic complexes which have no vertices of valence 3 (see in [19]). One of them the double 8-pyramid and the other one has four valence 4 vertices and four valence 5 vertices, and therefore it is the medial complex (see on Fig. 1). It has been shown that if this latter has Property Z then P is uniquely determined up to congruence and its volume is $\sqrt{\left[\frac{475+29\sqrt{145}}{250}\right]}$ giving the maximal volume polyhedron with eight vertices.

As concluding remarks Berman and Hanes raised the following questions:

Problem 2.1 *For which types of polyhedra does Property Z determine a unique polyhedron. More generally, for each isomorphism class of polyhedra is there one and only one polyhedron (up to congruence) which gives a relative maximum for the volume?*

Problem 2.2 *For $n = 4, \ldots, 7$ the duals of the polyhedra of maximum volume are just those polyhedra with n faces circumscribed about the unit sphere of minimum volume. For $n = 8$ the dual of the maximal volume polyhedron (described above) is*

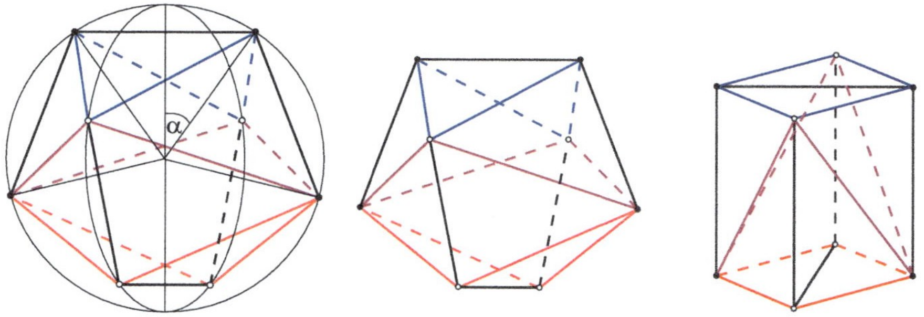

Fig. 1 The medial complex with 8 vertices and its two polyhedra, the maximal volume polyhedron and the cube

the best known solution to the isoperimetric problem for polyhedra with 8 faces. Is this true in general?

Recently there is no answer for these questions.

We have to mention a theorem of A. Florian ([22]) which immediately implies the inequalities in (3). Let P be a convex polyhedron with v vertices and volume $\mathrm{vol}(P)$. We consider an orthoscheme $T = OABC$ (where OA orthogonal to the plane ABC, and AB orthogonal to BC) with the properties:

(i) the radial projection of ABC onto the unit sphere with centre O is the spherical triangle $T' = A'B'C'$ given by

$$C'A'B'\sphericalangle = \frac{\pi}{3}, \quad A'B'C'\sphericalangle = \frac{\pi}{2}, \quad \mathrm{area}(T') = \frac{4\pi}{12(v-2)}$$

(ii) $\mathrm{vol}(T) = \frac{1}{12(v-2)} \mathrm{vol}(P)$.

Then we have:

Theorem 2.2 ([22]) *Let $K(\rho)$ be a ball with centre O and radius ρ. Let P be a convex polyhedron with v vertices and volume* $\mathrm{vol}(P)$*, and let the tetrahedron T be defined as above. Then*

$$\mathrm{vol}(P \cap K(\rho)) \leq 12(v-2)\,\mathrm{vol}(T \cap K(\rho)) \tag{4}$$

with equality if $v = 4, 6$ or 12 and P is a regular tetrahedron, octahedron or icosahedron with centre O. When $|OA| \leq \rho \leq |OC|$, these are the only cases of equality.

We recall the paper of Mutoh [12] who presented the results of a computer-aided search for optimal polyhedra with $4 \leq n \leq 30$ vertices. The solutions of the computation probably solved the mentioned cases, respectively, however there is no information in the paper either on the source code of the program or the algorithm which based the computation. Table 1 describes some of those polyhedra which suggested by Mutoh as the maximal volume one inscribed in the unit sphere. We refer here only a part of the complete table of Mutoh, for more information see the original paper [12].

Mutoh notes that it seems to be that the conjecture of Grace on medial polyhedron is false because the polyhedra found by computer in the cases $n = 11$ and $n = 13$ are not medial ones, respectively. Mutoh also listed the polyhedra circumscribed to the unit sphere with minimal volume and examined the dual conjecture of Goldberg (see also Problem 2.2). He said: "*Goldberg conjectured that the polyhedron of maximal volume inscribed to the unit sphere and the polyhedron of minimal volume circumscribed about the unit sphere are dual. A comparison of Table 1 and 3 shows that the number of vertices and the number of faces of the two class of polyhedra correspond with each other. The degrees of vertices of the polyhedra of maximal volume inscribed in the unit sphere correspond to the numbers of vertices of faces of the polyhedra of minimal volume circumscribed about the unit sphere. Indeed, the*

Table 1 Computer search results of polyhedra of maximal volume inscribed in the unit sphere

Number of vertices	Maximal volume	Number of facets	Valences of vertices
4	0.51320010	4	3×4
5	0.86602375	6	$3 \times 2, 4 \times 3$
6	1.33333036	8	4×6
7	1.58508910	10	$4 \times 5, 5 \times 2$
8	1.81571182	12	$4 \times 4, 5 \times 4$
9	2.04374046	14	$4 \times 3, 5 \times 6$
10	2.21872888	16	$4 \times 2, 5 \times 8$
11	2.35462915	18	$4 \times 2, 5 \times 8, 6 \times 1$
12	2.53614471	20	5×12
⋮	⋮	⋮	⋮
30	3.45322727	56	$5 \times 12, 6 \times 18$

volume of polyhedra whose vertices are the contact points of the unit sphere and the polyhedra circumscribed about the unit sphere differs only by 0.07299% from the volume of the polyhedra inscribed in the unit sphere."

We turn to a recent result that generalizes the triangle case of the inequality (1) of L. Fejes-Tóth. If A, B, C are three points on the unit sphere we can consider two triangles, one of the corresponding spherical triangle and the second one the rectilineal triangle with these vertices, respectively. Both of them are denoted by ABC. The angles of the rectilineal triangle are the half of the angles between those radius of the circumscribed circle which connect the center K of the rectilineal triangle ABC to the vertices A, B, C. Since K is also the foot of the altitude of the tetrahedron with base ABC and apex O, hence the angles α_A, α_B and α_C of the rectilineal triangle ABC, play an important role in our investigations, we refer to them as the *central angles* of the spherical edges BC, AC and AB, respectively. We call again the tetrahedron $ABCO$ the *facial tetrahedron* with base ABC and apex O.

Lemma 2.2 (See in [23]) *Let ABC be a triangle inscribed in the unit sphere. Then there is an isosceles triangle $A'B'C'$ inscribed in the unit sphere with the following properties:*

- *the greatest central angles and also the spherical areas of the two triangles are equal to each other, respectively;*
- *the volume of the facial tetrahedron with base $A'B'C'$ is greater than or equal to the volume of the facial tetrahedron with base ABC.*

From Lemma 2.2 it can be proved upper bound functions for the volume of the facial tetrahedron.

Proposition 2.1 *Let the spherical area of the spherical triangle ABC be τ. Let α_C be the greatest central angle of ABC corresponding to AB. Then the volume V of the facial tetrahedron $ABCO$ holds the inequality*

$$V \leq \frac{1}{3}\tan\frac{\tau}{2}\left(2-\frac{|AB|^2}{4}\left(1+\frac{1}{(1+\cos\alpha_C)}\right)\right). \tag{5}$$

In terms of τ and $c := AB$ we have

$$V \leq v(\tau, c) := \frac{1}{6}\sin c\frac{\cos\frac{\tau-c}{2}-\cos\frac{\tau}{2}\cos\frac{c}{2}}{1-\cos\frac{c}{2}\cos\frac{\tau}{2}}. \tag{6}$$

Equality holds if and only if $|AC| = |CB|$.

Observe that the function $v(\tau, c)$ is concave in the parameter domain $\mathcal{D} := \{0 < \tau < \pi/2, \tau \leq c < \min\{f(\tau), 2\sin^{-1}\sqrt{2/3}\}\}$ with certain concave (in τ) function $f(\tau)$ defined by the zeros of the Hessian; and non-concave in the domain $\mathcal{D}' = \{0 < \tau \leq \omega, f(\tau) \leq c \leq 2\sin^{-1}\sqrt{2/3}\} = \{0 < \tau \leq c \leq \pi/2\} \setminus D$, where $f(\omega) = 2\sin^{-1}\sqrt{2/3}$.

Assume now that the triangular star-shaped polyhedron P with f face inscribed in the unit sphere. Let $c_1, \ldots, c_f$ be the arc-lengths of the edges of the faces $F_1, \ldots, F_f$ corresponding to their maximal central angles, respectively. Denote by τ_i the spherical area of the spherical triangle corresponding to the face F_i for all i. We note that for a spherical triangle which edges a, b, c hold the inequalities $0 < a \leq b \leq c < \pi/2$, also holds the inequality $\tau \leq c$. In fact, for fixed τ the least value of the maximal edge length attend at the case of regular triangle. If $c < \pi/2$ then we have

$$\tan\frac{\tau}{4} = \left(\tan\frac{c}{4}\sqrt{\tan\frac{3c}{4}\tan\frac{c}{4}}\right) = \left(\tan\frac{c}{4}\sqrt{1-\frac{\tan\frac{3c}{4}+\tan\frac{c}{4}}{\tan c}}\right) < \tan\frac{c}{4},$$

and if $c = \pi/2$ then $\tau = 8\pi/4 = \pi/2$ proving our observation.

The following theorem gives an upper bound on the volume of the star-shaped polyhedron corresponding to the given spherical tiling in question.

Theorem 2.3 (See in [23]) *Assume that $0 < \tau_i < \pi/2$ holds for all i. For $i = 1, \ldots, f'$ we require the inequalities $0 < \tau_i \leq c_i \leq \min\{f(\tau_i), 2\sin^{-1}\sqrt{2/3}\}$ and for all j with $j \geq f'$ the inequalities $0 < f(\tau_j) \leq c_j \leq 2\sin^{-1}\sqrt{2/3}$, respectively. Let denote $c' := \frac{1}{f'}\sum_{i=1}^{f'} c_i$, $c^\star := \frac{1}{f-f'}\sum_{i=f'+1}^{f} f(\tau_i)$ and $\tau' := \sum_{i=f'+1}^{f} \tau_i$, respectively. Then we have*

$$v(P) \leq \frac{f}{6}\sin\left(\frac{f'c'+(f-f')c^\star}{f}\right)\frac{\cos\left(\frac{4\pi-f'c'-(f-f')c^\star}{2f}\right)-\cos\frac{2\pi}{f}\cos\left(\frac{f'c'+(f-f')c^\star}{2f}\right)}{1-\cos\frac{4\pi}{2f}\cos\left(\frac{f'c'+(f-f')c^\star}{2f}\right)}. \tag{7}$$

2.2 *The Cases of Higher Dimensions*

As we saw in the previous subsection, even the 3-dimensional case is completely proved only when the number of vertices less or equal to eight. This shows that in higher dimensions we cannot expect such complete results as was published by L. Fejes-Tóth, A. Flórian or Berman and Hanes in the second half on the last century, respectively. As Flórian said in [22]: "Several extremum properties of the regular triangle and the regular tetrahedron may be generalized to regular simplices in all dimensions.... Little is known in this respect about the general cross polytope, the hypercube and the nontrivial regular convex polytopes in 4-space". "Some extremum properties of these polytopes were established by comparing them with the topologically isomorphic convex polytopes.... But no methods are available for proving inequalities analogous to (3)." We now extract the method of Berman and Hanes to higher dimensions and using a combinatorial concept, the idea of Gale's transform solve some cases of few vertices. In this subsection we collect the results of the paper [24].

The first step is the generalization of Lemma 2.1 for arbitrary dimensions.

Lemma 2.3 *Consider a polytope $P \in \mathcal{P}_d(n)$ satisfying Property Z. For any $p \in V(P)$, let $\mathcal{F}_p$ denote the family of the facets of $\mathcal{C}(P)$ containing p. For any $F \in \mathcal{F}_p$, set*

$$A(F, p) = \mathrm{vol}_{d-1}\left(\mathrm{conv}\left((V(F) \cup \{o\}) \setminus \{p\}\right)\right),$$

and let $m(F, p)$ be the unit normal vector of the hyperplane, spanned by $(V(F) \cup \{o\}) \setminus \{p\}$, pointing in the direction of the half space containing p.

(2.3.1) Then we have $p = m/|m|$, where $m = \sum_{F \in \mathcal{F}_p} A(F, p) m(F, p)$.
(2.3.2) Furthermore P is simplicial.

Remark 2.1 Assume that $P \in \mathcal{P}_d(n)$ satisfies Property Z, and for some $p \in V(P)$, all the vertices of P adjacent to p are contained in a hyperplane H. Then the supporting hyperplane of $\mathbb{S}^{d-1}$ at p is parallel to H, or in other words, p is a normal vector to H. Thus, in this case all the edges of P, starting at p, are of equal length.

Lemma 2.4 *Let $P \in \mathcal{P}_d(n)$ satisfy Property Z, and let $p \in V(P)$. Let $q_1, q_2 \in V(P)$ be adjacent to p. Assume that any facet of P containing p contains at least one of q_1 and q_2, and for any $S \subset V(P)$ of cardinality $d-2$, $\mathrm{conv}(S \cup \{p, q_1\})$ is a facet of P not containing q_2 if, and only if $\mathrm{conv}(S \cup \{p, q_2\})$ is a facet of P not containing q_1. Then $|q_1 - p| = |q_2 - p|$.*

Corollary 2.1 is a straightforward consequence of Lemma 2.4 or, equivalently, Remark 2.1.

Corollary 2.1 *If $P \in \mathcal{P}_d(d+1)$ and $\mathrm{vol}_d(P) = v_d(d+1)$, then P is a regular simplex inscribed in $\mathbb{S}^{d-1}$.*

We note that this statement can be considered as a folklore. The analogous statement in d-dimensional spherical geometry (for simplices inscribed in a sphere of $\mathcal{S}^d$ with radius less than $\pi/2$) was proved by Böröczky in [25]. The method of Böröczky is based on the fact that Steiner's symmetrization is a volume-increasing transformation of the spherical space and so it can not be transformed immediately to the hyperbolic case. In hyperbolic spaces the investigations concentrated only to the simplices with ideal vertices. In dimension two every two triangles with ideal vertices are congruent to each other implying that they have the same area which value is maximal one among the triangles. On the other hand it was proved by Milnor (see in [26] or in [27]) that in hyperbolic 3-space, a simplex is of maximal volume if and only if it is ideal and regular. The same d-dimensional statement has been proved by Haagerup and Munkholm in [28]. This motivates the following:

Problem 2.3 *Prove or disprove that in hyperbolic d-space a simplex is of maximal volume inscribed in the unit sphere if and only if it is a regular one.*

Before the next corollary recall that if K is a $(d-1)$-polytope in $\mathbb{R}^d$, and $[p_1, p_2]$ is a segment intersecting the relative interior of K at a singleton different from p_1 and p_2, then $\operatorname{conv}(K \cup [p_1, p_2])$ is a *d-bipyramid* with base K and apexes p_1, p_2 (cf. [29]). In the literature the terminology "bipyramid" is more prevalent as of the nomenclature "double-pyramid" of Berman and Hanes. In the rest of this paper we use bipyramid.

Corollary 2.2 *Let $P \in \mathcal{P}_d(n)$ be combinatorially equivalent to a d-bipyramid. Assume that P satisfies Property Z. Then P is a d-bipyramid, its apexes p_1, p_2 are antipodal points, its base K and $[p_1, p_2]$ lie in orthogonal linear subspaces of $\mathbb{R}^d$, and K satisfies Property Z in the hyperplane* aff K.

Corollary 2.2 implies the following one:

Corollary 2.3 *If $P \in \mathcal{P}_d(2d)$ has maximal volume in the combinatorial class of cross-polytopes inscribed in $\mathbb{S}^{d-1}$, then it is a regular cross-polytope.*

The first non-trivial case is when the number of points is equal to $n = d + 2$. It has been proved:

Theorem 2.4 ([24]) *Let $P \in \mathcal{P}_d(d+2)$ have maximal volume over $\mathcal{P}_d(d+2)$. Then $P = \operatorname{conv}(P_1 \cup P_2)$, where P_1 and P_2 are regular simplices of dimensions $\lfloor \frac{d}{2} \rfloor$ and $\lceil \frac{d}{2} \rceil$, respectively, inscribed in $\mathbb{S}^{d-1}$, and contained in orthogonal linear subspaces of $\mathbb{R}^d$. Furthermore,*

$$v_d(d+2) = \frac{1}{d!} \cdot \frac{(\lfloor d/2 \rfloor + 1)^{\frac{\lfloor d/2 \rfloor + 1}{2}} \cdot (\lceil d/2 \rceil + 1)^{\frac{\lceil d/2 \rceil + 1}{2}}}{\lfloor d/2 \rfloor^{\frac{\lfloor d/2 \rfloor}{2}} \cdot \lceil d/2 \rceil^{\frac{\lceil d/2 \rceil}{2}}}$$

In the proof of the results on d-polytopes with $d + 2$ or $d + 3$ vertices, we use extensively the properties of the so-called Gale transform of a polytope (cf. [29, 30]). Since the application of this combinatorial theory leads to a new method in the investigation of our problem we review it.

Consider a d-polytope P with vertex set $V(P) = \{p_i : i = 1, 2, \dots, n\}$. Regarding $\mathbb{R}^d$ as the hyperplane $\{x_{d+1} = 1\}$ of $\mathbb{R}^{d+1}$, we can represent $V(P)$ as a $(d+1) \times n$ matrix M, in which each column lists the coordinates of a corresponding vertex in the standard basis of $\mathbb{R}^{d+1}$. Clearly, this matrix has rank $d+1$, and thus, it defines a linear mapping $L : \mathbb{R}^n \to \mathbb{R}^{d+1}$, with $\dim \ker L = n - d - 1$. Consider a basis $\{w_1, w_2, \dots, w_{n-d-1}\}$ of $\ker L$, and let $\bar{L} : \mathbb{R}^{n-d-1} \to \mathbb{R}^n$ be the linear map mapping the ith vector of the standard basis of $\mathbb{R}^{n-d-1}$ into w_i. Then the matrix $\bar{M}$ of $\bar{L}$ is an $n \times (n-d-1)$ matrix of (maximal) rank $n-d-1$, satisfying the equation $M\bar{M} = O$, where O is the matrix with all entries equal to zero. Note that the rows of $\bar{M}$ can be represented as points of $\mathbb{R}^{n-d-1}$. For any vertex $p_i \in V(P)$, we call the ith row of $\bar{M}$ the *Gale transform of* p_i, and denote it by $\bar{p}_i$. Furthermore, the n-element multiset $\{\bar{p}_i : i = 1, 2, \dots, n\} \subset \mathbb{R}^{n-d-1}$ is called the *Gale transform of* P, and is denoted by $\bar{P}$. If $\operatorname{conv} S$ is a face of P for some $S \subset V(P)$, then the (multi)set of the Gale transform of the points of S is called a face of $\bar{P}$. If $\bar{S}$ is a face of $\bar{P}$, then $\bar{P} \setminus \bar{S}$ is called a *coface* of $\bar{P}$.

Let $V = \{q_i : i = 1, 2, \dots, n\} \subset \mathbb{R}^{n-d-1}$ be a (multi)set. We say that V is a *Gale diagram* of P, if for some Gale transform P' the conditions $o \in \operatorname{relint} \operatorname{conv}\{q_j : j \in I\}$ and $o \in \operatorname{relint} \operatorname{conv}\{\bar{p}_j : j \in I\}$ are satisfied for the same subsets of $\{1, 2, \dots, n\}$. If $V \subset \mathbb{S}^{n-d-2}$, then V is a *normalized Gale diagram* (cf. [31]). A *standard Gale diagram* is a normalized Gale diagram in which the consecutive diameters are equidistant. A *contracted Gale diagram* is a standard Gale diagram which has the least possible number of diameters among all isomorphic diagrams. We note that each d-polytope with at most $d+3$ vertices may be represented by a contracted Gale diagram (cf. [29] or [30]). An important tool of the proofs the following theorem from [29] or also from [30].

Theorem 2.5 ([29, 30])

(i) A multiset $\bar{P}$ of n points in $\mathbb{R}^{n-d-1}$ is a Gale diagram of a d-polytope P with n vertices if and only if every open half-space in $\mathbb{R}^{n-d-1}$ bounded by a hyperplane through o contains at least two points of $\bar{V}$ (or, alternatively, all the points of $\bar{P}$ coincide with o and then $n = d+1$ and P is a d-simplex).

(ii) If F is a facet of P, and Z is the corresponding coface, then in any Gale diagram $\bar{V}$ of P, $\bar{Z}$ is the set of vertices of a (non-degenerate) set with o in its relative interior.

(iii) A polytope P is simplicial if and only if, for every hyperplane H containing $o \in \mathbb{R}^{n-d-1}$, we have $o \notin \operatorname{relint} \operatorname{conv}(\bar{V} \cap H)$.

(iv) A polytope P is a pyramid if and only if at least one point of $\bar{V}$ coincides with the origin $o \in \mathbb{R}^{n-d-1}$.

We note that (ii) can be stated in a more general form: F is a face of P if, and only if, for the corresponding co-face $\bar{F}$ of P, we have $o \in \operatorname{int} \operatorname{conv} \bar{Z}$.

Before stating the result on $n = d+3$ vertices, recall that a d-polytope with n vertices is *cyclic*, if it is combinatorially equivalent to the convex hull of n points on the moment curve $\gamma(t) = (t, t^2, \dots, t^d)$, $t \in \mathbb{R}$.

Theorem 2.6 ([24]) *Let $P \in \mathcal{P}_d(d+3)$ satisfy Property Z. If P is even, assume that P is not cyclic. Then $P = \text{conv}\{P_1 \cup P_2 \cup P_3\}$, where P_1, P_2 and P_3 are regular simplices inscribed in $\mathbb{S}^{d-1}$ and contained in three mutually orthogonal linear subspaces of $\mathbb{R}^d$. Furthermore:*

- *If d is odd and P has maximal volume over $\mathcal{P}_d(d+3)$, then the dimensions of P_1, P_2 and P_3 are $\lfloor d/3 \rfloor$ or $\lceil d/3 \rceil$. In particular, in this case we have*

$$(v_d(d+3) =) \operatorname{vol}_d(P) = \frac{1}{d!} \cdot \prod_{i=1}^{3} \frac{(k_i+1)^{\frac{k_i+1}{2}}}{k_i^{\frac{k_i}{2}}},$$

  where $k_1 + k_2 + k_3 = d$ and for every i, we have $k_i \in \left\{\lfloor \frac{d}{3} \rfloor, \lceil \frac{d}{3} \rceil\right\}$.
- *The same holds if d is even and P has maximal volume over the family of* not cyclic *elements of $\mathcal{P}_d(d+3)$.*

Remark 2.2 Let $d = 2m$ be even and $P \in \mathcal{P}_d(d+3)$ be a cyclic polytope satisfying Property Z. Then we need to examine the case that $\bar{P}$ is the vertex set of a regular $(2m+3)$-gon. Let the vertices of $\bar{P}$ be $\bar{p}_i, i = 1, 2, \ldots, 2m+3$ in counterclockwise order. Applying the method of the proof of Theorem 2.6, one can deduce that for every i, we have that $|p_{i-m-1} - p_i| = |p_{i+m+1} - p_i|$. On the other hand, for any other pair of vertices the conditions of Lemma 2.4 are not satisfied.

In the light of Theorem 2.6, it seems interesting to find the maximum volume cyclic polytopes in $\mathcal{P}_d(d+3)$, with d even. With regard to Remark 2.2, it is not unreasonable to consider the possibility that the answer for this question is a polytope $P = \text{conv}\{p_i : i = 1, 2, \ldots, d+3\}$ having a certain cyclic symmetry (if at all it is possible), namely that for any integer k, the value of $|p_{i+k} - p_i|$ is independent from i.

The following observation can be found both in [32], or, as an exercise, in [30].

Remark 2.3 Let $d \geq 2$ be even, and $n \geq d+3$. Let

$$C_d(n) = \sqrt{\frac{2}{d}} \operatorname{conv}\left\{\left(\cos\frac{i\pi}{n}, \sin\frac{i\pi}{n}, \cos\frac{2i\pi}{n}, \ldots, \cos\frac{di\pi}{2n}, \sin\frac{di\pi}{2n}\right) : i = 0, 1, \ldots, n-1\right\}.$$

Then $C(n,d)$ is a cyclic d-polytope inscribed in $\mathbb{S}^{d-1}$, and $\text{Sym}(C_d(n)) = D_n$.

It can be shown that for $d = 4, 6$ the only "symmetric" representations of a cyclic d-polytope with d even and $n = d+3$ are those congruent to $C_d(d+3)$. Using the concepts of *Löwner ellipsoid* it can be proved the following theorem:

Theorem 2.7 ([24]) *Let $P \in \mathcal{C}_d(d+3)$ be a cyclic polytope, where $d = 4$ or $d = 6$, and let $V(P) = \{p_i : i = 1, 2, \ldots, d+3\}$. If, for every value of k, $|p_{i+k} - p_i|$ is independent of the value of i, then P is congruent to $C_d(d+3)$.*

The above investigations raised a lot of questions and problems without answers. We collect some of them in the rest of this section.

Problem 2.4 *Prove or disprove that in $\mathcal{P}_d(d+3)$, the cyclic polytopes with maximal volume are the congruent copies of $C_d(d+3)$. In particular, is it true for $C_4(7)$? Is it true that any cyclic polytope in $\mathcal{P}_d(d+3)$ satisfying Property Z is congruent to $C_d(d+3)$?*

A straightforward computation shows that $C_4(7)$ satisfies Property Z. We note that in $\mathcal{P}_d(d+2)$, polytopes with maximal volume are cyclic, whereas in $\mathcal{P}_d(d+3)$, where d is odd, they are not. This leads to the following:

Problem 2.5 *Is it true that if $P \in \mathcal{P}_d(d+3)$, where d is even, has volume $v_d(d+3)$, then P is not cyclic?*

Remark 2.4 Let $P_4 \in \mathcal{P}_4(7)$ be the convex hull of a regular triangle and two diameters of $\mathbb{S}^3$, in mutually orthogonal linear subspaces. Furthermore, let $P_6 \in \mathcal{P}_6(9)$ be the convex hull of three regular triangles, in mutually orthogonal linear subspaces. One can check that

$$\mathrm{vol}_4(P_4) = \frac{3}{4} = 0.43301\ldots > \mathrm{vol}_4(C_4(7)) = \frac{49}{192}\left(\cos\frac{\pi}{7} + \cos\frac{2\pi}{7}\right) = 0.38905\ldots.$$

In addition,

$$\mathrm{vol}_6(C_6(9)) = \frac{7}{576}\sin\frac{\pi}{9} - \frac{7}{2880}\sin\frac{4\pi}{9} + \frac{7}{1152}\sin\frac{2\pi}{9} = 0.01697\ldots$$

and

$$\mathrm{vol}_6(P_6) = \frac{9\sqrt{3}}{640} = 0.02435\ldots > \mathrm{vol}_6(C_6(9)).$$

This suggests that the answer for Problem 2.5 is yes.

Remark 2.5 Using the idea of the proof of Theorem 2.7, for any small value of n, it may identify the polytopes having D_n as a subgroup of their symmetry groups. Nevertheless, it were unable to apply this method for general n, due to computational complexity. The authors [24] carried out the computations for $5 \le n \le 9$, and obtained the following polytopes, up to homothety:

- regular $(n-1)$-dimensional simplex in $\mathbb{R}^{n-1}$ for every n,
- regular n-gon in $\mathbb{R}^2$ for every n,
- $C_4(n)$ with $n = 6, 7, 8, 9$ and $C_6(n)$ with $n = 8, 9$,
- regular cross-polytope in $\mathbb{R}^3$ and $\mathbb{R}^4$,
- the polytope P_6 in $\mathbb{R}^6$, defined in Remark 2.4,
- the 3-polytope P with

$$V(P) = \left\{(1,0,0), \left(-\frac{2}{3}, -\frac{2}{3}, \frac{1}{3}\right), (0,1,0), \left(\frac{1}{3}, -\frac{2}{3}, -\frac{2}{3}\right), (0,0,1), \left(-\frac{2}{3}, \frac{1}{3}, -\frac{2}{3}\right)\right\}.$$

We note that, for d odd, the symmetry group of a cyclic d-polytope with $n \ge d+3$ vertices is $\mathbb{Z}_2 \times \mathbb{Z}_2$ (cf. [32]). Thus, the only cyclic polytopes in the above list are

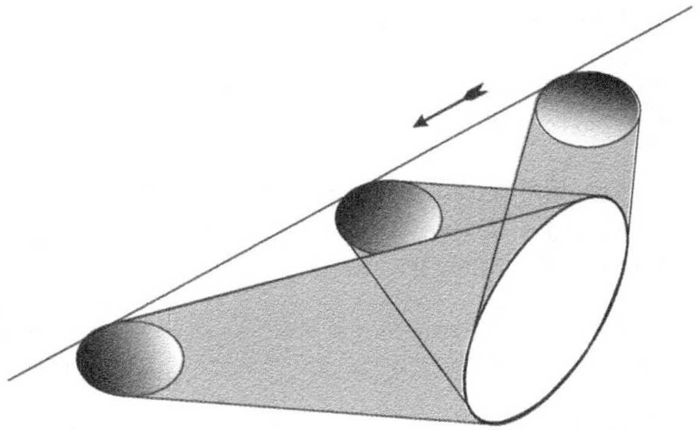

Fig. 2 The change of the convex hull

simplices and those homothetic to $C_d(n)$ for some values of n and d. This leads to the following question.

Problem 2.6 *Is it true that if, for some $n \geq d+3 \geq 5$, a cyclic polytope $P \in \mathcal{P}_d(n)$ satisfies $\mathrm{Sym}(P) = D_n$, then P is congruent to $C_d(n)$?*

3 Volume of the Convex Hull of Two Connecting Bodies

3.1 On the Volume Function of the Convex Hull of Two Convex Body

Following the chronology, we have to start here with a result of Fáry and Rédei from 1950 ([33]). They investigated the volume function defined on the convex hull of two convex bodies (Fig. 2). He proved that if one of the bodies moves on a line with constant velocity then the volume of the convex hull is a convex function of the time (see Satz.4 in [33]). It was also proved later in [15], and for convex polyhedra of dimension three in [34].

Theorem 3.1 ([15, 33, 34]) *The real valued function g of the real variable x defined by the fixed vector t and the formula*

$$g(x) := \mathrm{Vol}(\mathrm{conv}(K \cup (K' + t(x))), \textit{ where } t(x) := xt,$$

is convex.

The nice proof in [34] is based on the observation that the volume change function (by a translation in the direction of a line) can be calculated and it is an increasing function. Since it is also the derivative of g we get that g is convex. This calculation for the volume change can be done in the general case, too. Consider the shadow boundary of the convex hull $\mathrm{conv}(K \cup (K' + t))$ with respect to the line of translation t. This is an $(d-2)$-dimensional topological manifold separating the boundary of $\mathrm{conv}(K \cup (K' + t))$ into two domains, the front and back sides of it, respectively. (The translation t can be considered as a motion, hence the respective concepts of front and back sides can be regarded with respect to the direction of it.) Regarding a hyperplane H orthogonal to t the front side and back side are graphs of functions over the orthogonal projection X of $\mathrm{conv}(K \cup (K' + t))$ onto H. Thus the volume change in t can be calculated by the formula

$$g'(t) = \lim_{\varepsilon \to 0} \int_X (f^{t+\varepsilon}(X) - f^t(X)) + \int_X (b^{t+\varepsilon}(X) - b^t(X)),$$

where, at the moment t, f^t and b^t are the graphs of the front ad back sides, respectively. Since X is independent from t and for fixed X the functions

$$f^{t+\varepsilon}(x) - f^t(x) \text{ and } b^{t+\varepsilon}(x) - b^t(x)$$

in t are increasing and decreasing, respectively, we get that g' is also increasing in t implying that g is convex.

As a corollary we get the following:

Corollary 3.1 (see in [18]) *If we have two convex, compact bodies K and K' of the Euclidean space of dimension n and they are moving uniformly on two given straight lines then the volume of their convex hull is a convex function of the time.*

Remark 3.1 We emphasize that the statement of Theorem 3.1 is not true in hyperbolic space: Let K be a segment and K' be a point which goes on a line in the pencil of the rays ultraparallel to the line of the segment. Since the area function of the triangle defined by the least convex hull of K and K' is bounded (from below and also from above) it cannot be a convex function.

Note that if the "bodies" are points the statement simplified to a proposition of absolute geometry which implies e.g. the existence of the normal transversal of two skew lines in the hyperbolic space.

There are several applications of Theorem 3.1. In the paper of Hee-Kap Ahn, Peter Brass and Chan-Su Shin (see [34]) the following result, based on Lemma 3.1 appears.

Theorem 3.2 (See Theorem 3 in [34]) *Given two convex polyhedra P and Q in three-dimensional space, we can compute the translation vector t of Q that minimizes* $\mathrm{vol}(\mathrm{conv}(P \cup (Q + t)))$ *in expected time $O(n^3 \log^4 n)$. The d-dimensional problem can be solved in expected time $O(n^{d+1-3/d}(log n)^{d+1})$.*

In [33], Fáry and Rédey introduced the concepts of inner symmetricity (or outer symmetricity) of a convex body with the ratio (or inverse ratio) of the maximal (or minimal) volumes of the centrally symmetric bodies inscribed in (or circumscribed about) the given body. Using the mentioned Theorem 3.1 (and also its counterpart on the concavity of the volume function of the intersection of two bodies one of them from which moving on a line with constant velocity), they determined the inner symmetricity (and also the outer symmetricity) of a simplex (see Satz 5., resp. Satz 6. in [33]). It has been proved that if S is a simplex of dimension n then its inner symmetricity $c_\star(S)$ is equal to

$$c_\star(S) = \frac{1}{(n+1)^n} \sum_{0 \le \nu \le \frac{n+1}{2}} (-1)^\nu \binom{n+1}{\nu} (n+1-2\nu)^n. \tag{8}$$

On outer symmetricity $c^\star(S)$ of a simplex they proved that it is equal to

$$c^\star(S) = \frac{1}{\binom{n}{n_0}}, \tag{9}$$

where $n_0 = n/2$ if n is even and $n_0 = (n-1)/2$ if n is an odd number. The above values attain when we consider the volume of the intersection (or the convex hull of the union) of S with its centrally reflected copy S_O (taking the reflection at the centroid O of S).

Horváth and Lángi in [17] introduced the following quantity.

Definition 3.1 For two convex bodies K and L in $\mathbb{R}^d$, let

$$c(K, L) = \max\left\{\mathrm{vol}_d(\mathrm{conv}(K' \cup L')) : K' \cong K, L' \cong L \text{ and } K' \cap L' \neq \emptyset\right\}.$$

Furthermore, if $\mathcal{S}$ is a set of isometries of $\mathbb{R}^d$, we set

$$c(K|\mathcal{S}) = \frac{1}{\mathrm{vol}(K)} \max\left\{\mathrm{vol}_d(\mathrm{conv}(K \cup K')) : K \cap K' \neq \emptyset, K' = \sigma(K) \text{ for some } \sigma \in \mathcal{S}\right\}.$$

A quantity similar to $c(K, L)$ was defined by Rogers and Shephard [15], in which congruent copies were replaced by translates. It has been shown that the minimum of $c(K|\mathcal{S})$, taken over the family of convex bodies in $\mathbb{R}^d$, is its value for a d-dimensional Euclidean ball, if $\mathcal{S}$ is the set of translations or that of reflections about a point. Nevertheless, their method, approaching a Euclidean ball by suitable Steiner symmetrizations and showing that during this process the examined quantities do not increase, does not characterize the convex bodies for which the minimum is attained; they conjectured that, in both cases, the minimum is attained only for ellipsoids (cf. p. 94 of [15]). We note that the method of Rogers and Shephard [15] was used also in [13]. The results of the mentioned work based on the concept of *linear parameter system of convex sets* and such a generalization of Theorem 3.1 which has interest on its own-right, too.

Definition 3.2 ([15]) Let I be an arbitrary index set, with each member i of which is associated a point a_i in d-dimensional space, and a real number λ_i, where the sets $\{a_i\}_{i\in I}$ and $\{\lambda_i\}_{i\in I}$ are each bounded. If e is a fixed point and t is any real number, $A(t)$ denotes the set of points

$$\{a_i + t\lambda_i e\}_{i\in I},$$

and $C(t)$ is the least convex cover of this set of points, then the system of convex sets $C(t)$ is called a *linear parameter system*.

The authors proved (see Lemma 1 in [15]) that the volume $V(t)$ of the set $C(t)$ of a linear parameter system is a convex function of t. They noted that this result should be contrasted with that for a linear system of convex bodies as defined by Minkowski, where the d-th root of the volume of the body with parameter t is a concave function of t in its interval of definition.

In this paper we prove the following results:

Theorem 3.3 *Let H and K be two bodies and denote by $C(H, K)$ the least convex cover of the union of H and K. Furthermore let $V^\star(H, K)$ denote the maximum, taken over all point x for which the intersection $H \cap (K + x)$ is not empty, of the volume $\mathrm{vol}_d(C(H, K + x))$ of the set $C(H, K + x)$. Then $V^\star(H, K) \geq V^\star(SH, SK)$, where SH denotes the closed d-dimensional sphere with centre at the origin and with volume equal to that of H.*

Theorem 3.4 *If K is a convex body in d-dimensional space, then*

$$1 + \frac{2J_{d-1}}{J_d} \leq \frac{\mathrm{vol}_d(R^\star K)}{\mathrm{vol}_d(K)} \leq 2^d,$$

where J_d is the volume of the unit sphere in d-dimensional space, $R^\star K$ is the number to maximize with respect to a point a of K the volumes of the least centrally symmetric convex body with centre a and containing K. Equality holds on the left, if K is an ellipsoid; and on the right, if, and only if, K is a simplex.

Theorem 3.5 *If K is centrally symmetric body in d-dimensional space, then*

$$1 + \frac{2J_{d-1}}{J_d} \leq \frac{\mathrm{vol}_d(R^\star K)}{\mathrm{vol}_d(K)} \leq 1 + d$$

Equality holds on the left if K is an ellipsoid, and on the right if K is any centrally symmetric double-pyramid on a convex base.

Theorem 3.6 *If K is a convex body in d-dimensional space, then*

$$1 + \frac{2J_{d-1}}{J_d} \leq \frac{\mathrm{vol}_d(T^\star K)}{\mathrm{vol}_d(K)} \leq 1 + d,$$

where $T^\star K$ denotes the so-called translation body *of K. This is the body for which the volume of $K \cap (K + x) \neq$ and the volume of $C(K, K + x)$ is maximal one. Equality holds on the left if K is an ellipsoid, and on the right if K is a simplex.*

Theorem 3.7 *Let K be a convex body in d-dimensional space. Then there is a direction such that the volume of each cylinder Z, circumscribed to K, with its generators in the given direction, satisfies*

$$\frac{\mathrm{vol}_d(Z)}{\mathrm{vol}_d(K)} \geq \frac{2J_{d-1}}{J_d}.$$

It can be seen that these statements connect with the problem to determine the number $c(K|\mathcal{S})$ defined in Definition 3.1. In fact, Horváth and Lángi (in [17]) treated these problems in a more general setting. Let $c_i(K)$ be the value of $c(K|\mathcal{S})$, where $\mathcal{S}$ is the set of reflections about the i-flats of $\mathbb{R}^d$, and $i = 0, 1, \ldots, d-1$. Similarly, let $c^{tr}(K)$ and $c^{co}(K)$ be the value of $c(K|\mathcal{S})$ if $\mathcal{S}$ is the set of translations and that of all the isometries, respectively. In [17] the authors examined the minima of these quantities. In particular, in Theorem 3.8, was given another proof that the minimum of $c^{tr}(K)$, over the family of convex bodies in Re^n, is its value for Euclidean balls, and it was shown also that the minimum is attained if, and only if, K is an ellipsoid. This verifies the conjecture in [15] for translates.

Presented similar results about the minima of $c_1(K)$ and $c_{d-1}(K)$, respectively. In particular, the authors proved that, over the family of convex bodies, $c_1(K)$ is minimal for ellipsoids, and $c_{n-1}(K)$ is minimal for Euclidean balls. The first result proves the conjecture of Rogers and Shephard for copies reflected about a point.

During the investigation, $\mathcal{K}_d$ denotes the family of d-dimensional convex bodies. For any $K \in \mathcal{K}_d$ and $u \in \mathbb{S}^{n-1}$, $K|u^{\perp}$ denotes the orthogonal projection of K into the hyperplane passing through the origin o and perpendicular to u. The *polar* of a convex body K is denoted by K°. The denotation J_d of the paper [15] we are changing to the more convenient one v_d.)

The propositions are the followings:

Theorem 3.8 *For any $K \in \mathcal{K}_d$ with $d \geq 2$, we have $c^{tr}(K) \geq 1 + \frac{2v_{d-1}}{v_d}$ with equality if, and only if, K is an ellipsoid.*

We remark that a theorem related to Theorem 3.8 can be found in [35]. More specifically, Theorem 11 of [35] states that for any convex body $K \in \mathcal{K}_d$, there is a direction $u \in \mathbb{S}^{d-1}$ such that, using the notations of Theorem 3.8, $d_K(u)\,\mathrm{vol}_{d-1}(K|u^{\perp}) \geq \frac{2v_{d-1}}{v_d}$, and if for any direction u the two sides are equal, then K is an ellipsoid.

If, for a convex body $K \in \mathcal{K}_d$, we have that $\mathrm{vol}_d(\mathrm{conv}((v+K) \cup (w+K)))$ has the same value for any touching pair of translates, let us say that K satisfies the *translative constant volume property*. The characterization of the plane convex bodies with this property can be found also in this paper. Before formulating the result, we recall that a 2-dimensional o-symmetric convex curve is a Radon curve, if, for the convex hull K of a suitable affine image of the curve, it holds that K° is a rotated copy of K by $\frac{\pi}{2}$ (cf. [36]).

Theorem 3.9 *For any plane convex body $K \in \mathcal{K}_2$ the following are equivalent.*

(1) K satisfies the translative constant volume property.

(2) *The boundary of* $\frac{1}{2}(K - K)$ *is a Radon curve.*
(3) K *is a body of constant width in a Radon norm.*

In two situations we have more precise results, respectively. The first case is when the examined body is a centrally symmetric one, and the other one when it is symmetric with respect to a hyperplane. The authors proved the following two theorems:

Theorem 3.10 *For any* $K \in \mathcal{K}_d$ *with* $d \geq 2$, $c_1(K) \geq 1 + \frac{2v_{d-1}}{v_d}$, *with equality if, and only if,* K *is an ellipsoid.*

Theorem 3.11 *For any* $K \in \mathcal{K}_d$ *with* $d \geq 2$, $c_{d-1}(K) \geq 1 + \frac{2v_{d-1}}{v_d}$, *with equality if, and only if,* K *is a Euclidean ball.*

Finally, let $\mathcal{P}_m$ denote the family of convex m-gons in the plane $\mathbb{R}^2$. It is a natural question to ask about the minima of the quantities defined in the introduction over $\mathcal{P}_m$. More specifically, we set

$$\begin{aligned} t_m &= \min\{c^{tr}(P) : P \in \mathcal{P}_m\}; \\ p_m &= \min\{c_0(P) : P \in \mathcal{P}_m\}; \\ l_m &= \min\{c_1(P) : P \in \mathcal{P}_m\}. \end{aligned}$$

On these numbers the following results were shown:

Theorem 3.12 *(1)* $t_3 = t_4 = 3$ *and* $t_5 = \frac{25+\sqrt{5}}{10}$. *Furthermore,* $c^{tr}(P) = 3$ *holds for any triangle and quadrangle, and if* $c^{tr}(P) = t_5$ *for some* $P \in \mathcal{P}_5$, *then* P *is affine regular pentagon.*
(2) $p_3 = 4$, $p_4 = 3$ *and* $p_5 = 2 + \frac{4 sin \frac{\pi}{5}}{5}$. *Furthermore, in each case, the minimum is attained only for affine regular polygons.*
(3) $l_3 = 4$ *and* $l_4 = 3$. *Furthermore, among triangles, the minimum is attained only for regular ones, and among quadrangles for rhombi.*

Conjecture 3.1 *Let* $d \geq 2$ *and* $0 < i < d - 1$. *Prove that, for any* $K \in \mathcal{K}_d$, $c_i(K) \geq 1 + \frac{2v_{d-1}}{v_d}$. *Is it true that equality holds only for Euclidean balls?*

The maximal values of $c^{tr}(K)$ and $c_0(K)$, for $K \in \mathcal{K}_d$, and the convex bodies for which these values are attained, are determined in [15]. Using a suitable simplex as K, it is easy to see that the set $\{c_i(K) : K \in \mathcal{K}_n\}$ is not bounded from above for $i = 1, \ldots, n - 1$. This readily yields the same statement for $c^{co}(K)$ as well. On the other hand, from Theorem 3.11 we obtain the following.

Remark 3.2 For any $K \in \mathcal{K}_n$ with $n \geq 2$, we have $c^{co}(K) \geq 1 + \frac{2v_{n-1}}{v_n}$, with equality if, and only if, K is a Euclidean ball.

In Theorem 3.9 it was proved that in the plane, a convex body satisfies the translative equal volume property if, and only if, it is of constant width in a Radon plane. It is known (cf. [37] or [36]) that for $d \geq 3$, if every planar section of a normed space is Radon, then the space is Euclidean; that is, its unit ball is an ellipsoid. This motivates the conjecture:

Conjecture 3.2 *Let $d \geq 3$. If some $K \in \mathcal{K}_d$ satisfies the translative equal volume property, then K is a convex body of constant width in a Euclidean space.*

Furthermore, we remark that the proof of Theorem 3.9 can be extended, using the Blaschke-Santaló inequality, to prove Theorems 3.8 and 3.10 in the plane. Similarly, Theorem 3.11 can be proved by a modification of the proof of Theorem 3.8, in which we estimate the volume of the polar body using the width function of the original one, and apply the Blaschke-Santaló inequality.

Like in [15], Theorems 3.8 and 3.11 yield information about circumscribed cylinders. Note that the second corollary is a strengthened version of Theorem 5 in [15].

Corollary 3.2 *For any convex body $K \in \mathcal{K}_d$, there is a direction $u \in \mathbb{S}^{d-1}$ such that the right cylinder $H_K(u)$, circumscribed about K and with generators parallel to u has volume*

$$\operatorname{vol}(H_K(u)) \geq \left(1 + \frac{2v_{d-1}}{v_d}\right) \operatorname{vol}_d(K). \tag{10}$$

Furthermore, if K is not a Euclidean ball, then the inequality sign in (10) is a strict inequality.

Corollary 3.3 *For any convex body $K \in \mathcal{K}_d$, there is a direction $u \in \mathbb{S}^{d-1}$ such that any cylinder $H_K(u)$, circumscribed about K and with generators parallel to u, has volume*

$$\operatorname{vol}(H_K(u)) \geq \left(1 + \frac{2v_{d-1}}{v_d}\right) \operatorname{vol}_d(K). \tag{11}$$

Furthermore, if K is not an ellipsoid, then the inequality sign in (11) is a strict inequality.

Let P_m be a regular m-gon in $\mathbb{R}^2$.

Problem 3.1 *Prove or disprove that for any $m \geq 3$,*

$$t_m = c^{tr}(P_m), \quad p_m = c_0(P_m), \quad \textit{and} \quad l_m = c_1(P_m).$$

Is it true that for t_m and p_m, equality is attained only for affine regular m-gons, and for l_m, where $m \neq 4$, only for regular m-gons?

3.2 Simplices in the 3-Space

Horváth in [16] examined $c(K, K)$ in the special case that K is a regular tetrahedron and the two congruent copies have the same centre. It has been proved the following theorem.

Theorem 3.13 *The volume of the convex hull of two congruent regular triangles with a common center is maximal if and only if their planes are orthogonal to each* *other and one of their vertices are opposite position with respect to the common center O.*

The proof is based on some exact formulas, which can be extended to the non-regular case of triangles, too.

On regular tetrahedra was proved a theorem in that case when all of the spherical triangles contain exactly one from the vertices of the other tetrahedron and changing the role of the tetrahedra we also get it (so when the two tetrahedra are in *dual position*). We remark that in a dual position the corresponding spherical edges of the two tetrahedra are crossing to each other, respectively. In this case it has been proved that

Theorem 3.14 *The value $v = \frac{8}{3\sqrt{3}}r^3$ is an upper bound for the volume of the convex hull of two regular tetrahedra are in dual position. It is attained if and only if the eight vertices of the two tetrahedra are the vertices of a cube inscribed in the common circumscribed sphere.*

This paper considered the proof of that combinatorial case when two domains contain two vertices, respectively. The following statement were proved:

Statement 3.1 *Assume that the closed regular spherical simplices $S(1, 2, 3)$ and $S(4, 2, 3)$ contains the vertices $2'$, $4'$ and $1'$, $3'$, respectively. Then the two tetrahedra are the same.*

In the paper [23] the author closed this problem using a generalization of the icosahedron inequality of L. Fejes-Tóth. It has been shown the general statement:

Theorem 3.15 *Consider two regular tetrahedra inscribed in the unit sphere. The maximal volume of the convex hull P of the eight vertices is the volume of the cube C inscribed in the unit sphere, so*

$$\mathrm{vol}_3(P) \leq \mathrm{vol}_3(C) = \frac{8}{3\sqrt{3}}.$$

The paper [18] investigates also connecting simplices. It is assumed that the used set of isometries $\mathcal{S}$ consists only reflections at such hyperplanes H which intersect the given simplex S. (Hence the convex hull function is considered only on the pairs of the simplex and its copy at a hyperplane intersecting it.) They gave an explicit expression the relative volume of the convex hull of the simplices and gave upper bounds on it. The number $c(S, S^H)$ for the regular simplex is determined explicitly. The following lemma plays a fundamental role in the investigations.

Lemma 3.1 *If K and K' give a maximal value for $c_{K,K'}$ then the intersection $K \cap K'$ is an extremal point of each of the bodies.*

To formulate the results we introduce some new notation. Assume that the intersecting simplices S and S_H are reflected copies of each other at the hyperplane H.

Then H intersects each of them in the same set. By the Lemma 3.1 we have that the intersection of the simplices in an optimal case is a common vertex. Let $s_0 \in H$ and $s_i \in H^+$ for $i \geq 1$. We imagine that H is horizontal and H^+ is the upper half-space. Define the *upper side of* S as the union of those facets in which a ray orthogonal to H and terminated in a far point of H^+ is first intersecting with S. The volume of the convex hull is the union of those prisms which are based on the orthogonal projection of a facet of the simplex of the upper side. Let denote $F_{i_1}, \dots, F_{i_k}$ the facet-simplex of the upper side, $F'_{i_1}, \dots, F'_{i_k}$ its orthogonal projections on H and $u_{i_1}, \dots, u_{i_k}$ its respective unit normals, directed outwardly. We also introduce the notation $s = \sum_{i=0}^{d} s_i = \sum_{i=1}^{d} s_i$. Now we have

Statement 3.2

$$\frac{1}{\mathrm{vol}_d(S)}\mathrm{vol}_d(\mathrm{conv}(S, S^H)) = 2d \sum_{l=1}^{k} \frac{\langle u_{i_l}, u\rangle\langle u, s - s_{i_l}\rangle}{|\langle u_{i_l}, (d+1)s_{i_l} - s\rangle|}.$$

It can be solved the original problem in the case of the regular simplex. Denote the Euclidean norm of a vector x by $\|x\|$.

Theorem 3.16 *If S is the regular simplex of dimension n, then*

$$c(S, S^H) := \frac{1}{\mathrm{vol}_d(S)}\mathrm{vol}_d(\mathrm{conv}(S, S^H)) = 2d,$$

attained only in the case when $u = u_0 = \frac{s}{\|s\|}$.

We note that the result of the case of reflection at a hyperplane gives an intermediate value between the results corresponding to translates and point reflections. The part of the previous proof corresponding to the case of a single upper facet can be extended to a general simplex, too. Let G denote the Gram matrix of the vector system $\{s_1, \dots, s_n\}$, defined by the product $M^T M$, where $M = [s_1, \dots, s_n]$ is the matrix with columns s_i. In the following theorem we use the notation $\|\cdot\|_1$ for the l_1 norm of a vector or a matrix, respectively.

Theorem 3.17 *If the only upper facet is F_0 with unit normal vector u_0, then we have the inequality*

$$\frac{1}{\mathrm{vol}_d(S)}\mathrm{vol}_d(\mathrm{conv}(S, S^H)) \leq d\left(1 + \frac{\|s\|}{\langle u_0, s\rangle}\right) =$$

$$= \left(d + \sqrt{\left\|(1, \dots, 1)G^{-1}\right\|_1 \|M(1, \dots, 1)\|}\right).$$

Equality is attained if and only if the normal vector u of H is equal to $\frac{u_0+s'}{\|u_0+s'\|}$, where $s' = \frac{s}{\|s\|}$ is the unit vector of the direction of s.

We remark that for a regular simplex we get back the previous theorem [38–43], since

$$G = \begin{pmatrix} 1 & \frac{1}{2} & \cdots & \frac{1}{2} \\ \frac{1}{2} & 1 & \cdots & \frac{1}{2} \\ \vdots & \vdots & \vdots & \vdots \\ \frac{1}{2} & \cdots & \frac{1}{2} & 1 \end{pmatrix} \text{ and } G^{-1} = \begin{pmatrix} \frac{2d}{d+1} & -\frac{2}{d+1} & \cdots & -\frac{2d}{d+1} \\ -\frac{2}{d+1} & \frac{2d}{d+1} & \cdots & -\frac{2}{d+1} \\ \vdots & \vdots & \vdots & \vdots \\ -\frac{2}{d+1} & \cdots & -\frac{2}{d+1} & \frac{2d}{d+1} \end{pmatrix},$$

implying that

$$d + \sqrt{\left\|(1, \ldots, 1)G^{-1}\right\|_1} \, \|M(1, \ldots, 1)\| = d + \sqrt{\frac{2d}{d+1}}\sqrt{\frac{d(d+1)}{2}} = 2d.$$

References

1. L. Lindelöf, Propriétés générales des polyèdres qui, sous une étendue superficielle donnée, renferment le plus grand volume. Bull. Acad. Sci. St. Petersburg 14, 258–269 (1869). Math. Ann. **2**, 150–159 (1870)
2. P. Gruber, *Convex and Discrete Geometry, Grundlehren der mathematischen Wissenschaften 336* (Springer, Berlin, Heidelberg, 2007)
3. K. Böröczky Jr., K. Böröczky, Isoperimetric problems for polytopes with a given number of vertices. Mathematika **43**, 237–254 (1996)
4. H.T. Croft, K.J. Falconer, R.K. Guy, *Unsolved Problems in Geometry*, vol. 2 (Springer, New York, 1991)
5. Florian, A., in *Handbook of Convex Geometry*, ed. by P.M. Gruber, J.M. Wills. Extremum Problems for Convex Discs and Polyhedra (North-Holland Publishing Co., Amsterdam, 1993)
6. P. Brass, W. Moser, J. Pach, *Research Problems in Discrete Geometry* (Springer, New York, 2005)
7. B. Kind, P. Kleinschmidt, On the maximal volume of convex bodies with few vertices. J. Combin. Theory Ser. A **21**, 124–128 (1976)
8. A. Klein, M. Wessler, The largest small n-dimensional polytope with $n+3$ vertice. J. Combin. Theory Ser. A **102**, 401–409 (2003)
9. A. Klein, M. Wessler, A correction to "The largest small n-dimensional polytope with $n+3$ vertices" [J. Combin. Theory Ser. A **102**, 401–409 (2003)]. J. Combin. Theory Ser. A **112**, 173–174 (2005)
10. L. Fejes-Tóth, *Regular Figures* (The Macmillan Company, New York, 1964)
11. J.D. Berman, K. Hanes, Volumes of polyhedra inscribed in the unit sphere in E^3. Math. Ann. **188**, 78–84 (1970)
12. N. Mutoh, The polyhedra of maximal volume inscribed in the unit sphere and of minimal volume circumscribed about the unit sphere. JCDCG, Lect. Notes Comput. Sci. **2866**, 204–214 (2002)
13. A.M. Macbeath, An extremal property of the hypersphere. Proc. Cambridge Philos. Soc. **47**, 245–247 (1951)
14. C.A. Rogers, G.C. Shephard, Convex bodies associated with a given convex body. J. London Math. Soc. **33**, 270–281 (1958)
15. C.A. Rogers, G.C. Shephard, Some extremal problems for convex bodies. Mathematika **5**(2), 93–102 (1958)
16. Á.G. Horváth, Maximal convex hull of connecting simplices. Stud. Univ. Zilina **22/1**, 7–19 (2008)

17. Á.G. Horváth, Z. Lángi, On the volume of the convex hull of two convex bodies. Monatshefte für Mathematik **174**(2), 219–229 (2014)
18. Á.G. Horváth, On an extremal problem connected with simplices. Beiträge zur Geometrie und Algebra **55**(2), 415–428 (2014)
19. R. Bowen, S. Fisk, Generation of triangulations of the sphere. Math. Comp. **21**, 250–252 (1967)
20. M. Goldberg, The isoperimetric problem for polyhedra. Tóhoku Math. J. **40**, 226–236 (1935)
21. D.W. Grace, Search for largest polyhedra. Math. Comp. **17**, 197–199 (1963)
22. A. Florian, Integrale auf konvexen polyedern period. Math. Hungar. **1**, 243–278 (1971)
23. Á.G. Horváth, On the icosahedron inequality of László Fejes-Tóth. J. Math. Inequalities **10/2**, 521–539 (2016)
24. Á.G. Horváth, Z. Lángi, Maximum volume polytopes inscribed in the unit sphere. Monaschefte für Math. **181/2**, 341–354 (2016)
25. Böröczky, K., On an extremum property of the regular simplex in S^d. *Colloq. Math. Soc. János Bolyai***48** Intuitive Geometry, pp. 117–121. Siófok (1985)
26. J.W. Milnor, in *The geometry and Topology of 3-Manifolds*, ed. by W.P. Thurston, Computation of volume (Lecture notes at Princeton University, 1977–1978)
27. J.W. Milnor, Hyperbolic geometry: the first 150 years. Bull. Amer. Math. Soc. (N.S.) **6/1**, 9–24 (1982)
28. U. Haagerup, H.J. Munkholm, Simplices of maximal volume in hyperbolic n-space. Acta. Math. **147**, 1–12 (1981)
29. B. Grünbaum, *Convex Polytopes*, 2nd edn. (Springer, New York, 2003)
30. G.M. Ziegler, *Lectures on Polytopes, Graduate Texts in Mathematics 152* (Springer, New York, 1995)
31. C.W. Lee, Regular triangulation of convex polytopes. *Dimacs Ser. Discrete Math. Theor. Comput. Sci.***4**, 443–456 (1991)
32. V. Kaibel, A. Wassmer, *Automorphism Groups of Cyclic Polytopes, Chapter 8 of F. Lutz, Triangulated Manifolds with Few Vertices, Algorithms and Combinatorics* (Springer, New York, to appear)
33. I. Fáry, L.Der Rédei, zentralsymmetrische Kern und die zentralsymmetrische Hülle von konvexen Körpern. Math. Annalen. **122**, 205–220 (1950)
34. H.-K. Ahn, P. Brass, C.-S. Shin, Maximum overlap and minimum convex hull of two convex polyhedra under translations. Comput. Geom. **40**, 171–177 (2008)
35. H. Martini, Z. Mustafaev, Some applications of cross-section measures in Minkowski spaces. Period. Math. Hungar. **53**, 185–197 (2006)
36. H. Martini, K. Swanepoel, Antinorms and Radon curves. Aequationes Math. **72**, 110–138 (2006)
37. J. Alonso, C. Benítez, Orthogonality in normed linear spaces: a survey, II. Relat. Between Main Orthogonalities, Extracta Math. **4**, 121–131 (1989)
38. K. Ball, Ellipsoids of maximal volume in convex bodies. Geom. Dedicata **41**(2), 241–250 (1992)
39. R.J. Gardner, *Geometric Tomography* (Cambridge University Press, Cambridge, 1995)
40. P. Gronchi, M. Longinetti, Affine regular polygons as extremals of area functionals. Discrete Comput. Geom. **39**, 273–297 (2008)
41. M. Henk, Löwner-John ellipsoids. *Doc. Math.* Extra volume ISMP (2012), 95–106
42. F. John, in *Studies and Essays Presented to R. Courant on his 60th Birthday, January 8, 1948*. Extremum Problems with Inequalities as Subsidiary Conditions (Interscience Publishers, Inc., New York, 1948), pp. 187–204
43. A.C. Thompson, *Minkowski Geometry, Encyclopedia of Mathematics and its Applications 63* (Cambridge University Press, Cambridge, 1996)

Integers, Modular Groups, and Hyperbolic Space

Norman W. Johnson

Abstract In each of the normed division algebras over the real field $\mathbb{R}$—namely, $\mathbb{R}$ itself, the complex numbers $\mathbb{C}$, the quaternions $\mathbb{H}$, and the octonions $\mathbb{O}$—certain elements can be characterized as *integers*. An integer of norm 1 is a *unit*. In a *basic system* of integers the units span a 1-, 2-, 4-, or 8-dimensional lattice, the points of which are the vertices of a regular or uniform Euclidean honeycomb. A *modular group* is a group of linear fractional transformations whose coefficients are integers in some basic system. In the case of the octonions, which have a nonassociative multiplication, such transformations form a *modular loop*. Each real, complex, or quaternionic modular group can be identified with a subgroup of a Coxeter group operating in hyperbolic space of 2, 3, or 5 dimensions.

The relationship between the *modular group* $PSL_2(\mathbb{Z})$ of linear fractional transformations over the ring of rational integers and the regular hyperbolic tessellation $\{3, \infty\}$ has been known since the nineteenth century. When $\mathbb{Z}$ is replaced by the ring $\mathbb{G} = \mathbb{Z}[i]$ of Gaussian integers, the analogous *Picard group* is similarly related to the regular honeycomb $\{3, 4, 4\}$ of hyperbolic 3-space. Recent results of Egon Schulte, Barry Monson, Asia Ivić Weiss, and the author have extended these connections to modular groups over other systems of complex and quaternionic integers and other regular or uniform honeycombs of hyperbolic space.

1 Linear Fractional Transformations

When each point of a projective line FP^1 over a field F is identified uniquely either with an element $x \in F$ or with the extended value ∞, a projectivity (i.e., a permutation of the points of FP^1 that preserves cross ratios) can be expressed as a *linear fractional*

Norman W. Johnson—Deceased

N. W. Johnson (✉)
Norton, USA

M. D. E. Conder et al. (eds.), *Discrete Geometry and Symmetry*,
Springer Proceedings in Mathematics & Statistics 234,
https://doi.org/10.1007/978-3-319-78434-2_12

transformation of the extended field $F \cup \{\infty\}$, defined for four given field elements $a, b, c, d (ad - bc \neq 0)$ by

$$x \mapsto \frac{ax + c}{bx + d},$$

with $-d/b \mapsto \infty$ and $\infty \mapsto a/b$ if $b \neq 0$ and with $\infty \mapsto \infty$ if $b = 0$. Such mappings can also be represented by 2×2 invertible matrices over F, modulo scalars, constituting the *projective general linear* group $\mathrm{PGL}_2(\mathrm{F})$.

The complex projective line $\mathbb{C}\mathrm{P}^1$ with one point fixed, represented in $\mathbb{R}^2$ by the familiar Argand diagram, provides a conformal model for the hyperbolic plane H^2. Points in the "upper half-plane" $\mathrm{Im}\, z > 0$ are the *ordinary* points of H^2, and the real axis, together with $\{\infty\}$, represents the *absolute circle*. The isometry group of H^2 is the *projective pseudo-orthogonal* group $\mathrm{PO}_{2,1}$, isomorphic to the group of linear fractional transformations

$$\cdot\langle A\rangle : \mathbb{C} \cup \{\infty\} \to \mathbb{C} \cup \{\infty\}$$

with A real and $\det A \neq 0$, i.e., the (real) projective general linear group PGL_2. The subgroup $\mathrm{P}^+\mathrm{O}_{2,1}$ of direct isometries is isomorphic to the (real) *projective special linear* group PSL_2 (with $\det A > 0$).

A convex polytope P of finite content in H^n whose dihedral angles are all submultiples of π (or zero if two adjacent facets are parallel) is the fundamental region for a hyperbolic *Coxeter group*, generated by reflections in the facets of P. The group is *compact* if P is an ordinary simplex, with $n + 1$ ordinary vertices;*paracompact* if P is an asymptotic simplex, with one or more vertices at infinity; or *hypercompact* if P is not a simplex. When P is a right triangle with acute (or zero) angles π/p and π/q, the group is the symmetry group $[p, q]$ of a regular tessellation $\{p, q\}$ of p-gons, q at a vertex.

The set of 2×2 matrices A over the rational integers $\mathbb{Z}$ with $\det A = \pm 1$ forms the *unit linear* group $\bar{\mathrm{S}}\mathrm{L}_2(\mathbb{Z})$. The subgroup of matrices A with $\det A = 1$ is the *special linear* group $\mathrm{SL}_2(\mathbb{Z})$. The corresponding group $\mathrm{PSL}_2\mathbb{Z}$ of linear fractional transformations $\cdot\langle A\rangle$ is the (rational) *modular group*, with $\mathrm{P}\bar{\mathrm{S}}\mathrm{L}_2(\mathbb{Z})$ being the *extended modular group*.

Felix Klein showed in 1879 (cf. [20]) that the modular group is isomorphic to the rotation group of the regular hyperbolic tessellation $\{3, \infty\}$. This is the direct subgroup of the paracompact Coxeter group $[3, \infty]$:

$$\mathrm{PSL}_2(\mathbb{Z}) \cong [3, \infty]^+.$$

The absolute $(n - 1)$-sphere ("hypersphere at infinity") of hyperbolic n-space H^n has the geometry of *inversive* $(n - 1)$-space I^{n-1}, i.e., the Möbius $(n - 1)$-sphere. For $n > 1$ the group $\mathrm{PO}_{n,1}$ of isometries of H^n is isomorphic to the group of circularities (homographies and antihomographies) of I^{n-1}. When $n = 2$, this is the group $\mathrm{PGL}_2 \cong$

$\mathrm{PGL}_2(\mathbb{R})$ of linear fractional transformations $\cdot\langle A\rangle$ with A real. When $n = 3$, it is the group $\mathrm{PGL}_2(\mathbb{C})$ of linear fractional transformations

$$\cdot\langle A\rangle : \mathbb{C} \cup \{\infty\} \to \mathbb{C} \cup \{\infty\}$$

with A complex, i.e., the group of projectivities of the complex projective line $\mathbb{C}\mathrm{P}^1$.

2 Complex Modular Groups

For any square-free integer $d \neq 1$, the *quadratic field* $\mathbb{Q}(\sqrt{d})$ has elements $r + s\sqrt{d}$, where r and s belong to the rational field $\mathbb{Q}$. The *conjugate* of $a = r + s\sqrt{d}$ is $\tilde{a} = r - s\sqrt{d}$ its *trace* tr a is $a + \tilde{a} = 2r$, and its *norm* $N(a)$ is $a\tilde{a} = r^2 - s^2 d$. The elements a with both tr a and $N(a)$ in $\mathbb{Z}$ are *quadratic integers* and constitute an integral domain, a two-dimensional algebra $\mathbb{Z}^2(d)$ over $\mathbb{Z}$, whose invertible elements, or *units*, have norm ± 1.

For $d > 0$ the elements of $\mathbb{Z}^2(d)$ are real and there are infinitely many units. When $d < 0$, $\mathbb{Z}^2(d)$ has both real and imaginary elements, the conjugate of z is its complex conjugate $\bar{z}$, tr $z = 2$ Re z, $N(z) = |z|^2$, and (with two exceptions) the only units are ± 1.

Complex numbers of the form $g = g_0 + g_1\mathrm{i}$, where $(g_0, g_1) \in \mathbb{Z}^2$ and $\mathrm{i} = \sqrt{-1}$, belong to the ring $\mathbb{G} = \mathbb{Z}[i] = \mathbb{Z}^2(-1)$ of *Gaussian integers*. There are four units in all: ± 1 and $\pm\mathrm{i}$. When the complex field $\mathbb{C}$ is regarded as a two-dimensional vector space over $\mathbb{R}$, the Gaussian integers constitute a two-dimensional lattice C_2 spanned by the units 1 and i, as shown in Fig. 1. The points of C_2 are the vertices of a regular tessellation $\{4, 4\}$ of the Euclidean plane.

Just as restricting the coefficients of linear fractional transformations $\cdot\langle A\rangle$ to rational integers defines the rational modular group $\mathrm{PSL}_2(\mathbb{Z})$, so restricting them to Gaussian integers defines the *Gaussian modular group* $\mathrm{PSL}_2(\mathbb{G})$. This group was first described by Émile Picard in 1884 and is commonly known as the "Picard group."

In 1897 Fricke and Klein identified $\mathrm{PSL}_2(\mathbb{G})$ with a subgroup of the rotation group of the regular hyperbolic honeycomb $\{3, 4, 4\}$ (cf. [17], pp. 60, 196). Schulte and Weiss [21] showed that it is a subgroup of index 2 in $[3, 4, 4]^+$, and Monson and Weiss [18] exhibited it as a subgroup of index 2 in the hypercompact Coxeter group $[\infty, 3, 3, \infty]$. The five mirrors for the latter group are the bounding planes of a quadrangular pyramid whose apex lies on the absolute sphere of H^3.

Complex numbers of the form $e = e_0 + e_1\omega$, where e_0 and e_1 are rational integers and $\omega = -1/2 + 1/2\sqrt{-3}$, belong to the ring $\mathbb{E} = \mathbb{Z}[\omega] = \mathbb{Z}^2(-3)$ of *Eisenstein integers*. There are six units: ± 1, $\pm\omega$, $\pm\omega^2$. When the complex field $\mathbb{C}$ is regarded as a two-dimensional vector space over $\mathbb{R}$, the Eisenstein integers constitute a two-dimensional lattice A_2 spanned by the units 1 and ω, as shown in Fig. 2. The points of A_2 are the vertices of a regular tessellation $\{3, 6\}$ of the Euclidean plane.

Bianchi [2, 3] showed that if D is an imaginary quadratic integral domain, the group $\mathrm{PSL}_2(\mathrm{D})$ acts discontinuously on hyperbolic 3-space. Though Fricke and Klein

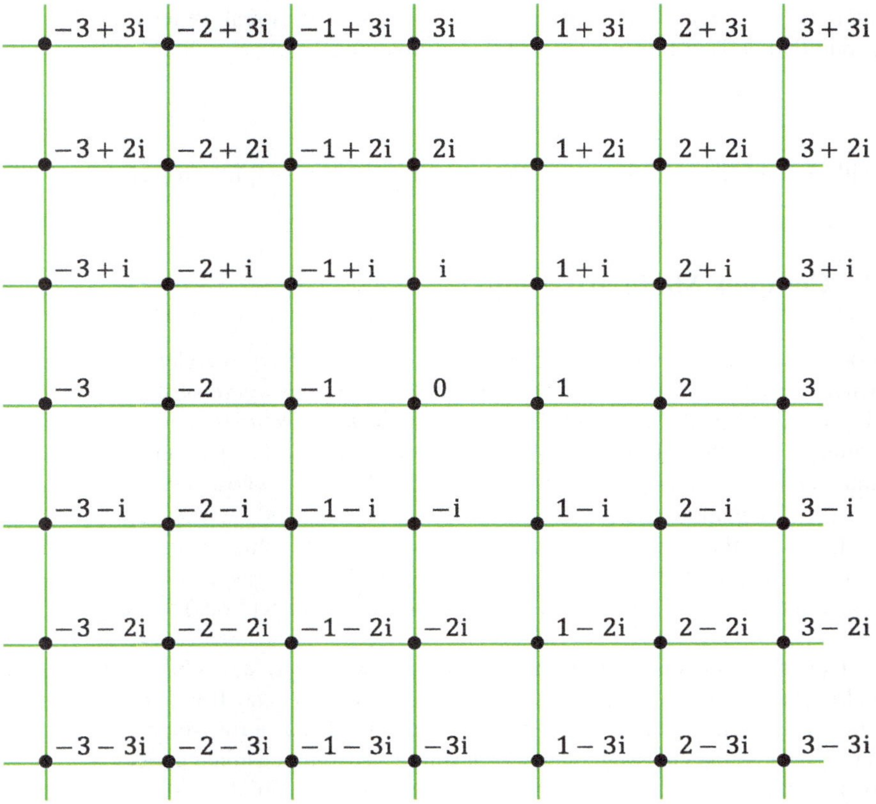

Fig. 1 The Gaussian integers

applied this to the Gaussian integers $\mathbb{G} = \mathbb{Z}[\mathrm{i}]$, the Eisenstein integers $\mathbb{E} = \mathbb{Z}[\omega]$ were generally ignored. It was not until 1994 that Schulte and Weiss cf. [18], Monson and Weiss [19] related the *Eisenstein modular group* $\mathrm{PSL}_2(\mathbb{E})$ to the regular hyperbolic honeycomb {3, 3, 6}, showing that $\mathrm{PSL}_2(\mathbb{E})$ is isomorphic to a subgroup of the rotation group $[3, 3, 6]^+$.

The ring $\mathbb{Z}$ of rational integers can be identified with the points of a lattice C_1 spanned by the units ± 1, the vertices of a regular partition $\{\infty\}$. The modular group $\mathrm{PSL}_2(\mathbb{Z})$ is isomorphic to the rotation group $[3, \infty]^+$ of the regular hyperbolic tessellation $\{3, \infty\}$. Similarly, the rings $\mathbb{G}$ and $\mathbb{E}$ of Gaussian and Eisenstein integers correspond to lattices C_2 and A_2, whose points are the vertices of the regular tessellations $\{4, 4\}$ and $\{3, 6\}$. As shown by Johnson and Weiss [14], the respective modular groups are isomorphic to "ionic" subgroups of the paracompact Coxeter groups [3, 4, 4] and [3, 3, 6], the symmetry groups of the regular hyperbolic honeycombs {3, 4, 4} and {3, 3, 6}:

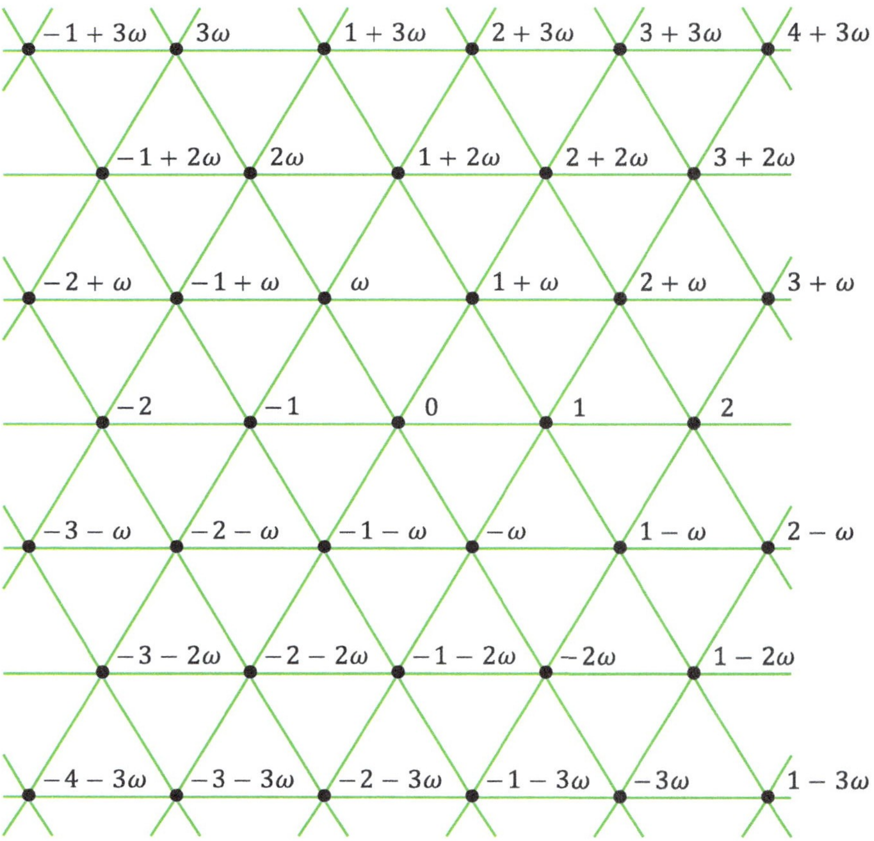

Fig. 2 The Eisenstein integers

$$\mathrm{PSL}_2(\mathbb{G}) \cong [3, 4, 1^+, 4]^+,$$
$$\mathrm{PSL}_2(\mathbb{E}) \cong [(3, 3)^+, 6, 1^+].$$

Such subgroups of a Coxeter group are obtained by replacing certain of the generating reflections by their pairwise products or by conjugates of other generators (cf. [7, §4.4]). If there are k superscript plus signs, the subgroup is of index 2^k.

3 Quaternionic Modular Groups

The division ring $\mathbb{H}$ of *quaternions* is a four-dimensional vector space over $\mathbb{R}$ with basis 1, i, j, k having an associative multiplication of vectors satisfying Hamilton's famous equations

$$\mathrm{i}^2 = \mathrm{j}^2 = \mathrm{k}^2 = \mathrm{ijk} = -1.$$

The multiplication so defined is noncommutative; e.g., ij = k = −ji. Each quaternion $\mathrm{Q} = t + x\mathrm{i} + y\mathrm{j} + z\mathrm{k}$ has a *conjugate* $\tilde{\mathrm{Q}} = t - x\mathrm{i} - y\mathrm{j} - z\mathrm{k}$, a *trace* $\mathrm{tr}\,\mathrm{Q} = \mathrm{Q} + \tilde{\mathrm{Q}} = 2t$, and a *norm* $N(\mathrm{Q}) = \mathrm{Q}\tilde{\mathrm{Q}} = t^2 + x^2 + y^2 + z^2$.

Vahlen [22] showed that homographies of inversive $(n-1)$-space I^{n-1} can be represented by linear fractional transformations over a Clifford algebra of dimension 2^{n-2} (cf. [1]). The cases $n = 2$, 3, and 4 correspond to the real field $\mathbb{R}$, the complex field $\mathbb{C}$, and the division ring $\mathbb{H}$.

Wilker [23] showed how a homography of I^4, or a direct isometry of H^5, is represented by a linear fractional transformation $\cdot\langle A\rangle$, determined by nonzero real scalar multiples of a 2×2 invertible matrix over $\mathbb{H}$. Thus the special projective pseudo-orthogonal group $\mathrm{P}^+\mathrm{O}_{5,1}$ is isomorphic to the quaternionic projective general linear group $\mathrm{PGL}_2(\mathbb{H})$.

William Rowan Hamilton, who discovered the quaternions in 1843, later investigated the ring $\mathbb{Z}[\mathrm{i}, \mathrm{j}]$ of quaternionic integers

$$\mathrm{G} = g_0 + g_1\mathrm{i} + g_2\mathrm{j} + g_3\mathrm{k},$$

where the g's are rational integers. Lipschitz [16] devoted a whole book to this system, which I denote by $\mathbb{H}$am and call the *Hamilton integers*. The ring $\mathbb{H}$am has eight invertible elements, or units:

$$\pm 1, \pm\mathrm{i}, \pm\mathrm{j}, \pm\mathrm{k}.$$

As points of Euclidean 4-space, these are the vertices of a regular 16-cell {3, 3, 4}.

In 1896 Adolf Hurwitz described the ring $\mathbb{Z}[u, v]$ of quaternionic integers

$$\mathrm{H} = h_0 + h_1 u + h_2 v + h_3 w,$$

where the h's are rational integers and where

$$u = \frac{1}{2} - \frac{1}{2}\mathrm{i} - \frac{1}{2}\mathrm{j} + \frac{1}{2}\mathrm{k} \quad \text{and} \quad v = \frac{1}{2} + \frac{1}{2}\mathrm{i} - \frac{1}{2}\mathrm{j} - \frac{1}{2}\mathrm{k},$$

with $w = \frac{1}{2} - \frac{1}{2}\mathrm{i} + \frac{1}{2}\mathrm{j} - \frac{1}{2}\mathrm{k} = (uv^{-1})$. This system will be denoted by $\mathbb{H}$ur and called the *Hurwitz integers*. The ring $\mathbb{H}$ur has 24 units, consisting of the eight Hamilton units and 16 others of the type

$$\pm\frac{1}{2} \pm \frac{1}{2}\mathrm{i} \pm \frac{1}{2}\mathrm{j} \pm \frac{1}{2}\mathrm{k}.$$

As points of Euclidean 4-space, these are the vertices of a regular 24-cell {3, 4, 3}.

Still another system of quaternionic integers is the ring $\mathbb{Z}[\omega, \mathrm{j}]$ of quaternions

$$\mathrm{E} = e_0 + e_1\omega + e_2\mathrm{j} + e_3\omega\mathrm{j},$$

where the e's are rational integers and where

$$\omega = -\frac{1}{2} + \frac{1}{2}\sqrt{3}\mathrm{i} \quad \text{and} \quad \omega\mathrm{j} = -\frac{1}{2}\mathrm{j} + \frac{1}{2}\sqrt{3}\mathrm{k}.$$

This system will be denoted by $\mathbb{H}$yb and called the *hybrid integers*. The ring $\mathbb{H}$yb has 12 units:

$$\pm 1, \pm\omega, \pm\omega^2, \pm\mathrm{j}, \pm\omega\mathrm{j}, \pm\omega^2\mathrm{j}.$$

As points of Euclidean 4-space, these are the vertices of a hexagonal double fusil $\{6\} + \{6\}$, the dual of a hexagonal double prism $\{6\} \times \{6\}$.

When $\mathbb{H}$ is taken as a four-dimensional vector space over $\mathbb{R}$, each of the rings of integral quaternions constitutes a four-dimensional lattice spanned by the units. For the Hamilton integers, points of the lattice C_4 are vertices of a regular honeycomb $\{4, 3, 3, 4\}$ of E^4. For the Hurwitz integers, points of the lattice D_4, which contains C_4 as a sublattice, are vertices of a regular honeycomb $\{3, 3, 4, 3\}$ of E^4. For the hybrid integers, points of the lattice $\mathrm{A}_2 \oplus \mathrm{A}_2$ are vertices of a prismatic honeycomb $\{3, 6\} \times \{3, 6\}$ of E^4, the rectangular product of two regular tessellations of E^2.

When the coefficients of a linear fractional transformation $\cdot\langle A\rangle$ are restricted to elements of a ring of integral quaternions, we have one of the *quaternionic modular groups* $\mathrm{PSL}_2(\mathbb{H}\mathrm{am})$, $\mathrm{PSL}_2(\mathbb{H}\mathrm{ur})$, or $\mathrm{PSL}_2(\mathbb{H}\mathrm{yb})$. These groups were investigated by Johnson and Weiss [15]. Each of them is a subgroup (or an extension of a subgroup) of a Coxeter group operating in H^5:

$$\begin{aligned}\mathrm{PSL}_2(\mathbb{H}\mathrm{am}) &\cong [3, 4, (3, 3)^{\Delta}, 4]^+,\\ \mathrm{PSL}_2(\mathbb{H}\mathrm{ur}) &\cong [(3, 3, 3)^+, 4, 3^+],\\ \mathrm{PSL}_2(\mathbb{H}\mathrm{yb}) &\cong 4[1^+, 6, (3, 3, 3, 3)^+, 6, 1^+].\end{aligned}$$

These are, respectively, a subgroup of index 3 in the ionic subgroup $[3, 4, (3, 3)^+, 4]^+$ of the paracompact group $[3, 4, 3, 3, 4]$, the commutator subgroup of the paracompact group $[3, 3, 3, 4, 3]$, and an extension of the commutator subgroup of the hypercompact group $[6, 3, 3, 3, 3, 6]$.

4 Integral Octonions

The division algebra $\mathbb{O}$ of *octonions* was discovered by John Graves in 1843 and rediscovered by Arthur Cayley in 1845. It constitutes an eight-dimensional vector space over $\mathbb{R}$, and (like $\mathbb{R}$, $\mathbb{C}$, and $\mathbb{H}$) has a multiplicative norm. Whereas both $\mathbb{R}$ and

$\mathbb{C}$ are fields and $\mathbb{H}$ is a skew-field, multiplication in $\mathbb{O}$ is neither commutative nor associative. The nonzero octonions form a multiplicative *Moufang loop* GM($\mathbb{O}$).

The notion of *integer*can be applied to any normed division algebra. Dickson [9, pp. 141–142] proposed criteria for a set of complex, quaternionic, or octonionic integers. Some of Dickson's requirements seem too strict, others not strict enough. In our theory a *basic system* of integers is a set with the following properties:

(1) the trace and the norm of each element are rational integers;
(2) the elements form a subring of $\mathbb{C}$, $\mathbb{H}$, or $\mathbb{O}$ with a set of invertible units (elements of norm 1) closed under multiplication;
(3) when $\mathbb{C}$, $\mathbb{H}$, or $\mathbb{O}$ is taken as a vector space over $\mathbb{R}$, the elements are the points of a two, four, or eight-dimensional lattice spanned by the units.

The only basic system of real integers is the ring $\mathbb{Z}$ of rational integers, with two units. The rings $\mathbb{G}$ and $\mathbb{E}$ of Gaussian and Eisenstein integers, the only domains of quadratic integers with both real and imaginary units (four for $\mathbb{G}$, six for $\mathbb{E}$), are the two basic systems of complex integers. Using results of Du Val [11], Johnson and Weiss [15] showed that the units of a basic system of integral quaternions must form a binary dihedral group $2D_2$ or $2D_3$ or the binary tetrahedral group $2A_4$ and hence that the only basic systems are the rings $\mathbb{H}$am (8 units), $\mathbb{H}$yb (12 units), and $\mathbb{H}$ur (24 units).

Conway and Smith [5] investigated rings of real, complex, quaternionic, and octonionic integers, which fall into four distinct families, labeled $\mathcal{G}$, $\mathcal{E}$, $\mathcal{H}$, and $\mathcal{D}$. There are just four basic systems of integral octonions, one in each family ([12, pp. 58–59]; cf. [4, 8]).

To the systems $\mathcal{G}^1 = \mathbb{Z}$ (2 units), $\mathcal{G}^2 = \mathbb{G}$ (4 units), and $\mathcal{G}^4 = \mathbb{H}$am (8 units) we can add the system $\mathcal{G}^8 = \mathbb{O}$cg of *Cayley–Graves integers* (or "Gravesian octaves") with 16 units spanning a lattice C_8, the points of which are the vertices of a regular honeycomb $\{4, 3^6, 4\}$ of E^8.

Along with systems $\mathcal{E}^2 = \mathbb{E}$ (6 units) and $\mathcal{E}^4 = \mathbb{H}$yb (12 units) we have the system $\mathcal{E}^8 = \mathbb{O}$ce of *compound Eisenstein integers* (or "Eisenstein octaves") with 24 units that span a lattice $4A_2 = A_2 \oplus A_2 \oplus A_2 \oplus A_2$, the points of which are the vertices of a prismatic honeycomb $\{3, 6\}^4$, the rectangular product of four regular tessellations of E^2.

Two systems $\mathcal{H}^4 = \mathbb{H}$ur (24 units) can be combined to produce the system $\mathcal{H}^8 = \mathbb{O}$ch of *coupled Hurwitz integers* (or "Hurwitzian octaves") with 48 units that span a lattice $2D_4 = D_4 \oplus D_4$, the points of which are the vertices of a prismatic honeycomb $\{3, 3, 4, 3\}^2$, the rectangular product of two regular honeycombs of E^4.

Dickson [9, pp. 319–325] showed that certain sets of octonions having coordinates in $\mathbb{Z}$ or $\mathbb{Z} + 1/2$ form a system of octonionic integers. In fact, he obtained three such systems. Coxeter [6] found that there are in all *seven* of these systems, one corresponding to each of the unit octonions that, together with 1, span $\mathbb{O}$. Each system $\mathcal{D}^8 = \mathbb{O}$cd of *Coxeter–Dickson integers* (or "Dicksonian octaves") has 240 units that span a lattice E_8, the points of which are the vertices of Thorold Gosset's uniform honeycomb 5_{21}. The lattice E_8 contains C_8, $4A_2$, and $2D_4$ as sublattices, and the ring $\mathbb{O}$cd contains $\mathbb{O}$cg, $\mathbb{O}$ce, and $\mathbb{O}$ch as subrings.

Rings of octonionic integers cannot be used to define modular groups. First, the division algebra $\mathbb{O}$ is nonassociative, satisfying only the weaker alternative laws $(aa)b = a(ab)$ and $(ab)b = a(bb)$. Second, the connection between linear fractional transformations and hyperbolic geometry runs through the family of Clifford algebras, including $\mathbb{R}$, $\mathbb{C}$, and $\mathbb{H}$ but not $\mathbb{O}$.

Though not associative, invertible 2×2 matrices over one of the basic systems of octonionic integers form a *special Moufang loop* $\mathrm{SM}_2(\mathbb{O}\mathrm{cg})$, $\mathrm{SM}_2(\mathbb{O}\mathrm{ce})$, $\mathrm{SM}_2(\mathbb{O}\mathrm{ch})$, or $\mathrm{SM}_2(\mathbb{O}\mathrm{cd})$. Identifying the matrices $\pm A$, we obtain an *octonionic modular loop* $\mathrm{PSM}_2(\mathbb{O}\mathrm{cg})$, $\mathrm{PSM}_2(\mathbb{O}\mathrm{ce})$, $\mathrm{PSM}_2(\mathbb{O}\mathrm{ch})$, or $\mathrm{PSM}_2(\mathbb{O}\mathrm{cd})$.

5 Summary

The ten basic systems of real, complex, quaternionic, or octonionic integers fall into four families:

$$\begin{array}{llll} \mathcal{G}^1 = \mathbb{Z}, & \mathcal{G}^2 = \mathbb{G} & \mathcal{G}^4 = \mathbb{H}\mathrm{am}, & \mathcal{G}^8 = \mathbb{O}\mathrm{cg}, \\ & \mathcal{E}^2 = \mathbb{E}, & \mathcal{E}^4 = \mathbb{H}\mathrm{yb}, & \mathcal{E}^8 = \mathbb{O}\mathrm{ce}, \\ & & \mathcal{H}^4 = \mathbb{H}\mathrm{ur}, & \mathcal{H}^8 = \mathbb{O}\mathrm{ch}, \\ & & & \mathcal{D}^8 = \mathbb{O}\mathrm{cd}, \end{array}$$

The elements of each basic system are the points of a lattice in E^1, E^2, E^4, or E^8. The real, complex, and quaternionic systems define modular groups related to Coxeter groups operating in H^2, H^3, or H^5. The four octonionic systems define modular loops. Further details may be found in the author's forthcoming book ([13, Chaps. 14–15]).

References

1. L.V. Ahlfors, Möbius transformations and Clifford numbers, in *Differential Geometry and Complex Analysis*, H. E. Rauch memorial volume, ed. by I. Chavel, H. M. Farkas (Springer, Berlin, 1985), pp. 65–73
2. L. BIanchi, Geometrische Darstellung der Gruppen linearer Substitutionen mit ganzen complexen Coefficienten nebst Anwendungen auf die Zahlentheorie," *Math. Ann.* **38**, 313–333 (1891). Reprinted in *Opere*, vol. I, pt. 1 (Edizione Cremonese, Rome, 1952), pp. 233–258
3. L. BIanchi, Sui gruppi de sostituzioni lineari con coeficienti appartenenti a corpi quadratici imaginari," *Math. Ann.* **40**, pp. 332–412 (1892). Reprinted in *Opere*, vol. I, pt. 1 (Edizione Cremonese, Rome, 1952), pp. 270–373
4. P. Boddington, D. Rumynin, On Curtis' theorem about finite octonionic loops. Proc. Amer. Math. Soc. **135**, 1651–1657 (2007)
5. J.H. Conway, D.A. Smith, *On Quaternions and Octonions: Their Geometry, Arithmetic, and Symmetry* (AK Peters, Natick, Mass., 2003)
6. H.S.M. Coxeter, Integral Cayley numbers, *Duke Math. J.* **13**, pp. 561–578. Reprinted in *Twelve Geometric Essays* (Southern Illinois Univ. Press, Carbondale, and Feffer & Simons, London-Amsterdam, 1968) or *The Beauty of Geometry: Twelve Essays* (Dover, Mineola, N.Y., 1999), pp. 21–39

7. H.S.M. Coxeter, W.O.J. Moser *Generators and Relations for Discrete Groups*, Ergebnisse der Mathematik und ihrer Grenzgebiete (N.F.), Bd. 14. (Springer-Verlag, Berlin-Göttingen-Heidelberg, 1957; 4th ed., Springer-Verlag, Berlin-Heidelberg-New York, 1980)
8. R.T. Curtis, Construction of a family of Moufang loops. Math. Proc. Cambridge Philos. Soc. **142**, 233–237 (2007)
9. L.E. Dickson, A new simple theory of hypercomplex integers," *J. Math. Pures Appl.* (9) **2**, pp. 281–326. Reprinted in *Collected Mathematical Papers*, Vol. VI (Chelsea, Bronx, N.Y., 1983), pp. 531–576
10. L.E. Dickson, *Algebras and Their Arithmetics*. Univ. of Chicago Press, Chicago, 1923; G.E. Stechert, New York, 1938. Trans. by J.J. Burckhardt, E. Schubarth as *Algebren und ihre Zahlentheorie* (Orell Füssli, Zurich-Leipzig, 1927)
11. P. Du Val, *Homographies, Quaternions, and Rotations* (Clarendon Press/Oxford Univ. Press, Oxford, 1964), p. 1964
12. N.W. Johnson, Integers. Math. Intelligencer **35**, 52–59 (2013). (No. 2)
13. N.W. Johnson, *Geometries and Transformations* (Press, Cambridge-New York, Cambridge Univ, 2017). (to appear)
14. N.W. Johnson, A.I. Weiss, Quadratic integers and Coxeter groups. Can. J. Math. **51**, 1307–1336 (1999a)
15. N.W. Johnson, A.I. Weiss, Quaternionic modular groups. Linear Algebra Appl. **295**, 159–189 (1999b)
16. R. Lipschitz, *Untersuchungen über die Summen von Quadraten* (Bonn, M. Cohen, 1886), p. 1886
17. W. Magnus, *Noneuclidean Tesselations and Their Groups* (Academic Press, New York-London, 1974), p. 1974
18. B.R. Monson, A.I. Weiss, Polytopes related to the Picard group. Linear Algebra Appl. **218**, 185–204 (1995)
19. B.R. Monson, A.I. Weiss, Eisenstein integers and related C-groups. Geom. Dedicata. **66**, 99–117 (1997)
20. H. PoincarÉ, Theorie des groupes fuchsiens, *Acta Math.* **1**, pp. 1–62. Reprinted in *Œuvres*, t. II (Gauthier-Villars, Paris, 1952), pp. 108–168. First two sections trans. by J. Stillwell as "Theory of Fuchsian groups" in *Sources of Hyperbolic Geometry*, History of Mathematics, Vol. 10 (American Math. Soc. and London Math. Soc., Providence, 1996), pp. 123–129 (1882)
21. E. Schulte, A.I. Weiss, Chirality and projective linear groups. Discrete Math. **131**, 221–261 (1994)
22. K.T. Vahlen, Ueber Bewegungen und complexe Zahlen. Math. Ann. **55**, 585–593 (1902)
23. J.B. Wilker, The quaternion formalism for Möbius groups in four or fewer dimensions. Linear Algebra Appl. **190**, 99–136 (1993)

Monge Points, Euler Lines, and Feuerbach Spheres in Minkowski Spaces

Undine Leopold and Horst Martini

Dedicated to Károly Bezdek and Egon Schulte on the occasion of their 60th birthdays

Abstract It is surprising, but an established fact that the field of elementary geometry referring to normed spaces (= Minkowski spaces) is not a systematically developed discipline. There are many natural notions and problems of elementary and classical geometry that were never investigated in this more general framework, although their Euclidean subcases are well known and this extended viewpoint is promising. An example is the geometry of simplices in non-Euclidean normed spaces; not many papers in this direction exist. Inspired by this lack of natural results on Minkowskian simplices, we present a collection of new results as non-Euclidean generalizations of well-known fundamental properties of Euclidean simplices. These results refer to Minkowskian analogues of notions like Euler line, Monge point, and Feuerbach sphere of a simplex in a normed space. In addition, we derive some related results on polygons (instead of triangles) in normed planes.

Keywords Birkhoff orthogonality · Vertex centroid · Circumsphere · Euler line · Feuerbach sphere · Isosceles orthogonality · Mannheim's theorem · Minkowskian simplex · Monge point · Normality · Normed space

2010 Mathematics Subject Classification 46B20 · 51M05 · 51M20 · 52A10 · 52A20 · 52A21 · 52B11

U. Leopold · H. Martini (✉)
Fakultät für Mathematik, Technische Universität Chemnitz,
09107 Chemnitz, Germany
e-mail: undine.leopold@googlemail.com

H. Martini
e-mail: horst.martini@mathematik.tu-chemnitz.de

M. D. E. Conder et al. (eds.), *Discrete Geometry and Symmetry*,
Springer Proceedings in Mathematics & Statistics 234,
https://doi.org/10.1007/978-3-319-78434-2_13

1 Introduction

Looking at basic literature on the geometry of finite dimensional real Banach spaces (see, e.g., the monograph [52] and the surveys [34, 42]), the reader will observe that there is no systematic representation of results in the spirit of elementary and classical geometry in such spaces (in other words, the field of elementary geometry is not really developed in normed spaces, also called Minkowski spaces). This is not only meant in the sense that a classifying, hierarchical structure of theorems is missing. Also, it is already appealing to find the way of correctly defining analogous notions. An example of such a non-developed partial field is the geometry of simplices in non-Euclidean Minkowski spaces. Inspired by this indicated lack of natural results on Minkowskian simplices, we derive a collection of new results which reflect non-Euclidean analogues and extensions of well known properties of Euclidean simplices. These results are based on, or refer to, generalizations of notions like Euler lines, Monge points, and Feuerbach spheres of simplices in Minkowski spaces. It should be noticed that some of these topics are not even established for Minkowski planes; most of our results are derived immediately for simplices in Minkowski spaces of arbitrary finite dimension.

In plane Euclidean geometry, the *Euler line* of a given triangle is a well-studied object which contains many interesting points besides the circumcenter and the vertex centroid of this triangle. Other special points on the Euler line include the orthocenter and the center of the so-called *nine-point-* or *Feuerbach circle*. Notions like this can be extended to simplices in higher dimensional Euclidean space, and the respective results can sometimes be sharpened for important subfamilies of general simplices, like, for example, the family of orthocentric simplices. Using new methods developed by Grassmann for studying the d-dimensional Euclidean space, this was done already in the 19th century. Two early related references are [45, 47]. Deeper results were obtained later; the concept of Euler line and some related notions have been generalized to Euclidean higher dimensional space in [10, 16–19, 23, 24, 28, 30, 46] for orthocentric simplices, and in [11, 15, 20, 33, 46, 49] for general simplices. Other interesting generalizations in Euclidean geometry refer to Euler lines of cyclic polygons, see [25]. For a few results in Minkowski planes and spaces we refer to [6, 8, 12, 37, 43]. The Feuerbach circle of a triangle in the Euclidean plane passes through the feet of the three altitudes, the midpoints of the three sides, and the midpoints of the segments from the three vertices to the orthocenter of that triangle. Beautiful generalizations of the Feuerbach circle to d-dimensional Euclidean space for orthocentric simplices have been obtained in [10, 19, 24, 28], and for general simplices in [11, 20, 46]. Minkowskian analogues have so far only been discussed in normed planes, see [8, 37, 48]. One should also mention that the concepts discussed here are certainly interesting for other non-Euclidean geometries; see, e.g., [26, 27].

A d-dimensional (*normed* or) *Minkowski space* $(\mathbb{R}^d, \|\cdot\|)$ is the vector space $\mathbb{R}^d$ equipped with a norm $\|\cdot\|$. A norm can be given implicitly by its *unit ball* $B(O, 1)$, which is a convex body centered at the origin O; its boundary $S(O, 1)$ is the *unit sphere* of the normed space. Any homothet of the unit ball is called a *Minkowskian ball* and denoted by $B(X, r)$, where X is its center and $r > 0$ its radius; its boundary is the *Minkowskian sphere* $S(X, r)$. Two-dimensional Minkowski spaces are *Minkowski planes*, and for an overview on what has been done in the geometry of normed planes and spaces we refer to the book [52], and to the surveys [34, 42].

The fundamental difference between non-Euclidean Minkowski spaces and the Euclidean space is the absence of an inner product, and thus the notions of angles and orthogonality do not exist in the usual sense. Nevertheless, several *types of orthogonality* can be defined (see [1, 2, 5] for an overview), with *isosceles* and *Birkhoff orthogonalities* being the most prominent examples. We say that y is *isosceles orthogonal* to x, denoted $x \perp_I y$, when $\|x + y\| = \|x - y\|$. Isosceles orthogonality is thus the orthogonality of diagonals in a parallelogram with equal side lengths (a rhombus in Euclidean space). It is also the orthogonality of chords over a diameter. By contrast, y is *Birkhoff orthogonal* to x, denoted $x \perp_B y$, when $\|x\| \leq \|x + \alpha y\|$ for any $\alpha \in \mathbb{R}$. Thus Birkhoff orthogonality is the (unsymmetric) orthogonality of a radius x and corresponding tangent vector y of some ball centered at the origin O. For hyperplanes and lines, there is the notion of *normality*. A direction (vector) v is *normal* to a hyperplane E if there exists a radius $r > 0$, such that E supports the ball $B(O, r)$ at a multiple of v. Equivalently, v is normal to E if any vector parallel to E is Birkhoff orthogonal to v.

For any two distinct points P, Q, we denote by $[PQ]$ the *closed segment*, by $\langle PQ \rangle$ the *spanned line* (affine hull), and by $[PQ\rangle$ the *ray* $\{P + \lambda(Q - P) \mid \lambda \geq 0\}$; we write $\|[PQ]\|$ for the *length* of $[PQ]$. We will use the usual abbreviation conv for the convex hull of a set.

In this article, we focus on the geometry of simplices in d-dimensional Minkowski spaces. A nice contribution to this topic, but with different aims, is the paper [9]. As usual, a *d-simplex* is the convex hull of $d + 1$ points in general linear position, or the bounded, non-empty intersection of $d + 1$ closed half-spaces in general position. We underline that by *circumcenters of simplices* we mean the *centers of circumspheres* (or *-balls*) *of simplices*, i.e., of Minkowskian spheres containing all the vertices of the respective simplex (see, e.g., [3]). A related, but different notion is that of *minimal enclosing spheres of simplices*, sometimes also called circumspheres (cf., e.g., [4]); this notion is not discussed here. In the two-dimensional situation, circumspheres and -balls are called *circumcircles* and *-discs*. In Minkowski spaces, simplices may have several, precisely one, or no circumcenter at all, depending on the shape of the unit ball, see Fig. 1. Examples without circumcenters may only be constructed for

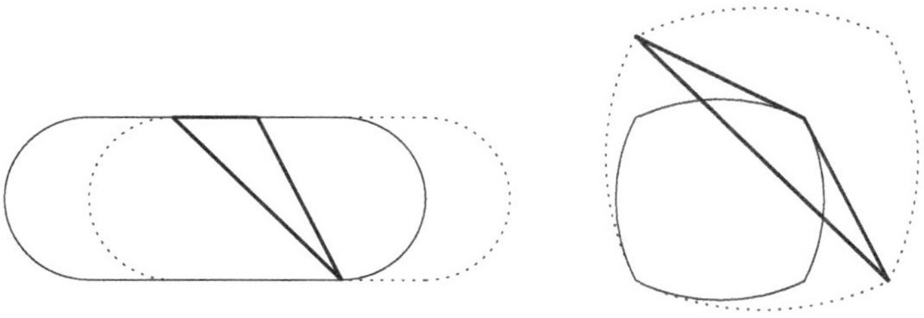

Fig. 1 A triangle with several circumcenters (left), and a triangle without a circumcenter (right), as illustrated by suitable homothets of the unit ball

non-smooth norms, as all smooth norms allow inscription into a ball [21, 31]. *We focus on the case where there is at least one circumcenter.*

2 Orthocentric Simplices and the Monge Point in Euclidean Space

We begin with a short survey on results related to orthocentricity in Euclidean space. In Euclidean geometry, not every simplex in dimension $\boldsymbol{d \geq 3}$ possesses an orthocenter, i.e., a point common to all the altitudes. However, if such a point H exists, the simplex is called *orthocentric* and possesses a number of special properties (compare the survey contained in [16, 23]). The following proposition is well known (see again [16]).

Proposition 2.1 *A d-simplex T in Euclidean space is orthocentric if and only if the direction of every edge is perpendicular to the affine hull of the vertices not in that edge (i.e., the affine hull of the opposite $(d-2)$-face). Equivalently, a d-simplex in Euclidean space is orthocentric if and only if any two disjoint edges are perpendicular.*

The $(d-2)$-faces of a d-polytope are sometimes called *ridges*, see [44]. The following fact (see also the survey in [16]) can be proved in many ways, and has been posed as a problem in the American Mathematical Monthly [29]. Note that orthocenters are not defined for an edge or a point.

Proposition 2.2 *In an orthocentric Euclidean d-simplex ($d \geq 3$), the foot of every altitude is the orthocenter of the opposite facet.*

In absence of a guaranteed orthocenter, the literature on Euclidean geometry (e.g. [7, 14] for three dimensions, [11, 16, 23] for the general case) defines the *Monge point* of a tetrahedron or higher dimensional simplex as the intersection of so-called *Monge (hyper-)planes*. The Monge point coincides with the Euclidean orthocenter if the latter exists [7, 11, 14]. From this, theorems about the Euler line, the Feuerbach circle, etc. can be generalized to higher dimensional simplices, see all the references given in the Introduction, and see Sect. 4 for Minkowskian analogues. We recall the definition and the following theorems from [11].

Definition 2.1 Let T be a d-simplex in Euclidean d-space. A *Monge hyperplane* is a hyperplane which is perpendicular to an edge of the simplex and which passes through the vertex centroid of the opposite $(d-2)$-face (ridge).

Theorem 2.1 (Monge Theorem) *The Monge hyperplanes of a Euclidean d-simplex have precisely one point in common, which is called the* Monge point N *of the simplex.*

Theorem 2.2 (Orthocenter Theorem) *In an orthocentric Euclidean d-simplex, the Monge point N coincides with the orthocenter H.*

Theorem 2.3 (Mannheim Theorem, see [7, 14] for $d = 3$, and [11] for arbitrary d) *For any d-simplex, the $d + 1$ planes, each determined by an altitude of a d-simplex and the Monge point (for $d = 3$, the orthocenter) of the corresponding facet, pass through the Monge point of the d-simplex.*

Regular simplices are orthocentric. Regular simplices are also *equilateral*, i.e., all their edges have equal length, as well as *equifacetal*, which means that all their facets are isometric (congruent). Furthermore, the circumcenter M, vertex centroid G, orthocenter H, and *incenter* I, i.e., the center of the unique inscribed sphere touching all facets, coincide. Conversely, we have the following statement, see [16].

Theorem 2.4 *A Euclidean d-simplex T is regular, if and only if any of the following conditions are fulfilled:*

1. *T is equilateral.*
2. *T is orthocentric and any two of the centers M, G, I, H coincide.*
3. *T is orthocentric and equifacetal.*

As we will see in the next Section, the concept of Monge point generalizes to arbitrary Minkowski spaces, at least for simplices with a circumcenter.

3 The Monge Point of Simplices in Minkowski Spaces

In this section, we generalize the definition of Monge point and its construction by Monge hyperplanes to Minkowski spaces of arbitrary (finite) dimension $\boldsymbol{d \geq 2}$.

Definition 3.1 Let $(\mathbb{R}^d, \|\cdot\|)$ be a d-dimensional Minkowski space, and let T be a d-simplex with a circumcenter M. For each pair (F, E_F) of a ridge F and opposite edge E_F, and if M is not the midpoint of E_F, define the *associated Monge line* as the line through the vertex centroid of F which is parallel to the line through M and the midpoint of E_F.

Theorem 3.1 *Let $(\mathbb{R}^d, \|\cdot\|)$ be a d-dimensional Minkowski space, and let T be a d-simplex with a circumcenter M. Then the Monge lines of T are concurrent in a single point N_M, called the* Monge point of T.

Remark 3.1 Before we proceed with the proof of Theorem 3.1, we remind the reader of the following well-know fact: the centroid G of $d + 1$ points is the weighted average

$$\frac{(d-k)G' + (k+1)G''}{d+1}$$

of the centroid G' of $k + 1$ of the points and the centroid G'' of the remaining $d - k$ points. Thus, it is possible to obtain G as the intersection of all *k-medians* (where $k = 0, \dots, \lfloor \frac{d+1}{2} \rfloor - 1$) between the centroid of a subset of $k + 1$ points and the centroid of the remaining $d - k$ points. For $d + 1$ points in general linear position and $k = 0$ we obtain the usual medians.

Proof (Proof of the theorem) The proof is similar to, but more general than, the one in [11] for Euclidean space. First, for each $(d-2)$-face F denote its vertex centroid $G(F)$, and let $G(E_F)$ be the midpoint or vertex centroid of the opposite edge E_F. Since a d-simplex possesses $\binom{d+1}{2}$ edges (ridges) and M can be located at the midpoint of at most one of them, the auxiliary lines $\langle M G(E_F)\rangle$ are well-defined for at least $\binom{d+1}{2} - 1$ pairs (F, E_F). The auxiliary line $\langle M G(E_F)\rangle$, if well-defined, is parallel to the associated Monge line $\langle G(F)L(F)\rangle$ of (F, E_F), where we define $L(F) := G(F) + G(E_F) - M$. Second, if $M = G$, then G and $G(F)$ both lie on $\langle M G(E_F)\rangle$, i.e., each auxiliary line coincides with the associated Monge line, and all these lines intersect in $N_M := M = G$ (and this is the only point, since different edge midpoints define different lines $\langle M G(E_F)\rangle$). If $M \neq G$, then auxiliary line and Monge line are distinct. Observe that each 1-median $[G(F)G(E_F)]$ connects a Monge line and the corresponding auxiliary line. The vertex centroid G of the simplex T divides each 1-median in the ratio $2 : (d-1)$, so the same division ratio holds true for the segment $[MN(F)]$ which passes through the given circumcenter M, the vertex centroid G of T, and ends at the point $N(F)$ on $[G(F)L(F)\rangle$, see Fig. 2. As a consequence of this common ratio, all points $N(F)$ are indeed the same point N_M, solely dependent on the chosen circumcenter (and the given simplex), and all rays $[G(F)L(F)\rangle$ meet at N_M. □

In keeping with the tradition in Euclidean space, we want to reformulate the theorem in terms of hyperplanes.

Definition 3.2 Let $(\mathbb{R}^d, \|\cdot\|)$ be a d-dimensional Minkowski space, and let T be a d-simplex with a circumcenter M. Suppose M is not the midpoint of an edge E_F opposite a $(d-2)$-face F of the simplex. For the pair (F, E_F) define the *auxiliary pencil* of hyperplanes through M and the midpoint of E_F. Furthermore, define the associated *Monge hyperplane pencil* for the pair (F, E_F) as the translate of the auxiliary pencil such that all hyperplanes go through the vertex centroid of F.

Corollary 3.2 *Let $(\mathbb{R}^d, \|\cdot\|)$ be a d-dimensional Minkowski space, and let T be a d-simplex with a circumcenter M. Then the hyperplanes of all (well-defined) Monge hyperplane pencils of T intersect in a single point, namely the Monge point of T.*

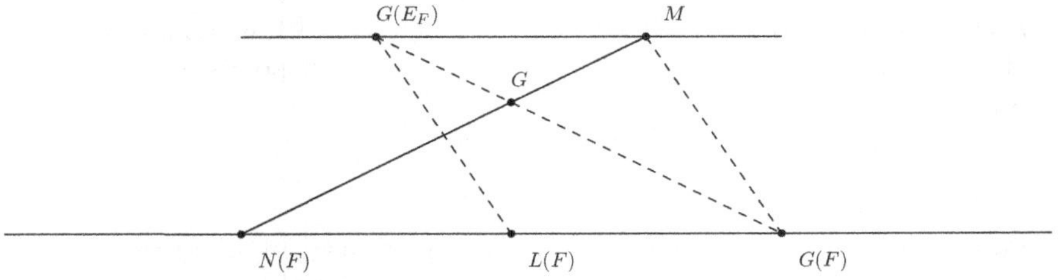

Fig. 2 Location of the Monge point

The following corollary tells us the precise location of the Monge point with respect to the vertices of the simplex and the given circumcenter.

Corollary 3.3 *Let* $T = \operatorname{conv}\{A_0, \ldots, A_d\}$ *be a* d*-simplex in* d*-dimensional Minkowski space, possessing a circumcenter* M. *Then the associated Monge point is determined as*

$$N_M = M + \frac{\sum_{i=0}^{d}(A_i - M)}{d-1}.$$

Proof Let F be a ridge of the simplex, opposite the edge E_F, such that $G(E_F) \neq M$ (i.e., the edge midpoint is distinct from M; such an edge must exist). From the proof of Theorem 3.1 we deduce for $M \neq G$ that

$$\|[M\,G(E_F)]\| : \|[G(F)N_M]\| = \|[M\,G]\| : \|[GN_M]\| = (d-1) : 2.$$

Thus

$$N_M = M + (d+1)\frac{G-M}{d-1} = M + \frac{(d+1)\frac{\sum_{i=0}^{d}(A_i-M)}{d+1}}{d-1} = M + \frac{\sum_{i=0}^{d}(A_i - M)}{d-1}.$$

For $M = G$ we obtain $N_M = M = G$. □

Remark 3.4 In Euclidean context, each *Monge hyperplane* passes through the vertex centroid of a $(d-2)$-face F and is perpendicular to the opposite edge E_F (here *opposite edge* means the edge between the two vertices not in the ridge F). However, we see that perpendicularity is not necessary for the construction, and any hyperplane containing the associated Monge line as per our definition is suitable (provided the Monge line is well-defined). Therefore, while our Minkowskian Monge pencils contain the correct Monge hyperplanes in Euclidean context, we have the confirmation that *orthogonality of lines and hyperplanes* need not necessarily play a role when finding the Monge point. The concept of Monge point is even an *affine* concept, as the circumcenter property of M is used nowhere (i.e., *any* point M can be used to construct "Monge lines" intersecting at N_M with the analytical expression given above).

In particular, we obtain the following corollary, which appears to give a new kind of construction also for the Euclidean case.

Corollary 3.5 *Let* $(\mathbb{R}^d, \|\cdot\|)$ *be a* d*-dimensional Minkowski space, and let* T *be a* d*-simplex with a circumcenter* M. *For each ridge* F *and the opposite edge* E_F *with midpoint* $G(E_F)$, *if* $M \neq G(E_F)$ *and* $\langle M\,G(E_F)\rangle$ *is* not *parallel to* F, *define an* M-hyperplane *as the hyperplane containing* F *and being parallel to* $\langle M\,G(E_F)\rangle$. *Then all defined* M*-hyperplanes intersect in the Monge point* N_M.

Remark 3.6 We summarize that the Monge point can be constructed in at least two concrete ways, once by Monge lines (Theorem 3.1), and once by M-hyperplanes (Corollary 3.5). Additionally, in Euclidean space, we have the usual construction via hyperplanes and orthogonality, which is a different specialization of Corollary 3.2.

Proof (Proof of the corollary) Let $A_0, \ldots, A_d$ denote the vertices of T. Observe that, since the medial hyperplanes of T are in general position, M lies in at most d of the $d+1$ medial hyperplanes. Without loss of generality, M does not lie in the medial hyperplane between A_0 and its opposite facet. Since $G([A_0A_i])$ lies in that medial hyperplane for $i = 1, \ldots, d$, and the ridge $F_{0,i}$ opposite $[A_0A_i]$ is parallel to that medial hyperplane, we conclude that $[M\,G([A_0A_i])]$ is not parallel to $F_{0,i}$, and the M-hyperplanes are defined at least for the d pairs $(F_{0,i}, [A_0A_i])$. Consider the $(d-1)$-simplex

$$T_0 := \operatorname{conv}\{G([A_0A_i]), i = 1, \ldots, d\},$$

which is a homothet of the *facet* F_0 of T opposite A_0 with homothety center A_0 and factor $\frac{1}{2}$. The related $(d-1)$-simplex T_0' is obtained by homothety of T_0 in $G(T_0) = \frac{A_0+G(F_0)}{2}$ and with homothety factor $-(d-1)$. Observe that the $(d-2)$-dimensional facets of T_0' pass through the vertices of T_0 and are parallel to the $(d-2)$-dimensional facets of T_0.

Now, the d M-hyperplanes previously considered are parallel to the hyperplanes defined by the facets of the d-simplex

$$\operatorname{conv}\{M \cup T_0'\}$$

through the vertex M. Therefore, these M-hyperplanes are in general position, intersecting only in the Monge point N_M which, by definition, is contained in every defined M-hyperplane. □

Another theorem concerning the Monge point in Euclidean space is the Mannheim theorem, see [14] for the three-dimensional case and [11] for generalizations. It is stated in Theorem 2.3 above, and it presents an example of a statement that cannot be extended to Minkowski spaces. The simple reason is that hyperplane sections of Minkowskian balls need not be centrally symmetric. Therefore, in general the concept of Monge point of a d-simplex cannot be transferred to its facets.

4 Euler Lines and Generalized Feuerbach Spheres of Minkowskian Simplices

We define as *Euler line associated to a circumcenter M* the straight line connecting M with the vertex centroid G. Thus, in the case of the vertex centroid being a circumcenter, the associated Euler line is not well-defined (see also Corollary 4.6). We now consider the situation in d-dimensional Minkowski space for $\boldsymbol{d \geq 2}$.

Definition 4.1 For a d-simplex $T := \operatorname{conv}\{A_0, \ldots, A_d\}$ with circumcenter M, define the *complementary line of a facet with respect to M* as the translate of the line between the circumcenter M of the simplex and the vertex centroid of the facet, passing through the opposite vertex. If $A_1, \ldots, A_d$ are the vertices of the chosen facet with vertex centroid G_0, then the complementary line is $A_0 + t \cdot (G_0 - M)$, $t \in \mathbb{R}$.

Remark 4.1 As in the planar case, for smooth norms such a circumcenter always exists (see [32], and [42, §7.1]). For a non-smooth norm, simplices without a circumcenter may exist (see again Fig. 1 (right) for the planar situation, and it is easy to construct examples also for general d).

The following theorem is an easy consequence of the definition of the vertex centroid.

Theorem 4.1 *The complementary lines of the facets of a d-simplex T with respect to a fixed circumcenter M connect all the vertices to the same point, the* complementary point *P_M associated to M.*

Proof Let $T = \mathrm{conv}\{A_0, \ldots, A_d\}$, and let G_j denote the vertex centroid of the facet opposite vertex A_j. Then the point

$$P_M = M + \sum_{i=0}^{d}(A_i - M) = A_j + d\left(\frac{\sum\limits_{\substack{i=0\\ i\neq j}}^{d} A_i}{d} - M\right)$$
$$= A_j + d(G_j - M) \text{ for each } j = 0, \ldots d,$$

lies on each complementary line. For each $j = 0, \ldots, d$, we have $\|[P_M A_j]\| = d\,\|[M G_j]\|$. □

Various useful types of orthogonalities have been defined in Minkowski spaces for pairs of vectors, all coinciding with the usual orthogonality in Euclidean space, yet we only have normality as a concept for vectors and (hyper-)planes. We call each segment $[P_M A_j]$ on a complementary line the *complementary segment* associated to the opposite facet. As such, a complementary segment is not orthogonal to a hyperplane in any known sense. However, in dimension two we obtain the familiar isosceles orthogonality between an edge of a simplex (triangle side) and the corresponding complementary segment (orthogonality if we are in the Euclidean plane!), and the complementary point is the C-*orthocenter* [8, 37]. Unlike the C-orthocenter, the notion of complementary point generalizes to any higher dimension.

Remark 4.2 The complementary point is even an affine notion (and so is the Monge point, see Remark 3.4), as we only used division ratios of segments on a line. In addition, the point P_M can be constructed for any point M (circumcenter or not) in the following way: take the line connecting M to the vertex centroid of a simplex facet (if distinct from M), and then consider the translated line passing through the vertex opposite the chosen facet. All lines of the latter kind intersect in a point (denoted P_M in the present article), which has already been observed by Snapper [49].

The complementary point and Monge point associated to a simplex with circumcenter M possess the following properties.

Theorem 4.2 *Let T be a d-simplex ($d \geq 2$) in Minkowskian space $(\mathbb{R}^d, \|\cdot\|)$, with a circumcenter M distinct from its vertex centroid G.*

(a) The associated complementary point P_M and the Monge point N_M lie on the Euler line $\langle MG \rangle$.
(b) The vertex centroid G divides the segment $[MP_M]$ internally in the ratio $1 : d$.
(c) The associated Monge point N_M divides the segment $[MP_M]$ internally in the ratio $1 : (d-2)$.
(d) The vertex centroid G divides the segment $[MN_M]$ internally in the ratio $(d-1) : 2$.

Proof Let $T = \mathrm{conv}\{A_0, \ldots, A_d\}$. That the Euler line $\langle GM \rangle$ associated to M passes through N_M and P_M can be seen from the following equations:

$$\begin{aligned} G &= \frac{\sum_{i=0}^{d} A_i}{d+1} = M + \frac{\sum_{i=0}^{d}(A_i - M)}{d+1}, \\ N_M &= M + \frac{\sum_{i=0}^{d}(A_i - M)}{d-1}, \\ P_M &= M + \sum_{i=0}^{d}(A_i - M). \end{aligned}$$

Thus (a) is proved. The above equations also immediately prove (b) and (c). Proving (d) is an easy exercise in arithmetic:

$$\begin{aligned} \|G - M\| : \|N_M - G\| &= \left\| \frac{\sum_{i=0}^{d}(A_i - M)}{d+1} \right\| : \left\| \frac{\sum_{i=0}^{d}(A_i - M)}{d-1} - \frac{\sum_{i=0}^{d}(A_i - M)}{d+1} \right\| \\ &= \left\| \frac{\sum_{i=0}^{d}(A_i - M)}{d+1} \right\| : \left\| \frac{2\sum_{i=0}^{d}(A_i - M)}{(d-1)(d+1)} \right\| \\ &= (d-1) : 2. \end{aligned}$$

□

Remark 4.3 We see that N_M can be obtained from M by homothety in G, with homothety ratio $-\frac{2}{d-1}$. Moreover, recall the M-hyperplanes from Corollary 3.5 which intersect in N_M. The above homothety takes each M-hyperplane to a certain parallel hyperplane through M. It turns out that these *central planes* (through the circumcenter M) encompass the supporting hyperplanes through M of the auxiliary simplex $\mathrm{conv}\{M \cup T_0'\}$ in the proof of Corollary 3.5.

Considering the points of interest in Theorem 4.2, one may ask whether the point $M + \frac{\sum_{i=0}^{d}(A_i - M)}{d}$ on the Euler line, dividing $[MP_M]$ internally in the ratio $1 : (d-1)$, holds any special meaning. It turns out that it is the center of a sphere analogous to the well-known Feuerbach circle of a triangle in the Euclidean plane. The extension to higher dimensional normed spaces for the case $M \neq G$ is as follows (for the "degenerate" case $M = G$ we refer to Corollary 4.6).

Theorem 4.3 *(The $2(d+1)$- or Feuerbach sphere of a d-simplex) In an arbitrary Minkowski d-space, let $T = \text{conv}\{A_0, \ldots, A_d\}$ be a d-simplex with a circumcenter M and circumradius R, and let $G(\neq M)$ be its vertex centroid. The sphere with center $F_M := M + \frac{\sum (A_i - M)_{i=0}^d}{d}$ on the Euler line and of radius $r := \frac{R}{d}$ passes through the following $2(d+1)$ points:*

(a) the vertex centroids G_i, $i = 0, \ldots, d$, of the facets F_i of T (F_i is opposite vertex A_i), and
(b) the points L_i^M dividing the segments connecting the Monge point N_M to the vertices A_i of T, $i = 0, \ldots, d$, in the ratio $1 : (d-1)$. Moreover, $S(F_M, r)$ is a homothet of the circumsphere $S(M, R)$ with respect to the vertex centroid G and homothety ratio $\frac{-1}{d}$, i.e., G divides the segment $[F_M M]$ internally in the ratio $1 : d$, and F_M divides the segment $[N_M M]$ internally in the ratio $1 : (d-1)$.

Remark 4.4 In analogy to the Feuerbach circle in the plane centered at the nine-point-center, we call F_M the $2(d+1)$*-center* of the simplex with respect to the circumcenter M, and $S(F_M, \frac{R}{d})$ its *Feuerbach* or $2(d+1)$*-point-sphere*.

Proof (Proof of the theorem) The vertex centroid of a facet opposite vertex A_j is $G_j = \frac{\sum_{\substack{i=0\\ i\neq j}}^{d} A_i}{d}$. We have $R = \|A_j - M\|$ for any $j = 0, \ldots, d$, and thus

$$\|G_j - F_M\| = \left\| \frac{\sum_{\substack{i=0\\ i\neq j}}^d A_i}{d} - M - \frac{\sum_{i=0}^d (A_i - M)}{d} \right\| = \left\| \frac{M - A_j}{d} \right\| = \frac{R}{d},$$

which proves that $S(F_M, \frac{R}{d})$ passes through the points in (a).

The Monge point is $N_M = M + \frac{\sum_{i=0}^d (A_i - M)}{d-1}$, thus

$$\begin{aligned} L_j^M &:= M + \frac{\sum_{i=0}^d (A_i - M)}{d-1} + \frac{A_j - M - \frac{\sum_{i=0}^d (A_i - M)}{d-1}}{d} \\ &= M + \frac{(d-1)\sum_{i=0}^d (A_i - M)}{d(d-1)} - \frac{M - A_j}{d} \\ &= M + \frac{\sum_{i=0}^d (A_i - M)}{d} - \frac{M - A_j}{d}. \end{aligned}$$

Therefore,

$$\|L_j^M - F_M\| = \left\| M + \frac{\sum_{i=0}^{d}(A_i - M)}{d} - \frac{M - A_j}{d} - M - \frac{\sum_{i=0}^{d}(A_i - M)}{d} \right\|$$
$$= \left\| -\frac{M - A_j}{d} \right\| = \frac{R}{d},$$

which proves that $S(F_M, \frac{R}{d})$ passes through the points in (b). We also have

$$\|F_M - G\| : \|G - M\| = \frac{\left\| M + \frac{\sum_{i=0}^{d}(A_i - M)}{d} - M - \frac{\sum_{i=0}^{d}(A_i - M)}{d+1} \right\|}{\left\| M + \frac{\sum_{i=0}^{d}(A_i - M)}{d+1} - M \right\|}$$
$$= \left\| \frac{\sum_{i=0}^{d}(A_i - M)}{d(d+1)} \right\| : \left\| \frac{\sum_{i=0}^{d}(A_i - M)}{d+1} \right\| = 1 : d$$

and

$$\|N_M - F_M\| : \|F_M - M\| = \frac{\left\| M + \frac{\sum_{i=0}^{d}(A_i - M)}{d-1} - M - \frac{\sum_{i=0}^{d}(A_i - M)}{d} \right\|}{\left\| M + \frac{\sum_{i=0}^{d}(A_i - M)}{d} - M \right\|}$$
$$= \left\| \frac{\sum_{i=0}^{d}(A_i - M)}{d(d-1)} \right\| : \left\| \frac{\sum_{i=0}^{d}(A_i - M)}{d} \right\| = 1 : (d-1),$$

proving the remaining statements. □

Remark 4.5 As noted in the Introduction, the sphere construction has been done for the Euclidean case in several earlier works, giving a $3(d+1)$-point-sphere. In Minkowski space, we "lose" the $(d+1)$ points which are orthogonal projections of the L_i^M onto the facets F_i. In the planar case, this has already been pointed out in [8, 37].

The following corollary is an immediate consequence of the affine nature of both the points mentioned in Theorem 4.2 and the $2(d+1)$-center introduced in Theorem 4.3.

Corollary 4.6 *In a d-simplex in Minkowskian d-space, the points M, G, F_M, N_M, P_M are either collinear (on the Euler line), or they all coincide. In the latter case, instead of speaking of the Euler line not being well-defined, sometimes the term* collapsing Euler line *is used.*

Remark 4.7 Based on the affine underpinning of our setting, we may consider the $(d+1)$-dimensional spatial representation of this configuration where the segments between M and the vertices of our simplex are projections of some segments spanning a $(d+1)$-dimensional parallelepiped. Then, the segment $[MP_M]$ on the Euler line corresponds to the projection of the main diagonal of the parallelepiped, and the points dividing the main diagonal in the ratio $1:d$, $1:(d-1)$, and $1:(d-2)$ project to the vertex centroid, the center of the Feuerbach-$2(d+1)$-point-sphere, and the Monge point, respectively.

Since it can be shown that N_M divides the segment $[F_M M]$ externally in the ratio $1:d$, i.e., $[F_M M]$ is divided harmonically by G and N_M, we obtain the following corollary, the second statement of which has been noted in [10] for Euclidean orthocentric simplices and the orthocenter. For a strictly convex normed plane ($d=2$), the second statement can be found in [37, Theorem 4.6].

Corollary 4.8 *The Monge point N_M associated to a circumcenter M of a d-simplex T is the center of homothety between the Feuerbach-$2(d+1)$-point-sphere centered at F_M and the circumsphere centered at M, with homothety ratio $1:d$. For any line from N_M meeting the associated circumsphere of T in Q, the point P dividing $[N_M Q]$ internally in the ratio $1:(d-1)$ is located on the Feuerbach sphere; conversely, for any line from N_M meeting the associated Feuerbach sphere in P, the point Q dividing the segment $[N_M P]$ externally in the ratio $d:(d-1)$ is located on the circumsphere of T.*

5 Generalizations for Polygons in the Plane

Generalizations of the concept of Euler line and Feuerbach circle have not just focused on raising the dimension of the space; there have also been attempts to generalize to polygons. We will now see that easy generalizations arise if we consider such polygons as projections of higher dimensional simplices or sections of parallelepipeds. This relates to *descriptive geometry* (see also Remark 4.7).

Herrera Gómez [25] and Collings [13] have written about *remarkable circles* in connection with *cyclic polygons* in the Euclidean plane. Their definition of *cyclic polygon* as a polygon possessing a circumcircle is directly extendable to any normed plane. Necessarily, cyclic polygons are convex.

Let $P=\operatorname{conv}\{A_0,\dots,A_d\}$, $d\geq 3$, be a cyclic polygon with circumcenter M in the normed plane $(\mathbb{R}^2,\|\cdot\|)$. We may view the vertices of P as the images under projection of certain vertices of a $(d+1)$-dimensional parallelepiped Q in $(d+1)$-dimensional space to an affine plane (which we then endow with the norm $\|\cdot\|$), namely the vertices adjacent to M' where M' projects to M (compare Remark 4.7). This makes P the projection of that hyperplane section P' of Q which is defined by all the vertices adjacent to M'. Alternatively, we may view P as the shadow of a d-simplex T, which itself is a projection of the hyperplane section P' of Q to an affine d-subspace.

We now define the points P_M (*complementary point*), N_M (*Monge point*), G (*vertex centroid*), F_M ($2(d+1)$-*center*) of the polygon to be the respective parallel projections of the following distinguished points on the main diagonal of the parallelepiped, which would have the corresponding meaning for the d-simplex T when M' projects to a circumcenter of T, see Sect. 4. That is,

$$G = \frac{\sum_{i=0}^{d} A_i}{d+1} = M + \frac{\sum_{i=0}^{d}(A_i - M)}{d+1} \quad \text{is called the } \textit{vertex centroid} \text{ of the polygon } P,$$

$$F_M = M + \frac{\sum_{i=0}^{d}(A_i - M)}{d} \quad \text{is called the } 2(d+1) - \textit{center} \text{ of the polygon } P,$$

$$N_M = M + \frac{\sum_{i=0}^{d}(A_i - M)}{d-1} \quad \text{is called the } \textit{Monge point} \text{ of the polygon } P,$$

$$P_M = M + \sum_{i=0}^{d}(A_i - M) \quad \text{is called the } \textit{complementary point} \text{ of the polygon } P.$$

These points either coincide or are collinear on the *Euler line* of the polygon P (compare Corollary 4.6), with the division ratios given in Theorem 4.2. We can then easily deduce the following relationships.

Theorem 5.1 *Let* $P = \text{conv}\{A_0, \ldots, A_d\}$, $d \geq 3$ *be a cyclic polygon with circumcenter* M *and circumradius* R *in the normed plane* $(\mathbb{R}^2, \|\cdot\|)$. *Then:*

(a) The complementary point P_M *is common to all the circles* $S(P_M^i, R)$, $i = 0, \ldots, d$, *where* P_M^i *is the complementary point of the subpolygon* $P_i = \text{conv}(\{A_0, \ldots, A_d\} \setminus \{A_i\})$ *with respect to the circumcenter* M.

(b) The lines $\langle A_i P_M^i \rangle$ *are concurrent in* C_M, *where* $C_M := M + \frac{1}{2}\sum_{i=0}^{d}(A_i - M)$ *is the midpoint of* $[MP_M]$ *and called the* spatial center *of* P *with respect to* M.

(c) The midpoints E_i *of the segments joining the vertices* A_i, $i = 0, \ldots, d$, *with the complementary point* P_M *are concyclic in the circle* $S(C_M, \frac{R}{2})$.

(d) The point C_M *is common to all the circles* $S(C_M^i, \frac{R}{2})$, *where* C_M^i *is the spatial center of the subpolygon* P_i *with respect to the circumcenter* M, $i = 0, \ldots, d$, *and the points* C_M^i *also lie on the circle* $S(C_M, \frac{R}{2})$.

Proof We have

$$P_M = M + \sum_{j=0}^{d}(A_j - M) = M + \sum_{\substack{j=0 \\ j\neq i}}^{d}(A_j - M) + (A_i - M) = P_M^i + (A_i - M).$$

Since $(A_i - M)$ is a radius of any translate of the circle $S(M, R)$, we obtain the statement in (a). In the spatial representation in $(d+1)$-dimensional space, the vertex projecting to the complementary point P_M is the endpoint opposite M' of the main diagonal of the parallelepiped Q (i.e., the line which projects to the Euler line), whereas the pre-images of the points P_M^i are vertices adjacent to the pre-image of P_M.

Thus the pre-images of each point P_M^i and A_i, $i = 0, \ldots, d$, together span another main diagonal of the parallelepiped Q. The main diagonals of the parallelepiped intersect in one point C' (the vertex centroid of the parallelepiped), and this point halves each main diagonal. The projection of this point is the point C_M by definition, which proves part (b). Note that at most $d - 1$ of the lines $\langle A_i P_M^i \rangle$ may not be well-defined, and precisely when their pre-images are parallel to the null space of the projection, but at least 2 lines remain to determine the point C_M. Part (d) is similar to part (a), in that

$$C_M = M + \frac{1}{2}\sum_{j=0}^{d}(A_j - M) = M + \frac{1}{2}\sum_{\substack{j=0\\ j\neq i}}^{d}(A_j - M) + \frac{1}{2}(A_i - M) = C_M^i + \frac{1}{2}(A_i - M).$$

The second statement in (d) follows trivially. Finally, for part (c), consider Fig. 3 and observe that the line $\langle C_M E_i \rangle$ is parallel to $\langle MA_i \rangle$ for each $i = 0, \ldots, d$. □

Remark 5.1 For the Euclidean plane and $d = 3$, part (d) is well known [53, pp. 22–23]. For all $d \geq 3$, the statements (a)–(c) have been established in [25] where P_M is called the orthocenter, and $S(C_M, \frac{R}{2})$ the Feuerbach circle of the polygon. For strictly convex normed planes part (d) has been shown in [37, Theorem 4.18], calling the point C_M the center of the Feuerbach circle $S(C_M, \frac{R}{2})$, and the circles $S(C_M^i, \frac{R}{2})$ the Feuerbach circles of the subpolygons. The motivation in either case was to observe a radius half as long as the radius of the original circumcircle. We see that the statements extend in some way to *all* Minkowski planes, though one has to

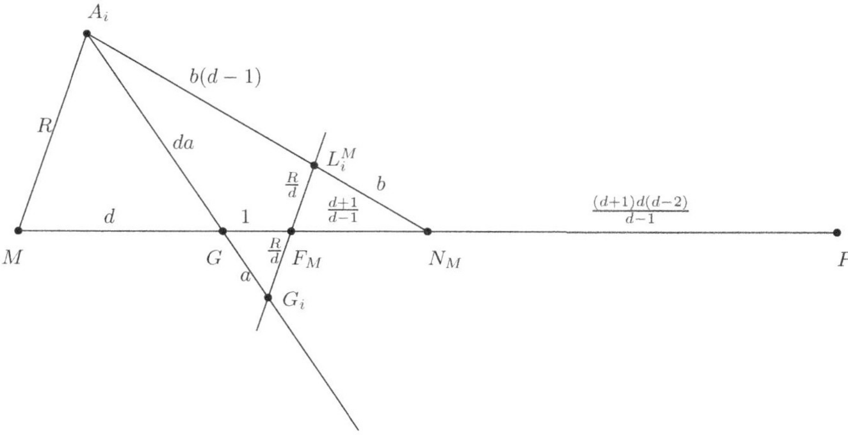

Fig. 3 Points on the Euler line and Feuerbach sphere, and ratios of line segments

be careful in their formulation; recall that in planes which are not strictly convex, we cannot necessarily speak of *the (unique)* circumcircle, or *the (unique)* intersection of several circles.

Remark 5.2 Note that M is a circumcenter of P, and also a circumcenter for each of its sub-polygons with $d \geq 3$ vertices. The analogous statement for a d-simplex in d-space is wrong, i.e., a circumcenter of a d-simplex T is *not* a circumcenter for each of its facets, which is the reason for the lack of analogous higher dimensional statements involving the complementary points of facets of T in Sect. 4.

An alternative, equally plausible definition of (orthocenter and) Feuerbach circle of a polygon in the Euclidean plane was given by Collings [13]. This, too, generalizes to normed (Minkowski) planes, and is easily provable using the spatial representation given above. Both concepts of Feuerbach circles are illustrated in Fig. 4, for cyclic pentagons in the ℓ_1-norm.

Theorem 5.2 *Let* $P = \operatorname{conv}\{A_0, \ldots, A_d\}$, $d \geq 3$, *be a cyclic polygon with circumcenter* M *and circumradius* R *in the normed plane* $(\mathbb{R}^2, \|\cdot\|)$.

(a) The Monge point N_M *is the point of intersection of the lines* $\langle A_i N_M^i \rangle$, $i = 0, \ldots, d$, *where* N_M^i *is the Monge point of the subpolygon* $P_i = \operatorname{conv}(\{A_0, \ldots, A_d\} \setminus \{A_i\})$.

(b) The vertex centroids G_i *of the subpolygons* $P_i = \operatorname{conv}(\{A_0, \ldots, A_d\} \setminus \{A_i\})$, $i = 0, \ldots, d$, *are concyclic on* $S(F_M, \frac{R}{d})$, *where* F_M *is the* $2(d+1)$*-center of the polygon. Furthermore, the circle* $S(F_M, \frac{R}{d})$ *passes through the* $(d+1)$ *points* L_i^M *dividing the segments* $[N_M A_i]$ *in the ratio* $1 : (d-1)$.

(c) The Monge points N_M^i *of the subpolygons are concyclic on the circle*

$$S\left(M + \frac{1}{d-2}\sum_{j=0}^{d}(A_j - M), \frac{R}{d-2}\right)$$

with its center on the Euler line.

Proof We have

$$\begin{aligned} N_M &= M + \frac{1}{d-1}\sum_{j=0}^{d}(A_j - M) \\ &= A_i + \frac{d-2}{d-1}\left(\left(M + \frac{1}{d-2}\sum_{\substack{j=0 \\ j\neq i}}^{d}(A_j - M)\right) - A_i\right) \\ &= A_i + \frac{d-2}{d-1}\left(N_M^i - A_i\right), \end{aligned}$$

Fig. 4 Comparison of different definitions of the Feuerbach circle for a pentagon in the ℓ_1-norm. For **a** see Remark 5.1, and for **b** see Remark 5.4. Respective radii are marked by dotted line segments, and the relevant part of the Euler line $[MP_M]$ is marked in bold. One pair of special points on the Feuerbach circle is constructed in each case (with thin solid auxiliary lines)

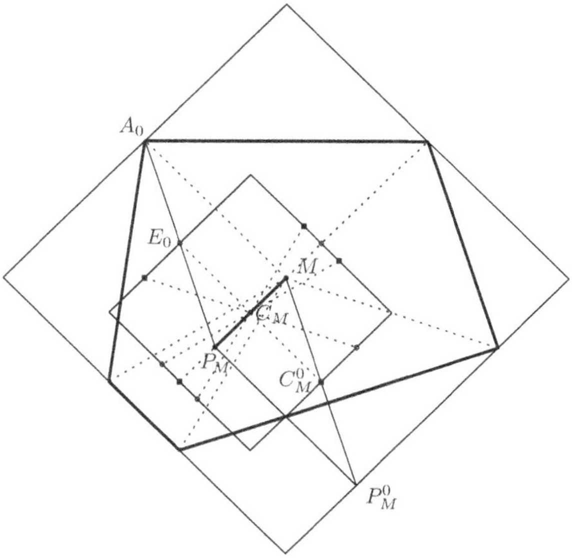

(a) The Feuerbach circle of half size.

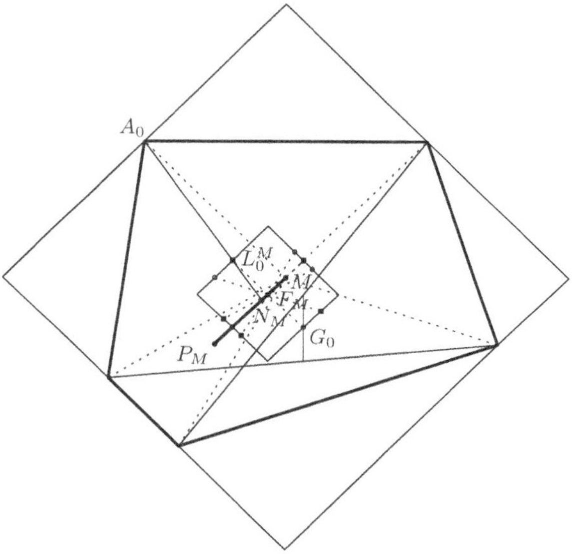

(b) The Feuerbach circle of Collings.

which proves part (a). Part (b) is clear with Theorem 4.3 and the fact that the segments $[F_M G_i]$ and $[F_M L_i^M]$ have equal length and are homothets of $[MA_i]$ for each $i = 0, \dots, d$ (with factor $\frac{1}{d}$ and homothety center N_M). For part (c), observe that for each $i = 0, \dots, d$, N_M^i is the intersection of the lines $\langle M G_i \rangle$ and $\langle A_i N_M \rangle$, see also Fig. 3. Since the above equation shows that N_M divides the segment $[A_i N_M^i]$ in the ratio $(d-2)\colon 1$, the homothet of the circumsphere with respect to homothety center N_M and homothety ratio $-\frac{1}{d-2}$ passes through the N_M^i. Thus the corresponding center can be calculated as $M + \frac{1}{d-2}\sum_{j=0}^{d}(A_j - M)$ (on the Euler line), and the radius is $\frac{R}{d-2}$. □

Remark 5.3 Collings [13] proved a variant of part (a) for the Euclidean plane and called the point N_M differently, namely the *orthocenter* of the polygon. In fact, Collings' orthocenter (per our definition, the Monge point N_M) was defined inductively, using the base case $d = 2$, i.e., starting at sub-triangles of P, whose Monge point, complementary point, and C-orthocenter coincide. Note that an inductive definition of the Monge point as such necessitates that M is the circumcenter at each stage of the recursion (otherwise the resulting points at each stage would not correspond to our definition of Monge point), and thus only works in the plane. In the context of d-simplices, we did not consider this recursion for precisely this reason (although of course, the respective lines exist in higher dimensional space, and they are concurrent at the corresponding points!).

Remark 5.4 Part (b) was also proved for the Euclidean plane in [13], and in analogy with the nine-point-circle of a triangle, the circle $S(F_M, \frac{R}{d})$ was named the (generalized) *nine-point-circle*, although it was only observed to pass through the $(d+1)$ vertex centroids G_i. B. Herrera Gómez [25] extended the statements, for example by proving (c) for the Euclidean plane, and by investigating related infinite families of circles.

6 Concluding Remarks and Open Problems

Solutions to questions from elementary geometry in normed spaces often yield an interesting tool and form the first step for attacking problems in the spirit of Discrete and Computational Geometry in such spaces (see, e.g., [3, 4] for the concepts of circumballs and minimal enclosing balls, or [34, Section 4] referring to bisectors as basis of an approach to Minkowskian Voronoi diagrams). And of course it is an interesting task for geometers to generalize notions like orthogonality (see [1, 2, 5]), orthocentricity (cf. [8, 37, 43, 48]), isometries (see [39, 41]), and regularity (see [40]) in absence of an inner product. In case of regularity, we may ask which figures are special, and what are useful concepts to describe their degree of symmetry in normed planes and spaces? For Minkowski spaces nothing really satisfactory is done in this direction, and it is clear that a corresponding hierarchical classification of types of simplices would yield the first step here. Thus, it would be an interesting research program to extend the generalizable parts of the concepts investigated in [15–17] to normed spaces: what particular types of simplices are obtained if special points of them, called "centers" (like circum- and incenters, vertex centroids, Monge points, Fermat-Torricelli points etc.), coincide or lie, in cases where this is not typical (e.g., in case of the incenter), on the Euler line? In view of [39, 41], a related interesting task might be the development of symmetry concepts based on Minkowskian isometries.

Another interesting point of view comes in with the field of geometric configurations which is summarized by the recent monograph [22]. Namely, the Three-Circles-Theorem and Miquel's Theorem can be successfully extended to normed planes (see [8, 37, 50] and thus have acquired some recent popularity. Clifford's circle configuration, for circles of equal radii also called Clifford's Chain of Theorems (see [36, 54]), is a direct generalization of the Three-Circles-Theorem and also part of the collection of theorems which nicely ties to visualizations of the Euler line and the Feuerbach circle in the spirit of descriptive geometry (see our discussion at the beginning of Sect. 5 above). Based on [8, 37], Martini and Spirova extend in [38] the Clifford configuration for circles of equal radii to strictly convex normed planes, and prove properties of the configuration as well as characterizations of the Euclidean plane among Minkowski planes. Using our terminology from Sect. 5 above, one may easily color the vertices of the parallelepiped Q alternatingly red and blue, with M' being blue. Then the projected blue vertices are centers of circles of the Clifford configuration, whereas the projected red vertices are in the intersection of certain subsets of the circles. Due to the successful extension of these topics to normed planes and spaces one might hope that further configuration concepts can be generalized this way. E.g., one can check whether the comprehensive geometry of n-lines (which are the natural extensions of complete quadrilaterals; see Section 4 of the survey [35]) and systems of circles corresponding with them contain parts which are generalizable this way.

As basic notions like isoperimetrix (see [52, §4.4 and §5.4]) demonstrate, *duality* (of norms) plays an essential role in the geometry of normed spaces. This concept should also be used in that part of Minkowski Geometry discussed here. It should be checked how far this important concept can be applied to get, in correspondence with already obtained results, also "dual results", such that, for example, results on notions like "circumball" and "inball" might be dual to each other.

Finally we mention that even for the Euclidean plane there are new generalizations of notions, such as generalized Euler lines in view of so-called circumcenters of mass etc. (see [51]), which could, a fortiori, also be studied for normed planes and spaces.and minimal enclosing balls

Acknowledgements The authors are grateful to Emil Molnár for several hints and remarks which helped to improve the presentation in the final version of this paper.

References

1. J. Alonso, C. Benítez, Orthogonality in normed linear spaces: a survey. I. Main properties. Extracta Math. **3**(1), 1–15 (1988)
2. J. Alonso, C. Benítez, Orthogonality in normed linear spaces: a survey. II. Relations between main orthogonalities. Extracta Math. **4**(3), 121–131 (1989)
3. J. Alonso, H. Martini, M. Spirova, Minimal enclosing discs, circumcircles, and circumcenters in normed planes (Part I). Comput. Geom. **45**(5–6), 258–274 (2012)
4. J. Alonso, H. Martini, M. Spirova, Minimal enclosing discs, circumcircles, and circumcenters in normed planes (Part II). Comput. Geom. **45**(7), 350–369 (2012)
5. J. Alonso, H. Martini, S. Wu, On Birkhoff orthogonality and isosceles orthogonality in normed linear spaces. Aequationes Math. **83**(1–2), 153–189 (2012)
6. C. Alsina, M.S. Tomás, Orthocenters in real normed spaces and the Euler line. Aequationes Math. **67**(1–2), 180–187 (2004)
7. N. Altshiller-Court, *Modern Pure Solid Geometry*, 2nd edn. (Chelsea Publishing Co., New York, 1964)
8. E. Asplund, B. Grünbaum, On the geometry of Minkowski planes. Enseignement Math. (2), **6**, 299–306 (1961), 1960
9. G. Averkov, On the geometry of simplices in Minkowski spaces. Stud. Univ. Žilina Math. Ser. **16**(1), 1–14 (2003)
10. M. Buba-Brzozowa, Analogues of the nine-point circle for orthocentric n-simplexes. J. Geom. **81**(1–2), 21–29 (2004)
11. M. Buba-Brzozowa, The Monge point and the $3(n + 1)$ point sphere of an n-simplex. J. Geom. Graph. **9**(1), 31–36 (2005)
12. G.D. Chakerian, M.A. Ghandehari, The Fermat problem in Minkowski spaces. Geom. Dedicata **17**(3), 227–238 (1985)
13. S.N. Collings, Cyclic polygons and their Euler lines. Math. Gaz. **51**, 108–114 (1967)
14. R.A. Crabbs, Gaspard Monge and the Monge point of the tetrahedron. Math. Mag. **76**(3), 193–203 (2003)
15. A.L. Edmonds, M. Hajja, H. Martini, Coincidences of simplex centers and related facial structures. Beiträge Algebra Geom. **46**(2), 491–512 (2005)
16. A.L. Edmonds, M. Hajja, H. Martini, Orthocentric simplices and their centers. Results Math. **47**(3–4), 266–295 (2005)
17. A.L. Edmonds, M. Hajja, H. Martini, Orthocentric simplices and biregularity. Results Math. **52**(1–2), 41–50 (2008)
18. E. Egerváry, On orthocentric simplexes. Acta Litt. Sci. Szeged **9**, 218–226 (1940)
19. E. Egerváry, On the Feuerbach-spheres of an orthocentric simplex. Acta Math. Acad. Sci. Hungar. **1**, 5–16 (1950)
20. R. Fritsch, "Höhenschnittpunkte" für n-Simplizes. Elem. Math. **31**(1), 1–8 (1976)
21. M.L. Gromov, Simplexes inscribed on a hypersurface. Mat. Zametki **5**, 81–89 (1969)
22. B. Grünbaum, Configurations of points and lines, vol. 103, Graduate Studies in Mathematics (American Mathematical Society, Providence, RI, 2009)
23. M. Hajja, H. Martini, Orthocentric simplices as the true generalizations of triangles. Math. Intelligencer **35**(2), 16–28 (2013)
24. G. Hajós, Über die Feuerbachschen Kugeln mehrdimensionaler orthozentrischer Simplexe. Acta Math. Acad. Sci. Hungar. **2**, 191–196 (1951)
25. B.H. Gómez, The Feuerbach circle and the other remarkable circles of the cyclic polygons. Elem. Math. **69**(1), 1–10 (2014)
26. Á.G. Horváth, Hyperbolic plane geometry revisited. J. Geom. **106**(2), 341–362 (2015)
27. Á.G. Horváth, On the hyperbolic triangle centers. Stud. Univ. Žilina Math. Ser. **27**(1), 11–34 (2015)
28. S. Iwata, The problem in closure in the n-dimensional space. Sci. Rep. Fac. Lib. Arts Ed. Gifu Univ. Natur. Sci. **3**(1), 28–32 (1962)

29. M.S. Klamkin, K. Hanes, M.D. Meyerson, Problems and solutions: solutions: orthocenters uponOrthocenters: 10453. Amer. Math. Monthly **105**(6), 563–565 (1998)
30. A.O. Konnully, Orthocentre of a simplex. J. London Math. Soc. **39**, 685–691 (1964)
31. V.V. Makeev, The degree of a mapping in some problems of combinatorial geometry. Ukrain. Geom. Sb. **30**, 62–66 (1987). ii
32. V.V. Makeev, Inscribed simplices of a convex body. Ukrain. Geom. Sb. **35**, 47–49 (1992). 162
33. S.R. Mandan. Altitudes of a simplex in n-space. J. Austral. Math. Soc. **2**, 403–424 (1961/1962)
34. H. Martini, K.J. Swanepoel, The geometry of Minkowski spaces–a survey. II. Expo. Math. **22**(2), 93–144 (2004)
35. H. Martini, Recent results in elementary geometry. II, in *Proceedings of the 2nd Gauss Symposium. Conference A: Mathematics and Theoretical Physics (Munich, 1993)*, Sympos. Gaussiana, (de Gruyter, Berlin, 1995), pp. 419–443
36. H. Martini, The three-circles theorem, Clifford configurations, and equilateral zonotopes. In: *Proceedings of the 4th International Congress of Geometry (Thessaloniki, 1996)* (Giachoudis-Giapoulis,Thessaloniki, 1997), pp 281–292
37. H. Martini, M. Spirova, The Feuerbach circle and orthocentricity in normed planes.Enseign. Math. (2), 53(3–4), 237–258 (2007)
38. H. Martini, M. Spirova, Clifford's chain of theorems in strictly convex Minkowski planes. Publ. Math. Debrecen **72**(3–4), 371–383 (2008)
39. H. Martini, M. Spirova, Reflections in strictly convex Minkowski planes. Aequationes Math. **78**(1–2), 71–85 (2009)
40. H. Martini, M. Spirova, Regular tessellations in normed planes. Symmetry: Culture Sci. **1–2**, 223–235 (2011)
41. H. Martini, M. Spirova, K. Strambach, Geometric algebra of strictly convex Minkowski planes. Aequationes Math. **88**(1–2), 49–66 (2014)
42. H. Martini, K.J. Swanepoel, G. Weiß, The geometry of Minkowski spaces–a survey. I. Expo. Math. **19**(2), 97–142 (2001)
43. H. Martini, S. Wu, On orthocentric systems in strictly convex normed planes. Extracta Math. **24**(1), 31–45 (2009)
44. P. McMullen, E. Schulte, *Abstract Regular Polytopes*, vol. 92, Encyclopedia of Mathematics and its Applications (Cambridge University Press, Cambridge, 2002)
45. R. Mehmke, Ausdehnung einiger elementarer Sätze über das ebene Dreieck auf Räume von beliebig viel Dimensionen. Arch. Math. Phys. **70**, 210–219 (1884)
46. F. Molnár, Über die Eulersche Gerade und Feuerbachsche Kugel des n-dimensionalen Simplexes. Mat. Lapok **11**, 68–74 (1960)
47. H.W. Richmond, On extensions of the property of the orthocentre. Quart. J. **32**, 251–256 (1900)
48. T.R. Soto, W.P. Redondo, On orthocentric systems in Minkowski planes. Beitr. Algebra Geom. **56**(1), 249–262 (2015)
49. E. Snapper, An affine generalization of the Euler line. Amer. Math. Monthly **88**(3), 196–198 (1981)
50. M. Spirova, On Miquel's theorem and inversions in normed planes. Monatsh. Math. **161**(3), 335–345 (2010)
51. S. Tabachnikov, E. Tsukerman, Remarks on the circumcenter of mass. Arnold Math. J. **1**(2), 101–112 (2015)
52. A.C. Thompson, *Minkowski Geometry*, vol. 63, Encyclopedia of Mathematics and its Applications (Cambridge University Press, Cambridge, 1996)
53. I.M. Yaglom.*Geometric transformations. II.* (Translated from the Russian by Allen Shields) (Random House, New York; The L. W. Singer Co., Syracuse, N.Y., 1968)
54. P. Ziegenbein, Konfigurationen in der Kreisgeometrie. J. Reine Angew. Math. **183**, 9–24 (1940)

An Algorithm for Classification of Fundamental Polygons for a Plane Discontinuous Group

Zoran Lučić, Emil Molnár and Nebojša Vasiljević

Abstract Based on the procedure given in [15] we describe an algorithm, implemented in a computer program, for complete enumeration of combinatorial equivalence classes of fundamental polygons for any fixed plane discontinuous group given by its signature. This is a solution of a long standing problem, we call it Poincaré-Delone problem to honour of Henry Poincaré and Boris Nikolaevich Delone (Delaunay). We give examples and computations together with some complete lists of combinatorially different polygons which serve as fundamental domains for the groups with the Macbeath signatures, e.g. $(2, +, [\ ]; \{\ \})$,$(3, -, [\ ]; \{\ \})$ and $(3, +, [\ ]; \{\ \})$.

Keywords Plane discontinuous group · Fundamental polygon · Macbeath signature · Poincaré-Delone problem

2010 Mathematics Subject Classification 57S30 · 52C20 · 51M10 · 68R99

Supported by Hungarian Nat. Found. for Sci. Res. (OTKA) No. 20498 (1996).

Z. Lučić · N. Vasiljević
Department of Mathematics, University of Belgrade,
Studentski trg 16, P.B. 550 Belgrade, Serbia
e-mail: zlucic@matf.bg.ac.rs

N. Vasiljević
e-mail: nebojsa@matf.bg.ac.rs

E. Molnár (✉)
Institute of Mathematics, Department of Geometry,
Budapest University of Technology and Economics,
Egry József út. 1. H. II 22, Budapest 1521, Hungary
e-mail: emolnar@math.bme.hu
URL: http://www.math.bme.hu/~emolnar

M. D. E. Conder et al. (eds.), *Discrete Geometry and Symmetry*,
Springer Proceedings in Mathematics & Statistics 234,
https://doi.org/10.1007/978-3-319-78434-2_14

1 Introduction

Suppose that $\widetilde{M}$ is a 2-dimensional, closed, compact manifold, with possible singular points and boundaries, i.e. that $\widetilde{M} = \Pi/G$ is a good orbifold that serves as a compact fundamental domain F of an isometry group G acting discontinuously on a classical plane Π of constant curvature. For omitting the too long introduction, we refer to [11] and Sect. 4, where the 46 fundamental domains of the well-known 17 Euclidean plane crystallographic groups are displayed. Hence, we consider $\widetilde{M} = F$ as a polygon with side pairing identifications, i.e. with piecewise linear (PL) presentation on the affine plane A^2 (see [34]).

By the Macbeath signature (see [17, 20, 33] or [34])

$$(g, \pm; [h_1, \ldots, h_l]; \{(h_{11}, \ldots, h_{1l_1}), \ldots, (h_{q1}, \ldots, h_{ql_q})\}),$$

where $\pm$ is $+$ or $-$, and where g, h_i $(1 \le i \le l)$ and h_{ij} $(1 \le i \le q, 1 \le j \le l_i)$ are integers such that $g \ge 0$, $h_i \ge 2$ and $h_{ij} \ge 2$, we express the following features of $\widetilde{M}$:

(i) If $\pm = +$ and $g > 0$, then $\widetilde{M}$ is an orientable surface of genus g, which means that $\widetilde{M}$ is a connected sum of g tori (Fig. 1a); if $g = 0$, then $\widetilde{M}$ is homeomorphic to a sphere; if $\pm = -$ and $g > 0$, then $\widetilde{M}$ is non-orientable surface of genus g, which means that $\widetilde{M}$ is a connected sum of g projective planes.

(ii) There are l *singular points* on $\widetilde{M}$, with *periods* $h_1, h_2, \ldots, h_l$. If the set of periods is empty, i.e. if empty square brackets appear in a signature, then $\widetilde{M}$ has no singular points.

(iii) There are q disjoint closed (Jordan) curves $\gamma_1, \gamma_2, \ldots, \gamma_q$ (called *boundary components*) on $\widetilde{M}$ and l_i $(1 \le i \le q)$ *dihedral points* on the curve γ_i, of *periods* h_{ij} $(1 \le i \le q, 1 \le j \le l_i)$. If $q = 0$ then empty curly brackets denote that there are no boundary components.

Equivalent to the Macbeath signature is Conway's orbifold notation (see [5] or [1]):

$$\circ\circ\ldots\circ \quad h_1, \ldots, h_l \quad * h_{11}, \ldots, h_{1l_1} \;\ldots\; * h_{q1}, \ldots, h_{ql_q} \quad \times\times\ldots\times,$$

either with g initial circles which represent tori (for orientable case) or g final crosses which represent projective planes (for non-orientable case). The absence of circles and crosses indicates that the base manifold M of the orbifold $\widetilde{M}$ is a sphere.

If G is a finitely generated isometry group, with fundamental domain $F = \Pi/G = \widetilde{M}$, which acts discontinuously on a complete, simply connected 2-dimensional manifold Π of constant curvature 0, $+1$ or -1 (i. e. Π is the Euclidean plane $\boldsymbol{E}^2$, or the 2-sphere $\boldsymbol{S}^2$, or the Bolyai-Lobachevskian hyperbolic plane $\boldsymbol{H}^2$), then a Macbeath signature may serve as a *signature of* G in order to indicate the orientability $(\pm)$ of Π/G, its genus (g), the orders $h_1, \ldots, h_l$ of the rotation centers and the stabilizers of the orders $2h_{ij}$ $(1 \le i \le q, 1 \le j \le l_i)$ associated with the dihedral centers on the i-th boundary component.

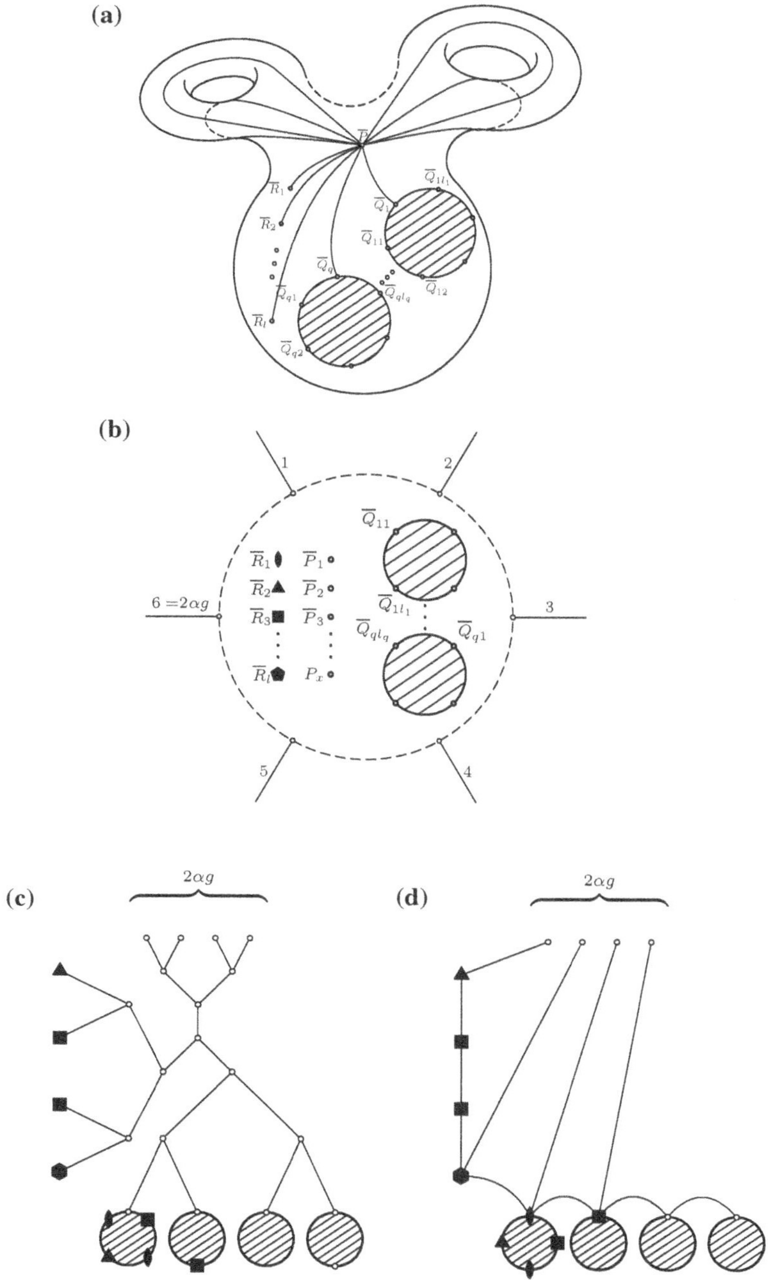

Fig. 1 **a**, **b** Scheme of an orientable surface. Graph **c** for a maximal, **d** for a minimal fundamental domain, respectively

Identifying points X from the orbit X^G of G, by a covering map

$$\kappa : \Pi \to \Pi/G, \qquad X \mapsto \overline{X} := X^G,$$

we obtain a surface $\widetilde{M} = \Pi/G$ which is a *good orbifold* [26, p. 87] (compact surface) if all the rotation and dihedral centers of G are of finite order.

The map κ is a local homeomorphism almost everywhere on Π except at points with non-trivial stabilizers, hence, at the rotation centers and the points mapped onto the boundary of $\widetilde{M}$.

We give the well known necessary and sufficient conditions for topological (homeomorphically equivariant) or geometrical isomorphism of plane discontinuous groups, where a topological mapping ϕ of Π induces the group isomorphism $\Upsilon : G \to G'$, $g \mapsto g' = \phi^{-1} g \phi$ (for the proof see [17] or [34, th. 4.6.3-4]).

Two plane discontinuous groups G and G' are topologically isomorphic (equivariant) if and only if:

(a) The surfaces $\widetilde{M} = \Pi/G$ and $\widetilde{M}' = \Pi/G'$ are homeomorphic.
(b) The numbers of the non-equivalent rotation centers are the same and the orders of the rotations are the same, i.e. up to their permutation.
(c) On each boundary curve γ_i of $\widetilde{M}$, and γ_i' of $\widetilde{M}'$, respectively, there is a cycle of dihedral centers with corresponding orders $2h_{i1}, \ldots, 2h_{il_i}$. If $\widetilde{M}$ and $\widetilde{M}'$ are orientable, then either both have the same cycles or all those of $\widetilde{M}'$ are inverse to those of $\widetilde{M}$. If $\widetilde{M}$ and $\widetilde{M}'$ are non-orientable, then the cycles of $\widetilde{M}$ may be put in bijective correspondence with those of $\widetilde{M}'$, where image and pre-image may have the same or opposite orientation.

By a formal contraction of the q boundary disks into q singular points of the compact surface $\widetilde{M}$, we obtain a compact surface M^* without boundary, with q additional singular points and with the same rotation centers, the same genus and orientability as the starting $\widetilde{M}$.

If G_X is a stabilizer of a point $X \in \Pi$, we define the *indicator function*

$$S(G_X) = \begin{cases} h^+, & \text{if } X \text{ is rotation center of order } h; \\ h, & \text{if } X \text{ is dihedral center of order } 2h, \end{cases}$$

where $h \in \{1, 2, \ldots\}$. If $h^+ = 1^+$, stabilizer G_X is trivial ($h = 1$ will not be indicated at the polygon symbol later on, $^+$ can also be omitted).

For the above plane discontinuous group G there is a simply connected bounded closed set F, called a *fundamental domain of* G, whose G-images cover Π without any interior point in common (see [34, p. 115]). Moreover, as a fundamental domain of G may serve a *generalized polygon*, i.e. a topological disk F whose boundary is divided by a finite set of *vertices* into piecewise linear sides. The polygon F is called *fundamental polygon*. The sides of F are identified (or κ-paired) by isometries of Π which generate G. Generalized polygon F which serves as a fundamental domain of G, together with the set of identifications defined by G, is said to be κ*-paired polygon* of G. The vertices of F (where at least three G-images of F meet) fall into

G-equivalence classes with G-conjugate stabilizers such that the indicator function takes the same value on them.

If Y is the midpoint of an edge such that $S(G_Y) = 2^+$, this point is (exceptionally) considered as a vertex of F, although only two G-images of F meets around Y. This is the point where our method differs from the D-symbol method given in [1, 12].

If a line reflection appears as a generator in G, a side on that line is identified with itself and this side appears on a boundary cycle of $\widetilde{M}$.

In order to characterize polygons which serve as fundamental domains for a given discontinuous group, we provide the following theorem proved in [15] (see Fig. 1a–d).

Theorem 1 *A κ-paired polygon F serves as a fundamental domain for a plane discontinuous group G given by a Macbeath signature above, if and only if the κ-images of its sides form a graph C on a surface $\widetilde{M} = \Pi/G$ with the following properties:*

1. *$\widetilde{M} \setminus C$ is an open disk.*
2. *The graph C can be contracted on the surface $\widetilde{M}$ with genus g into a graph $\widetilde{C}$ with one vertex of valency $v = 2\alpha g$, and αg loops ($\alpha = 2$ if M is orientable, $\alpha = 1$ otherwise).*
3. *A singular point $\overline{R}_i$, that is a κ-image of a rotation center R_i, is a vertex of C with valency $v(\overline{R}_i) \geq 1$.*
4. *A subgraph C_i of C which belongs to the boundary component $\overline{b}_i$ can be contracted (as $\widetilde{M} \to M^*$ is a formal contraction indicated above) into a point $\overline{Q}_i$ with valency $v(\overline{Q}_i) \geq 1$.*
5. *A vertex $\overline{P}$ of C, that is a κ-image of a vertex P of F with trivial stabilizer, has a valency of at least 3.*

□

Two fundamental (or κ-paired) polygons are said to be *combinatorially equivalent* if there is a bijection mapping one onto the other which preserves the relation of incidence of vertices and edges, their cyclic order, and the G-equivalence of vertices and the directed edges together with their stabilizers (see [14, p. 511]).

To determine all combinatorially different polygons which serve as fundamental domains for a given plane discontinuous group we describe the following procedure given in [14, 15].

Theorem 2 *Suppose that G is a finitely generated discontinuous isometry group acting on Π with a compact fundamental domain. Suppose further that G is given with a fixed good Macbeath signature, different from 4 types of bad orbifolds [26, p. 87]*

$$(0, +; [u]; \{\ \}), \quad (0, +; [\ ]; \{(u)\}), \quad 2 \leq u,$$

$$(0, +; [u, v]; \{\ \}), \quad (0, +; [\ ]; \{(u, v)\}, \quad 2 \leq u < v,$$

for which such a group G does not exist, and different from three further types of orbifolds

$$S_1 = (0, +; [\ ]; \{\ \}), \quad S_{\overline{1}} = (1, -; [\ ]; \{\ \}),$$

$$R = (0, +; [\ ]; \{(h_{11}, \ldots, h_{1l_1})\}), \quad 0 < l_1,$$

with combinatorially unique fundamental domains. The set of all combinatorially different polygons, that are fundamental domains for G, is obtained by the following procedure:

(a) On $\widetilde{M}$ we determine a finite set (up to combinatorial equivalence) of all possible non-contractible graphs with one vertex and αg loops ($\alpha = 2$ if $\widetilde{M}$ is orientable, $\alpha = 1$ otherwise), such that for any $\widetilde{C}$ from that set, $\widetilde{M} \setminus \widetilde{C}$ is a disk.

(b) We associate the graph $\widetilde{C}$ with a disconnected graph $\widetilde{C}'$ which consists of αg disjoint paths belonging to the loops of $\widetilde{C}$ ($\widetilde{C}'$ can be obtained by cutting a (small) disk D on $\widetilde{M}$ around an added vertex, Fig. 1b).

(c) We determine a finite set of all possible trees on $\widetilde{M}$ (in D), each of them meets $\widetilde{C}'$ only at the set of its $2\alpha g$ vertices (on the boundary of D), such that the set of vertices of each of these trees consists of:

(i) $2\alpha g$ vertices of $\widetilde{C}'$, each is of valency one.

(ii) l rotation centers $\overline{R}_1, \ldots, \overline{R}_l$.

(iii) Points $\overline{Q}_1, \ldots, \overline{Q}_q$ obtained by contractions of the boundary components of $\widetilde{M} \to M^*$.

(iv) Some additional points $\overline{P}_1, \overline{P}_2, \ldots, \overline{P}_x$ on $\widetilde{M}$, each is of valency at least three, whence $x \leq 2\alpha g + l + q - 2$.

(d) We join each of these trees with $\widetilde{C}'$ and replace $\overline{Q}_1, \ldots, \overline{Q}_q$ by the boundary components $\overline{b}_1, \ldots, \overline{b}_q$ of M^* with dihedral centers on them as new vertices, to obtain a new graph C on $\widetilde{M}$.

(e) To every disk $\widetilde{M} \setminus C$ we correspond a polygon F which serves as a fundamental domain for G.

(f) Among all the polygons F we select the combinatorially different ones.

□

The inequality $x \leq 2\alpha g + l + q - 2$ (that was omitted in [15] but already mentioned in [14]) is justified by a well known property of a tree, that the number of its vertices that are of a degree at least 3, is equal to the number of vertices of degree 1, subtracted by 2. In our case the number of vertices of degree 1 is equal to the sum of $2\alpha g$, and of the number of vertices of a degree 1 among $\overline{R}_1, \ldots, \overline{R}_l$, and $\overline{Q}_1, \ldots, \overline{Q}_q$, which is at most $l + q$.

By comparing the angle sum of the polygon that is a fundamental domain of the plane discontinuous group G, given by its Macbeath signature, with the angle sum of the corresponding Euclidean polygon, we conclude that G is realizable as isometry group, acting discontinuously on $\boldsymbol{S}^2(<)$, in $\boldsymbol{E}^2(=)$ or $\boldsymbol{H}^2(>)$, if and only if

$$0 \gtreqless 4 - 2\alpha g - 2\sum_{i=1}^{l}(1 - 1/h_i) - 2q - \sum_{j=1}^{q}\sum_{k=1}^{l_j}(1 - 1/h_{jk}),$$

where $\alpha = 2$ if $\widetilde{M}$ is orientable, or $\alpha = 1$ otherwise.

We give sharp estimates for the number $\mathfrak{n}$ of sides of fundamental polygon F obtained by the procedure described in Theorem 2 (see [14]):

Theorem 3 *If* $\mathfrak{n}$ *is the number of edges (and vertices) of a fundamental polygon of finite area for our group G given by its Macbeath signature, different from the groups* S_1, $S_{\overline{1}}$ *and R with combinatorially unique fundamental domains, then*

$$\mathfrak{n}_{\min} \le \mathfrak{n} \le \mathfrak{n}_{\max},$$

where

$$\mathfrak{n}_{\min} = 2\alpha g \quad \text{if} \quad l = q = 0,$$

or

$$\mathfrak{n}_{\min} = q_0 + \sum_{k=1}^{q} l_k + 2\alpha g + 2l + 2q - 2$$

otherwise, and

$$\mathfrak{n}_{\max} = \sum_{k=1}^{q} l_k + 6\alpha g + 4l + 5q - 6,$$

where $\alpha = 2$ *if* $\widetilde{M}$ *is orientable, or* $\alpha = 1$ *otherwise, and* q_0 *is the number of the boundary components of M containing no dihedral centers. Moreover, for a given G there exist fundamental domains with* $\mathfrak{n}_{\min}$ *and* $\mathfrak{n}_{\max}$ *edges.* □

For the group with signature

$$(1, +, [3, 4, 4, 6]; \{(2, 3, 2, 4), (4), (\,), (\,)\}),$$

we illustrate these facts in Fig. 1c, d (see [14, p. 516]).

Theorem 4 *For each plane discontinuous group G, given by its signature and described in Theorems 2 and 3, there are finitely many combinatorially different fundamental polygons. There exists an algorithm that enumerates all fundamental polygons of G.* □

Remark The above procedure is based on the enumeration of trees with $2\alpha g + l + q$ fixed vertices and x additional vertices each of valency at least 3, where $x \le 2\alpha g + l + q - 2$. This procedure can be applied even in the case when extended rotations or dihedral centers of infinite order exist.

Namely, within the Macbeath signature we allow that $h_i = \infty$ $(1 \le i \le l)$ for extended rotation center, which is an end in $\boldsymbol{H}^2$ determined by a horocyclic rotation, or the ideal point of two parallel lines in $\boldsymbol{E}^2$ determined by an Euclidean translation. Furthermore, we allow that $h_{ij} = \infty$ $(1 \le i \le q, 1 \le j \le l_i)$ for extended dihedral center defined by parallel reflection lines in $\boldsymbol{H}^2$, or an ideal point defined by two parallel reflection lines in the Euclidean plane $\boldsymbol{E}^2$. Moreover, at a boundary component could be more than one extended dihedral centers.

As we see these extended centers cause some differences in the above procedure of determining fundamental polygons for the above group G. Difficulties appear also in the metric realization of the corresponding fundamental domain with ideal vertices.

Fortunately our existence Theorem 5, based on [15, Prop. 3.2], allows unified formulation, but this will not be discussed here.

Theorem 5 *[15, Prop. 3.2] Among all convex polygons in $\boldsymbol{S}^2$ or in $\boldsymbol{H}^2$ (resp. in $\boldsymbol{E}^2$) with given angles $\alpha_1, \alpha_2, \ldots, \alpha_m$, $m \ge 3$, there exists exactly one up to isometry (resp. similarity) polygon, respecting the order of angles, circumscribed around a circle.* □

In Sect. 3 we will describe a particular algorithm whose existence is explained in Theorem 4. It is based on the procedure from Theorem 2 and facts from Theorem 3.

The computer implementation of that algorithm was developed by the third author in his B.Sc. thesis [32]. The product of this implementation is program COMCLASS (see Sect. 4). The complexity of the procedure, which is clearly super-exponential (note that the number of labeled trees with n vertices is n^{n-2}, see [3]), will be discussed elsewhere (see e.g. [30]).

Particular problems have independently been solved in [1, 12, 20, 30], partially by different methods. This gives us an opportunity to illustrate some of the steps in the procedure only by examples and figures.

Thus, the long standing Poincaré-Delone problem has been completely solved in dimensions 2 by Theorems 1, 2, 3, 4 and 5, and the algorithm described in Sect. 3. Our procedure also completes the classification of plane discontinuous groups with fundamental domains of finite area, finalized in [17, 33, 34]. The corresponding 2-orbifolds have also been completely described (see [26]).

Poincaré's classification of compact 2-dimensional surfaces is obviously related to our topic. This is based in principle also on an algorithm of super-exponential complexity (by the genus). Poincaré [28] initiated finding orientation preserving hyperbolic plane groups (so-called Fuchsian groups) via finding their possible fundamental domains (at least one domain for each group), and he characterized these domains (see e.g. [18, 19]). Delone (see e.g. [7, 8]) classified all Euclidean planigons (onto 93 combinatorial "marked" classes, among them 46 fundamental domains, see Sect. 4), and asked for the general stereohedron problem, i.e. for finding "all" combinatorial polyhedra (polytopes) which can be fundamental domains of groups acting discontinuously on a space of constant curvature (only particular cases are solved for dimensions greater than 2).

Our program is available in a source code for on-line execution (see Sect. 4). Its output is a list of all combinatorially different fundamental domains (represented by canonized polygon descriptors list) for a group given by its Macbeath signature.

2 Paired Polygon

A κ-paired polygon $F = V_0V_1 \ldots V_{n-1}$ of a group G given by a (good) Macbeath's signature, whose edges are

$$a_0 = V_0V_1,\ a_1 = V_1V_2, \ldots,\ a_{n-1} = V_{n-1}V_0$$

is well described by the relations $\uparrow\uparrow$ and $\uparrow\downarrow$ among edges, and the relation $\leftrightarrow$ among vertices (induced by the previous two relations), defined on the set $\{0, 1, \ldots, n-1\}$, $n \in \boldsymbol{N}$, and by the finite sequence

$$m = (m_0, m_1, \ldots, m_{n-1}), \quad m_i \in \boldsymbol{N}$$

such that $i \uparrow\uparrow j$ if edges a_i and a_j are κ-paired by a direct isometry from G different from identity,

$i \uparrow\downarrow j$ if edges a_i and a_j are κ-paired by an indirect isometry from G (if $i = j$ that isometry is a reflection in a line which contains a_i),

$i \leftrightarrow j$ if vertices V_i and V_j are κ-paired by an isometry from G, such that $m_i = S(G_{V_i})$.

Let $\not\uparrow\uparrow$ mean *not* $\uparrow\uparrow$, and $\not\uparrow\downarrow$ *not* $\uparrow\downarrow$. We define a *paired polygon* as a finite sequence

$$(n, \uparrow\uparrow, \uparrow\downarrow, \leftrightarrow, m), \quad n \in \boldsymbol{N}, \quad m = (m_0, m_1, \ldots, m_{n-1}) \in \boldsymbol{N}^n$$

where $\uparrow\uparrow$, $\uparrow\downarrow$ and $\leftrightarrow$ are binary relations defined on the set $\{0, 1, \ldots, n-1\}$ that

1. $\uparrow\uparrow$ and $\uparrow\downarrow$ are symmetric,
2. if $i \uparrow\uparrow j$ then $i \not\uparrow\downarrow k$, for every k, if $i \uparrow\downarrow j$ then $i \not\uparrow\uparrow k$, for every k,
3. if $i \uparrow\uparrow j$ and $i \uparrow\uparrow k$ then $j = k$, if $i \uparrow\downarrow j$ and $i \uparrow\downarrow k$ then $j = k$,
4. $\leftrightarrow$ is a minimal equivalence relation such that if $i \uparrow\uparrow j$ then $i \leftrightarrow j +_n 1$ and $i +_n 1 \leftrightarrow j$ (Fig. 2a), if $i \uparrow\downarrow j$ then $i \leftrightarrow j$ and $i +_n 1 \leftrightarrow j +_n 1$ (Fig. 2b, wiht $+_n$ as mod n operation),
5. if $i \leftrightarrow j$ then $m_i = m_j$,
6. if $m_i = 1^+$ then $|i/\!\leftrightarrow| > 2$, i.e. the equivalence class of V_i consists of more than 2 vertices.

We see that a paired polygon $(n, \uparrow\uparrow, \uparrow\downarrow, \leftrightarrow, m)$ serves as a *combinatorial description* of a given κ-paired polygon $F = V_0V_1 \ldots V_{n-1}$ of a plane discontinuous group G.

If $F = V_0V_1 \ldots V_{n-1}$ and $F' = V_0'V_1' \ldots V_{n-1}'$ are two κ-paired polygons and $(n, \uparrow\uparrow_1, \uparrow\downarrow_1, \leftrightarrow_1, m^1)$ and $(n, \uparrow\uparrow_2, \uparrow\downarrow_2, \leftrightarrow_2, m^2)$ their combinatorial descriptions, respectively, then F and F' are combinatorially equivalent if and only if there is a permutation $\delta \in D_n$ (D_n is the dihedral group defined on the set $\{0, 1, \ldots, n-1\}$), induced by

$$V_i \mapsto V'_{\delta(i)}, i = 0, 1, \ldots, n-1,$$

such that

$$i \uparrow\uparrow_1 j \Leftrightarrow \overline{\delta}(i) \uparrow\uparrow_2 \overline{\delta}(j), \qquad i \uparrow\downarrow_1 j \Leftrightarrow \overline{\delta}(i) \uparrow\downarrow_2 \overline{\delta}(j)$$

$$\text{and} \quad m_i^1 = m_{\delta(i)}^2$$

where $\overline{\delta}$ denotes the edge permutation, naturally derived from the vertex permutation δ.

3 Discrete Structures and the Algorithm

In this section, we will describe discrete structures and the algorithm implemented in the computer program COMCLASS (see Sect. 4).

3.1 Descriptor of Paired Polygon

Suppose that $\mathcal{A}$ is an order alphabet which consists of lower and upper case letters (say, of English alphabet), of ten digits, and of the hyphen (–). On the set of lower and upper case letters we define operation $^{-1}$ of *case changing*: If α is a letter, α^{-1} is the opposite case letter. By $\mathcal{A}^*$ we denote a set of sequences (words) of elements from $\mathcal{A}$.

To a paired polygon $p = (n, \uparrow\uparrow, \uparrow\downarrow, \leftrightarrow, m)$ we associate the word

$$v_0 e_0 v_1 e_1 \cdots v_{n-1} e_{n-1}$$

from $\mathcal{A}^*$ called a *polygon descriptor*, such that

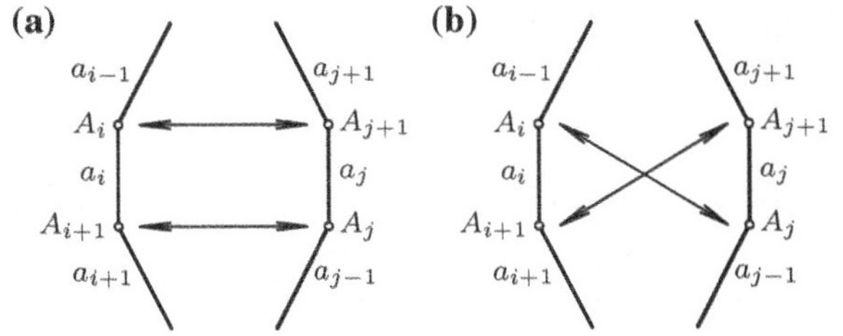

Fig. 2 For side pairing a polygon

$$v_i = \begin{cases} \varepsilon & \text{if } m_i = 1 \text{ or } m_i = 1^+, \\ \delta_h & \text{otherwise, where } m_i = h^+ \text{ or } m_i = h, \end{cases} \tag{1}$$

$$e_i = e_j^{-1} \text{ if } i \uparrow\uparrow j \tag{2}$$

$$e_i = e_j \text{ if } i \uparrow\downarrow j \text{ and } i \neq j \tag{3}$$

$$e_i \text{ is} - \text{(hyphen) if } i \uparrow\downarrow i \tag{4}$$

where ε is the empty word and δ_h is a sequence of digits which serves as a decimal notation of number $h \in \boldsymbol{N}$.

A paired polygon may have many different polygon descriptors. We introduce *a normal form* of a polygon descriptor by a rule of *"choosing next lower case letter"*:

$$\text{if } (i \uparrow\uparrow j \text{ or } i \uparrow\downarrow j) \text{ and } i < j \text{ then } e_i = \lambda_{\max\{k|\lambda_k \in \{e_0,\ldots,e_{i-1}\}\}+1}$$

where λ_k denotes k-th lower case letter of the alphabet. By the rule of "choosing next lower case letter", to every polygon descriptor we correspond its normal form. This procedure is called *normalization*. We introduce two more operations on the set of polygon descriptors:

$$\text{rot}(p) = v_{n-1}e_{n-1}v_0e_0v_1e_1 \cdots v_{n-2}e_{n-2} \tag{5}$$

$$\text{inv}(p) = v_0e_{n-1}v_{n-1}e_{n-2}v_{n-2} \cdots e_1v_1e_0 \tag{6}$$

where $p = v_0e_0v_1e_1 \cdots v_{n-1}e_{n-1}$.

Two normalized polygon descriptors are said to be *equivalent* if one can be produced from the other by applying rot, inv and normalization. Two polygon descriptors are said to be *equivalent* if their normalized polygon descriptors are equivalent. Clearly, all descriptors of a given κ-paired polygon are equivalent, and two combinatorially equivalent κ-paired polygons have the same set of descriptors.

Let us introduce an order in our alphabet: digits are lesser then letters, hyphen is lesser then any lower case letter, a lower case letter is lesser then the corresponding capital letter. This order implies lexicographic order of words. In the class of equivalent descriptors there is finitely many normalized descriptors (at most $2n$). Among normalized descriptors there is the minimal one, which is said to be the *canonical form*. We obtain canonical form of a given normalized descriptor by the following **canonization procedure** (see also [20]):

1. Starting with a given descriptor and applying composition of rot and normalization, and composition of inv and normalization as many times as we are getting new values, we obtain a set of all normalized descriptors in an equivalence class.
2. The minimal descriptor in this set is the canonical form.

Hence, canonical form of polygon descriptor represents a class of equivalent descriptors as well as a class of equivalent κ-paired polygons.

If $v_0 e_0 v_1 e_1 \cdots v_{n-1} e_{n-1}$ is a polygon descriptor then

$$v_i e_i v_{i+_n 1} e_{i+_n 1} \cdots v_{i+_n (k-1)} e_{i+_n (k-1)} v_{i+_n k} \text{ where } i, k \in \{0 \ldots n-1\}$$

is said to be a *descriptor segment*.

3.2 Starting Descriptor

Suppose that the Macbeath signature of our group G is of the form

$$(g, \pm; [\ ]; \{\ \}).$$

By step (a) of the procedure described in Theorem 2 we will get a fundamental polygon with minimal number ($2\alpha g$) of edges. If

$$e_0 e_1 \cdots e_{2\alpha g-1}$$

is a descriptor of that polygon ($v_i = \varepsilon$ because its vertices have trivial stabilizers) we consider it as a *starting descriptor* in determining a descriptor for a fundamental polygon of G with the Macbeath signature

$$(g, \pm; [h_1, \ldots, h_l]; \{\ \}).$$

In order to decide if $e_0 e_1 \cdots e_n$ serves as a starting descriptor, we use the following **procedure**:

1. Check if $n = 2\alpha g$.
2. Check that $\uparrow\downarrow$ is empty if and only if orientability is positive.
3. Check if $\uparrow\uparrow$ and $\uparrow\downarrow$ fulfils the rules in the definition of a paired polygon.
4. Check if $\leftrightarrow$ is defined according to the rules in the definition of a paired polygon.
5. Check if $\leftrightarrow$ is a full relation (this means that all vertices of the polygon are identified).

3.3 Tree Decomposition

Suppose that G is of Macbeath signature

$$(g, \pm; [h_1, \ldots, h_l]; \{\ \}),$$

and T is a tree in the disk D, described by step (c) of Theorem 2. Let

$$n = 2\alpha g.$$

We denote by $L_0, L_1, \dots L_{n-1}$ ordered vertices on the boundary of D (see Fig. 3a, notice that L_i's are not vertices of the fundamental polygon), and associate letters to edges of T where we avoid first αg letters from the alphabet. We denote by $a_0, a_1, \dots, a_n$ the edges that contain $L_0, L_1, \dots L_{n-1}$, respectively, and those edges need not to have associated letters.We chose one of two possible directions of every edge, to be positive.

Let

$$l_i = L_i a_i V_{i,0} t_{i,0} \cdots V_{i,k_i-2} t_{i,k_i-2} V_{i,k_i-1} a_{i+_n 1} L_{i+_n 1}$$

be a directed path on T connecting L_i to $L_{i+_n 1}$, which divides disk D into two parts such that the whole T belongs to the closure of one of these two parts.Suppose that

$$v_{i,j} = \begin{cases} \varepsilon & \text{if } S(V_{i,j}) = 1^+, \\ \delta_h & \text{otherwise, where } S(V_{i,j}) = h^+. \end{cases} \tag{7}$$

(δ_h is a sequence of digits which serves as a decimal notation of number $h \in \boldsymbol{N}$) and let $e_{i,j}$ be a letter λ associated to $t_{i,j}$, if the oriented path passes the edge $t_{i,j}$ in the positive direction, or λ^{-1} otherwise. If

$$s_i := v_{i,0} e_{i,0} \dots v_{i,k_i-2} e_{i,k_i-2} v_{i,k_i-1},$$

(the descriptor segment determined by the oriented path l_i), then the sequence

$$(s_0, s_1, \dots s_{n-1})$$

is said to be a *decomposition* of tree T. We illustrate this procedure in Fig. 3a where the decomposition of the given tree is

$$(s_0, s_1, s_2, s_3, s_4, s_5) = (\texttt{x4}, \texttt{4y2}, \texttt{2}, \texttt{2}, \texttt{2Y4X}, \varepsilon).$$

So, by the process of discretization of the algorithm given in Theorem 2, graphs from the step (a) are represented by starting descriptors and trees from (c) are represented by tree decompositions.

We are now ready to give **rules** for calculation of the polygon descriptor in a form

$$\overline{s}_0 e_0 \overline{s}_1 e_1 \cdots \overline{s}_{2\alpha g-1} e_{2\alpha g-1} \tag{8}$$

where $e_0 e_1 \cdots e_{2\alpha g-1}$ is starting descriptor and $\overline{s}_0, \dots, \overline{s}_{2\alpha g-1}$ are descriptor segments.

If $(s_0, s_1, \dots s_{n-1})$ is a tree decomposition then $\overline{s}_i = s_{q_i}^{z_i}$, i.e. the polygon descriptor is

$$s_{q_0}^{z_0} e_0 s_{q_1}^{z_1} e_1 \cdots s_{q_{n-1}}^{z_{n-1}} e_{q_{n-1}},$$

where $z_i \in \{-1, 1\}$ and

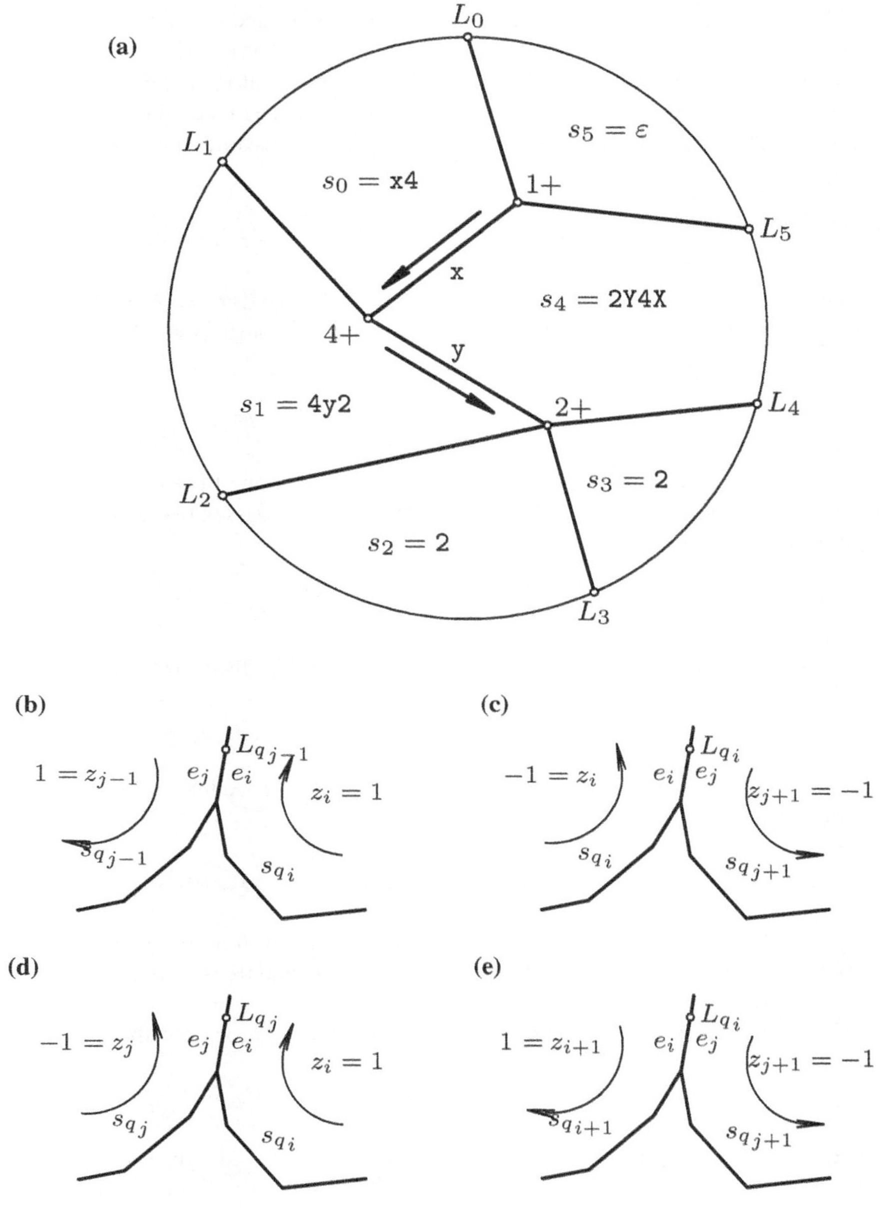

Fig. 3 For tree decomposition, blank derivation and qualifier

1. if $e_i = e_j^{-1}$ $(i \uparrow\uparrow j)$ then $q_{j+1} = q_i + z_i$ and $z_{j+1} = z_i$ (Fig. 3b, c),

1'. if $e_i = e_j^{-1}$ $(i \uparrow\uparrow j)$ then $q_j = q_{i+1} - z_{i+1}$ and $z_j = z_{i+1}$ (derived from (2) in definition of the polygon descriptor),

2. if $e_i = e_j$ $(i \uparrow\downarrow j)$ then $q_j = q_i + z_i$ and $z_j = -z_i$ (Fig. 3d, d with swaped i and j),

2'. if $e_i = e_j$ $(i \uparrow\downarrow j)$ then $q_{j+1} = q_{i+1} - z_i$ and $z_j = -z_i$ (Fig. 3e, e with swaped i and j).

Additions and subtractions in rules are modulo n (see also Fig. 3b–e). In order to determine unique polygon descriptor by our rules, it is necessary to determine q_0 and z_0. If we compare our rules with the property 2 in Theorem 1, we see that the set of all k which define q_k and z_k (if q_0 and z_0 are known) is $0/_{\leftrightarrow}$ defined by the class of $2\alpha g$ vertices of the starting polygon descriptor. Procedure in Sect. 2 gives us that $\leftrightarrow$ has to be full relation. We choose q_0 and z_0 arbitrary since we always obtain the same polygon.

If we choose starting descriptor to be abcBac and $q_0 = 0$ and $z_0 = 1$, our rules applied to the illustration given in Fig. 3a determine $q_1 = 2, z_1 = -1, q_2 = 4, z_2 = -1, q_3 = 5, z_3 = -1, q_4 = 1, z_4 = -1, q_5 = 3, z_5 = 1$. Hence, the polygon descriptor is

$$s_0\mathtt{a}s_2^{-1}\mathtt{b}s_4^{-1}\mathtt{c}s_5^{-1}\mathtt{B}s_1^{-1}\mathtt{a}s_3\mathtt{c} = \mathtt{x4a2bx4y2cB2Y4a2c}.$$

3.4 Blank Derivation and Qualifier

On the tree given in Fig. 3a we denote by A a vertex with trivial stabilizer, by B a rotation center of order 4, by C a rotation center of order 2. Let us consider the tree as a rooted tree where

1. Root of the tree is: L_0,
2. L_0 has 1 child: A,
3. A has 2 children: B and L_5,
4. B has 2 children: L_1 and C,
5. C has 3 children: L_2, L_3 and L_4,
6. Each L_1, L_2, L_3, L_4, L_5 has no children and it is on the border of the disk D.

We used the following listing rule for simple description of a tree: among the vertices with new defined children we always chose first the left one. If we neglect the type of stabilizer, the tree rooted at L_0 in Fig. 3a we describe by a finite sequence $(1, 2, 2, 0', 0', 3, 0', 0', 0')$, where we express only the number of children in a row. Notice that there can exist a vertex with no children, not on a border of the disk D (in that case necessarily with nontrivial stabilizer), so we use denotation $0'$ to make difference with 0. An n-tuple above is called the *blank derivation* of a tree.

If $(b_0, b_1, \ldots, b_{k-1})$ is blank derivation then each element of a k-tuple is corresponded to a vertex. Suppose that $V_0, V_1, \ldots, V_{k-1}$ are corresponding vertices. In our example we have that $V_0 = L_0$, $V_1 = A$, etc. Notice that $b_0 = 1$, and for

$0 < i < k$ valency of V_i is $b_i + 1$. Types of stabilizers we describe' by a k-tuple $(S(V_0), S(V_1), \ldots, S(V_{k-1}))$ which we call a *qualifier*. Hence

$$b_i = 0' \Rightarrow S(V_i) = 1^+ \tag{9}$$

$$b_i \neq 0' \wedge b_i < 2 \Rightarrow S(V_i) > 1^+ \tag{10}$$

Blank derivation describes a tree with possible vertices not on the border of D, such that their valencies are less then 3, i.e. $b_i \neq 0'$ and $b_i < 2$. The number of such vertices is said to be the *defect* of a blank derivation.

Blank derivation together with its qualifier has all relevant information about our tree.

3.5 Algorithm

Suppose that G is our above group with a given (good) Macbeath signature:

$$(g, \pm; [h_1, \ldots, h_l]; \{\}), g > 0.$$

The set of all canonized descriptors of fundamental polygons for G are obtained by the following **algorithm**:

1. Using the procedure given in Sect. 3.2, determine all possible starting descriptors of genus g and orientability $\pm$.
2. Determine all blank derivations with $2\alpha g - 1$ 0''s in the descriptor and of a defect less or equal l.
3. For each blank derivation determine a set of all qualifiers which are permutations of
$$(h_1^+, \ldots, h_l^+, \underbrace{1^+, \ldots, 1^+}_{k-l}) \tag{11}$$
(where k is the size of a blank derivation) and fulfils (9) and (10).
4. For every blank derivation and qualifier of this descriptor, determine a tree decomposition. This procedure is in a domain of routine computer science problems.
5. For each tree decomposition and each starting descriptor, using the rules given in Sect. 3.3, with $q_0 = 0$ and $z_0 = 1$, determine a polygon descriptor. It is enough to use only one pair of values for q_0 and z_0 (not necessarily 0 and 1).
6. We canonize each determined polygon descriptor in the previous step, by the given canonization procedure.

3.6 Boundaries and Genus 0

If our group G admits reflections, i.e. if its Macbeath signature is of the form

$$(g, \pm; [h_1, \ldots, h_l]; \{(h_{1,1}, \ldots, h_{1,l_1}), \ldots, (h_{q,1} \ldots, h_{q,l_q})\}),$$

the procedure given in Sect. 3.5 is essentially the same but more bulky.

At the beginning we suppose that $g > 0$.

Contracted boundary components $\overline{Q}_1, \ldots, \overline{Q}_q$ in a graph C (Theorem 2) gives us new vertices, hence in a tree qualifier there are new elements which we denote by $1^*, \ldots, q^*$, and because of that instead of (11) we have

$$(h_1^+, \ldots, h_l^+, 1^*, \ldots, q^*, \underbrace{1^+, \ldots, 1^+}_{k-l-q})$$

((9) and (10) are fulfilled).

We extend our alphabet with asterisk and in (7) we add a possibility for a vertex to be a contracted component of a boundary component, hence

$$v_{i,j} = \delta_x * \delta_y \quad \text{if } V_{i,j} = Q_x, x = 1, 2, \ldots q, \text{ and } v_{i,j} \text{ represents } y\text{-th occurrence of boundary component } Q_x$$

where δ_x and δ_y are decimal notations of x and y, respectively. Step 5 of the procedure given in Sect. 3.5 is essentially the same. Obtained words, which are not necessarily polygon descriptors, may have subwords from the previous formula.

Between steps 5 and 6 we add step 5': word obtained in step 5 is changed by adding q boundary components together with the valencies of vertices obtained by their contractions, since we have to replace $\delta_x * \delta_y$ by y-th segment of x-th *boundary decomposition* analogous to a tree decomposition.

If $g = 0$ we use simplified procedure to obtain all possible trees on a sphere with the given properties, and to cut a sphere along each of these trees. For this procedure we established all necessary techniques: cutting a sphere along a tree is a decomposition in one segment, and the tree generation is as usual.

4 Program COMCLASS

Based on the given algorithm computer program COMCLASS, written in the programming language C, for a given group signature determine all canonized polygon descriptors. Closely related to COMCLASS is computer program FUNDAMENTAL developed by Daniel Huson [12].

COMCLASS is available at http://comclass.math.rs in source code and for online execution.

At the end we give some group examples.

Discrete groups of Euclidean plane		
*p*1	(1, +, [ ]; { })	abAB abcABC
*p*2	(0, +, [2, 2, 2, 2]; { })	2a2b2B2c2C2A 2a2b2c2C2B2A 2ab2Bc2Cd2DA 2a2bc2Cd2DB2A 2abc2Cd2DBe2EA
pm	(0, +, [ ]; {( ), ( )})	a-A-
pg	(2, −, [ ]; { })	abaB abba abcbaC abccBa
cm	(1, −, [ ]; {( )})	a-a a-a- ab-Ba
pmm	(0, +, [ ]; {(2, 2, 2, 2)})	2-2-2-2-
pmg	(0, +, [2, 2]; {( )})	2ab2B-A 2a-b2B-A 2a2b-B2A 2ab-Bc2CA
pgg	(1, −, [2, 2]; { })	2a2b2B2a 2a2b2a2b ab2Bac2C ab2Bc2Ca 2ab2Bc2ac ab2c2C2Ba ab2Bcd2Dac abc2Cd2DBa
cmm	(0, +, [2]; {(2, 2)})	2a-2-2-A 2a2-2-2A
*p*4	(0, +, [2, 4, 4]; { })	4a2b4B2A 4a4b2B4A 4ab4Bc2CA
*p*4*m*	(0, +, [ ]; {(2, 4, 4)})	4-4-2-
*p*4*g*	(0, +, [4]; {(2)})	4a-2-A 4a2-2A

$p3$	$(0, +, [3, 3, 3]; \{\})$	3a3b3B3A 3ab3Bc3CA
$p3m1$	$(0, +, [\ ]; \{(3, 3, 3)\})$	3-3-3-
$p31m$	$(0, +, [3]; \{(3)\})$	3a-3-A 3a3-3A
$p6$	$(0, +, [2, 3, 6]; \{\})$	3a6b2B6A 6a2b3B2A 6a3b2B3A 6ab3Bc2CA
$p6m$	$(0, +, [\ ]; \{(2, 3, 6)\})$	6-3-2-

(see also [7, 12, pp. 518–519]).

If

$$(g, \pm, [\ ]; \{\})$$

is the Macbeath signature of our group then the result is purely topological. As an example, for the Macbeath signature

$$(2, -, [\ ]; \{\})$$

(of our pg), there are four topological gluings of a disk which all give Klein bottle:

abaB, abba, abcbaC, abccBa

(see e.g. [1, p. 310]).

Combinatorially different polygons of group with the Macbeath signature

$$(2, +, [\ ]; \{\})$$

are:

1 abcdABCD	2 abcdBCAD	3 abcdBDAC
4 abcdCDAB	5 abcdBeCEAD	6 abcdBeDAEC
7 abcdBeDEAC	8 abcdCeBDAE	9 abcdCeBEAD
10 abcdCeDEAB	11 abcdeABCDE	12 abcdeBCADE
13 abcdeBCDAE	14 abcdeBCEAD	15 abcdeBDACE
16 abcdeBDAEC	17 abcdeBDEAC	18 abcdeCDABE
19 abcdeCDEAB	20 abcdeDBCAE	21 abcdeDBEAC
22 abcdeDECAB	23 abcBdeDfCFAE	24 abcBdeDfEFAC
25 abcdBCDefEAF	26 abcdBeDfEAFC	27 abcdBefCDFAE
28 abcdBefCEADF	29 abcdBefDEACF	30 abcdCefBEFAD
31 abcdCefEBDAF	32 abcdCefEBFAD	33 abcdeBDfEACF
34 abcdeBDfEAFC	35 abcdeBfCDFAE	36 abcdeBfCEADF
37 abcdeBfCEAFD	38 abcdeBfCEFAD	39 abcdeBfDEFAC
40 abcdeBfECFAD	41 abcdeCfBDEAF	42 abcdeCfDEFAB

43 abcdeDfBCFAE	44 abcdeDfEFCAB	45 abcdefBCDAEF
46 abcdefBCFADE	47 abcdefBDAECF	48 abcdefBDEACF
49 abcdefBDEAFC	50 abcdefBECFAD	51 abcdefCDFABE
52 abcdefDBCAEF	53 abcdefDBEFAC	54 abcdefDEFABC
55 abcdefDEFCAB	56 abcdefEBCDAF	57 abcdBeCDEfgFAG
58 abcdBefEgCDGAF	59 abcdCeDEBfgFAG	60 abcdCefEBgDGAF
61 abcdCefEBgFGAD	62 abcdCefgEBFGAD	63 abcdeBfDgEFACG
64 abcdeBfDgEFAGC	65 abcdeBfEgCFAGD	66 abcdeBfgCEFADG
67 abcdeBfgCEFAGD	68 abcdeCDEBfgFAG	69 abcdeCfBgDEGAF
70 abcdeCfgBDEAFG	71 abcdeCfgDEGABF	72 abcdeCfgFBDEAG
73 abcdeDfBgFCGAE	74 abcdeDfgBFCGAE	75 abcdefBDgEACFG
76 abcdefBEgCFADG	77 abcdefBEgCFAGD	78 abcdefBgCDGAEF
79 abcdefBgCFADGE	80 abcdefBgCFDGAE	81 abcdefBgDEGACF
82 abcdefBgECFGAD	83 abcdefBgEFGACD	84 abcdefCgDFABEG
85 abcdefDBgEFGAC	86 abcdefDgBCGAEF	87 abcdefDgBEFACG
88 abcdefDgBEFAGC	89 abcdefDgEFGABC	90 abcdefDgEFGCAB
91 abcdefgBDEAFCG	92 abcdefgEBCDAFG	93 abcdefgEBFGACD
94 abcdefgEFGDABC	95 abcdBefgEhCDHAFG	96 abcdBefgEhFGHACD
97 abcdCefgEBhFGHAD	98 abcdeCfDEFBghGAH	99 abcdeCfgFBhDEHAG
100 abcdeCfghFBDEAGH	101 abcdeCfghFBGHADE	102 abcdeDfgBhFCGHAE
103 abcdefBgChDGAEHF	104 abcdefBgDhEGACFH	105 abcdefBgDhEGAHCF
106 abcdefBgEhCFGADH	107 abcdefBgEhCFGAHD	108 abcdefBgEhFGACHD
109 abcdefDgBhEFHAGC	110 abcdefDghBGCHAEF	111 abcdefEgBhCGDHAF
112 abcdefgBhCGDHAEF	113 abcdefgBhFCGHADE	114 abcdefgEhFGHDABC
115 abcdeCfDEFBghiGAHI	116 abcdeCfghFBiDEIAGH	117 abcdeCfghFBiGHIADE
118 abcdefBgDhiEGAHCFI	119 abcdefDghBiGCHIAEF	120 abcdefgBhCGiDHAEIF
121 abcdefgBhFiCGHADIE	122 abcdefgEhBiCHDIAFG	

We found 65 combinatorially different polygons for group with Macbeath signature

$$(3, -, [\ ]; \{\ \}),$$

see also [24].

There are 82 polygons with minimal number (12) of edges for the group

$$(3, +, [\ ]; \{\ \}).$$

Program has an option to limit the number of edges for fundamental polygons. It is also possible to obtain only the number of fundamental polygons (see also [24, 30]). Thus for group with the Macbeath signature $(4, -, [\ ]; \{\ \})$ there are 2.498 combinatorially different fundamental polygons, and for group with the Macbeath signature $(4, +, [\ ]; \{\ \})$ there are 7.258 fundamental polygons with a minimal number (16) of edges. For group $(5, -, [\ ]; \{\ \})$ there are 473, and for group $(6, -, [\ ]; \{\ \})$ there are 7.190 fundamental polygons with a minimal number (10 for first one and 12 for the last one) of edges.

5 Closing Remarks

We hope that this conference paper will conclude our research which looks back to a long history. The GEOSYM 2015 Conference organized in Veszprém, Hungary (http://geosym.mik.uni-pannon.hu/) gave us further insights into the development of the subject of this meeting. The second named author presented the results and its relevance became immediately apparent during numerous other talks. Thus, we extend our references to include additional publications (with no pretending of completeness): [2, 4, 6, 9, 10, 13, 16, 21–23, 25, 27, 29, 31].

An Oberwolfach seminar "Kombinatorische Geometrie" organized by Andreas W. M. Dress and Jörg M. Wills in 1984, 23–29. September, brought to the attention of the participants, which had opened new directions which had been settled during this conference.

We would like to remember our charismatic teachers, recently passed away, among them Professor Stanko Bilinski, the "Father" of Yugoslavian and Hungarian geometers in addition to the authors; Professor Ludwig Danzer, who was the supervisor of Egon Schulte, and an enthusiastic master photographer in his free time.

We would like to thank the excellent initiators and our colleagues at the University of Pannonia for organizing this wonderful Geometry Festival. Together with all participants we offer our best wishes to our celebrated colleagues and friends, Károly Bezdek and Egon Schulte.

References

1. L. Balke, D. Huson, Two-dimensional groups, orbifolds and tilings. Geom. Dedicata **60**, 89–106 (1996)
2. S. Bilinski, Die quasiregulären Polyeder vom Geschlecht 2, Österreich. Akad. Wiss. Math. Natur. Kl. Sitzungsber **II**(194), 63–78 (1985)
3. A. Cayley, A theorem on trees. Quart. J. Math **23**, 376–378 (1889)
4. M. Conder, P. Dobcsányi, Determination of all regular maps of small genus. J. Combin. Theory Ser. B **81**, 224–242 (2001)
5. J.H. Conway, The orbifold notation for surface groups (In Groups, combinatorics, and geometry). LMS Lect. Notes **165**, 438–447 (1992). Cambridge
6. L. Danzer, E. Schulte, Reguläre Inzidenzkomplexe I. Geometriae Dedicata **13**, 295–308 (1982)
7. B.N. Delone, Theory of planigons (in Russian). Izv. Akad. Nauk SSSR Ser. Mat. **23**(3), 365–386 (1959)
8. B.N. Delone, N.P. Dolbilin, M.I. Shtogrin, Combinatorial and metrical theory of planigons, Tr. Mat. Inst. Steklov Akad. Nauk SSSR, **148**, 109–140 (in Russian); English trans: Proceedings of the Steklov Institute of Mathematics **4**(1980), 111–141 (1978)
9. A.W.M. Dress, Presentation of discrete groups, acting on simply connected manifolds in terms of parametrized systems of Coxeter matrices. Adv. Mathe. **63**, 198–212 (1987)
10. G. Gévay, E. Schulte, J.M. Wills, The regular Grünbaum polyhedron of genus 5. Adv. Geomet. **14**, 465–482 (2014)
11. D. Hilbert, S. Cohn-Vossen, *Geomet. Imagination* (Chelsea, New York, 1952)
12. D. Huson, The generation and classification of tile-k-transitive tilings of the euclidean plane, the sphere and the hyperbolic plane. Geom. Dedicata **47**, 269–296 (1993)

13. U. Leopold, *Vertex-Transitive Polyhedra in Three-Space*, Ph.D. Dissertation to The Faculty of the College of Science of Northeastern University, Boston 2014
14. Z. Lučić, E. Molnár, Combinatorial classification of fundamental domains of finite area for planar discontinuous isometry groups. Arch. Math **54**, 511–520 (1990)
15. Z. Lučić, E. Molnár, Fundamental domains for planar discontinuous groups and uniform tilings. Geom. Dedicata **40**, 125–143 (1991)
16. Z. Lučić, E. Molnár, M. Stojanović, The 14 infinite families of isotoxal tilings in the planes of constant curvature. Periodica Math. Hungar. **29**(2), 177–193 (1994)
17. A.M. Macbeath, The classification of non-euclidean plane crystallographic groups. Can. J. Math. **19**, 1192–1205 (1967)
18. B. Maskit, On Poincaré's theorem for fundamental polygons. Adv. In Math. **7**, 219–230 (1971)
19. E. Molnár, Polyhedron complexes with simply transitive group actions and their realizations. Acta Math. Hung. **59**(1–2), 175–216 (1992)
20. E. Molnár, Discontinuous groups in homogeneous Riemannian space by classification of D-symbols. Publ. Math. Debrecen **49**(3-4), 265–294 (1996)
21. E. Molnár, On projective models of Thurston geometries, some relevant notes on Nil orbifolds and manifolds, Siberian Electron. Mathe. Rep. **7**, 491–498 (2010). http://semr.math.nsc.ru
22. E. Molnár, On D-Symbols and Orbifolds in an Algorithmic Way. Atti Semin. Mat. Fis. Univ. Modena Reggio Emilia, **58**, 263–276 (2011)
23. E. Molnár, I. Prok, J. Szirmai, Classification of tile-transitive simplex tilings and their realizations in homogeneous spaces, *Non-Euclidean Geometries, János Bolyai Memorial Volume*, Ed. András Prékopa and Emil Molnár, Mathematics and Its Applications, vol. 581 (Springer 2006), pp. 321–363
24. E. Molnár, E. Stettner, Symmetry groups and fundamental tilings for the compact surface of genus 3− (2. The normalizer diagram with classification). Beiträge zur Algebra und Geometrie (Contr. Alg. Geom.), **46/1**, 19–42 (2005)
25. B. Monson, T. Pisanski, E. Schulte, A.I. Weiss, Semisymmetric Graphs from Polytopes, Manuscript arXiv 2006
26. J.M. Montesinos, *Classical Tessellations and Three-Manifolds* (Springer, Berlin, 1987)
27. A. Orbanić, D. Pellicer, T. Pisanski, T.W. Tucker, Edge-transitive maps of low genus. Ars Math. Contemp. **4**, 385–402 (2011)
28. H. Poincaré, Théorie des groupes fuchsiens. Acta Mathematica **1**(1), 1–62 (December 1882)
29. J. Širáň, T.W. Tucker, M. Watkins, Realizing finite edge-transitive orientable maps. J. Graph. Theory. **37**, 1–34 (2001)
30. E. Stettner, Die computergestützte Klassifizierung der Flächeneinwickelungen in einem Vieleck vorgegebener Seitenanzahl. Annales Univ. Sci. Budapest. Sect. Math. **41** (1998), 103–115, Manuskript (1989)
31. O. Šuch, Vertex transitive maps on the Klein bottle. Ars Math. Contemp. **4**, 363–374 (2011)
32. N. Vasiljević, *Combinatorial Structure of Fundamental Polygons* (in Serbo-Croat), B.Sc. thesis, Faculty of Mathematics, University of Belgrade, 1993
33. H.C. Wilkie, On non-Euclidean Crystallographic Groups. Math. Z. **91**, 87–102 (1966)
34. H. Zieschang, E. Vogt, H. Coldewey, *Surfaces and Planar Discontinuous Groups* (Springer, Berlin, 1980)

Self-inscribed Regular Hyperbolic Honeycombs

Peter McMullen

Abstract This paper describes ways that certain regular honeycombs of non-finite type in d-dimensional hyperbolic space $\mathbb{H}^d$ for $d = 2, 3$ and 5 can be inscribed in others, in particular showing that some can be inscribed properly in copies of themselves.

Keywords Coxeter group · Simplex dissection · Hyperbolic space · Regular honeycomb · Self-inscribed · Compound

MSC (2010): Primary 51M20 · Secondary 51M10

1 Introduction

The purpose of this paper is to draw attention to some curious properties of certain families of regular hyperbolic honeycombs with ideal vertices. The existence of similarities in euclidean spaces enables some regular honeycombs to be inscribed in smaller copies of themselves, by which we mean that the vertices of one form a subset of the vertices of the other. The d-dimensional cubic tilings exemplify this property, in infinitely many different ways. We shall see here that the same behaviour is exhibited in four families of regular hyperbolic honeycombs, one in $\mathbb{H}^2$, two in $\mathbb{H}^3$ and one in $\mathbb{H}^5$.

At the instigation of one of the referees of an earlier version of the paper, we have shown that certain subgroups of Coxeter groups that we employ are themselves Coxeter groups; these connexions are closely related to simplex dissections of Debrunner [1] which formalized two folkloristic results. At the suggestion of the other, we have added two more families of honeycombs. As a consequence, the paper has been substantially rewritten.

P. McMullen (✉)
University College London, Gower Street, London WC1E 6BT, UK
e-mail: p.mcmullen@ucl.ac.uk

M. D. E. Conder et al. (eds.), *Discrete Geometry and Symmetry*,
Springer Proceedings in Mathematics & Statistics 234,
https://doi.org/10.1007/978-3-319-78434-2_15

We have been told by Asia Weiss that Donald Coxeter had come up with similar ideas to those in this note, although nothing was ever published.

2 Regular Polytopes and Automorphism Groups

In this section, we briefly set the scene. As shown in, for example, [3, Chap. 2], a regular n-polytope $\mathcal{P}$ and its automorphism group $\boldsymbol{G}$ can be identified in a natural way. The group $\boldsymbol{G}$ of $\mathcal{P}$ has distinguished generators $\boldsymbol{r}_0, \ldots, \boldsymbol{r}_{n-1}$ satisfying – possibly among others – the relations $(\boldsymbol{r}_j\boldsymbol{r}_k)^{p_{jk}} = \boldsymbol{e}$ for $0 \leqslant j \leqslant k \leqslant n-1$, where

$$p_{jk} = \begin{cases} 1, & \text{if } j = k, \\ p_k & \text{if } j = k-1, \\ 2, & \text{if } j \leqslant k-2. \end{cases} \tag{2.1}$$

We always assume here that $p_k \geqslant 3$ for each k, and that $p_k = \infty$ is allowed. In addition, the $\boldsymbol{r}_i$ also have the intersection property:

$$\langle \boldsymbol{r}_i \mid i \in \mathsf{J}\rangle \cap \langle \boldsymbol{r}_i \mid i \in \mathsf{K}\rangle = \langle \boldsymbol{r}_i \mid i \in \mathsf{J} \cap \mathsf{K}\rangle \tag{2.2}$$

for $\mathsf{J}, \mathsf{K} \subseteq \{0, \ldots, n-1\}$. Conversely, such a group $\boldsymbol{G}$ is the automorphism group of a regular polytope $\mathcal{P}$, in which case $\{p_1, \ldots, p_{n-1}\}$ is called the Schläfli type of $\mathcal{P}$. Combinatorially, for $0 \leqslant k \leqslant n-1$ a k-face of $\mathcal{P}$ is identified with a right coset of the distinguished subgroup $\boldsymbol{G}_k := \langle \boldsymbol{r}_i \mid i \neq k\rangle$, with the incidence relation given by

$$\boldsymbol{G}_j\boldsymbol{a} \leqslant \boldsymbol{G}_k\boldsymbol{b} \iff j \leqslant k \text{ and } \boldsymbol{G}_j\boldsymbol{a} \cap \boldsymbol{G}_k\boldsymbol{b} \neq \emptyset$$

for $\boldsymbol{a}, \boldsymbol{b} \in \boldsymbol{G}$ turning $\mathcal{P}$ into a poset.

If $\boldsymbol{G}$ is specified solely by the relations implied by (2.1), then it is a (string) Coxeter group, and is denoted $[p_1, \ldots, p_{n-1}]$. The corresponding polytope $\mathcal{P}$ is universal, by which we mean that any regular polytope of Schläfli type $\{p_1, \ldots, p_{n-1}\}$ is a quotient of $\mathcal{P}$. It is this situation that prevails throughout the paper; $\{p_1, \ldots, p_{n-1}\}$ will henceforth mean the universal polytope.

Associated with the Coxeter group $\boldsymbol{G}$ is its contragredient representation G, say. We need little from this, except to know that G acts faithfully on a certain convex cone – the Tits cone – in $\mathbb{R}^n$. Its generators R_k corresponding to the involutions $\boldsymbol{r}_k$ are linear reflexions in hyperplane mirrors, which bound a fundamental chamber C; copies of C under $\boldsymbol{G}$ fit together face-to-face to form the chamber complex, and their union is the Tits cone. The fact that $\boldsymbol{G}$ is a Coxeter group means that the local relations – how chambers fit together around their $(n-2)$-faces – determine the whole structure of the chamber complex. See [3, Sect. 3A] for a brief exposition, as well as further references.

Remark 2.3 We adopt the convention that heavy braces denote an abstract regular polytope, as in the Schläfli type $\{p_1, \ldots, p_{n-1}\}$ of $\mathcal{P}$. Light braces indicate a geometric regular polytope in euclidean or hyperbolic space. Thus **{4, 3}** is an abstract 3-cube; {4, 3} is the ordinary 3-cube in $\mathbb{E}^4$. We similarly use R_k to denote a geometric reflexion corresponding to the involutory automorphism $\boldsymbol{r}_k$.

3 The Coxeter Group $[3^{n-2}, 2r]$

This section treats the first of the subgroup relationships among Coxeter groups. We begin with something that should be obvious to which we shall appeal twice.

Lemma 3.1 *For $k = 0, \ldots, n-2$ and $r \geqslant 3$, the mapping $\boldsymbol{r}_{n-1} \mapsto \boldsymbol{e}$ and $\boldsymbol{r}_j \mapsto \boldsymbol{r}_j$ for $j = 0, \ldots, n-2$ induces a homomorphism on $[3^{n-2}, 2r]$ with quotient $[3^{n-2}] \cong S_n$, the symmetric group.*

The notation p^m stands for $p, \ldots, p$, with m occurrences of p.

We then have

Theorem 3.2 *For $k = 0, \ldots, n-2$ and $r \geqslant 3$, the Coxeter group $[3^{k-1}, 2r, r, 2r, 3^{n-k-3}]$ is a subgroup of $[3^{n-2}, 2r]$ of index $\binom{n}{k+1}$.*

Proof The conventions for extreme values of r should be obvious; just think of the block $2r, r, 2r$ as migrating through a sequence of 3s. The generators $\boldsymbol{s}_0, \ldots, \boldsymbol{s}_{n-1}$ of the subgroup $\boldsymbol{G}_k$ (say) are given by

$$\boldsymbol{s}_j := \begin{cases} \boldsymbol{r}_j, & \text{if } j = 0, \ldots, k-1, \\ \boldsymbol{r}_k \boldsymbol{r}_{k+1} \cdots \boldsymbol{r}_{n-2} \boldsymbol{r}_{n-1} \boldsymbol{r}_{n-2} \cdots \boldsymbol{r}_{k+1} \boldsymbol{r}_k, & \text{if } j = k, \\ \boldsymbol{r}_{n+k-j}, & \text{if } j = k+1, \ldots, n-1. \end{cases} \tag{3.3}$$

In the language of [3, Chap. 7], this defines a mixing operation $\boldsymbol{\nu}_k : (\boldsymbol{r}_0, \ldots, \boldsymbol{r}_{n-1}) \mapsto (\boldsymbol{s}_0, \ldots, \boldsymbol{s}_{n-1})$. The indexing of $\boldsymbol{\nu}_k$ is chosen to indicate that $\boldsymbol{r}_k$ is the only generator which changes, although the order of $\boldsymbol{r}_{k+1}, \ldots, \boldsymbol{r}_{n-1}$ is reversed. We can extend the range of k in a natural way by $\boldsymbol{\nu}_{n-1} = \boldsymbol{\iota}$ (the identity), and $\boldsymbol{\nu}_{-1} = \boldsymbol{\delta}$ (the duality operation – which reverses the order of all the $\boldsymbol{r}_j$).

It is a routine matter (which we leave to the reader) to verify that $\boldsymbol{s}_0, \ldots, \boldsymbol{s}_{n-1}$ do generate a group satisfying the relations of $[3^{k-1}, 2r, r, 2r, 3^{n-k-3}]$. Bear in mind that, if $\boldsymbol{a}^2 = \boldsymbol{b}^2 = (\boldsymbol{ab})^3 = \boldsymbol{e}$, then $\boldsymbol{aba} = \boldsymbol{bab}$; then appeal to conjugacy. For example,

$$\begin{aligned} \boldsymbol{s}_{k-1}\boldsymbol{s}_k &= \boldsymbol{r}_{k-1} \cdot \boldsymbol{r}_k \boldsymbol{r}_{k+1} \cdots \boldsymbol{r}_{n-2} \boldsymbol{r}_{n-1} \boldsymbol{r}_{n-2} \cdots \boldsymbol{r}_{k+1} \boldsymbol{r}_k \\ &\sim \boldsymbol{r}_k \boldsymbol{r}_{k-1} \boldsymbol{r}_k \cdot \boldsymbol{r}_{k+1} \cdots \boldsymbol{r}_{n-2} \boldsymbol{r}_{n-1} \boldsymbol{r}_{n-2} \cdots \boldsymbol{r}_{k+1} \\ &= \boldsymbol{r}_{k-1} \boldsymbol{r}_k \boldsymbol{r}_{k-1} \cdot \boldsymbol{r}_{k+1} \cdots \boldsymbol{r}_{n-2} \boldsymbol{r}_{n-1} \boldsymbol{r}_{n-2} \cdots \boldsymbol{r}_{k+1} \\ &\sim \boldsymbol{r}_k \cdot \boldsymbol{r}_{k+1} \cdots \boldsymbol{r}_{n-2} \boldsymbol{r}_{n-1} \boldsymbol{r}_{n-2} \cdots \boldsymbol{r}_{k+1} \\ &\cdots \\ &\sim \boldsymbol{r}_{n-2} \boldsymbol{r}_{n-1}, \end{aligned}$$

and so on. We must therefore check that no additional relations are acquired.

To see what $\boldsymbol{\nu}_k$ does geometrically, we look at the contragredient representation G of $\boldsymbol{G}$. The fundamental chamber C is a cone over a simplex. If, as in Sect. 2, we let R_k be the linear reflexion corresponding to $\boldsymbol{r}_k$, then a consequence of Lemma 3.1 is that the conjugates of R_{n-1} under $\langle R_0, \ldots, R_{n-2}\rangle$ generate a subgroup of G whose fundamental region is the cone F corresponding to the initial simplex facet of $\mathcal{P} := \{3^{n-2}, 2r\}$.

There are $n!$ copies of C in F, which are the images of C under the subgroup $\mathrm{H} := \langle R_0, \ldots, R_{n-2}\rangle$ (this is a symmetric group). Regarding R_j interchangeably as a linear reflexion and its mirror, the hyperplane R_j slices F into two halves, which the reflexion R_j swaps. The fundamental cone C_k of the representation G_k corresponding to $\boldsymbol{G}_k$ is similarly cut out of F by the hyperplanes S_j with $j \neq k, n-1$. The images of C_k in F are those under the subgroup $\langle S_0, \ldots, S_{k-1}\rangle\langle S_{k+2}, \ldots, S_{n-1}\rangle = \langle R_0, \ldots, R_{k-1}\rangle\langle R_{k+1}, \ldots, R_{n-2}\rangle$, of order $(k+1)!(n-k-1)!$.

It should now be clear that the local geometric structure around C_k is inherited from that around C, and thus that this suffices to determine G_k, and hence $\boldsymbol{G}_k$. In other words, the latter is also a Coxeter group. □

As we have just pointed out, Lemma 3.1 says that $\mathcal{P} = \{3^{n-2}, 2r\}$ collapses onto its initial facet, which is an $(n-1)$-simplex. Consequently, lifting this collapse back into $\mathcal{P}$ implies the first part of

Theorem 3.4 *The vertices of the universal regular polytope $\mathcal{P} = \{3^{n-2}, 2r\}$ can be n-coloured. Moreover, the polytope $\mathcal{P}_k := \{3^{k-1}, 2r, r, 2r, 3^{n-k-3}\}$ can be inscribed in $\mathcal{P}$, using (any) $k+1$ of the colour-classes of its vertices.*

Proof If $\mathcal{P}$, $\mathcal{Q}$ are regular polytopes, we write $\mathcal{Q} \prec \mathcal{P}$ to mean that $\mathcal{Q}$ is inscribed in $\mathcal{P}$; that is, vert $\mathcal{Q} \subset$ vert $\mathcal{P}$, with vert $\mathcal{P}$ the vertex-set of $\mathcal{P}$. The crucial fact is the subgroup relationship between the groups of the vertex-figures: in the previous notation, $\langle \boldsymbol{s}_1, \ldots, \boldsymbol{s}_{n-1}\rangle \leqslant \langle \boldsymbol{r}_1, \ldots, \boldsymbol{r}_{n-1}\rangle$. This means that $\mathcal{P}_k$ has the same initial vertex v (say) as $\mathcal{P}$; indeed, it has the same initial j-face for $j = 0, \ldots, k$.

We now appeal to induction on n. Replacing n by $n-1$ implies replacing k by $k-1$, which means that we first have to establish the case $k = 0$. In this case, $\boldsymbol{s}_0$ swaps the initial facet $\{3^{n-2}\}$ of $\mathcal{P}$ with the one that shares the ridge opposite v; then $v\boldsymbol{s}_0$ has the same colour 1 as v. Moreover, since $\langle \boldsymbol{s}_1, \ldots, \boldsymbol{s}_{n-1}\rangle = \langle \boldsymbol{r}_1, \ldots, \boldsymbol{r}_{n-1}\rangle$, all such antipodal vertices in facets through the initial vertex are vertices of $\mathcal{P}_0$, and this quickly leads to vert $\mathcal{P}_0$ consisting of the whole of colour-class 1 of $\mathcal{P}$.

If $\mathcal{Q}$ is a regular polytope (or honeycomb), then we denote by $\mathcal{Q}^{\boldsymbol{v}}$ its broad vertex-figure; that is, the vertices of $\mathcal{Q}^{\boldsymbol{v}}$ consist of those vertices of $\mathcal{Q}$ that are joined to its initial vertex by an edge. For $k > 0$, we may now assume that $\mathcal{P}_k^{\boldsymbol{v}}$ consists of all vertices of $\mathcal{P}^{\boldsymbol{v}}$ in colour-classes $2, \ldots, k+1$, that is, those adjacent to the initial vertex coloured 1. The claim of the theorem quickly follows from the action of $\boldsymbol{G}_k$ and the symmetry of the colour-classes. □

We next observe that the polytopes $\mathcal{P}_k$ occur in dual pairs. More specifically, since we can freely permute colour-classes, we deduce

Theorem 3.5 *For each $n \geqslant 4$, $r \geqslant 3$ and $k = 0, \ldots, n-2$, $\mathcal{P}_k$ and $\mathcal{P}_{n-k-2}$ are dual polytopes. Moreover, they can be inscribed in $\mathcal{P}$ so that their vertex-sets are complementary colour-classes.*

The following illustrates Theorems 3.2 and 3.4.

Example 3.6 The first theorem yields subgroups [6, 3] and [3, 6] of index 3 in the Coxeter group [3, 6]. By Theorem 3.4, the vertices of the planar tessellation {3, 6} of $\mathbb{E}^3$ by triangles can be 3-coloured; we do not need the general theory to see this. Two out of the three colour classes form the vertices of an inscribed copy of a tessellation {6, 3}, while the third then forms the vertex-set of the dual copy of {3, 6}, now scaled up from the original by $\sqrt{3}$, we have

$$\{3, 6\} \succ \{6, 3\} \succ \{3, 6\};$$

we can iterate the process and extend it to

$$\cdots \succ \{3, 6\} \succ \{3, 6\} \succ \{3, 6\} \succ \cdots$$

with each copy having a third of the vertices of the one before; there is a similar infinite sequence with {6, 3} replacing {3, 6}.

4 The Tessellation {3, ∞}

The vertices of the tessellation $\{3, \infty\}$ in the hyperbolic plane $\mathbb{H}^2$ can be 3-coloured (again, we do not really need the general discussion to see this). Either two out of three, or one out of three of the colour classes yields a tessellation $\{\infty, \infty\}$, and so we can strictly inscribe one copy in another using half the vertices. This process can be repeated to inscribe a copy using a quarter of the vertices, and then an eighth, so on. We can clearly treat this easy case by hand.

5 Honeycombs Inscribed in {3, 3, 6}

We now apply the resuts of Sect. 3 to $\{3^{n-2}, 6\}$. We looked at the first case {3, 6} in Example 3.6, and so here we concentrate on the honeycomb {3, 3, 6} with ideal vertices in $\mathbb{H}^3$. We can take its symmetry group to have generators R_j for $j = 0, \ldots, 3$ given by

$$xR_j := \begin{cases} (\eta, \zeta_2, \zeta_1, \zeta_3), & \text{if } j = 0, \\ (\eta, \zeta_1, \zeta_3, \zeta_2), & \text{if } j = 1, \\ (\eta, -\zeta_2, -\zeta_1, \zeta_3), & \text{if } j = 2, \\ \frac{1}{4}(5\eta - \sqrt{3}\langle z, u\rangle, \sqrt{3}\eta u + 4z - 3\langle z, u\rangle u), & \text{if } j = 3, \end{cases} \tag{5.1}$$

where $u := (1, 1, 1)$. The initial vertex is $(\sqrt{3}, -1, 1, 1)$, and – after rationalization – the vertices in general can be written in the form $x = (\sqrt{3}\eta, z)$, with $\eta, \zeta_1, \zeta_2, \zeta_3 \in \mathbb{Z}$ having no common factor, and $\eta > 0$ such that $3\eta^2 = \zeta_1^2 + \zeta_2^2 + \zeta_3^2$. In fact, considering congruences modulo 8, it is easy to see that $\eta, \zeta_1, \zeta_2, \zeta_3$ must all be odd, if the expression for x is in lowest terms. Note that R_3 does not preserve such expressions, because of the factor $\frac{1}{4}$.

If we successively apply $R_0, R_1, R_2, R_3, R_2, R_1, R_0$ to $(\sqrt{3}, -1, 1, 1)$, then we obtain all $(\sqrt{3}, z)$ with

$$\begin{aligned} z = {} & (1, -1, 1), (1, 1, -1), (-1, -1, -1), (1, 1, 1), \\ & (-1, -1, 1), (-1, 1, -1), (1, -1, -1); \end{aligned}$$

in other words, we have all vertices of the form $(\sqrt{3}, \pm 1, \pm 1, \pm 1)$. This reflects the following fact.

Lemma 5.2 *The two honeycombs* $\{3, 3, 6\}$ *and* $\{4, 3, 6\}$ *have the same vertices.*

Indeed, $[3, 3, 6]$ and $[4, 3, 6]$ have a common subgroup, of index 5 in the first and 2 in the second; see [3, Sect. 11G] for the details. This further implies that all changes of sign of the coordinates of z are allowed, as well as all permutations, and so leads us to

Theorem 5.3 *The vertices of* $\{3, 3, 6\}$ *can be taken to be all points of the form* $(\sqrt{3}\eta, z)$ *just described.*

Proof It is clear that vertices of $\{3, 3, 6\}$ are all of the required form. The form of R_3 gets in the way of showing the converse immediately. However, let $(\sqrt{3}\eta, z)$ be of the given form in its lowest terms; permuting the coordinates and changing signs allows us to assume that $\zeta_1 \geqslant \zeta_2 \geqslant \zeta_3 > 0$, with at least one more strict inequality if $\eta > 1$; in particular, $\zeta_1 > \eta$ and $\langle z, u\rangle - \eta > 0$. If $\langle z, u\rangle - \eta \equiv 0 \mod 4$, then we see at once that $(\sqrt{3}\eta', z') = (\sqrt{3}\eta, z)R_3$ has integer entries $\eta', \ldots, \zeta_3'$ with $\eta' < \eta$. Otherwise, observe that $\zeta_1 + \zeta_2 - \zeta_3 - \eta \equiv 0 \mod 4$, since $2\zeta_3 \equiv 2 \mod 4$. We thus change the sign of ζ_3; since η remains the same, while the new $\langle z, u\rangle - \eta$ is still positive, we can apply R_3 as before, yielding a new integral $\eta' < \eta$. This completes the argument. □

Remark 5.4 Something that we cannot explain is that every odd positive η seems to occur in such a reduced expression $3\eta^2 = \|z\|^2$ (we have only checked this for $\eta \leqslant 19$)

As we have seen, the vertices of $\{3, 3, 6\}$ can be 4-coloured; its facets are tetrahedra. Inscribed in $\{3, 3, 6\}$, using 3, 2 and 1 of its colour classes in turn, we find

$$\{3, 3, 6\} \succ \{3, 6, 3\} \succ \{6, 3, 6\} \succ \{3, 6, 3\}.$$

From this, it follows that $\{3, 6, 3\}$ can be inscribed in itself, using just a third of its vertices. Of course, we have the same pattern as before; the two copies of $\{3, 6, 3\}$, and the two copies of {6, 3, 6}, can be inscribed in {3, 3, 6} using complementary colour classes.

This leads to families of compounds. For example, iterating {3, 6, 3}[*3*{3, 6, 3}] leads to {3, 6, 3}[3^k{3, 6, 3}] for each $k \geqslant 1$. But we actually have more: Lemma 5.2 and duality show that we have compounds like

$$\begin{aligned} &\{4, 3, 6\}[2\{6, 3, 6\}]\{6, 3, 4\}, \\ 3&\{4, 3, 6\}[4\{3, 6, 3\}]\{6, 3, 4\}, \\ &\{4, 3, 6\}[4\{3, 6, 3\}]3\{6, 3, 4\}. \end{aligned}$$

Once again, we leave further details to the interested reader.

6 The Coxeter Group $[3^{n-3}, 4, q]$

For the other families, we again begin with a subsidiary remark; compare Lemma 3.1.

Lemma 6.1 *For $k = 0, \ldots, n-2$ and $q \geqslant 3$, the mapping $\boldsymbol{r}_{n-2}, \boldsymbol{r}_{n-1} \mapsto \boldsymbol{e}$ and $\boldsymbol{r}_j \mapsto \boldsymbol{r}_j$ for $j = 0, \ldots, n-3$ induces a homomorphism on $[3^{n-3}, 4, q]$ with quotient $[3^{n-3}] \cong S_{n-1}$, the symmetric group.*

The main result of the section is

Theorem 6.2 *For $k = 0, \ldots, n-2$ and $q \geqslant 3$, the Coxeter group $[3^{k-1}, 4, q, q, 4, 3^{n-k-4}]$ is a subgroup of $[3^{n-3}, 4, q]$ of index $\binom{n-1}{k+1}$.*

Proof Let $\boldsymbol{r}_0, \ldots, \boldsymbol{r}_{n-1}$ be the distinguished generators of $\boldsymbol{G} = [3^{n-3}, 4, q]$. For $k = 0, \ldots, n-3$, we define the mixing operation $\boldsymbol{\mu}_k \colon (\boldsymbol{r}_0, \ldots, \boldsymbol{r}_{n-1}) \mapsto (\boldsymbol{t}_0, \ldots, \boldsymbol{t}_{n-1})$ by

$$\boldsymbol{t}_j := \begin{cases} \boldsymbol{r}_j, & \text{if } j = 0, \ldots, k-1, \\ \boldsymbol{r}_k \boldsymbol{r}_{k+1} \cdots \boldsymbol{r}_{n-3} \boldsymbol{r}_{n-2} \boldsymbol{r}_{n-3} \cdots \boldsymbol{r}_{k+1} \boldsymbol{r}_k, & \text{if } j = k, \\ \boldsymbol{r}_{n+k-j}, & \text{if } j = k+1, \ldots, n-1. \end{cases} \tag{6.3}$$

Again, the indexing of $\boldsymbol{\mu}_k$ is chosen to indicate that $\boldsymbol{r}_k$ is the only generator which changes, although the order of $\boldsymbol{r}_{k+1}, \ldots, \boldsymbol{r}_{n-1}$ is reversed. As before, we can extend the range of k in a natural way by $\boldsymbol{\mu}_{n-2} = \boldsymbol{\iota}$ (the identity), and $\boldsymbol{\mu}_{-1} = \boldsymbol{\delta}$ (the duality operation).

The proof follows the lines of that of Theorem 3.2 quite closely, in particular in verifying that the required relations are satisfied. In the present case, $2^{k-1}(k-1)!$ copies of the fundamental cone of $\boldsymbol{G}$ in the contragredient representation fit together to form the cone over an $(n-1)$-cross-polytope. We perform the same construction as before, but in a facet of this cross-polytope; Lemma 6.1 ensures that the construction is compatible with the whole group. We leave it to the reader to fill in the details. □

Remark 6.4 The operation $\boldsymbol{\mu}_{n-3}$ coincides with the halving operation η applied to the 3-coface $\{4, q\}$; compare [3, (10E2)].

Corresponding to the group $\boldsymbol{G}_k$ of Theorem 6.2 is the (universal) abstract regular polytope

$$\mathcal{P}_k := \{3^{k-1}, 4, q, q, 4, 3^{n-k-4}\},$$

with the same conventions as in the proposition; in particular, $\mathcal{P} := \mathcal{P}_{n-2} = \{3^{n-3}, 4, q\}$. Exactly analogous to Theorem 3.4, we have

Theorem 6.5 *The vertices of the universal regular polytope $\mathcal{P} = \{3^{n-3}, 4, q\}$ can be $(n-1)$-coloured. Moreover, the polytope $\mathcal{P}_k := \{3^{k-1}, 4, q, q, 4, 3^{n-k-4}\}$ can be inscribed in $\mathcal{P}$ using (any) $k+1$ of the colour-classes of its vertices.*

Proof As before, we appeal to induction on n, noting that the initial vertex always stays the same. Replacing n by $n-1$ implies replacing k by $k-1$, which means that we first have to establish the case $k = 0$. In this case, $\boldsymbol{t}_0$ takes the initial vertex into the opposite vertex of the initial cross-polytopal facet of $\mathcal{P}$; this vertex has the same colour 1 as the initial one. Moreover, since $\langle \boldsymbol{t}_1, \ldots, \boldsymbol{t}_{n-1} \rangle = \langle \boldsymbol{r}_1, \ldots, \boldsymbol{r}_{n-1} \rangle$, all such antipodal vertices in facets through the initial vertex are vertices of $\mathcal{P}_0$, and this quickly leads to vert $\mathcal{P}_0$ consisting of the whole colour-class 1 of $\mathcal{P}$.

For $k > 0$, we may now assume that $\mathcal{P}_k^v$ consists of all vertices of $\mathcal{P}^v$ in colour-classes $2, \ldots, k+1$, that is, those adjacent to the initial vertex coloured 1. The claim of the theorem quickly follows from the action of $\boldsymbol{G}_k$ and the symmetry of the colour-classes. □

Again as before, the polytopes $\mathcal{P}_k$ occur in dual pairs, and we deduce

Theorem 6.6 *For each $n \geqslant 4$, $q \geqslant 3$ and $k = 0, \ldots, n-3$, $\mathcal{P}_k$ and $\mathcal{P}_{n-k-3}$ are dual polytopes. Moreover, they can be inscribed in $\mathcal{P}$ so that their vertex-sets are complementary colour-classes.*

For the moment, we just illustrate Theorems 6.2 and 6.5 by two familiar cases. We have expressed them in terms of abstract polytopes, but of course they are isomorphic to the geometric ones in $\mathbb{E}^4$.

Example 6.7 When $n = 4$ and $q = 3$, we have

$$\{3, 4, 3\} \succ \{4, 3, 3\} \succ \{3, 3, 4\}.$$

Example 6.7 tells us that the vertices of the 24-cell $\{3, 4, 3\}$ are 3-colourable, and that the vertices of the 4-cross-polytope $\{3, 3, 4\}$ and 4-cube $\{4, 3, 3\}$ comprise one and two of the colour-classes, respectively. Moreover, they can be re-arranged so that $\{3, 3, 4\}$ and $\{4, 3, 3\}$ have complementary vertex-sets in vert $\{3, 4, 3\}$; they are then in dual position.

Example 6.8 When $n = 5$ and $q = 3$, we have

$$\{3, 3, 4, 3\} \succ \{3, 4, 3, 3\} \succ \{4, 3, 3, 4\} \succ \{3, 3, 4, 3\}.$$

We have the same pattern in Example 6.8. The vertices of the last copy of $\{3, 3, 4, 3\}$ form one out of four colour-classes of the first; the complementary three colour-classes make up the vertex-set of the dual $\{3, 4, 3, 3\}$. Similarly, we can inscribe two dual copies of the cubic tiling $\{4, 3, 3, 4\}$ in $\{3, 3, 4, 3\}$ with complementary vertex-sets (or colour-classes).

Familiar coordinates for the vertices of the geometric honeycombs graphically illustrate all this. For the original copy, we have

$$\text{vert}\{3, 3, 4, 3\} = \{(\xi_1, \ldots, \xi_4) \in \tfrac{1}{2}\mathbb{Z}^4 \mid \xi_1 \equiv \cdots \equiv \xi_4 \mod 1\}.$$

This actually identifies vert$\{3, 3, 4, 3\}$ with the integer quaternions $\xi_1 + \xi_2\mathbf{i} + \xi_3\mathbf{j} + \xi_4\mathbf{k}$.

The obvious splitting

$$\text{vert}\{3, 3, 4, 3\} = \mathbb{Z}^4 \cup \left(\mathbb{Z}^4 + \tfrac{1}{2}(1, 1, 1, 1)\right)$$

into two congruence classes modulo 1 gives the vertices of two dual copies of $\{4, 3, 3, 4\}$. Finally, the other copy of $\{3, 3, 4, 3\}$ has vertex-set

$$\{(\xi_1, \ldots, \xi_4) \in \mathbb{Z}^4 \mid \xi_1 + \cdots + \xi_4 \equiv 0 \mod 2\}.$$

7 Honeycombs Inscribed in {3, 4, 4}

We now have two applications of the results of Sect. 6 which yield information that we do not recall having seen before. For $n = q = 4$, our pattern is

$$\{3, 4, 4\} \succ \{4, 4, 4\} \succ \{4, 4, 4\}; \tag{7.1}$$

the vertex-sets of the two copies of $\{4, 4, 4\}$ form two or one of the three colour-classes of vertices of $\{3, 4, 4\}$, respectively. Indeed, the two copies can be regarded as duals, and so re-arranged to have complementary vertex-sets. However, a striking consequence is that one copy of $\{4, 4, 4\}$ can be inscribed in another using half its vertices; this leads to a doubly-infinite sequence

$$\cdots \succ \{4, 4, 4\} \succ \{4, 4, 4\} \succ \{4, 4, 4\} \succ \cdots,$$

with each copy having half the vertices of the one before.

The universal polytopes are realizable as regular honeycombs in hyperbolic space $\mathbb{H}^3$. For the latter, we adopt the standard model

$$\mathbb{H}^n = \{(\xi_0, \ldots, \xi_n) \in \mathbb{R}^{n+1} \mid \xi_0 > 0,\ \xi_0^2 = \xi_1^2 + \cdots + \xi_n^2 + 1\}.$$

The symmetry group of $\{3, 4, 4\}$ can be taken to have generators R_j (corresponding to $\boldsymbol{r}_j$) as follows. With $x = (\eta, z)$, where $z = (\zeta_1, \zeta_2, \zeta_3) \in \mathbb{E}^3$, we have

$$xR_j := \begin{cases} (\eta, \zeta_2, \zeta_1, \zeta_3), & \text{if } j = 0, \\ (\eta, \zeta_1, \zeta_3, \zeta_2), & \text{if } j = 1, \\ (\eta, \zeta_1, \zeta_2, -\zeta_3), & \text{if } j = 2, \\ \big(2\eta - \langle z, u\rangle, z + (\eta - \langle z, u\rangle)u\big), & \text{if } j = 3, \end{cases} \tag{7.2}$$

where $u = (1, 1, 1)$. Thus R_0, R_1, R_2 generate the symmetry group of the octahedron in a natural way. We write R_3 and points of $\mathbb{H}^3$ in this way for future computational convenience. Observe as well that each R_j preserves the set $\mathbb{Z}^4$ of integer vectors.

The vertices of $\{3, 4, 4\}$ are ideal, and so are to be thought of as rays $\{\lambda(\eta, z) \mid \lambda > 0\}$, with $\eta > 0$ and $\eta^2 = \|z\|^2$. We can normalize these in two ways, either by taking $\eta = 1$ and thus $z \in \mathbb{S}^2$ (the unit sphere), or $(\eta, z) \in \mathbb{Z}^4$ with $\gcd(\eta, \zeta_1, \zeta_2, \zeta_3) = 1$. With the latter representation, we have

Theorem 7.3 *The vertex-set of* $\{3, 4, 4\}$ *is*

$$\text{vert}\{3, 4, 4\} = \{(\eta, z) \in \mathbb{Z}^4 \mid \eta > 0,\ \eta^2 = \|z\|^2\}.$$

Proof We first note that the assumed condition $\gcd(\eta, \zeta_1, \zeta_2, \zeta_3) = 1$ implies that η is odd since, if $\zeta \in \mathbb{Z}$, then $\zeta^2 \equiv 0$ or $1 \mod 4$; thus we cannot have η even and at least one of $\zeta_1, \zeta_2, \zeta_3$ odd. It follows that exactly one of $\zeta_1, \zeta_2, \zeta_3$ is odd; hence $\langle z, u\rangle$ must also be odd, and it is then easy to see that each R_j takes one vector of the given form into another.

To see that every vector of that form occurs, we begin by noting that the initial vertex of $\{3, 4, 4\}$ is $(1, 1, 0, 0)$. We next observe that R_0, R_1, R_2 allow us freedom to permute the coordinates of z and change their signs. If $\eta > 1$, then we change signs so that z is a non-negative vector. From $\gcd(\eta, \zeta_1, \zeta_2, \zeta_3) = 1$ we infer that $\eta < \langle z, u\rangle = \zeta_1 + \zeta_2 + \zeta_3$ (just compare η^2 and $\langle z, u\rangle^2$); if $(\eta, z)R_3 =: (\eta', z')$, then we deduce that $\eta' < \eta$. Induction on η leads at once to the claim of the theorem. □

Remark 7.4 In the alternative normalization, we can identify vert$\{3, 4, 4\}$ with $\mathbb{S}^2 \cap \mathbb{Q}^3$.

In fact, we can say rather more. We have seen that exactly one of $\zeta_1, \zeta_2, \zeta_3$ is odd; moreover, which coordinate is odd is preserved by R_2 and R_3 (for the latter, note that $\eta - \langle z, u\rangle$ is even – of course, R_0 and R_1 permute the colour classes). As a consequence, we have

Proposition 7.5 *With the previous notation, the vertex* (η, z) *of* $\{3, 4, 4\}$ *is coloured* j *just when the* j*th coordinate of* z *is odd.*

Remark 7.6 As a matter of interest, in the given coordinate system the isometry Φ of $\mathbb{H}^3$ with matrix

$$\Phi = \frac{1}{\sqrt{2}} \begin{bmatrix} 2 & 1 & 1 & 0 \\ 0 & 1 & -1 & 0 \\ -1 & -1 & -1 & 1 \\ -1 & -1 & -1 & -1 \end{bmatrix}$$

is such that $\mathcal{P}_0 \prec \mathcal{P}_0\Phi = \mathcal{P}_1$, with the indexing introduced in Sect. 6; we do not give the details of the calculation. Thus the powers of Φ (positive and negative) induce the sequence of copies of $\{4, 4, 4\}$, each properly inscribed in the next.

As we have seen, we can inscribe $\{4, 4, 4\}$ in $\{3, 4, 4\}$, using either two or one of its three colour-classes. As a result, we obtain two, dual, regular compounds (on the abstract level as well)

$$2\{3, 4, 4\}[3\{4, 4, 4\}]\{4, 4, 3\}, \ \ \{3, 4, 4\}[3\{4, 4, 4\}]2\{4, 4, 3\}.$$

(The fact that the copies P_0 and P_1 of $\{4, 4, 4\}$ can be arranged to have complementary vertex-sets in vert$\{3, 4, 4\}$ accounts for the numbers *2* and *3*.) Thus, even though $\{4, 4, 4\}$ is self-dual, perhaps surprisingly its compounds in $\{3, 4, 4\}$ are not.

8 Honeycombs Inscribed in {3, 3, 3, 4, 3}

For $n = 6$ and $q = 3$, our pattern is

$$\{3, 3, 3, 4, 3\} \succ \{3, 3, 4, 3, 3\} \succ \{3, 4, 3, 3, 4\} \succ \{4, 3, 3, 4, 3\} \succ \{3, 3, 4, 3, 3\}. \tag{8.1}$$

There are five colour-classes of vertices of $\{3, 3, 3, 4, 3\}$, and the inscribed polytopes use four, three, two or one of these, respectively. The two copies of $\{3, 3, 4, 3, 3\}$ can be regarded as duals, with complementary vertex-sets in those of $\{3, 3, 3, 4, 3\}$; the dual polytopes $\{3, 4, 3, 3, 4\}$ and $\{4, 3, 3, 4, 3\}$ can be viewed similarly.

Remark 8.2 Note that we have an alternative picture of $\{3, 3, 4, 3, 3\} \prec \{4, 3, 3, 4, 3\}$, in the form

$$\{3, 3, 4, 3, 3\} = \left\{3, {3, 4, 3 \atop 3}\right\}$$

with vertices alternate vertices of $\{4, 3, 3, 4, 3\}$ (of course, as in the pattern here).

As in Sect. 7, we obtain a doubly-infinite sequence

$$\cdots \succ \{3, 3, 4, 3, 3\} \succ \{3, 3, 4, 3, 3\} \succ \{3, 3, 4, 3, 3\} \succ \cdots ;$$

in this case, each copy has a quarter of the vertices of its predecessor. Here, though, we can interpolate copies of {4, 3, 3, 4, 3} and {3, 4, 3, 3, 4} between successive ones of {3, 3, 4, 3, 3}, as in the first pattern.

Geometrically, we have a realization {3, 3, 3, 4, 3} as a regular honeycomb in $\mathbb{H}^5$. Of its symmetry group $\langle R_0, \dots, R_5 \rangle$ acting on vectors (η, z), $R_0, \dots, R_4$ fix η and act on $z = (\zeta_1, \dots, \zeta_5)$ in the standard way as symmetries of the 5-cross-polytope (that is, R_j interchanges ζ_{j+1} and ζ_{j+2} for $j = 0, \dots, 3$, while R_4 changes the sign of ζ_5 – compare (7.2)). Further,

$$(\eta, z)R_5 = \tfrac{1}{2}\big(3\eta - \langle z, u\rangle, 2z + (\eta - \langle z, u\rangle)u\big), \tag{8.3}$$

where $u = (1, 1, 1, 1, 1)$. In analogy to the case {3, 4, 4}, we can represent a vertex of {3, 3, 3, 4, 3} by a vector $(\eta, z) \in \mathbb{Z}^6$ with $\eta > 0$ and $\gcd(\eta, \zeta_1, \dots, \zeta_5) = 1$. Exactly the same arguments as deployed in Sect. 7 lead to

Theorem 8.4 *The vertex-set of* {3, 3, 3, 4, 3} *is*

$$\operatorname{vert}\{3, 3, 3, 4, 3\} = \{(\eta, z) \in \mathbb{Z}^6 \mid \eta > 0,\ \eta^2 = \|z\|^2\}.$$

Remark 8.5 In the alternative normalization, we can identify vert{3, 3, 3, 4, 3} with $\mathbb{S}^4 \cap \mathbb{Q}^5$.

In a similar way, we have

Proposition 8.6 *With the same convention as before, the vertex* (η, z) *of* {3, 3, 3, 4, 3} *is coloured* j *just when the* j*th coordinate* ζ_j *of* z *has the same parity as* η.

Proof If η is even, then the assumed condition that $\gcd(\eta, \zeta_1, \dots, \zeta_5) = 1$ implies that at least one of $\zeta_1, \dots, \zeta_5$ must be odd; since $\eta^2 \equiv 0 \mod 4$ and $\zeta^2 \equiv 1 \mod 4$ if $\zeta \in \mathbb{Z}$ is odd, we see that exactly four of them are odd. If η is odd, then $\eta^2, \zeta^2 \equiv 1 \mod 8$ (for odd ζ) similarly implies that exactly one of $\zeta_1, \dots, \zeta_5$ is odd. A final observation that R_4 and R_5 preserve the parity condition completes the proof; once again, the fact that $\eta - \langle z, u\rangle$ is even is the key for R_5. □

The discussion shows that, for example, we can inscribe four copies of {3, 3, 4, 3, 3} in itself (with disjoint vertex-sets); this leads to geometric vertex-regular compounds of the form

$$\{3, 3, 4, 3, 3\}[4^k\{3, 3, 4, 3, 3\}]$$

for each k. Since, as remarked earlier, we can interpolate copies of {4, 3, 3, 4, 3} and {3, 4, 3, 3, 4} between successive ones of {3, 3, 4, 3, 3}, a consequence is that completely classifying possible regular compounds of hyperbolic honeycombs may be far from straightforward. So, what we shall do is point out that even some simple compounds do not behave as one might expect.

As in the [3, 4, 4]-family, we have a pair of compounds

$$4\{3,3,3,4,3\}[5\{3,3,4,3,3\}]\{3,4,3,3,3\},$$
$$\{3,3,3,4,3\}[5\{3,3,4,3,3\}]4\{3,4,3,3,3\}$$

where a self-dual honeycomb is inscribed in non-self-dual compounds. Of course, we also have the dual pair

$$3\{3,3,3,4,3\}[5\{3,4,3,3,4\}]2\{3,4,3,3,3\},$$
$$2\{3,3,3,4,3\}[5\{4,3,3,4,3\}]3\{3,4,3,3,3\};$$

once again, the numbers are explained by the fact that $\{3, 4, 3, 3, 4\}$ and $\{4, 3, 3, 4, 3\}$ can be taken to have complementary subsets of vertices of $\{3, 3, 3, 4, 3\}$ (that is, counting colour-classes). Last, though, note that we have further compounds such as

$$3\{3,3,4,3,3\}[4\{3,4,3,3,4\}],\ \{3,3,4,3,3\}[2\{4,3,3,4,3\}],$$
$$2\{3,4,3,3,4\}[3\{4,3,3,4,3\}],$$

which are only vertex-regular; the interested reader will easily be able to derive many others.

9 Quotients

The regular hyperbolic honeycombs with ideal vertices have quotients which are locally toroidal, in that their facets and vertex-figures are either spherical or toroidal; these are discussed in considerable detail in [3, Chaps. 10–12]; we also mention [4]. But on passing to the quotients, it is usually the case that subgroup relationships are not preserved.

However, among the locally toroidal regular polytopes described in [3, Chap. 10] are

$$\{\{3,4\},\{4,4:2s\}\} \succ \{\{4,4:2s\},\{4,4\mid s\}\} \succ \{\{4,4\mid s\},\{4,4:2s\}\},$$

for each $s \geqslant 2$; recall that the torus components$\{4, 4 : 2s\} = \{4, 4\}_{(s,s)}$ and $\{4, 4 \mid s\} = \{4, 4\}_{(s,0)}$ (in the notation of the monograph) are determined by their Petrie polygons $\{2s\}$ and holes $\{s\}$, respectively. Exactly the same pattern of colour-classes of vertices carries over to the quotients. Only $s = 2$ gives a finite case; for $s = 3$ the polytopes are naturally realizable in $\mathbb{E}^5$.

In [3, Chap. 11] other polytopes than those arising from quotients of [3, 3, 6] and its subgroups are considered. In that family, the quotients do not preserve indices of subgroups; indeed, the same group may occur. In no case do the inscriptions carry over.

Though there are far from degenerate finite quotients of $\{3, 3, 3, 4, 3\}$, the discussion of [3, Chap. 12] shows that these do not induce nice inscriptions of locally toroidal regular polytopes like those in the previous family. However, the operation μ_k (with different indices) was employed in a different context in [2] (see also [3, Sect. 14A]) to produce a family of locally projective regular polytopes.

References

1. H.E. Debrunner, Dissecting orthoschemes into orthoschemes. Geom. Ded. **33**, 123–152 (1990)
2. P. McMullen, Locally projective regular polytopes. J. Combin. Theory A **65**, 1–10 (1994)
3. P. McMullen and E. Schulte, *Abstract Regular Polytopes* (*Encyclopedia of Mathematics and its Applications* **92**). Cambridge University Press (Cambridge, 2002)
4. B.R. Monson, E. Schulte, Reflection groups and polytopes over finite fields. III. Adv. Appl. Math. **41**, 76–94 (2008)

Sphere-of-Influence Graphs in Normed Spaces

Márton Naszódi, János Pach and Konrad Swanepoel

Dedicated to Károly Bezdek and Egon Schulte on the occasion of their 60th birthdays

Abstract We show that any k-th closed sphere-of-influence graph in a d-dimensional normed space has a vertex of degree less than $5^d k$, thus obtaining a common generalization of results of Füredi and Loeb (Proc Am Math Soc 121(4):1063–1073, 1994 [1]) and Guibas et al. (Sphere-of-influence graphs in higher dimensions, Intuitive geometry [Szeged, 1991], 1994, pp. 131–137 [2]).

Toussaint [8] introduced the sphere-of-influence graph of a finite set of points in Euclidean space for applications in pattern analysis and image processing (see [7] for a recent survey). This notion was later generalized to so-called closed sphere-of-influence graphs [3] and to k-th closed sphere-of-influence graphs [4]. Our setting

---

Márton Naszódi acknowledges the support of the János Bolyai Research Scholarship of the Hungarian Academy of Sciences, and the Hung. Nat. Sci. Found. (OTKA) grant PD104744. Part of this paper was written when Swanepoel visited EPFL in April 2015. Research by János Pach was supported in part by Swiss National Science Foundation grants 200020-144531 and 200020-162884.

M. Naszódi
Department of Geometry, Lorand Eötvös University,
Pazmány Péter Sétany 1/C, Budapest 1117, Hungary
e-mail: marton.naszodi@math.elte.hu

J. Pach
EPFL Lausanne and Rényi Institute, Budapest, Hungary
e-mail: pach@cims.nyu.edu

K. Swanepoel (✉)
Department of Mathematics, London School of Economics
and Political Science, Houghton Street, London WC2A 2AE, UK
e-mail: k.swanepoel@lse.ac.uk

M. D. E. Conder et al. (eds.), *Discrete Geometry and Symmetry*,
Springer Proceedings in Mathematics & Statistics 234,
https://doi.org/10.1007/978-3-319-78434-2_16

will be a d-dimensional normed space $\mathcal{N}$ with norm $\|\cdot\|$. We denote the ball with center $c \in \mathcal{N}$ and radius r by $B(c, r)$.

Definition 1 Let $k \in \mathbb{N}$ and let $V = \{c_i : i = 1, \dots, m\}$ be a set of points in the d-dimensional normed space $\mathcal{N}$. For each $i \in \{1, \dots, m\}$, let $r_i^{(k)}$ be the smallest r such that

$$\{j \in \mathbb{N} : j \neq i, \|c_i - c_j\| \leq r\}$$

has at least k elements. Define the *k-th closed sphere-of-influence graph on V* by setting $\{c_i, c_j\}$ an edge whenever $B(c_i, r_i^{(k)}) \cap B(c_j, r_j^{(k)}) \neq \emptyset$.

Füredi and Loeb [1] gave an upper bound for the minimum degree of any closed sphere-of-influence graph in $\mathcal{N}$ in terms of a certain packing quantity of the space (see also [5, 6].)

Definition 2 Let $\vartheta(\mathcal{N})$ denote the largest cardinality of a subset A of the ball $B(o, 2)$ of the normed space $\mathcal{N}$ such that any two points of A are at distance at least 1, and the origin o is in A.

Füredi and Loeb [1] showed that any closed sphere-of-influence graph (that is, in our terminology, a first closed sphere-of-influence graph) in $\mathcal{N}$ has a vertex of degree smaller than $\vartheta(\mathcal{N}) \leq 5^d$. (It is clear that $\vartheta(\mathcal{N})$ is bounded above by the number of balls of radius $1/2$ that can be packed into a ball of radius $5/2$, which is at most 5^d by volume considerations.)

Guibas, Pach and Sharir [2] showed that any k-th closed sphere-of-influence graph in d-dimensional Euclidean space has a vertex of degree at most $c^d k$, for some universal constant $c > 1$. In this note we show the following more precise result, valid for all norms, and generalizing the result of Füredi and Loeb [1] mentioned above.

Theorem 3 *Every k-th sphere-of-influence graph on at least two points in a normed space* $\mathcal{N}$ *has at least two vertices of degree smaller than* $\vartheta(\mathcal{N})k \leq 5^d k$.

We note that the theorem still holds when there are repeated elements.

Corollary 4 *A k-th sphere-of-influence graph on n points in* $\mathcal{N}$ *has at most* $(\vartheta(\mathcal{N})k - 1)n \leq (5^d k - 1)n$ *edges.*

Proof of Theorem 3 Let $V = \{c_1, c_2, \dots, c_m\}$. Relabel the vertices $c_1, c_2, \dots, c_m$ such that $r_1^{(k)} \leq r_2^{(k)} \leq \dots \leq r_m^{(k)}$. We define an auxiliary graph H on V by joining c_i and c_j whenever $\|c_i - c_j\| < \max\{r_i^{(k)}, r_j^{(k)}\}$. Thus, if $\{c_i : i \in I\}$ is an independent set in H, then no ball in $\{B(c_i, r_i^{(k)}) : i \in I\}$ contains the center of another in its interior. We next bound the chromatic number of H.

Lemma 5 *The chromatic number of H does not exceed k.*

Proof Note that for each $i \in \{1, \dots, m\}$, the set

$$\{j < i : c_i c_j \in E(H)\} = \{j < i : \|c_i - c_j\| < r_i^{(k)}\}$$

has less than k elements. Therefore, we can greedily color H in the order $c_1, c_2, \ldots, c_m$ by k colors. $\square$

We next show that the degrees of c_1 and c_2 (corresponding to the two smallest values of $r_i^{(k)}$) are both at most $\vartheta(\mathcal{N})k$, which will complete the proof of Theorem 3. We first need the so-called "bow-and-arrow" inequality of [1]. For completeness, we include the proof from [1].

Lemma 6 *(Füredi–Loeb [1]) For any two non-zero elements a and b of a normed space,*

$$\left\| \frac{1}{\|a\|}a - \frac{1}{\|b\|}b \right\| \geq \frac{\|a-b\| - |\|a\| - \|b\||}{\|b\|}.$$

Proof Without loss of generality, we may assume that $\|a\| \geq \|b\| > 0$. Then

$$\begin{aligned}
\|a-b\| &= \left\| \|a\| \frac{1}{\|a\|}a - \|b\| \frac{1}{\|b\|}b \right\| \\
&= \left\| \|b\| (\frac{1}{\|a\|}a - \frac{1}{\|b\|}b) + (\|a\| - \|b\|)\frac{1}{\|a\|}a \right\| \\
&\leq \|b\| \left\| \frac{1}{\|a\|}a - \frac{1}{\|b\|}b \right\| + \|a\| - \|b\|.
\end{aligned}$$ $\square$

The next lemma is abstracted with minimal hypotheses from [5, Proof of Theorem 6] (see also [1, Proof of Theorem 2.1]).

Lemma 7 *Consider the balls $B(v_1, \lambda_1)$ and $B(v_2, \lambda_2)$ in the normed space $\mathcal{N}$, such that $\max\{\lambda_1, \lambda_2\} \geq 1$, $v_1 \notin \mathrm{int}(B(v_2, \lambda_2))$, $v_2 \notin \mathrm{int}(B(v_1, \lambda_1))$ and $B(v_i, \lambda_i) \cap B(o, 1) \neq \emptyset$ $(i = 1, 2)$. Define $\pi : \mathcal{N} \to B(o, 2)$ by*

$$\pi(x) = \begin{cases} x & \text{if } \|x\| \leq 2, \\ \frac{2}{\|x\|}x & \text{if } \|x\| \geq 2. \end{cases}$$

Then $\|\pi(v_1) - \pi(v_2)\| \geq 1$.

Proof In terms of the norm, we are given that $\|v_1 - v_2\| \geq \max\{\lambda_1, \lambda_2\} \geq 1$, $\|v_1\| \leq \lambda_1 + 1$, and $\|v_2\| \leq \lambda_2 + 1$. Without loss of generality, we may assume that $\|v_2\| \leq \|v_1\|$.

If $v_1, v_2 \in B(o, 2)$ then $\|\pi(v_1) - \pi(v_2)\| = \|v_1 - v_2\| \geq 1$.

If $v_1 \notin B(o, 2)$ and $v_2 \in B(o, 2)$, then

$$\begin{aligned}
\|\pi(v_1) - \pi(v_2)\| &= \left\| 2\frac{1}{\|v_1\|}v_1 - v_2 \right\| \geq \|v_1 - v_2\| - \left\| v_1 - 2\frac{1}{\|v_1\|}v_1 \right\| \\
&= \|v_1 - v_2\| - (\|v_1\| - 2) \geq \lambda_1 - (\lambda_1 + 1) + 2 = 1.
\end{aligned}$$

If $v_1, v_2 \notin B(o,2)$, then

$$\|\pi(v_1)-\pi(v_2)\| = \left\|2\frac{1}{\|v_1\|}v_1 - 2\frac{1}{\|v_2\|}v_2\right\| \geq 2\frac{\|v_1-v_2\|-\|v_1\|+\|v_2\|}{\|v_2\|} \quad \text{by Lemma 6}$$
$$\geq 2\left(\frac{\lambda_1-(\lambda_1+1)}{\|v_2\|}+1\right) = \frac{-2}{\|v_2\|}+2 \geq -1+2=1. \qquad \square$$

We can now finish the proof of Theorem 3. Let $i \in \{1,2\}$, and let $c := c_i$, that is, the radius corresponding to c is the smallest, or second smallest. By Lemma 5 we can partition the set of neighbors of c in the k-th closed sphere-of-influence graph on V into k classes $N_1, \dots, N_k$ so that each N_t is an independent set in H. We may assume that the radius $r_i^{(k)}$ corresponding to c is 1. Then for any $t \in \{1,\dots,k\}$, each ball in $\{B(c_j, r_j^{(k)})\colon c_j \in N_t\}$ intersects $B(c,1)$, and the center of no ball is in the interior of another ball. By Lemma 7, $\{\pi(p-c)\colon p \in N_t\}$ is a set of points contained in $B(o,2)$ with a distance of at least 1 between any two. That is, $|N_t \setminus \mathrm{int}(B(c,1))| \leq \vartheta(\mathcal{N})-1$ for each $t=1,\dots,k$. Since there are at most $k-1$ points in $V \cap \mathrm{int}(B(c,1)) \setminus \{c\}$, it follows that the degree of c is at most $\sum_{t=1}^{k}|N_t \setminus \mathrm{int}(B(c,1))| + k - 1 \leq (\vartheta(\mathcal{N})-1)k + k - 1 = \vartheta(\mathcal{N})k - 1$.

Acknowledgements We thank the referee for helpful suggestions that improved the paper.

References

1. Z. Füredi, P.A. Loeb, On the Best Constant for the Besicovitch Covering Theorem. Proc. Am. Math. Soc. **121**(4), 1063–1073 (1994). MR1249875 (95b:28003)
2. L. Guibas, J. Pach, M. Sharir, Sphere-of-influence graphs in higher dimensions, in *Intuitive Geometry* (Szeged, 1991) 1994, pp. 131–137. MR1383618 (97a:05183)
3. F. Harary, M.S. Jacobson, M.J. Lipman, F.R. McMorris, Abstract sphere-of-influence graphs. Math. Comput. Modelling **17**(11), 77–83 (1993). *Graph-Theoretic Models in Computer Science, II* (Las Cruces, NM, 1988–1990), p. 1236512
4. J. Klein, G. Zachmann, Point cloud surfaces using geometric proximity graphs. Comput. Graph. **28**(6), 839–850 (2004)
5. T.S. Michael, T. Quint, Sphere of influence graphs: edge density and clique size. Math. Comput. Model. **20**(7), 19–24 (1994). MR1299482
6. J.M. Sullivan, Sphere packings give an explicit bound for the Besicovitch covering theorem. J. Geom. Anal. **4**(2), 219–231 (1994). MR1277507
7. G.T. Toussaint, The sphere of influence graph: theory and applications. Int. J. Inf. Technol. Comput. Sci. **14**(2), 37–42 (2014)
8. G.T. Toussaint, A graph-theoretical primal sketch. Mach. Intell. Pattern Recognit. **6**, 229–260 (1988). *A Computational Geometric Approach to the Analysis of Form*, MR993994

On Symmetries of Projections and Sections of Convex Bodies

Dmitry Ryabogin

Abstract In this paper we discuss several questions of unique determination of convex (or star-shaped) bodies with projections (sections) satisfying a certain symmetry property.

Keywords Sections and projections of convex bodies

1 Introduction: Questions on Bodies with Congruent Projections and Sections

In 1932 Nakajima [1] and Süss [2] proved that two convex bodies in $\mathbb{R}^3$ are translates of each other, provided that their orthogonal projections onto every plane passing through the origin are translates of each other.

It is very natural to ask what happens if the group of translations is replaced by the group of isometries. The following problem is probably one of the oldest open problems of uniqueness in classical convexity (see, [3], p. 125, Problem 3.2).

Problem 1 *Let K and L be two convex bodies in $\mathbb{R}^n$, $n \geq 2$, such that the corresponding orthogonal projections $K|H$, $L|H$, onto all subspaces H of a fixed dimension k, $2 \leq k \leq n-1$, are congruent. Does it follow that K and L coincide up to translation and reflection in the origin?*

Here we say that two sets A and B in $\mathbb{R}^k$, $k \geq 2$, are congruent if there exists an orthogonal transformation $\varphi \in O(k)$ such that the sets $\varphi(A)$ and B are translates of each other.

The author is supported in part by U.S. National Science Foundation Grants DMS-0652684 and DMS-1101636.

D. Ryabogin (✉)
Department of Mathematics, Kent State University, Kent, OH 44242, USA
e-mail: ryabogin@math.kent.edu

M. D. E. Conder et al. (eds.), *Discrete Geometry and Symmetry*,
Springer Proceedings in Mathematics & Statistics 234,
https://doi.org/10.1007/978-3-319-78434-2_17

It is natural to consider an analogue of Problem 1 for sections (cf. [3], p. 289, Problem 7.3).

Problem 2 *Let K and L be star bodies in $\mathbb{R}^n$, $n \geq 2$, such that the section $K \cap H$ is congruent to $L \cap H$ for all subspaces H of a fixed dimension k, $2 \leq k \leq n-1$. Is K a translate of $\pm L$?*

Why do we expect the bodies in the ambient space to be different up to translation and reflection only, and not to be, say, congruent? Is it possible to simply describe a fairly large class of bodies for which the answer is affirmative? In this article we will attempt to give partial answers to some of these questions. On the way we will formulate several related problems of uniqueness, which, in our opinion, are interesting in their own right.

All necessary definitions and notions that will be used in the sequel may be found in the books of Gardner [3] and Schneider [4].

1.1 Notation

We will denote by $S^{n-1} = \{x \in \mathbb{R}^n : |x| = 1\}$ the unit sphere in $\mathbb{R}^n$, $n \geq 2$, and by $\xi^\perp = \{x \in \mathbb{R}^n : x \cdot \xi = 0\}$ the subspace orthogonal to a direction $\xi \in S^{n-1}$. Here the "dot" stands for the usual inner product in $\mathbb{R}^n$, $|x| = \sqrt{x_1^2 + \cdots + x_n^2}$ is the usual Euclidean norm. Given a spherical cap $U = U_\epsilon(\xi) = \{\theta \in S^{n-1} : \theta \cdot \xi \geq 1 - \epsilon\}$, $\epsilon \in (0, 1)$, we denote by $U^\perp = U_\epsilon^\perp(\xi)$ the equatorial neighborhood of $\xi^\perp \cap S^{n-1}$, i.e., $U^\perp = \{\theta \in S^{n-1} : |\theta \cdot \xi| < \epsilon\}$. The notation $SO(n)$ and $O(n)$ for the special orthogonal and orthogonal groups, acting on $\mathbb{R}^n$, is standard. We will say that a rotation $\varphi = \varphi_\xi$ in the subspace $\xi^\perp$ is in $SO(n-1, \xi^\perp)$ if there exists a rotation $\Phi_\xi \in SO(n)$ bringing $\xi^\perp$ into $e_n^\perp$ and such that $\varphi_\xi = \Phi_\xi^{-1} \varphi_{e_n} \Phi_\xi$ with $\varphi_{e_n} \in SO(n-1) = SO(n-1, e_n^\perp)$. Here the $(n-1)$dimensional subspace $e_n^\perp$ being identified with $\mathbb{R}^{n-1}$, and $e_n = (0, \ldots, 0, 1)$. A similar notation is used for $O(n-1, H)$, where H is a $k-$dimensional subspace of $\mathbb{R}^n$, $2 \leq k \leq n-1$. Let G be a fixed subgroup of $O(n-1, e_n^\perp)$. We will denote by $G_\xi = o_\xi G$, the corresponding subgroup acting in $\xi^\perp$, where $o_\xi \in SO(n)$ such that $o_\xi(e_n^\perp) = \xi^\perp$. The *reflection in the origin* is the group consisting of the identity map and the map $x \to -x$. We remark that this group is a subgroup of $SO(n-1)$ only for odd n. A *rigid motion* is an orthogonal transformation followed by a translation. A *direct rigid motion* is a rotation followed by a translation. We say that a subset A of $\mathbb{R}^n$, $n \geq 2$, has an $O(n)$-symmetry or a rigid motion symmetry if there exists a non-trivial orthogonal transformation $\varphi \in O(n)$ such that the sets $\varphi(A)$ and A are translates of each other.

We denote by $h_K(x)$ the support function of a convex body $K \subset \mathbb{R}^n$. For $x \in \mathbb{R}^n$ it is defined as $h_K(x) = \sup\limits_{y \in K} x \cdot y$, ([5], p. 37), and it is a homogeneous function of degree 1. The width of a set $A \subset \mathbb{R}^n$ in the direction $x \in \mathbb{R}^n$, is defined as $w_A(x) = h_A(x) + h_A(-x)$. We will repeatedly use the following well-known properties of the support function. For every convex body K,

$$h_{K|\xi^\perp}(x) = h_K(x) \text{ and } h_{\varphi_\xi(K|\xi^\perp)}(x) = h_{K|\xi^\perp}(\varphi_\xi^{-1}(x)), \quad \forall x \in \xi^\perp, \tag{1}$$

(see, for example, [3], (0.21), (0.26), pp. 17–18); here φ_ξ^{-1} stands for the inverse of $\varphi_\xi \in SO(n-1, \xi^\perp)$.

The notation $\rho_K(x) = \sup_{\lambda>0}\{\lambda x \in K\}$, $x \in \mathbb{R}^n \setminus \{0\}$, is used for the radial function of a star-shaped body with respect to the origin ([3], p. 18). It is a homogeneous function of degree -1.

We will write f_e and f_o for the even and odd parts of the function f,

$$f_e(x) = \frac{f(x) + f(-x)}{2}, \qquad f_o(x) = \frac{f(x) - f(-x)}{2}.$$

2 Translations Only

2.1 *Projections*

In the two-dimensional case, Problem 1 has a negative answer even in the particular case in which the projections are translates of each other. Indeed, considering the Reuleaux triangle R ([3], p. 108) and the disc D of the same width $w_R(x) = w_D(x)$, $\forall x \in \mathbb{R}^2$, we see that the corresponding projections (segments) have the same length (equal to the corresponding width in the direction orthogonal to the direction of the projection), hence these projections are translates of each other.

As we mentioned above, in the three-dimensional case the problem was formulated and solved independently by Nakajima [1] and Süss [2] in 1932. Their proof can be generalized to higher dimensions (see [6], Lemma I) for another proof obtained by I. Lieberman; an elementary proof can be also found in [7]. Besides, for Problem 1, with $k = n - 1$, H. Hadwiger established a more general result and showed that it is not necessary to consider projections onto all $(n-1)$-dimensional subspaces; the hypotheses need only be true for one fixed subspace H, together with all subspaces containing a line orthogonal to H. In other words, one requires only a "ground" projection on H and all corresponding "side" projections; see Fig. 1 in Sect. 3. Moreover, Hadwiger noted that in $\mathbb{R}^n$, $n \geq 4$, the ground projection might be dispensed with (see [8], and [3], pp. 126–127).

2.2 *Sections*

As in the case of projections, Problem 2 on the plane has a negative answer. This could be seen by considering the disc and the plane equichordal convex set. Here we say that a convex set containing the origin in its interior is equichordal if $\rho_K(\theta) + \rho_K(-\theta) =$

const for all $\theta \in S^1$. For the existence of these bodies we refer the reader to ([3], p. 255, Theorem 6.3.2).

In the three-dimensional case the problem was solved by C. A. Rogers, who showed that it has an affirmative answer, see ([3], p. 270, Theorem 7.1.1). We are not aware of an analogue of Hadwiger's result for sections.

3 Directly Congruent Projections

Since on the plane the answers to Problems 1 and 2 are negative, from now on we will assume that $n \geq 3$. We say that two sets A and B in $\mathbb{R}^{n-1}$ are *directly congruent* if there exists a rotation $\varphi \in SO(n-1)$ such that the sets $\varphi(A)$ and B are translates of each other.

If $\text{codim}(H) = 1$, in the case of direct congruence Problem 1 can be reformulated in terms of support functions as follows.

Problem 3 *Assume that for every $\xi \in S^{n-1}$ we have a non-trivial rotation $\varphi_\xi \in SO(n-1, \xi^\perp)$ such that*

$$h_{K|\xi^\perp}(\varphi_\xi^{-1}(\theta)) + a_\xi \cdot \theta = h_{L|\xi^\perp}(\theta) \qquad \forall \theta \in (\xi^\perp \cap S^{n-1}). \tag{2}$$

Does it follow that there exists $b \in \mathbb{R}^n$ such that $h_K(\theta) + b \cdot \theta = h_L(\theta)$ or $h_K(\theta) + b \cdot \theta = h_L(-\theta)$ for all $\theta \in S^{n-1}$?

3.1 Symmetric Bodies

We recall that a set $E \subset \mathbb{R}^n$, $n \geq 2$, is *centrally symmetric*, ([3], p. 3), if there exists a vector c such that the translate $E - c$ is *centered*, i.e., $x \in E - c$ if and only if $-x \in E - c$.

The class of symmetric bodies is a fairly large class for which the answer to Problem 1 is affirmative. We remark that, taking into account the aforementioned result of Nakajima and Süss, it is enough to consider the case of subspaces of codimension 1 (cf., for example, [9], proof of Theorem 2).

To answer Problems 1 and 2 in the centrally-symmetric case, the idea is to use the Funk transform Ff, which is defined on continuous functions f on the unit sphere as

$$Ff(\xi) = \int\limits_{\xi^\perp \cap S^{n-1}} f(\sigma) d\sigma.$$

It is well-known ([10], Chap. III, Sect. 1) that $Ff = Fg$ implies $f = g$ for *even* continuous functions on S^{n-1}.

Without loss of generality we may assume that the centers of bodies are located at the origin. Integrating the "even part" of Eq. (2) over $\xi^\perp \cap S^{n-1}$, i.e., the equation

$$h_K(\varphi_\xi^{-1}(\theta)) = (h_K)_e(\varphi_\xi^{-1}(\theta)) = (h_L)_e(\theta) = h_L(\theta) \qquad \forall \theta \in (\xi^\perp \cap S^{n-1}),$$

and using the fact that the Lebesgue measure is invariant under isometries, we get

$$(Fh_K)(\xi) = \int_{\xi^\perp \cap S^{n-1}} h_K(\sigma)d\sigma = \int_{\xi^\perp \cap S^{n-1}} h_K(\varphi_\xi^{-1}(\sigma))d\sigma =$$

$$\int_{\xi^\perp \cap S^{n-1}} h_L(\sigma)d\sigma = (Fh_L)(\xi) \qquad \forall \xi \in S^{n-1}.$$

We conclude that $h_K = h_L$. A similar argument related to Problem 2 can be applied to the corresponding equation in terms of the radial functions, see [11]; we also refer the reader to [12].

We remark that if we assume that only one of the bodies in Problem 1 is symmetric, then the other body must be symmetric as well. Indeed, if K is symmetric, then, all its projections are symmetric. Hence, all the projections of L are symmetric, and using the result from ([13], p. 132), we conclude that L is symmetric.

Since for symmetric bodies the answers to Problems 1 and 2 are affirmative, it is natural to conjecture the same result in the general case of non-symmetric bodies.

3.2 Golubyatnikov's Approach

The heuristic idea is that rotations in Problem 1 are trivial. To be more precise, let $\theta \in S^{n-1}$ and define its "orbit" under the action of the family of rigid motions as

$$\Gamma(\theta) = \bigcup_{\{\xi \in (\theta^\perp \cap S^{n-1})\}} (\varphi_\xi(\theta) + a_\xi),$$

where φ_ξ and a_ξ are as in Problem 3. Observe that if we use (1) and change $\xi \in (\theta^\perp \cap S^{n-1})$ in (2), the right-hand side $h_L(\theta)$ does not change. This means that the value of h_K on $\Gamma(\theta)$ is constant. Therefore, if we manage to show that for close points $\sigma, \theta \in S^{n-1}$, their orbits intersect, $\Gamma(\sigma) \cap \Gamma(\theta) \neq \emptyset$, then we would have $h_L(\sigma) = h_L(\theta)$. If the above equality is true for all σ and θ, then L must be an Euclidean ball and one can assume that the projections of K and L are translates of each other.

Thus, if this line of thought was possible, the group of rigid motions in Problem 1 could be reduced to a group of translations. In reality, for arbitrary bodies it is hard to prove (if it is true at all) that for close points on the sphere the orbits intersect. One obstacle stems from the fact that the orbit map $\mathcal{O}_\sigma : \sigma^\perp \cap S^{n-1} \to S^{n-1}$, $\mathcal{O}_\sigma(\xi) = \varphi_\xi(\sigma)$ is not well-defined (we could have several rotations and translations in one subspace $\xi^\perp$, $\xi \in (\sigma^\perp \cap S^{n-1})$). The second obstacle is that even if $\mathcal{O}_\sigma$ is well-defined, it is not, in general, continuous.

On the other hand, one can gain additional information about projections on subspaces $\xi^\perp$ for which $\mathcal{O}_\sigma(\xi)$ is not well-defined or is discontinuous. Indeed, by simple considerations (see [14], Lemma 2.1.1, p. 15) one can show that in both cases we have two conditions on projections onto $\xi^\perp$, i.e., $\forall \theta \in (S^{n-1} \cap \xi^\perp)$ we have

$$h_{K|\xi^\perp}(\varphi_\xi^{-1}(\theta)) + a_\xi \cdot \theta = h_{L|\xi^\perp}(\theta), \qquad h_{K|\xi^\perp}(\psi_\xi^{-1}(\theta)) + b_\xi \cdot \theta = h_{L|\xi^\perp}(\theta)$$

for $\psi_\xi \neq \varphi_\xi$. In other words,

$$h_{K|\xi^\perp}(\chi_\xi^{-1}(\theta)) + c_\xi \cdot \theta = h_{K|\xi^\perp}(\theta), \qquad \forall \theta \in (S^{n-1} \cap \xi^\perp),$$

with $\chi_\xi = \psi_\xi \circ \varphi_\xi^{-1}$, $c_\xi = a_\xi - \psi(b_\xi)$. This means that $K|\xi^\perp$ (and $L|\xi^\perp$) have an $SO(n-1)$-symmetry (here we say that a set $A \subset \mathbb{R}^n$, $n \geq 2$, has an $SO(n)$-symmetry if there exists a non-trivial rotation $\varphi \in SO(n)$ such that $\varphi(A)$ is a translate of A).

A general hope is that if the projections $K|\xi^\perp$ have a symmetry for all ξ belonging to an open set in S^{n-1}, then $K|\xi^\perp$ must degenerate into Euclidean $(n-1)$-dimensional balls. This will eliminate the rotations, reducing the problem to translations only.

Using these ideas one can get, in particular, an affirmative answer to Problem 1 for polytopes [15]. We refer the reader to the book of Vladimir Golubaytnkov for several results related to Problem 1, obtained using this method, [14]; see also Sect. 3.5 below.

To use the orbit approach in order to attack Problem 2 is harder, for the radial function does not behave well under translations (see [9] for some results, obtained by the above approach).

3.3 One Body, A Rotational Symmetry

To simplify the orbit motion, one can look at the symmetrals of K and L. This helps to "separate" translations and rotations. Using (2) for K and $-K$ we have

$$h_{(K-K)|\xi^\perp}(\varphi_\xi^{-1}(\theta)) = h_{(L-L)|\xi^\perp}(\theta) \quad \forall \theta \in \xi^\perp \quad \forall \xi \in S^{n-1}. \tag{3}$$

Applying the Funk transform on the sphere to h_{K-K} and h_{L-L}, and repeating the argument used in Sect. 3.1, we see that $K - K = L - L$. This shows that any time we have a non-trivial rotation in the case of congruence of projections of bodies onto

$\xi^\perp$, the projections of their symmetrals should have the corresponding rotational symmetry. In other words,

$$h_{(K-K)|\xi^\perp}(\varphi_\xi^{-1}(\theta)) = h_{(K-K)|\xi^\perp}(\theta) \quad \forall\theta \in \xi^\perp, \tag{4}$$

and a similar equality is valid for $L - L$. Thus, Problem 1 about *two* bodies leads naturally to a problem about *one* origin-symmetric body.

Problem 4 *Let K be an origin-symmetric convex body in $\mathbb{R}^n$, $n \geq 3$, such that for every $\xi \in S^{n-1}$, the projection $K|\xi^\perp$ has the symmetry of a subgroup $G_\xi \subset SO(n-1, \xi^\perp)$, where for n odd G_ξ is not a reflection in the origin. Does it follow that K is an Euclidean ball?*

In the three-dimensional case the problem has an affirmative answer. This could be proved by the methods that are similar to those in [11]. It is open for $n \geq 4$.

To simplify Problem 4 one can *fix* the symmetry in every subspace. Consider, for example, a case of symmetries of the cube. We say that a set $A \subset \mathbb{R}^n$, $n \geq 3$, has a symmetry of a cube, if there are n pairwise orthogonal axes in $\mathbb{R}^n$ such that A is invariant under any rotation by the angle $\pi/2$ around any of these axes, i.e., if A is invariant under the finite subgroup of $SO(n)$ of the symmetries of a cube.

Problem 5 *Let K be an origin-symmetric convex body in $\mathbb{R}^n$, $n \geq 4$, such that for every $\xi \in S^{n-1}$, the projection $K|\xi^\perp$ has the symmetry of a cube. Does it follow that K is an Euclidean ball?*

If the body K is in $\mathbb{R}^3$, the analogus question is not difficult. Indeed, let K have plane projections with square symmetries (here we say that a plane centered set A has a square symmetry if $\varphi^{\pi/2}(A) = A$ with $\varphi^{\pi/2}$ being a rotation by $\pi/2$). Fix any point $\sigma \in S^2$. Then the orbit of σ satisfies

$$\Gamma(\sigma) = \bigcup_{\{\xi\in(\sigma^\perp\cap S^2)\}} \varphi_\xi^{\pi/2}(\sigma) = (\sigma^\perp \cap S^2). \tag{5}$$

Since any two great circles of S^2 intersect, we see that h_K must be identically constant, and we are done.

The situation is much more complicated for $K \subset \mathbb{R}^n$, $n \geq 4$, since we do not have any information about invariant subspaces of a rotation in the corresponding subspace $\xi^\perp$. If, say, $n = 4$, to follow the orbit of a fixed point one has, probably, to consider all locally-continuous vector fields on S^3, generated by the axes of rotations, and the problem becomes quite hard.

There is nothing special about the cubic symmetry, and one can ask questions about any fixed symmetry related to a subgroup of $SO(n-1)$. One can also ask similar questions about sections of origin-symmetric star-bodies. Here the situation is the same as for projections: using the ideas from [11], in the three-dimensional case the problem can be answered affirmatively; it is open for $n \geq 4$.

3.4 One Body, A Direct Rigid Motion Symmetry

It is interesting to know what happens if the body is not symmetric, and projections have a *direct rigid motion* symmetry.

Problem 6 *Let K be a convex body in $\mathbb{R}^n$, $n \geq 3$, such that for every $\xi \in S^{n-1}$, the projection $K|\xi^\perp$ has a fixed direct rigid motion symmetry, which is different from reflection in the origin for n odd. Does it follow that K is an Euclidean ball?*

Here we say that a set $D \subset \xi^\perp$ has a direct rigid motion symmetry if $\varphi(D) = D + a$ for some vector $a \in \xi^\perp$ and some non-trivial rotation $\varphi \in SO(n-1, \xi^\perp)$.

It seems that convexity plays no role in Problem 6, and we could reformulate it for an arbitrary continuous function on S^{n-1}.

Problem 7 *Let G be a fixed subgroup of $SO(n-1)$, which is different from the subgroup of reflections in the origin for odd n, and let f be a continuous function on S^{n-1}. Assume that $\forall \xi \in S^{n-1}$ and $\forall \varphi_\xi \in G_\xi \subset SO(n-1, \xi^\perp)$ there exist $a_\xi \in \xi^\perp$ such that*

$$f(\varphi_\xi^{-1}(\theta)) + a_\xi \cdot \theta = f(\theta), \qquad \forall \theta \in (S^{n-1} \cap \xi^\perp). \tag{6}$$

Does it follow that $f(\theta) = const + b \cdot \theta$ for some $b \in \mathbb{R}^n$?

The idea of Fedor Nazarov is to use Harmonic Analysis to separate translations from rotations. We will show how this idea works in the three-dimensional case, with f restricted to $\xi^\perp$ having the direct rigid motion symmetry of the square for all $\xi \in S^2$, i.e., with f satisfying (6) for $n = 3$ and φ_ξ^{-1} being a rotation by $\pi/2$ for all $\xi \in S^2$.

Without loss of generality, we can assume that f is odd. Indeed, looking at the "even" part of (6), we have $f_e(\varphi_\xi^{-1}(\theta)) = f_e(\theta)$, $\forall \theta \in (S^2 \cap \xi^\perp)$ $\forall \xi \in S^2$. Using (5) we see that f_e is identically constant.

Parametrizing a large circle $\xi^\perp \cap S^2$ as $(\cos s, \sin s, 0)$, we rewrite (6) as

$$f_o(s + \frac{\pi}{2}) + (a_1, a_2) \cdot (\cos s, \sin s) = f_o(s), \qquad \forall s \in [0, 2\pi]. \tag{7}$$

It is enough to show that if the odd part of f satisfies (6) (or (7)), then it must be linear. To "separate" translations from rotations we look at the Fourier coefficients of both parts of (7). The point is that the linear term $a_1 \cos s + a_2 \sin s$ has only two non-trivial Fourier coefficients. Hence, taking the Fourier coefficients of both parts of (7) we have

$$(1 - e^{\frac{\pi}{2} in})\widehat{f_o}(n) = 0 \qquad \forall n \neq \pm 1,$$

where

$$\widehat{f}(n) = \frac{1}{2\pi} \int_0^{2\pi} f(s) e^{-ins} ds, \qquad n \in \mathbb{Z}.$$

In other words, $\widehat{f_o}(n) = 0$ for all odd $n \in \mathbb{Z}$, $n \neq \pm 1$. On the other hand, in our parametrization the oddness of f_o on $\xi^\perp \cap S^2$ is expressed as $f_o(s+\pi) = -f_o(s)$ $\forall s \in [0, 2\pi]$, and we have

$$(1 + e^{\pi i n})\widehat{f_o}(n) = 0 \qquad \forall n \neq \pm 1.$$

Hence, $\widehat{f_o}(n) = 0$ for all even $n \in \mathbb{Z}$. We see that the Fourier coefficients of the restriction of f_o onto every $\xi^\perp$ vanish, except $n = \pm 1$. In other words, the restrictions $f_o|_{\xi^\perp}$ are linear for all $\xi \in S^2$. If the restriction of a continuous function onto every subspace is linear, then the function must be linear, [7]. Thus, there exists $b \in \mathbb{R}^3$ such that $f_o(\theta) = b \cdot \theta \; \forall \theta \in S^2$. Finally, we obtain

$$f(\theta) = f_e(\theta) + f_o(\theta) = const + b \cdot \theta \qquad \forall \theta \in S^2,$$

which gives the desired result in the particular case of the square direct rigid motion symmetry.

The above problem is open for $n \geq 4$ for any fixed direct rigid motion symmetry. We conclude this section with

Problem 8 *Let K be a star-shaped body with respect to the origin in $\mathbb{R}^n$, and let $n \geq 3$. Assume that for every $\xi \in S^{n-1}$, the section $K \cap \xi^\perp$ has a fixed direct rigid motion symmetry, which is different from the reflection in the origin for n odd. Does it follow that K is an Euclidean ball?*

3.5 Main Results

Vladimir Golubyatnikov proved the following theorem (see Theorem 2.1.1, [14], p. 13).

Theorem 1 *Let K and L be two convex bodies in $\mathbb{R}^3$ such that their projections $K|\xi^\perp$, $L|\xi^\perp$ onto every subspace $\xi^\perp$ are directly congruent, and have no $SO(2)$-symmetries. Then K is a translate of $\pm L$.*

The assumption on bodies having no projections with $SO(2)$-symmetries yields the continuity of the map $\varphi : S^2 \to S^1$, defined as $\varphi(\xi) = \varphi_\xi$, where φ_ξ is the smallest (in absolute value) angle of rotation such that (2) holds with $n = 2$, (cf. [14], Lemma 2.1.1, p. 15).

Recently, Myroshnychenko [16] observed that the convexity assumption in Golubyatnikov's proof could be dispensed with, and it can be generalized to the case of hedgehogs, [17]. If we "separate" translations from rotations, it looks like the convexity assumption could also be dispensed with; the next question of Richard Gardner and Vladimir Golubyatnikov [14] is very interesting.

Problem 9 *Let f and g be two continuous functions on S^{n-1} such that for every $\xi \in S^{n-1}$ there exists a rotation $\varphi_\xi \in SO(n-1, \xi^\perp)$ verifying*

$$f(\varphi_\xi^{-1}(\theta)) = g(\theta), \qquad \forall \theta \in (S^{n-1} \cap \xi^\perp). \tag{8}$$

Does it follow that $f = g$ or $f(\theta) = g(-\theta)$ $\forall \theta \in S^{n-1}$?

If $n = 3$ one can answer this question [11] using ideas of Golubyatnikov [14], and of Schneider [4]. The case $n \geq 4$ are open. One can show that Problem 9 can be reduced to a "*local* question" about one function.

Problem 10 *Let G be a fixed subgroup of $SO(n-1)$, $n \geq 4$, which is different from reflection in the origin for n odd, and let f be a continuous function on $U^\perp$ for some spherical cap $U \subset S^{n-1}$. Assume that $\forall \xi \in U$ and $\forall \varphi_\xi \in G_\xi \subset SO(n-1, \xi^\perp)$ we have $f(\varphi_\xi^{-1}(\theta)) = f(\theta)$ $\forall \theta \in (S^{n-1} \cap \xi^\perp)$. Does it follow that $f \equiv const$ on $U^\perp$?*

Golubyatnikov [14] also obtained several interesting results related to Problem 1 in the case $k = 3$ ([14], Theorem 2.1.1, p. 13; Theorem 3.2.1, p. 48). Following the ideas from [14] and [11] one can obtain several Hadwiger-type results related to both Problems 1 and 2 in the case $k = 3$ [9]. In order to formulate one of these results we introduce some notation and definitions.

We denote by $d_K(\zeta)$ the diameter of a convex body K, which is parallel to the direction $\zeta \in S^{n-1}$. We also denote by $\mathcal{O} = \mathcal{O}_\zeta \in O(n)$ the orthogonal transformation satisfying $\mathcal{O}|_{\zeta^\perp} = -I|_{\zeta^\perp}$, and $\mathcal{O}(\zeta) = \zeta$.

In the case when D is a subset of a 3−dimensional subspace H, and $\xi \in (H \cap S^{n-1})$, $n \geq 4$, we say that D has a $(\xi, \alpha\pi)$-symmetry if $\varphi(D) = D + a$ for some vector $a \in H$ and some rotation $\varphi \in SO(3, H)$ by the angle $\alpha\pi$, $\alpha \in (0, 2)$, satisfying $\varphi(\xi) = \xi$. If, in particular, the angle of rotation is π, we say that D has a (ξ, π)-symmetry. We say that a set $D \subset H$ has a rigid motion symmetry if $\varphi(D) = D + a$ for some vector $a \in \xi^\perp$ and some non-identical orthogonal transformation $\varphi \in O(3, H)$.

Theorem 2 *Let K and L be two convex bodies in $\mathbb{R}^4$ having countably many diameters. Assume that there exists a diameter $d_K(\zeta)$, such that the "side" projections $K|w^\perp$, $L|w^\perp$ onto all subspaces $w^\perp$ containing ζ are directly congruent, see Fig. 1. Assume also that these projections have no (ζ, π)-symmetries and no (u, π)-symmetries for any $u \in (\zeta^\perp \cap w^\perp \cap S^3)$. Then $K = L + b$ or $K = \mathcal{O}L + b$ for some $b \in \mathbb{R}^4$.*

If, in addition, the "ground" projections $K|\zeta^\perp$, $L|\zeta^\perp$, are directly congruent and do not have rigid motion symmetries, then $K = L + b$ for some $b \in \mathbb{R}^4$.

One can obtain a similar result for sections of star-shaped bodies [9]. We are not aware of any results for $k \geq 4$.

It is interesting to note that the proof of the above statement is based on a certain analytic result with no convexity assumption (see Proposition 1, [9]). Taking into account some of the results mentioned above, it makes sense to reformulate Problem 1 in terms of continuous functions, satisfying no convexity conditions.

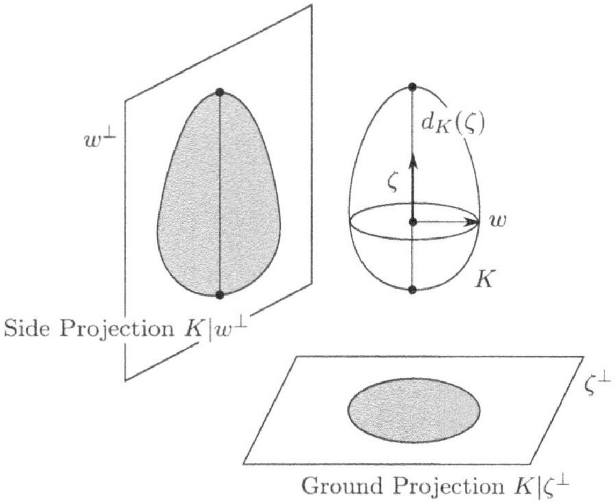

Fig. 1 Diameter $d_K(\zeta)$, side projection $K|w^\perp$ and ground projection $K|\zeta^\perp$

Problem 11 *Let f and g be two continuous functions on $\mathbb{R}^n$, $n \geq 2$, homogeneous of degree 1. Assume that for every subspace H of dimension k, $1 \leq k \leq n-1$, there exists an orthogonal transformation $\psi_H \in O(n, H)$ and a vector $a_H \in H$ such that*

$$f(\psi_H(x)) + a_H \cdot x = g(x) \qquad \forall x \in H. \tag{9}$$

Does it follow that $f(x) + b \cdot x = g(x)$ or $f(x) + b \cdot x = g(-x)$ for some $b \in \mathbb{R}^n$ and all $x \in \mathbb{R}^n$?

It is not clear how to formulate a functional analogue of Problem 2. The point is that the sections of a star body are usually described using the radial function, which does not behave well under translations.

4 Other Groups of Symmetries

4.1 Adding Reflections, Symmetries of $O(n)$

The plausible analogues of Problems 4, 6, 7, and 10 with $O(n-1)$ instead of $SO(n-1)$ seem to be more difficult. The most reasonable conjectures related to these analogues would be that the resulting bodies are bodies of revolution or ellipsoids. This is due to the fact that (unlike in the case of $SO(n-1)$, where an Euclidean ball seems to be the only body with projections/sections having a fixed rotational symmetry) there are many bodies (for example, ellipsoids or bodies of revolution) whose projections/sections have an axis of symmetry.

The next question of Karoly Bezdek seems to be open for a long time.

Problem 12 *Let K be a convex body in $\mathbb{R}^3$ such that every section of K has an axis of symmetry, i.e., for every $(t,\xi) \in \mathbb{R} \times S^2$ satisfying $(int(K) \cap (\xi^\perp + t\xi)) \neq \emptyset$ there exists a reflection $\psi_{t,\xi} \in O(2, \xi^\perp + t\xi)$ with respect to a line $l(t,\xi) \in (\xi^\perp + t\xi)$ such that*

$$\psi_{t,\xi}(K \cap (\xi^\perp + t\xi)) = K \cap (\xi^\perp + t\xi).$$

Does it follow that K is an ellipsoid or a body of revolution?

L. Montejano obtained some partial results related to Problem 12 [18]; in this connection see also [19].

4.2 Groups of Symmetries Containing $O(n)$

The next beautiful statement by Petty and MacKenney [20] shows that for pairs of bodies the results are, in general, negative.

Theorem 3 *There exist two different origin-symmetric bodies of revolution K and L in $\mathbb{R}^3$ such that for every $\xi \in S^2$ there exists a rotation φ_ξ by $\pi/2$ and a dilation $\lambda_\xi > 0$ such that*

$$\lambda_\xi(\varphi_\xi^{\pi/2}(K|\xi^\perp)) = L|\xi^\perp.$$

The corresponding result for sections can be obtained by duality.

However, the situation is different for one body.

Problem 13 *Let K be a convex origin-symmetric body in $\mathbb{R}^n$, $n \geq 3$, such that for any pair of $k-$dimensional subspaces H and G, $2 \leq k \leq n-1$, there exists a linear transformation $A_{H,G} \in GL(n)$ such that*

$$A_{H,G}(K \cap H) = K \cap G.$$

Does it follow that K is an ellipsoid?

This question of Banach goes back to 1930s and is not yet completely solved, one of the simplest open cases is $n = 4, k = 3$. An interested reader should consult the paper of Pelczyński [21] for information about this problem, see also ([3], pp. 128, 290). One has also to mention the results of R. Schneider, who proved that an Euclidean ball is the only convex body such that all its sections are congruent to each other [4].

5 Concluding Remarks

Most of the problems considered in this article could be formulated over the field of complex numbers. As far as we know, all of them are open except the results of

Gromov [22] related to Problem 13, and some partial results of V. Golubyatnikov related to Problem 1 ([14], Theorem 3.3.1, p. 53).

It would be interesting to formulate and solve analogous problems in spherical or hyperbolic geometry. Our list of references is very incomplete and we strongly recommend the reader, interested in these problems, to consult the books of Gardner [3] and Schneider [4].

References

1. S. Nakajima, Eine Kennzeichnung homothetischer Eiflächen. Tohoku Math. J. **35**, 285–286 (1932)
2. W. Süss, Zusammensetzung von Eikörpen und homothetische Eiflächen. Tohoku Math. J. **35**, 47–50 (1932)
3. R.J. Gardner, *Geometric Tomography*, 2nd edn. Encyclopedia of Mathematics and its Applications, vol. 58 (Cambridge University Press, Cambridge, 2006)
4. R. Schneider, Convex bodies with congruent sections. Bull. Lond. Math. Soc. **312**, 52–54 (1980)
5. R. Schneider, *Convex Bodies: The Brunn-Minkowski Theory*. Encyclopedia of Mathematics and its Applications, vol. 44 (Cambridge University Press, Cambridge, 1993)
6. A.D. Alexandrov, On the theory of mixed volumes of convex bodies. II. New inequalities between mixed volumes and their applications. Mat. Sbornik **2**, 1205–1238 (1937). (In Russian)
7. D. Ryabogin, On a Lemma of Nakajima and Süss. AMM **122**(9), 890–892 (2015)
8. H. Hadwiger, Seitenrisse konvexer Körper und Homothetie. Elem. Math. **18**, 97–8 (1963)
9. M.A. Alfonseca, M. Cordier, D. Ryabogin, On bodies with congruent projections. Isr. J. Math. **215**(2), 765–799 (2016)
10. S. Helgason, *Radon Transform*, 2nd edn. (Birkhäuser, 1999)
11. D. Ryabogin, On the continual Rubik's cube. Adv. Math. **231**, 3429–3444 (2012)
12. K.J. Falconer, Applications of a result on spherical integration to the theory of convex sets. Am. Math. Mon. **90**, 690–693 (1983)
13. T. Bonnesen, W. Fenchel, *Theory of Convex Bodies*, Moscow, Idaho, 1987
14. V.P. Golubyatnikov, *Uniqueness questions in reconstruction of multidimensional objects from tomography type projection data, Inverse and Ill-posed problems series*, Utrecht-Boston-Koln-Tokyo, 2000
15. S. Myroshnychenko, D. Ryabogin, *On Polytopes with Congruent Projections*, Preprint
16. S. Myroshnychenko, On a functional equation related to hedgehogs with congruent projections. JMAA **445**(2), 1492–1504 (2017)
17. Y. Martinez-Maure, *Hedgehogs*, Banff's lecture 2014, personal communication
18. L. Montejano, Two applications of topology to convex geometry. Proc. Steklov Inst. Math. **247**, 182–185 (2004)
19. L. Montejano, Recognizing sets by means of some of their sections. Manuscripta Math. **76**, 227–239 (1992)
20. C.M. Petty, J.R. McKinney, Convex bodies with circumscribing boxes of constant volume. Portugal Math. **44**, 447–455 (1987)
21. A. Pelczyński, On some problems of Banach. Russ. Math. Surv. **28**(6), 67–75 (1973). (In Russain)
22. M. Gromov, On one geometric hypothesis of Banach. Izv. AN SSSR, **31**(5), 1105–1114 (1967) (In Russian)

Regular Incidence Complexes, Polytopes, and C-Groups

Egon Schulte

Abstract Regular incidence complexes are combinatorial incidence structures generalizing regular convex polytopes, regular complex polytopes, various types of incidence geometries, and many other highly symmetric objects. The special case of abstract regular polytopes has been well-studied. The paper describes the combinatorial structure of a regular incidence complex in terms of a system of distinguished generating subgroups of its automorphism group or a flag-transitive subgroup. Then the groups admitting a flag-transitive action on an incidence complex are characterized as generalized string C-groups. Further, extensions of regular incidence complexes are studied, and certain incidence complexes particularly close to abstract polytopes, called abstract polytope complexes, are investigated.

Keywords Abstract polytope · Regular polytope · C-group · Incidence geometries

MSC 2010 Primary: 51M20 · Secondary: 52B15 · 51E24

1 Introduction

Regular incidence complexes are combinatorial incidence structures with very high combinatorial symmetry. The concept was introduced by Danzer [12, 13] building on Grünbaum's [17] notion of a polystroma. Regular incidence complexes generalize regular convex polytopes [7], regular complex polytopes [8, 42], various types of incidence geometries [4, 5, 21, 44], and many other highly symmetric objects. The terminology and notation is patterned after convex polytopes [16] and was ultimately inspired by Coxeter's work on regular figures [7, 8]. The first systematic study of incidence complexes from the discrete geometry perspective occurred in [33] and the related publications [13, 34–36].

E. Schulte (✉)
Department of Mathematics, Northeastern University, Boston, MA 02115, USA
e-mail: e.schulte@northeastern.edu

M. D. E. Conder et al. (eds.), *Discrete Geometry and Symmetry*,
Springer Proceedings in Mathematics & Statistics 234,
https://doi.org/10.1007/978-3-319-78434-2_18

The special case of abstract polytopes has recently attracted a lot of attention (see [23]). Abstract polytopes (or incidence polytopes, as they were called originally) are incidence complexes close to ordinary polytopes and are in a sense topologically real.

Incidence complexes can also be viewed as incidence geometries or diagram geometries with a linear diagram (see [4, 5, 21, 44]), although here we study these structures from the somewhat different discrete geometric and combinatorial perspective of polytopes and ranked partially ordered sets.

The present paper is organized as follows. In Sect. 2, we introduce incidence complexes following the original definition of [13] with minor amendments. Then in Sects. 3 and 4 we derive structure results for flag-transitive subgroups of regular incidence complexes and characterize these groups as what we will call here generalized C-groups, basically following [33, 34] (apart from minor changes inspired by [23]). Section 5 explains how abstract regular polytopes fit into the more general framework of regular incidence complexes. In Sect. 6 we discuss extensions of regular incidence complexes. Section 7 is devoted to the study of abstract polytope complexes, a particularly interesting class of regular incidence complexes which are not abstract polytopes but still relatively close to abstract polytopes. This section also describes a number of open research problems. Finally, Sect. 8 collects historical notes on incidence complexes and some personal notes related to the author's work.

2 Incidence Complexes

Following [13, 33], an *incidence complex* $\mathcal{K}$ *of rank* n, or simply an n*-complex*, is a partially ordered set (poset), with elements called *faces*, which has the properties (I1), ..., (I4) described below.

(I1) $\mathcal{K}$ has a least face F_{-1} and a greatest face F_n, called the *improper* faces. All other faces of $\mathcal{K}$ are *proper* faces of $\mathcal{K}$.

(I2) Every totally ordered subset, or *chain*, of $\mathcal{K}$ is contained in a (maximal) totally ordered subset of $\mathcal{K}$ with exactly $n+2$ elements, called a *flag* of $\mathcal{K}$.

The conditions (I1) and (I2) make $\mathcal{K}$ into a *ranked* partially ordered set with a strictly monotone rank function with range $\{-1, 0, \dots, n\}$. A face of rank i is called an i*-face*. A face of rank 0, 1 or $n-1$ is also called a *vertex*, an *edge* or a *facet*, respectively. The faces of $\mathcal{K}$ of ranks -1 and n are F_{-1} and F_n, respectively. The *type* of a chain of $\mathcal{K}$ is the set of ranks of faces in the chain. Thus each flag has type $\{-1, 0, \dots, n\}$; that is, each flag Φ of $\mathcal{K}$ contains a face of $\mathcal{K}$ of each rank i with $i = -1, 0, \dots, n$.

For an i-face F and a j-face G of $\mathcal{K}$ with $F \leq G$ we call

$$G/F := \{H \in \mathcal{K} \mid F \leq H \leq G\}$$

a *section* of $\mathcal{K}$. This will be an incidence complex in its own right, of rank $j - i - 1$. Usually we identify a j-face G of $\mathcal{K}$ with the j-complex G/F_{-1}. Likewise, if F is an i-face, the $(n - i - 1)$-complex F_n/F is called the *co-face* of F in $\mathcal{K}$, or the *vertex-figure* at F if F is a vertex.

A partially ordered set $\mathcal{K}$ with properties (I1) and (I2) is said to be *connected* if either $n \leq 1$, or $n \geq 2$ and for any two proper faces F and G of $\mathcal{K}$ there exists a finite sequence of proper faces $F = H_0, H_1, \ldots, H_{k-1}, H_k = G$ of $\mathcal{K}$ such that H_{j-1} and H_j are incident for $j = 1, \ldots, k$. We say that $\mathcal{K}$ is *strongly connected* if each section of $\mathcal{K}$ (including $\mathcal{K}$ itself) is connected.

(I3) $\mathcal{K}$ is *strongly connected.*

(I4) For each $i = 0, 1, \ldots, n - 1$, if F and G are incident faces of $\mathcal{K}$, of ranks $i - 1$ and $i + 1$ respectively, then there are *at least two* i-faces H of $\mathcal{K}$ such that $F < H < G$.

Thus, an n-complex $\mathcal{K}$ is a partially ordered set with properties (I1),...,(I4).

An *abstract n-polytope*, or briefly *n-polytope*, is an incidence complex of rank n satisfying the following condition (I4P), which is stronger than (I4):

(I4P) For each $i = 0, 1, \ldots, n - 1$, if F and G are incident faces of $\mathcal{K}$, of ranks $i - 1$ and $i + 1$ respectively, then there are *exactly two* i-faces H of $\mathcal{K}$ such that $F < H < G$.

We call two flags of $\mathcal{K}$ *adjacent* if one differs from the other in exactly one face; if this face has rank i, with $i = 0, \ldots, n - 1$, the two flags are *i-adjacent*. Then the conditions (I4) and (I4P) are saying that each flag has at least one or exactly one i-adjacent flag for each i, respectively. We refer to (I4P) as the *diamond condition* (for polytopes).

Though the above definitions of connectedness and strong connectedness are satisfactory from an intuitive point of view, in practice the following equivalent definitions in terms of flags are more useful.

A partially ordered set $\mathcal{K}$ with properties (I1) and (I2) is called *flag-connected* if any two flags Φ and Ψ of $\mathcal{K}$ can be joined by a sequence of flags $\Phi = \Phi_0, \Phi_1, \ldots, \Phi_{k-1}, \Phi_k = \Psi$ such that successive flags are adjacent. Further, $\mathcal{K}$ is said to be *strongly flag-connected* if each section of $\mathcal{K}$ (including $\mathcal{K}$ itself) is flag-connected. It can be shown that $\mathcal{K}$ is strongly flag-connected if and only if any two flags Φ and Ψ of $\mathcal{K}$ can be joined by a sequence of flags $\Phi = \Phi_0, \Phi_1, \ldots, \Phi_{k-1}, \Phi_k = \Psi$, all containing $\Phi \cap \Psi$, such that successive flags are adjacent.

It turns out that a partially ordered set $\mathcal{K}$ with properties (I1) and (I2) is strongly flag-connected if and only if $\mathcal{K}$ is strongly connected. Thus in place of (I3) we could have required the following equivalent condition

(I3′) $\mathcal{K}$ is *strongly flag-connected.*

A bijection $\varphi : \mathcal{K} \to \mathcal{L}$ from a complex $\mathcal{K}$ to a complex $\mathcal{L}$ is called an *isomorphism* if φ is order-preserving (in both directions); that is, $F \leq G$ in $\mathcal{K}$ if and only if $F\varphi \leq G\varphi$ in $\mathcal{L}$. An *automorphism* of a complex $\mathcal{K}$ is an isomorphism from $\mathcal{K}$ to itself. The group of all automorphisms $\Gamma(\mathcal{K})$ of a complex $\mathcal{K}$ is called the *automorphism group* of $\mathcal{K}$.

A complex $\mathcal{K}$ is said to be *regular* if $\Gamma(\mathcal{K})$ is transitive on the flags of $\mathcal{K}$. The automorphism group of a regular complex may or may not be simply transitive on the flags. However, if $\mathcal{K}$ is a polytope then $\Gamma(\mathcal{K})$ is simply transitive on the flags.

Lemma 2.1 *Let $\mathcal{K}$ be a regular n-complex. Then all sections of $\mathcal{K}$ are regular complexes, and any two sections which are defined by faces of the same ranks are isomorphic. In particular, $\mathcal{K}$ has isomorphic facets and isomorphic vertex-figures.*

Proof Let F be an i-face and G a j-face of $\mathcal{K}$ with $F < G$. Let Ω denote a chain of $\mathcal{K}$ of type $\{-1, 0, \ldots, i-1, i, j, j+1, \ldots, n\}$ containing F and G. Now consider the action of the stabilizer of Ω in $\Gamma(\mathcal{K})$ induced on the section G/F of $\mathcal{K}$. Since $\mathcal{K}$ is a regular complex and the flags of G/F are just the restrictions of flags of $\mathcal{K}$ to G/F, this stabilizer is a group that acts flag-transitively on G/F (but not necessarily faithfully). Thus the $(j-i-1)$-complex G/F is regular and its automorphism group $\Gamma(G/F)$ contains, as a flag-transitive subgroup, a quotient of the stabilizer of Ω in $\Gamma(\mathcal{K})$. (Unlike for regular polytopes this quotient may be proper.)

Let F' and G' be another pair of i-face and j-face with $F' < G'$, and let Ψ' be a flag of $\mathcal{K}$ containing F' and G'. Then each automorphism of $\mathcal{K}$ mapping Ψ to Ψ' induces an isomorphism from G/F to G'/F'. Thus G'/F' is isomorphic to G/F. ■

3 Flag-Transitive Subgroups of the Automorphism Group

In this section we establish structure results for flag-transitive subgroups Γ of the automorphism group $\Gamma(\mathcal{K})$ of a regular complex $\mathcal{K}$. We follow [34, Sect. 2] (and [33]) and show that any such group (including $\Gamma(\mathcal{K})$ itself) has a distinguished system of generating subgroups obtained as follows. For corresponding results for regular polytopes see [23, Chap. 2B].

Throughout this section let $\mathcal{K}$ be a regular n-complex, with $n \geq 1$, and let Γ be a flag-transitive subgroup of $\Gamma(\mathcal{K})$. Define $N := \{-1, 0, \ldots, n\}$ and for $J \subseteq N$ set $\overline{J} := N \setminus J$. Let $\Phi := \{F_{-1}, F_0, \ldots, F_n\}$ be a fixed, or *base flag*, of $\mathcal{K}$, where F_i designates the i-face in Φ for each $i \in N$. For each $\Omega \subseteq \Phi$ let Γ_Ω denote the stabilizer of Ω in Γ. In particular, Γ_Φ is the stabilizer of the base flag Φ, and $\Gamma_\emptyset = \Gamma$. For $i \in N$ define the subgroup R_i of Γ as

$$R_i := \Gamma_{\Phi \setminus \{F_i\}} = \langle \varphi \in \Gamma \mid F_j \varphi = F_j \text{ for all } j \neq i \rangle. \tag{1}$$

Then each R_i contains Γ_Φ as a subgroup, and $R_{-1} = \Gamma_\Phi = R_n$.

For $i = 0, \ldots, n-1$ let k_i denote the number of i-faces of $\mathcal{K}$ in a section G/F, where F is an $(i-1)$-face and G an $(i+1)$-face with $F < G$; since $\mathcal{K}$ is regular, this number is independent of the choice of F and G. Then

$$k_i = |R_i : R_{-1}| \quad (i = 0, \ldots, n-1). \tag{2}$$

Note that each flag of $\mathcal{K}$ has exactly $k_i - 1$ flags i-adjacent to it for each i.

If $\mathcal{K}$ is a regular polytope then R_i is generated by an involution ρ_i for $i = 0, \ldots, n-1$, and the subgroups R_{-1} and R_n are trivial. In this case $k_i = 2$ for each i.

Our first goal is to describe the stabilizers of the subchains of the base flag Φ. For $J \subseteq N$ set $\Phi_J := \{F_j \in \Phi \mid j \in J\}$.

Lemma 3.1 *For $J \subseteq N$ we have $\Gamma_{\Phi_J} = \langle R_j \mid j \in \overline{J} \rangle$.*

Proof Let $\Lambda := \langle R_j \mid j \in \overline{J} \rangle$. It is clear that Λ is a subgroup of Γ_{Φ_J}, since each subgroup R_j with $j \in \overline{J}$ stabilizes Φ_J. To prove equality of the two groups, note first that Γ_{Φ_J} acts transitively on the set of all flags Ψ of $\mathcal{K}$ with $\Phi_J \subseteq \Psi$. Hence, since the base flag stabilizer Γ_Φ lies in Γ_{Φ_J}, it suffices to show that Λ also acts transitively on these flags.

Let Ψ be a flag with $\Phi_J \subseteq \Psi$. We show that Ψ lies in the orbit of Φ under Λ. Choose a sequence of flags

$$\Phi = \Phi_0, \Phi_1, \ldots, \Phi_{k-1}, \Phi_k = \Psi,$$

all containing Φ_J, such that successive flags are adjacent. We proceed by induction on k, the case $k = 0$ being trivial. By the inductive hypothesis, there exists $\psi \in \Lambda$ such that $\Phi\psi = \Phi_{k-1}$. We know that Φ_{k-1} and $\Psi = \Phi_k$ are j-adjacent flags for some j, so $\Phi = \Phi_{k-1}\psi^{-1}$ and $\Phi_k\psi^{-1}$ are also j-adjacent. By the flag-transitivity of Γ there exists an element $\tau \in R_j$ such that $\Phi_k\psi^{-1} = \Phi\tau$ and hence $\Psi = \Phi\tau\psi$. But $j \notin J$, since $\Phi_J \subseteq \Phi_i$ for each i, so $\tau, \psi \in \Lambda$ and hence $\tau\psi \in \Lambda$. Thus Ψ lies in the orbit of Φ under Λ. ∎

As the subgroups R_{-1} and R_n lie in R_j for each j, the previous lemma with $J = \emptyset$ immediately implies

Lemma 3.2 $\Gamma = \langle R_{-1}, R_0, \ldots, R_n \rangle = \langle R_0, \ldots, R_{n-1} \rangle$.

The subgroups $R_{-1}, R_0, \ldots, R_n$ of Γ are called the *distinguished generating subgroups* of Γ (*with respect to* Φ).

For each $I \subseteq N$, $I \neq \emptyset$, define the subgroup $\Gamma_I := \langle R_i \mid i \in I \rangle$. For $I = \emptyset$ we set $\Gamma_\emptyset := R_{-1}$. Then by Lemma 3.1,

$$\Gamma_I = \Gamma_{\{F_i \mid i \in \overline{I}\}} = \Gamma_{\Phi_{\overline{I}}} \quad (I \subseteq N); \tag{3}$$

or equivalently,

$$\Gamma_\Omega = \Gamma_{\{i \mid F_i \notin \Omega\}} \qquad (\Omega \subseteq \Phi). \tag{4}$$

The subgroups Γ_I, with $I \subseteq N$, are called the *distinguished subgroups* of Γ (*with respect to* Φ). Note that the notation $\Gamma_\emptyset$ can have two meanings, namely as Γ_I with $I = \emptyset$, and as Γ_Ω with $\Omega = \emptyset$. The intended meaning should be clear from the context.

The distinguished subgroups satisfy the following important *intersection property* (with respect to the distinguished generating subgroups):

Lemma 3.3 *For $I, J \subseteq N$ we have $\Gamma_I \cap \Gamma_J = \Gamma_{I \cap J}$.*

Proof This follows from the fact that the subgroups involved are stabilizers of subchains of Φ, as expressed in Eq. (3). In fact,

$$\Gamma_I \cap \Gamma_J = \Gamma_{\Phi_{\overline{I}}} \cap \Gamma_{\Phi_{\overline{J}}} = \Gamma_{\Phi_{\overline{I}} \cup \Phi_{\overline{J}}} = \Gamma_{\Phi_{\overline{I \cap J}}} = \Gamma_{I \cap J}.$$

Thus the lemma follows. ■

When $n \geq 2$ we often omit R_{-1} and R_n from the system of generating subgroups and also refer to $R_0, \ldots, R_{n-1}$ as the *distinguished generating subgroups*. In fact, in this case Lemma 3.3 shows that

$$R_{-1} = R_n = R_0 \cap \ldots \cap R_{n-1},$$

so R_{-1} and R_n are completely determined by $R_0, \ldots, R_{n-1}$. However, for the system of generating subgroups to also permit a characterization of the combinatorial structure of $\mathcal{K}$ when $n = 1$, the two subgroups R_{-1} and $R_n = R_1$ (with $R_{-1} = R_1$) must be included in the system; that is, R_{-1} and R_1 are not determined by R_0 alone.

The distinguished generating subgroups have the following commuting properties, which hold at the level of groups, but not generally at the level of elements.

Lemma 3.4 *For $-1 \leq i < j - 1 \leq n - 1$ we have $R_i R_j = \langle R_i, R_j \rangle = R_j R_i$.*

Proof This is trivial when $i = -1$ or $j = n$. Now suppose $-1 < i < j - 1 < n - 1$. Clearly, it suffices to show that $\langle R_i, R_j \rangle = R_i R_j$. Here the inclusion $\supseteq$ is trivial.

To establish the opposite inclusion let $\varphi \in \langle R_i, R_j \rangle$. Then φ fixes F_k for each $k \neq i, j$, since both R_i and R_j fix F_k. Hence, since $i < j - 1$, there exists an element $\psi \in R_i$ such that $F_i \psi = F_i \varphi$. Then $F_k \varphi \psi^{-1} = F_k$ for each $k \neq j$. But then there also exists an element $\tau \in R_j$ such that $F_j \tau = F_j \varphi \psi^{-1}$. Hence $F_k \varphi \psi^{-1} \tau^{-1} = F_k$ for each k, and therefore $\varphi \psi^{-1} \tau^{-1} \in \Gamma_\Phi$. But Γ_Φ is a subgroup of R_j, so we have

$$\varphi \in \Gamma_\Phi \tau \psi \subseteq \Gamma_\Phi R_j R_i = R_j R_i,$$

as required. This completes the proof. ■

The commuting properties of Lemma 3.4 do not generally extend to the case when $j = i + 1$. It can be shown that if $\mathcal{K}$ is a lattice, then

$$R_i R_{i+1} \cap R_{i+1} R_i \; = \; R_i \cup R_{i+1} \quad (i = 0, \ldots, n-2). \tag{5}$$

Recall that a lattice is a partially ordered set in which every two elements have a supremum (a least upper bound) and an infimum (a greatest lower bound) [43].

We introduce some further notation. For each $i \in N$ we write

$$\Gamma_i := \Gamma_{N \setminus \{i\}} = \langle R_j \mid j \neq i \rangle$$

and

$$\begin{aligned}\Gamma_i^- &:= \Gamma_{\{-1,0,\dots,i\}} = \langle R_j \mid j \leq i \rangle, \\ \Gamma_i^+ &:= \Gamma_{\{i,\dots,n\}} \quad = \langle R_j \mid j \geq i \rangle.\end{aligned}$$

Note that $\Gamma_{-1} = \Gamma_n = \Gamma$. As an immediate consequence of the commutation rules of Lemma 3.4 we have

$$\Gamma_i^- \Gamma_j^+ = \Gamma_j^+ \Gamma_i^- \qquad (-1 \leq i < j-1 \leq n-1). \tag{6}$$

Further, for each $i \in N$,

$$\Gamma_i = \Gamma_{i-1}^- \Gamma_{i+1}^+ = \Gamma_{i+1}^+ \Gamma_{i-1}^-. \tag{7}$$

Observe that when $-1 \leq i \leq j \leq n$ the distinguished subgroup $\langle R_{i+1}, \dots, R_{j-1} \rangle$ of Γ acts flag-transitively (but generally not faithfully) on the section F_j/F_i of $\mathcal{K}$ between the base i-face and the base j-face. The quotient of $\langle R_{i+1}, \dots, R_{j-1} \rangle$ defined by the kernel of this action is a (generally proper) flag-transitive subgroup of $\Gamma(F_j/F_i)$. In particular, Γ_{i-1}^- acts flag-transitively on the base i-face F_i/F_{-1} of $\mathcal{K}$, and Γ_{i+1}^+ acts flag-transitively on the co-face F_n/F_i of the base i-face of $\mathcal{K}$.

Our next goal is the characterization of the structure of a regular complex $\mathcal{K}$ in terms of the distinguished generating subgroups $R_{-1}, R_0, \dots, R_n$ of the chosen flag-transitive subgroup Γ of $\Gamma(\mathcal{K})$. By the transitivity properties of Γ we can write each i-face of $\mathcal{K}$ in the form $F_i\varphi$ with $\varphi \in \Gamma$. We begin with a lemma.

Lemma 3.5 *Let $0 \leq i \leq j \leq n-1$, and let G_i be an i-face of $\mathcal{K}$. Then $G_i \leq F_j$ if and only if $G_i = F_i\gamma$ for some $\gamma \in \Gamma_j$.*

Proof If $G_i = F_i\gamma$ with $\gamma \in \Gamma_j$, then $G_i \leq F_j\gamma = F_j$, as claimed. For the converse, let Ψ be any flag of $\mathcal{K}$ such that $\{G_i, F_j\} \subseteq \Psi$. Then, by Lemma 3.1, $F_j \in \Phi \cap \Psi$ implies that $\Psi = \Phi\gamma$ for some $\gamma \in \Gamma_{\{F_j\}} = \Gamma_j$. Thus $G_j = F_j\gamma$, as required. ■

We now have the following characterization of the partial order in $\mathcal{K}$.

Lemma 3.6 *Let $0 \leq i \leq j \leq n-1$, and let $\varphi, \psi \in \Gamma$. Then the following three conditions are equivalent:*

(a) $F_i\varphi \leq F_j\psi$;
(b) $\varphi\psi^{-1} \in \Gamma_{i+1}^+ \Gamma_{j-1}^-$;
(c) $\Gamma_i\varphi \cap \Gamma_j\psi \neq \emptyset$.

Proof We shall prove the equivalence in the form (a) $\Rightarrow$ (c) $\Rightarrow$ (b) $\Rightarrow$ (a).

Assume that (a) holds. Then $F_i\varphi\psi^{-1} \leq F_j$, and thus $F_i\varphi\psi^{-1} = F_i\gamma$ for some $\gamma \in \Gamma_j$ by Lemma 3.5. In turn, this says that $\Gamma_i\gamma \cap \Gamma_j \neq \emptyset$, since γ lies in this intersection. But $(\varphi\psi^{-1})\gamma^{-1} \in \Gamma_{\{F_i\}} = \Gamma_i$, so $\Gamma_i\varphi\psi^{-1} = \Gamma_i\gamma$. Hence $\Gamma_i\varphi \cap \Gamma_j\psi \neq \emptyset$. Thus (c) holds.

If (c) holds, the commuting properties of Lemma 3.4 show that

$$
\begin{aligned}
\varphi\psi^{-1} \in \Gamma_i\Gamma_j &= \Gamma^+_{i+1}\Gamma^-_{i-1}\Gamma^-_{j-1}\Gamma^+_{j+1} \\
&= \Gamma^+_{i+1}\Gamma^-_{j-1}\Gamma^+_{j+1} \\
&= \Gamma^+_{i+1}\Gamma^+_{j+1}\Gamma^-_{j-1} \\
&= \Gamma^+_{i+1}\Gamma^-_{j-1},
\end{aligned} \tag{8}
$$

as required for (b).

Finally, suppose (b) holds. Then $\varphi\psi^{-1} = \alpha\beta$ for some $\alpha \in \Gamma^+_{i+1}$ and $\beta \in \Gamma^-_{j-1}$. We deduce that

$$F_i\varphi\psi^{-1} = F_i\alpha\beta = F_i\beta \leq F_j\beta = F_j,$$

so that $F_i\varphi \leq F_j\psi$, which is (a). This completes the proof. ■

The previous lemma has important consequences. In effect, it says that we may identify a face $F_i\varphi$ of a regular complex $\mathcal{K}$ with the right coset $\Gamma_i\varphi$ of the stabilizer $\Gamma_i = \Gamma_{\{F_i\}} = \langle R_k \mid k \neq i\rangle$ of the base i-face F_i in Γ.

We conclude this section with a remark about the flag stabilizers of arbitrary regular complexes. A priori only little can be said about their structure. However, there are bounds on the prime divisors of the group order. For a regular complex with a finite flag stabilizer $R_{-1} = \Gamma_\Phi$, the prime divisors of the order of R_{-1} are bounded by

$$\max(k_i - 1 \mid 0 \leq i \leq n-1).$$

In fact, an element of R_{-1} of prime order exceeding this number would necessarily have to fix all adjacent flags of a flag that it fixes. But a simple flag connectivity argument shows that in a regular complex only the trivial automorphism can have this property.

4 Regular Complexes from Groups

In the previous section we derived various properties of flag-transitive subgroups of the automorphism groups of regular complexes. In particular, in Lemma 3.6, we proved that the combinatorial structure of a regular n-complex $\mathcal{K}$ can be completely described in terms of the distinguished generating subgroups $R_{-1}, R_0, \ldots, R_n$ of any flag-transitive subgroup Γ of $\Gamma(\mathcal{K})$.

If $\mathcal{K}$ is a regular n-polytope, then $\Gamma(\mathcal{K})$ is simply flag-transitive and hence has no proper flag-transitive subgroup. In this case $\Gamma = \Gamma(\mathcal{K})$, the base flag stabilizer $R_{-1} = R_n = \Gamma_\Phi$ is trivial, and each subgroup R_i (with $i = 0, \ldots, n-1$) is generated by an involutory automorphism ρ_i which maps the base flag Φ to its unique i-adjacent flag. The group of $\mathcal{K}$ is then what is called a *string C-group* (see [23, Chap. 2E]), that is, the *distinguished generators* $\rho_0, \ldots, \rho_{n-1}$ satisfy both the commutativity relations typical of a Coxeter group with a string diagram, and the intersection property of Lemma 3.3, which now takes the form

$$\langle \rho_i \mid i \in I \rangle \cap \langle \rho_i \mid i \in J \rangle = \langle \rho_i \mid i \in I \cap J \rangle \qquad (I, J \subseteq \{0, \ldots, n-1\}). \tag{9}$$

In this section we characterize the groups that can occur as flag-transitive subgroups of the automorphism group of a regular complex as what we will call here generalized string C-groups (with trivial core). As one of the most important consequences of this approach, we may think of regular complexes and corresponding generalized string C-groups as being essentially the same objects. We follow [34, Sect. 3] (and [33]).

Let Γ be a group generated by subgroups $R_{-1}, R_0, \ldots, R_n$, where R_{-1} and R_n are proper subgroups of R_i for each $i = 0, \ldots, n-1$, and $R_n = R_{-1}$; we usually assume that $n \geq 1$. These subgroups are the *distinguished generating subgroups* of Γ, and along with Γ will be kept fixed during this section. As before we set $N := \{-1, 0, \ldots, n\}$. Further, the subgroups $\Gamma_I := \langle R_i \mid i \in I \rangle$ with $I \subseteq N$ are called the *distinguished subgroups* of Γ; here $\Gamma_\emptyset = R_{-1}$. Then Γ is called a *generalized C-group* if Γ has the following *intersection property (with respect to its distinguished generating subgroups)*:

$$\Gamma_I \cap \Gamma_J = \Gamma_{I \cap J} \quad (I, J \subseteq N). \tag{10}$$

It is immediate from the definition that the distinguished subgroups Γ_I are themselves generalized C-groups, with distinguished generating subgroups those R_i with $i \in I \cup \{-1, n\}$.

It also follows from the definition that, in a generalized C-group Γ, the subgroups Γ_I with $I \subseteq \{0, \ldots, n-1\}$ are pairwise distinct. To see this, first observe that by (10) and our assumption that R_{-1} $(= \Gamma_\emptyset)$ be a proper subgroup of R_i for each $i = 0, \ldots, n-1$, a group R_i cannot be a subgroup of a group Γ_I when $i \notin I \cup \{-1, n\}$. Consequently, if $I, J \subseteq \{0, \ldots, n-1\}$ and $\Gamma_I = \Gamma_J$, then (10) implies that $\Gamma_I = \Gamma_{I \cap J} = \Gamma_J$; hence it follows from what was said before that $I = I \cap J = J$, as required.

A generalized C-group Γ is called a *generalized string C-group* (strictly speaking, a *string generalized C-group*) if its generating subgroups satisfy

$$R_i R_j = R_j R_i \qquad (-1 \leq i < j - 1 \leq n - 1). \tag{11}$$

Thus, for the remainder of this section we assume that $\Gamma = \langle R_{-1}, R_0, \ldots, R_n \rangle$, with $n \geq 1$, is a generalized string C-group.

As in the previous section, for each $i \in N$ we write

$$\begin{aligned} \Gamma_i &:= \langle R_j \mid j \neq i \rangle, \\ \Gamma_i^- &:= \langle R_j \mid j \leq i \rangle, \\ \Gamma_i^+ &:= \langle R_j \mid j \geq i \rangle, \end{aligned} \tag{12}$$

so in particular, $\Gamma_{-1} = \Gamma_n = \Gamma$. Then (6) and (7) carry over, as before by the commuting properties (11). Observe that the subgroups $\Gamma_0, \ldots, \Gamma_{n-1}$ are mutually distinct, and distinct from Γ.

We now construct a regular n-incidence complex $\mathcal{K}$ from Γ. For $i \in N$, we take as the set of i-faces of $\mathcal{K}$ (that is, its faces of rank i) the set of all right cosets $\Gamma_i \varphi$ in Γ, with $\varphi \in \Gamma$. As improper faces of $\mathcal{K}$, we choose two copies of Γ, one denoted by Γ_{-1}, and the other by Γ_n; in this context, they are regarded as distinct. Then, for the right cosets of Γ_{-1} and Γ_n, we have $\Gamma_{-1}\varphi = \Gamma_{-1}$ and $\Gamma_n \varphi = \Gamma_n$ for all $\varphi \in \Gamma$. On (the set of all proper and improper faces of) $\mathcal{K}$, we define the following partial order:

$$\Gamma_i \varphi \leq \Gamma_j \psi \;:\Longleftrightarrow\; -1 \leq i \leq j \leq n,\; \varphi\psi^{-1} \in \Gamma^+_{i+1}\Gamma^-_{j-1}. \tag{13}$$

Then Γ acts on $\mathcal{K}$ in an obvious way as a group of order preserving automorphisms.

Alternatively the partial order on $\mathcal{K}$ can be defined by

$$\Gamma_i \varphi \leq \Gamma_j \psi \;:\Longleftrightarrow\; -1 \leq i \leq j \leq n,\; \Gamma_i\varphi \cap \Gamma_j \psi \neq \emptyset. \tag{14}$$

The equivalence of the two definitions is based on the commutation rules of (11). In fact, it follows as in Eq. (8) that $\Gamma^+_{i+1}\Gamma^-_{j-1} = \Gamma_i\Gamma_j$, so $\varphi\psi^{-1} \in \Gamma^+_{i+1}\Gamma^-_{j-1} = \Gamma_i\Gamma_j$ if and only of $\Gamma_i\varphi \cap \Gamma_j\psi \neq \emptyset$.

If the dependence of $\mathcal{K}$ on Γ and $R_{-1}, R_0, \ldots, R_n$ is to be emphasized, we write $\mathcal{K}(\Gamma)$ or $\mathcal{K}(\Gamma; R_{-1}, R_0, \ldots, R_n)$ for $\mathcal{K}$.

We first show that the condition (13) induces a partial order on $\mathcal{K}$. For reflexivity and antisymmetry of $\leq$ we can appeal to (14). Certainly, a coset $\Gamma_i\varphi$ is incident with itself, which is reflexivity. If $\Gamma_i\varphi$ and $\Gamma_j\psi$ are two cosets with $\Gamma_i\varphi \leq \Gamma_j\psi$ and $\Gamma_j\psi \leq \Gamma_i\varphi$, then $i = j$ and the cosets (for the same subgroup) must coincide as they intersect; this implies antisymmetry. Finally, if $-1 \leq i \leq j \leq k \leq n$, we have

$$\Gamma^+_{j+1}\Gamma^-_{i-1} \cdot \Gamma^+_{i+1}\Gamma^-_{k-1} = \Gamma^+_{j+1}\Gamma^+_{i+1}\Gamma^-_{i-1}\Gamma^-_{k-1} = \Gamma^+_{i+1}\Gamma^-_{k-1}. \tag{15}$$

Transitivity of $\leq$ then is an immediate consequence if we appeal to the original definition of $\leq$ in (13). Thus $\leq$ is a partial order.

Clearly, $\Phi := \{\Gamma_{-1}, \Gamma_0, \ldots, \Gamma_{n-1}, \Gamma_n\}$ is a flag of $\mathcal{K}$, which we naturally call the *base* flag; its faces are also called the *base* faces of $\mathcal{K}$. Since Φ is a flag, so is its image $\Phi\varphi = \{\Gamma_{-1}\varphi, \Gamma_0\varphi, \ldots, \Gamma_{n-1}\varphi, \Gamma_n\varphi\}$ for each $\varphi \in \Gamma$.

We next establish that Γ acts transitively on all chains of $\mathcal{K}$ of each given type $I \subseteq N$. When $I = N$ this shows that Γ acts transitively on the flags of $\mathcal{K}$. Now let $I \subseteq N$, and let $\{\Gamma_i\varphi_i \mid i \in I\}$ be a chain of type I. We proceed by induction. Suppose that, for some $k \in I$, we have already shown that there exists an element $\psi \in \Gamma$ such that $\Gamma_i\varphi_i = \Gamma_i\psi$ for each $i \in I$ with $i \geq k$. Let $j \in I$ be the next smaller number than k (assuming that there is one). Then $\Gamma_j\varphi_j \leq \Gamma_k\psi$ implies by (13) that $\varphi_j\psi^{-1} \in \Gamma^+_{k+1}\Gamma^-_{j-1}$, say $\varphi_j\psi^{-1} = \alpha\beta$, with $\alpha \in \Gamma^+_{k+1}$ and $\beta \in \Gamma^-_{j-1}$. It follows that $\alpha^{-1}\varphi_j = \beta\psi =: \chi$, say, and hence that

$$\begin{aligned} \Gamma_i\chi &= \Gamma_i\beta\psi = \Gamma_i\psi, \quad \text{for } i \in I, i \geq k, \\ \Gamma_j\chi &= \Gamma_j\alpha^{-1}\varphi_j = \Gamma_j\varphi_j, \end{aligned}$$

giving the same property with j instead of k (and ψ replaced by χ). This is the inductive step, and the transitivity follows.

If $I \subseteq N$ and Φ_I denotes the subchain of Φ of type I (consisting of the faces in Φ with ranks in I), then the stabilizer of Φ_I in Γ is the subgroup $\Gamma_{\bar{I}}$. In particular, the stabilizer of the base flag Φ itself is R_{-1}. In fact, an element $\varphi \in \Gamma$ stabilizes Φ_I if and only if $\Gamma_i\varphi = \Gamma_i$ for each $i \in I$. Equivalently, $\Phi_I\varphi = \Phi_I$ if and only if

$$\varphi \in \bigcap_{i\in I} \Gamma_i = \bigcap_{i\in I} \langle R_j \mid j \neq i\rangle = \Gamma_{\bar{I}},$$

by the intersection property (10) for Γ. Thus the stabilizer of Φ_I is $\Gamma_{\bar{I}}$.

We can now state the following theorem.

Theorem 4.1 *Let $n \geq 1$, and let $\Gamma = \langle R_{-1}, R_0, \dots, R_n\rangle$ be a generalized string C-group and $\mathcal{K} := \mathcal{K}(\Gamma)$ the corresponding partially ordered set. Then $\mathcal{K}$ is a regular n-complex on which Γ acts flag-transitively. In particular, $\mathcal{K}$ is finite, if Γ is finite.*

Proof For $\mathcal{K}$ we need to check the defining properties (I1), ..., (I4) of incidence complexes. The property (I1) is trivially satisfied with Γ_{-1} and Γ_n as the least and greatest face, respectively. In fact, by (13), $\Gamma_{-1} \leq \Gamma_i\varphi \leq \Gamma_n$ for all φ and all i.

Next, we exploit the fact that every chain Ω in $\mathcal{K}$ of type I can be expressed in the form $\Omega = \Phi_I\varphi$, for some $\varphi \in \Gamma$. In particular, Ω is contained in the flag $\Phi\varphi$, which gives (I2).

We then prove (I4). Now if we take $I = N \setminus \{i\}$ for any $i \in \{0, \dots, n-1\}$, we see that the stabilizer of $\Phi_{N\setminus\{i\}} = \{\Gamma_{-1}, \Gamma_0, \dots, \Gamma_{i-1}, \Gamma_{i+1}, \dots, \Gamma_n\}$ is $\Gamma_{\{i\}} = R_i$. On the other hand, the stabilizer of Φ itself is $\Gamma_\emptyset = R_{-1}$, which by assumption is a proper subgroup of R_i. Hence the number of flags of $\mathcal{K}$ containing $\Phi_{N\setminus\{i\}}$, which is given by $|R_i : R_{-1}|$, is at least 2. The transitivity of Γ on chains of type $N \setminus \{i\}$ then gives (I4). Thus, for each flag, the number of i-adjacent flags is at least 1 and is given by $(|R_i : R_{-1}| - 1)$ for each $i = 0, \dots, n-1$.

Finally, we demonstrate (I3), in the alternative form (I3′) of strong flag-connectedness. As Γ acts flag-transitively on $\mathcal{K}$, to prove (I3′), it suffices to consider the special case where one flag is the base flag Φ. If Ψ is another flag of $\mathcal{K}$, let $J \subseteq N$ be such that $\Phi \cap \Psi = \Phi_J$. Since $\Phi_J \subseteq \Psi$ and the stabilizer of Φ_J is $\Gamma_{\bar{J}}$, the flag-transitivity of Γ shows that $\Psi = \Phi\varphi$ for some $\varphi \in \Gamma_{\bar{J}}$. Suppose $\varphi = \varphi_1 \dots \varphi_k$ such that $\varphi_l \in R_{j_l}$, for some $j_l \in \bar{J}$, for $l = 1, \dots, k$. Define $\psi_l := \varphi_l \dots \varphi_k$ for $l = 1, \dots, k$. Then

$$\Phi, \Phi\psi_k, \Phi\psi_{k-1}, \dots, \Phi\psi_2, \Phi\psi_1 = \Phi\varphi = \Psi,$$

is a sequence of successively adjacent flags, all containing Φ_J, which connects Φ and Ψ. Note here that $\Phi\psi_{l+1}$ and $\Phi\psi_l$ are j_l-adjacent for each $l = 1, \dots, k-1$, since Φ and $\Phi\varphi_l$ are j_l-adjacent and so are $\Phi\psi_{l+1}$ and $\Phi\varphi_l\psi_{l+1} = \Phi\psi_l$. Thus $\mathcal{K}$ is strongly connected, and the proof of the theorem is complete. ■

Note that the action of Γ on $\mathcal{K} := \mathcal{K}(\Gamma)$ need not be faithful in general. The kernel of the action consists of the elements of Γ which act trivially on $\mathcal{K}$, or equivalently, on the set of flags of $\mathcal{K}$. The stabilizer of the base flag Φ in Γ is R_{-1}, and hence the stabilizer of a flag $\Phi\varphi$ with $\varphi \in \Gamma$ is $\varphi^{-1}R_{-1}\varphi$. Hence the kernel of the action of Γ of $\mathcal{K}$ is given by its subgroup

$$\mathrm{core}(R_{-1}) = \bigcap_{\varphi\in\Gamma} (\varphi^{-1}R_{-1}\varphi). \tag{16}$$

Recall that in a group B, the *core* of a subgroup A, denoted core(A), is the largest normal subgroup of B contained in A; that is, $\mathrm{core}(A) = \cap_{b\in B} b^{-1}Ab$ (see [2]). Clearly, Γ itself can be identified with a flag-transitive subgroup of the automorphism group of $\mathcal{K}(\Gamma)$ if and only if core(R_{-1}) is trivial.

Our next theorem describes the structure of the sections of the regular complex $\mathcal{K}(\Gamma)$. For the proof we require the following consequence of the intersection property (10):

$$\Gamma^+_{k+1}\Gamma^-_{l-1} \cap \Gamma_{\{i+1,\dots,j-1\}} = \Gamma_{\{k+1,\dots,j-1\}}\Gamma_{\{i+1,\dots,l-1\}} \quad (i \le k \le l \le j) \tag{17}$$

To prove this property, suppose φ is an element in the set on the left hand side, $\varphi = \alpha\beta$ (say), with $\alpha \in \Gamma^+_{k+1}$ and $\beta \in \Gamma^-_{l-1}$. Now apply (10) twice, bearing in mind that $i \le k \le l \le j$: first, with $I = \{-1, 0, \dots, l-1\}$ and $J = \{i+1, \dots, n\}$ to obtain

$$\beta = \alpha^{-1}\varphi \in \Gamma^-_{l-1} \cap \Gamma^+_{i+1} = \Gamma_{\{i+1,\dots,l-1\}},$$

and second, with $I = \{k+1, \dots, n\}$ and $J = \{-1, 0, \dots, j-1\}$ to obtain

$$\alpha = \varphi\beta^{-1} \in \Gamma^+_{k+1} \cap \Gamma^-_{j-1} = \Gamma_{\{k+1,\dots,j-1\}}.$$

Thus $\varphi \in \Gamma_{\{k+1,\dots,j-1\}}\Gamma_{\{i+1,\dots,l-1\}}$, as required. The opposite inclusion is clear, since the two groups $\Gamma_{\{k+1,\dots,j-1\}}$ and $\Gamma_{\{i+1,\dots,l-1\}}$ both lie in $\Gamma_{\{i+1,\dots,j-1\}}$, and are subgroups of Γ^+_{k+1} and Γ^-_{l-1} respectively.

Theorem 4.2 *Let $\mathcal{K} := \mathcal{K}(\Gamma)$ be the regular n-complex associated with the generalized string C-group $\Gamma = \langle R_{-1}, R_0, \dots, R_n\rangle$.*
(a) Let $-1 \le i < j-1 \le n-1$, and let F be an i-face and G a j-face of $\mathcal{K}$ with $F \le G$. Then the section G/F of $\mathcal{K}$ is isomorphic to

$$\mathcal{K}(\Gamma_{\{i+1,\dots,j-1\}}) = \mathcal{K}(\Gamma_{\{i+1,\dots,j-1\}}; R_{-1}, R_{i+1}, \dots, R_{j-1}, R_n).$$

(b) The facets and vertex-figures of $\mathcal{K}$ are isomorphic to the regular $(n-1)$-complexes $\mathcal{K}(\Gamma_{n-1})$ and $\mathcal{K}(\Gamma_0)$, respectively.
(c) Let $-1 \le i \le n-1$, and let F be an $(i-1)$-face and G an $(i+1)$-face of $\mathcal{K}$ with $F \le G$. Then the number of i-faces of $\mathcal{K}$ in G/F is $|R_i : R_{-1}|$.

Proof We already established part (c). Part (b) is a special case of part (a). Now for part (a) assume that $-1 \le i < j-1 \le n-1$. The transitivity of Γ on chains of type $\{i, j\}$ implies that it suffices to prove the result for the section $\mathcal{K}(i, j) := \Gamma_j/\Gamma_i$ of $\mathcal{K}$. Let $I := \{0, \ldots, i, j, \ldots, n-1\}$. There is a one-to-one correspondence between chains of $\mathcal{K}(i, j)$ and chains of $\mathcal{K}$ which contain Φ_I. In particular, appealing again to the transitivity of Γ on chains of a given type, we deduce that each face $\Gamma_k\varphi \in \mathcal{K}(i, j)$ (with $i \leq k \leq j$) admits a representation with φ in the stabilizer of Φ_I, namely $\Gamma_{\bar{I}} = \Gamma_{\{i+1,\ldots,j-1\}}$. It now follows from (17) that $\mathcal{K}(i, j)$ is isomorphic to $\mathcal{K}(\Gamma_{\{i+1,\ldots,j-1\}})$. Set $\Lambda := \Gamma_{\{i+1,\ldots,j-1\}}$. In fact, by (17), if $\varphi, \psi \in \Gamma_{\{i+1,\ldots,j-1\}} = \Lambda$ and $i \leq k \leq l \leq j$, then $\varphi\psi^{-1} \in \Gamma^+_{k+1}\Gamma^-_{l-1}$ if and only if

$$\varphi\psi^{-1} \in \Gamma_{\{k+1,\ldots,j-1\}}\Gamma_{\{i+1,\ldots,l-1\}} = \Lambda^+_{k+1}\,\Lambda^-_{l-1},$$

or equivalently, $\Gamma_k\varphi \leq \Gamma_l\psi$ in $\mathcal{K}(i, j)$ (that is, in $\mathcal{K}$) if and only if

$$\Lambda_k\,\varphi \leq \Lambda_l\,\psi$$

in $\mathcal{K}(\Lambda) = \mathcal{K}(\Gamma_{\{i+1,\ldots,j-1\}})$. Thus $\mathcal{K}(i, j)$ and $\mathcal{K}(\Gamma_{\{i+1,\ldots,j-1\}})$ are isomorphic complexes, and the proof of the theorem is complete. ■

There are two immediate consequences of the earlier results and the construction of $\mathcal{K}(\Gamma)$.

Corollary 4.1 *The generalized string C-groups $\Gamma = \langle R_{-1}, R_0, \ldots, R_n\rangle$ with a trivial group* core (R_{-1}) *are precisely the flag-transitive subgroups of the automorphism groups of regular complexes.*

Proof Clearly, the flag stabilizer in a flag-transitive subgroup of the automorphism group of a regular complex must have trivial core since only the identity automorphism fixes every face. This shows one direction. The converse was already addressed above. ■

Theorem 4.3 *Let $n \geq 1$, let $\mathcal{K}$ be a regular n-complex, let Γ be a flag-transitive subgroup of $\Gamma(\mathcal{K})$, and let $R_{-1}, R_0, \ldots, R_n$ be the distinguished generating subgroups of Γ associated with the base flag $\Phi = \{F_{-1}, F_0, \ldots, F_n\}$ of $\mathcal{K}$. Then the regular complexes $\mathcal{K}$ and $\mathcal{K}(\Gamma)$ (or more exactly, $\mathcal{K}(\Gamma; R_{-1}, R_0, \ldots, R_n)$) are isomorphic. In particular, the mapping $\mathcal{K} \to \mathcal{K}(\Gamma)$ given by*

$$F_i\varphi \to \Gamma_i\varphi \quad (-1 \leq i \leq n;\ \varphi \in \Gamma)$$

is an isomorphism.

5 Regular Polytopes and C-groups

The basic structure results for abstract regular polytopes and their automorphism groups can be derived from the results of Sects. 3 and 4 (see [23, 33, 34]). Abstract polytopes are incidence complexes in which every flag has exactly one i-adjacent flag for each $i = 0, \ldots, n-1$; that is, polytopes satisfy the diamond condition. Regular polytopes have a simply flag-transitive automorphism group, so all flag stabilizers are trivial and there are no proper flag-transitive subgroups.

Let $\mathcal{K}$ be a regular n-polytope, $\Phi = \{F_{-1}, F_0, \ldots, F_n\}$ be a base flag of $\mathcal{K}$, and let $\Gamma := \Gamma(\mathcal{K})$. Then $R_i = \langle \rho_i \rangle$ for each $i = 0, \ldots, n-1$, where ρ_i is the unique automorphism of $\mathcal{K}$ mapping Φ to its i-adjacent flag. The subgroups R_{-1} and R_n are trivial. Thus $\Gamma = \langle \rho_0, \ldots, \rho_{n-1} \rangle$. The involutions $\rho_0, \ldots, \rho_{n-1}$ are called the *distinguished generators* of Γ. Then the structure results of Sect. 3 for the distinguished generating subgroups of Γ translate directly into corresponding statements for the distinguished generators.

Conversely, let Γ be a group generated by involutions $\rho_0, \ldots, \rho_{n-1}$, called the *distinguished generators* of Γ. Then Γ is a group of the type discussed at the beginning of Sect. 4, with $R_i := \langle \rho_i \rangle$ for $i = 0, \ldots, n-1$, and $R_{-1} = R_n = \{1\}$. In this case it suffices to consider the distinguished subgroups $\Gamma_I := \langle \rho_i \mid i \in I \rangle$ with $I \subseteq \{0, \ldots, n-1\}$. We call Γ a *C-group* if Γ has the intersection property (10), that is,

$$\langle \rho_i \mid i \in I \rangle \cap \langle \rho_i \mid i \in J \rangle = \langle \rho_i \mid i \in I \cap J \rangle \qquad (I, J \subseteq \{0, \ldots, n-1\}). \tag{18}$$

A C-group Γ is called a *string C-group* if the distinguished generators also satisfy the relations

$$(\rho_i \rho_j)^2 = 1 \qquad (-1 \leq i < j-1 \leq n-1), \tag{19}$$

which is equivalent to requiring (11). The number of generators n is called the *C-rank*, or simply the *rank*, of Γ. Clearly, C-groups are generalized C-groups, and string C-groups are generalized string C-groups.

The regular n-complex $\mathcal{K} = \mathcal{K}(\Gamma)$ of Theorem 4.1 associated with a string C-group $\Gamma = \langle \rho_0, \ldots, \rho_{n-1} \rangle$ is a polytope, by Theorem 4.2(c). Thus $\mathcal{K}$ is a regular n-polytope, with partial order given by (13), or equivalently, (14). The relevant subgroups involved in describing the partial order are $\Gamma_i = \langle \rho_j \mid j \neq i \rangle$, $\Gamma_i^- = \langle \rho_j \mid j \leq i \rangle$, and $\Gamma_i^+ = \langle \rho_j \mid j \geq i \rangle$. The i-faces of $\mathcal{K}$ are the right cosets of Γ_i for each i.

Thus the string C-groups are precisely the groups of regular polytopes.

Abstract polytopes of rank 3 are also called *(abstract) polyhedra*. Regular polyhedra are regular maps on surfaces, and most regular maps on surfaces are regular polyhedra (see [6, 10]).

A regular n-polytope $\mathcal{K}$ is of *(Schläfli) type* $\{p_1, \ldots, p_{n-1}\}$ if its sections G/F of rank 2 defined by an $(i-2)$-face F and an $(i+1)$-face G with $F < G$ are isomorphic to p_i-gons (possibly $p_i = \infty$) for $i = 1, \ldots, n-1$; then p_i is the order of $\rho_{i-1}\rho_i$ in $\Gamma(\mathcal{K})$.

Coxeter groups are a particularly important class of C-groups (see [23, Chap. 3] and [20, 33, 34]). Let $p_1, \ldots, p_{n-1} \geq 2$, and let $\Gamma = \langle \rho_0, \ldots, \rho_{n-1} \rangle$ be the (string) Coxeter group defined by the relations

$$\begin{array}{ll} \rho_i^2 & = 1 \text{ for } 0 \le i \le n-1; \\ (\rho_i \rho_j)^2 & = 1 \text{ for } 0 \le i < j-1 \le n-2; \\ (\rho_{i-1}\rho_i)^{p_i} & = 1 \text{ for } 1 \le i \le n-1. \end{array} \qquad (20)$$

Then Γ is a string C-group. The corresponding regular n-polytope is called the *universal regular polytope of type* $\{p_1, \ldots, p_{n-1}\}$ and is denoted by the Schläfli symbol $\{p_1, \ldots, p_{n-1}\}$. This polytope covers every regular polytope of type $\{p_1, \ldots, p_{n-1}\}$. For combinatorial and geometric constructions of the universal regular polytopes from the Coxeter complexes of the underlying Coxeter groups Γ see [23, Sect. 3D] (or [33, 34]).

The regular convex polytopes and regular tessellations (or honeycombs) of spherical, Euclidean or hyperbolic spaces are particular instances of universal regular polytopes, with the type determined by the standard Schläfli symbol.

6 Extensions of Regular Complexes

A central problem in the classical theory of regular polytopes is the construction of polytopes with prescribed facets. In this section, we briefly investigate the corresponding problem for regular incidence complexes. We say that a regular complex $\mathcal{L}$ is an *extension* of a regular complex $\mathcal{K}$ if the facets of $\mathcal{L}$ are isomorphic to $\mathcal{K}$ and if all automorphisms of $\mathcal{K}$ are extended to automorphisms of $\mathcal{L}$. In conjunction with the former condition the latter condition means that the stabilizer of a facet of $\mathcal{L}$ in $\Gamma(\mathcal{L})$ contains $\Gamma(\mathcal{K})$ as a subgroup; or, less formally, $\Gamma(\mathcal{K}) \le \Gamma(\mathcal{L})$.

Let $\mathcal{K}$ be a regular n-complex with $n \ge 1$, let $\Phi = \{F_{-1}, F_0, \ldots, F_n\}$ be a base flag of $\mathcal{K}$, let Γ be a flag-transitive subgroup of $\Gamma(\mathcal{K})$, and let $R_{-1}, R_0, \ldots, R_n$ be the distinguished generating subgroups of Γ associated with Φ. In constructing extensions of $\mathcal{K}$ we consider certain groups Λ with distinguished systems of generating subgroups $R'_{-1}, R'_0, \ldots, R'_{n+1}$. We use similar notation for the distinguished subgroups of Λ as for Γ. The following theorem was proved in [33, 35] (see also [36]).

Theorem 6.1 *Let $\mathcal{K}$ be a regular n-complex with $n \ge 1$, and let $\Gamma = \langle R_{-1}, R_0, \ldots, R_n\rangle$ be a flag-transitive subgroup of $\Gamma(\mathcal{K})$, as above. Let Λ be a group generated by a system of subgroups $R'_{-1}, R'_0, \ldots, R'_{n+1}$ satisfying the following conditions (a), (b) and (c).*

(a) $R'_{-1} = R'_{n+1} \subset R'_n$, $\Lambda \ne \Lambda^-_{n-1}$;

(b) $R'_i R'_j = R'_j R'_i$ for $0 \le i < j-1 \le n-1$;

(c) There exists a surjective homomorphism $\pi : \Lambda^-_{n-1} \to \Gamma$ such that

(c1) $\pi^{-1}(R_i) = R'_i$ for $i = -1, 0, \ldots, n-1$;

(c2) $\Lambda^+_i \cap \Lambda^-_{n-1} = \pi^{-1}(\Gamma^+_i)$ for $i = -1, 0, \ldots, n$.

Then there exists a regular $(n+1)$-complex $\mathcal{L}$ with facets isomorphic to $\mathcal{K}$. In particular, Λ acts flag-transitively on $\mathcal{L}$, and $\mathcal{L}$ is finite if Λ is finite. If π is an isomorphism, then Λ is isomorphic to a flag-transitive subgroup of $\Gamma(\mathcal{L})$, the group Γ is a subgroup of Λ, and $\mathcal{L}$ is finite if and only if Λ is finite. Further, $\mathcal{L}$ is a lattice, if $\mathcal{K}$ is a lattice *and Λ satisfies the following condition:*

(d) Let $0 \le i \le j < k \le n$ and $\tau \in \Lambda_{k-1}^{-}$. If $\tau \notin \Lambda_{i+1}^{+}\Lambda_{j-1}^{-}$ and if further $\tau \notin \Lambda_{i+1}^{+}\Lambda_{l-1}^{-}\Lambda_{\{j+1,\dots,k-1\}}$ for each l with $j < l < k$, then $\Lambda_{j+1}^{+} \cap \Lambda_{n-1}^{-}\Lambda_{i+1}^{+}\tau \subseteq \Lambda_{n-1}^{-}\Lambda_{k+1}^{+}$.

Note that condition (d) of Theorem 6.1 can be reformulated as follows: if $0 \le i \le j < k \le n$, $\tau \in \Lambda_{n-1}^{-}$, and F_k is the supremum of F_j and $F_i(\tau\pi)$ in $\mathcal{K}$, then $\Lambda_{j+1}^{+} \cap \Lambda_{n-1}^{-}\Lambda_{i+1}^{+}\tau \subseteq \Lambda_{n-1}^{-}\Lambda_{k+1}^{+}$.

Theorem 6.1 translates the problem of finding an extension of a regular complex $\mathcal{K}$ into an embedding problem for its automorphism group $\Gamma(\mathcal{K})$ into a suitable group Λ. A regular complex has many possible extensions. However, it is much harder to find an extension which is a lattice, if $\mathcal{K}$ is a lattice.

The following result was proved in [33, 35] (see also [38]).

Theorem 6.2 *Let $\mathcal{K}$ be a finite regular n-complex, and let f denote the number of facets of $\mathcal{K}$. Then $\mathcal{K}$ admits an extension $\mathcal{L}$ whose automorphism group $\Gamma(\mathcal{L})$ contains a flag-transitive subgroup isomorphic to the symmetric group S_{f+1}. If $\mathcal{K}$ is a polytope, then $\mathcal{L}$ is a polytope and $\Gamma(\mathcal{L}) = S_{f+1}$.*

In the extension $\mathcal{L}$ of Theorem 6.2 the $(n-1)$-faces always lie in exactly two facets, regardless of whether or not $\mathcal{L}$ is a polytope. For lattices $\mathcal{K}$, this complex $\mathcal{L}$ is almost always again a lattice. A slightly modified construction for the group Λ, with S_{f+1} replaced by the larger group $S_{f+1} \times \Gamma(\mathcal{K})$, always guarantees that the corresponding extension of $\mathcal{K}$ is again a lattice if $\mathcal{K}$ is a lattice (see [37, 38]). For an extension of a regular complex $\mathcal{K}$ it usually is the lattice property which is the hardest to verify.

For regular polytopes, further extension results have been obtained in recent years (for example, see [30]). For these results, both $\mathcal{K}$ and $\mathcal{L}$ are regular polytopes. There are also good extension results for chiral polytopes (see [11, 40, 41]) and for hypertopes (see [15]).

There are also interesting infinite extensions $\mathcal{L}$ of regular n-complexes $\mathcal{K}$, which in the case of polytopes have certain universality properties. Let $k \ge 2$ be an integer, and let C_k denote the cyclic group of order k. Let Λ be the amalgamated free product of $\Gamma := \Gamma(\mathcal{K}) = \langle R_{-1}, R_0, \dots, R_n \rangle$ and the direct product $\Gamma_{n-2}^{-} \times C_k$, with amalgamation of the subgroups Γ_{n-2}^{-} and $\Gamma_{n-2}^{-} \times \{1\}$ under the isomorphism $\kappa : \Gamma_{n-2}^{-} \to \Gamma_{n-2}^{-} \times \{1\}$ defined by $\varphi \to (\varphi, 1)$. (For amalgamated free products see [22].) Then Λ is the quotient of the free product of the two groups Γ and $\Gamma_{n-2}^{-} \times C_k$ obtained by imposing on the free product the set of new relations

$$(\varphi)\kappa = \varphi \quad (\varphi \in \Gamma_{n-2}^{-}),$$

which in effect identify φ and $(\varphi, 1)$ for each $\varphi \in \Gamma_{n-2}^{-}$. Thus, in standard notation for amalgamated free products,

$$\Lambda = \Gamma \underset{\Gamma_{n-2}^{-}}{*} (\Gamma_{n-2}^{-} \times C_k).$$

We use slightly simpler notation and write

$$\Lambda = \Gamma *_{\mathcal{K}} (\Gamma_{n-2}^{-} \times C_k). \tag{21}$$

Then, with the distinguished generating system $R'_{-1}, R'_0, \ldots, R'_{n+1}$ given by $R'_i := R_i$ for $i \leq n-1$, $R'_n := R_{-1} \times C_k$, and $R'_{n+1} := R_{-1}$, this group Λ turns out to be a generalized C-group. More explicitly, we have the following result (see [36]).

Theorem 6.3 *Let $\mathcal{K}$ be a regular n-complex, and let $k \geq 2$. Then $\mathcal{K}$ admits an infinite extension $\mathcal{L}$ whose automorphism group $\Gamma(\mathcal{L})$ contains a flag-transitive subgroup isomorphic to $\Gamma *_{\mathcal{K}} (\Gamma_{n-2}^{-} \times C_k)$. In $\mathcal{L}$, each $(n-1)$-face lies in exactly k facets (the k facets containing the base $(n-1)$-face of $\mathcal{L}$ are cyclically permuted by the subgroup C_k of R'_n). Moreover, $\mathcal{L}$ is a lattice if $\mathcal{K}$ is a lattice.*

If $\mathcal{K}$ is a regular n-polytope and $k = 2$, then $\mathcal{L}$ has the following universality property: if $\mathcal{P}$ is any regular $(n+1)$-polytope with facets isomorphic to $\mathcal{K}$, then $\mathcal{P}$ is covered by $\mathcal{L}$. The polytope $\mathcal{L}$ was called the *universal extension* of $\mathcal{K}$ in [23, Chap. 4D]. For example, if $\mathcal{K}$ is the triangle $\{3\}$, then $\mathcal{L}$ is the regular hyperbolic tessellation $\{3, \infty\}$ by ideal triangles, whose automorphism group is the projective general linear group $PGL_2(\mathbb{Z})$.

Recently there has been a lot of progress in the study of combinatorial coverings of arbitrary abstract polytopes (see [19, 27, 28]). For example, in the paper [27], with Monson, it was shown that every finite abstract polytope is a quotient of a regular polytope of the same rank; that is, every finite abstract polytope has a finite regular cover.

For open questions related to extensions of regular complexes see Problem 7.2 in the next section.

7 Abstract Polytope Complexes

An incidence complex $\mathcal{K}$ of rank n is called an *$(n-1)$-polytope complex*, or simply a *polytope complex*, if all facets of $\mathcal{K}$ are abstract polytopes. If the rank n is 3 or 4 respectively, we also use the term *polygon complex* or *polyhedron complex*.

Let $\mathcal{K}$ be a regular polytope complex of rank n, and let $\Phi := \{F_{-1}, F_0, \ldots, F_n\}$ be a base flag of $\mathcal{K}$. Let $\Gamma = \langle R_{-1}, R_0, \ldots, R_n \rangle$ be a flag-transitive subgroup of $\Gamma(\mathcal{K})$, where as before $R_{-1}, R_0, \ldots, R_n$ are the distinguished generating subgroups of Γ. Recall that $R_{-1} = R_n = \Gamma_\Phi$, which is the stabilizer of Φ in Γ. Each subgroup R_i with $i = 0, \ldots, n-1$ acts transitively on the $k_i =: k_i(\mathcal{K})$ faces of $\mathcal{K}$ of rank i in F_{i+1}/F_{i-1}, and by (2) we know that $|R_i : R_{-1}| = k_i$. As $\mathcal{K}$ is a polytope complex, $k_i = 2$ for $i \leq n-2$ and $k := k_{n-1} \geq 2$. Thus $|R_i : R_{-1}| = 2$ if $i \leq n-2$, so R_{-1} is a normal subgroup of R_i in this case, and $|R_{n-1} : R_{-1}| = k$. It follows that R_{-1} is also a normal subgroup of $\Gamma_{n-1} = \langle R_{-1}, R_0, \ldots, R_{n-2} \rangle$, the stabilizer of F_{n-1} in Γ.

Moreover, Γ_{n-1} acts flag-transitively on the base facet F_{n-1}/F_{-1} of $\mathcal{K}$, and its subgroup R_{-1} is the stabilizer of the base flag $\{F_{-1}, F_0, \ldots, F_{n-1}\}$ of F_{n-1}/F_{-1} in

Γ_{n-1}. Now F_{n-1}/F_{-1} is a polytope since $\mathcal{K}$ is a polytope complex, so R_{-1} must be the kernel of the action of Γ_{n-1} on F_{n-1}/F_{-1}; in fact, in a flag-transitive action on a polytope, if a group element stabilizes a flag then it stabilizes every flag and thus also every face. Also, again because the facet is a polytope, the flag-transitive subgroup of the automorphism group $\Gamma(F_{n-1}/F_{-1})$ of the facet F_{n-1}/F_{-1} induced by Γ_{n-1} must be $\Gamma(F_{n-1}/F_{-1})$ itself. Thus Γ_{n-1}/R_{-1} is a string C-group and

$$\Gamma_{n-1}/R_{-1} \cong \Gamma(F_{n-1}/F_{-1}). \tag{22}$$

The skeletons of abstract polytopes provide interesting examples of polytope complexes. Let $\mathcal{L}$ be an abstract m-polytope, and let $n \leq m$. The $(n-1)$-*skeleton* of $\mathcal{L}$, denoted $skel_{n-1}(\mathcal{L})$, is the n-complex with faces those of $\mathcal{L}$ of rank at most $n-1$ or of rank m (the m-face of $\mathcal{L}$ becomes the n-face of $skel_{n-1}(\mathcal{L})$). Then $skel_{n-1}(\mathcal{L})$ is an $(n-1)$-polytope complex whose facets are the $(n-1)$-faces of $\mathcal{L}$. For example, the 2-complex $skel_1(\mathcal{L})$ can be viewed as a graph often called the *edge-graph* of $\mathcal{L}$.

Now suppose $\mathcal{L}$ is a regular m-polytope and $\Gamma(\mathcal{L}) = \langle \rho_0, \dots, \rho_{m-1} \rangle$, where $\rho_0, \dots, \rho_{m-1}$ are the distinguished involutory generators (with respect to a base flag of $\mathcal{L}$). Then the $(n-1)$-skeleton $\mathcal{K} := skel_{n-1}(\mathcal{L})$ is a regular polytope complex of rank n admitting a flag-transitive (but not necessarily faithful) action by $\Gamma := \Gamma(\mathcal{L})$. This action of Γ on $\mathcal{K}$ is faithful if $\mathcal{L}$ is a lattice (but weaker assumptions suffice); in fact, in this case $\Gamma(\mathcal{L})$ acts faithfully on the vertex set of $\mathcal{L}$ and hence also on the set of faces of $\mathcal{L}$ of rank smaller than n. In any case, the distinguished generating subgroups $R_{-1}, R_0, \dots, R_n$ of Γ for its action on $\mathcal{K}$ are given by $R_{-1} = R_n := \langle \rho_n, \dots, \rho_{m-1} \rangle$,

$$R_i := \langle \rho_i, \rho_n, \dots, \rho_{m-1} \rangle \;\; (\cong \langle \rho_i \rangle \times \langle \rho_n, \dots, \rho_{m-1} \rangle) \quad (i \leq n-2),$$

and $R_{n-1} := \langle \rho_{n-1}, \rho_n, \dots, \rho_{m-1} \rangle$. (Here we are in a slightly more general situation than discussed in Sect. 3, in that Γ may act on $\mathcal{K}$ with nontrivial kernel, that is, Γ may not be a subgroup of $\Gamma(\mathcal{K})$. However, the corresponding results of Sect. 3 carry over to this situation as well.) In particular, R_{n-1} must be a string C-group of rank $m-n+1$. Note that the parameter $k = k_{n-1}$ for $\mathcal{K}$ is given by the number of $(n-1)$-faces of $\mathcal{L}$ that contain a given $(n-2)$-face of $\mathcal{L}$ (or equivalently, by the number of vertices of the co-face of $\mathcal{L}$ at an $(n-2)$-face of $\mathcal{L}$). For example, if $m = n+1$ and $\mathcal{L}$ is a regular polytope of Schläfli type $\{p_1, \dots, p_n\}$, then $k = p_n$ and R_{n-1} is the dihedral group D_{p_n}.

There are a number of interesting open problems concerning the characterization of regular polytope complexes which are skeletons of regular polytopes of higher rank.

Problem 7.1 *Let $\mathcal{K}$ be a regular polytope complex of rank n with automorphism group $\Gamma(\mathcal{K}) = \langle R_{-1}, R_0, \dots, R_n \rangle$ and base flag $\{F_{-1}, F_0, \dots, F_n\}$, and let $k := k_{n-1}(\mathcal{K}) > 2$. Suppose R_{n-1} is isomorphic to a string C-group.*
(a) Is $\mathcal{K}$ always the $(n-1)$-skeleton of a regular polytope?
(b) Is $\mathcal{K}$ the $(n-1)$-skeleton of a regular $(n+l-1)$-polytope if R_{n-1} is isomorphic to a string C-group of rank l?

(c) Suppose $\mathcal{K}$ is a lattice. Is $\mathcal{K}$ always the $(n-1)$-skeleton of a regular polytope which is also a lattice?
(d) Let $\mathcal{L}$ denote a regular polytope with $(n-1)$-skeleton $\mathcal{K}$. What can be said about the structure of $\mathcal{L}$ in the interesting special cases when R_{n-1} acts on F_n/F_{n-2} as a dihedral group D_k, alternating group A_k, or symmetric group S_k?
(e) Under which conditions on $\mathcal{K}$ is there a unique regular polytope with $\mathcal{K}$ as its $(n-1)$-skeleton?

Similar questions can be asked when $\Gamma(\mathcal{K})$ is replaced by a flag-transitive subgroup Γ.

There are also a number of open problems for regular polytope complexes with preassigned type of (polytope) facets.

Problem 7.2 *Let $\mathcal{F}$ be a regular $(n-1)$-polytope, and let $k > 2$.*
(a) Among the regular polytope complexes $\mathcal{K}$ of rank n with facets isomorphic to $\mathcal{F}$ and $k_{n-1}(\mathcal{K}) = k$, when is there a "universal" polytope complex covering all these polytope complexes?
(b) Among the regular polytope complexes $\mathcal{K}$ of rank n with facets isomorphic to $\mathcal{F}$, with $k_{n-1}(\mathcal{K}) = k$, and with R_{n-1} acting on F_n/F_{n-2} as a cyclic group C_k, dihedral group D_k, alternating group A_k, or symmetric group S_k, is there a "universal" polytope complex covering all these polytope complexes?
(c) Among the regular polytope complexes $\mathcal{K}$ of rank n with facets isomorphic to $\mathcal{F}$ and $k_{n-1}(\mathcal{K}) = k$, to what extent can one preassign the transitive permutation action of R_{n-1} on F_n/F_{n-2}? In other words, which permutation groups on k elements arise as R_{n-1}?

Note that when Theorem 6.3 is applied to a regular $(n-1)$-polytope $\mathcal{F}$, interesting examples of regular polytope complexes $\mathcal{K}$ of rank n arise in which R_{n-1} acts on F_n/F_{n-2} as a cyclic group C_k. The underlying construction of these complexes from group amalgamations has a somewhat "universal flavor"; however, it is not known if these polytope complexes actually are universal among all polytope complexes with R_{n-1} acting as C_k, unless $k = 2$. When $k = 2$ the corresponding complex is indeed universal and is known as the universal (polytope) extension of $\mathcal{F}$ (see [23, Chap. 4D]).

Another interesting problem concerns the existence of simply flag-transitive subgroups.

Problem 7.3 *Describe conditions on a regular polytope complex $\mathcal{K}$ that guarantee that $\Gamma(\mathcal{K})$ contains a simply flag-transitive subgroup.*

The final set of problems we describe here concerns geometric realizations of regular polytope complexes or more general incidence complexes in real Euclidean spaces or unitary complex spaces. It is known that the set of all Euclidean realizations of a given finite abstract regular polytope, if not empty, has the structure of a convex cone called the *realization cone* of the given polytope (see [23, Chap. 5] or [24, 25]).

Problem 7.4 *Let $\mathcal{K}$ be a finite regular polytope complex of rank n with automorphism group $\Gamma(\mathcal{K}) = \langle R_{-1}, R_0, \ldots, R_n \rangle$ and facets isomorphic to $\mathcal{F}$, and let $k := k_{n-1}(\mathcal{K})$. Describe the realization space of $\mathcal{K}$ (that is, the set of all realizations of $\mathcal{K}$ in a Euclidean space) in terms of the realization cone of $\mathcal{F}$ and the permutation action of R_{n-1} on the k facets of $\mathcal{K}$ containing the base $(n-2)$-face (that is, the action of R_{n-1} on F_n/F_{n-2}).*

Our next problem deals with geometric realizations in a Euclidean space of a given dimension d.

Problem 7.5 *Classify all geometrically regular polytope complexes in a Euclidean space of dimension $d \geq 4$.*

The case $d = 3$ was solved in [31, 32], where all geometrically regular polygon complexes (of rank 3) in ordinary Euclidean 3-space were classified. The classification is quite involved.

There may be a nice theory of unitary complex realizations for certain types of incidence complexes (including perhaps the duals of regular polytope complexes). The well-known regular complex n-polytopes in unitary complex n-space $\mathbb{C}^n$ are examples of incidence complexes which are realized as affine complex subspace configurations on which the (unitary) geometric symmetry group acts flag-transitively (see [8]). These structures could serve as a guide to develop a complex realization theory for more general kinds of incidence complexes.

8 Notes

(1) *Regular incidence complexes* were introduced around 1977 by Ludwig Danzer as combinatorial generalizations of regular polytopes [7], regular complex polytopes [8], and other highly "regular" incidence structures. The notion built on Branko Grünbaum's work on *regular polystromata* (see [17]). The first systematic study of incidence complexes from the discrete geometry perspective occurred in my doctoral dissertation [33] (and the related publications [13, 34–36]), at about the same time when the concept of diagram geometries was introduced by Buekenhout [3] to find geometric interpretations for the sporadic simple groups. At the time of the writing of my dissertation, I was not aware of Grünbaum's paper [17], nor did I know about Buekenhout's work [3]. (These were times before Google!) I learnt about both papers in 1981. Starting with [38, 39], my own work focussed on the class of incidence complexes now called abstract polytopes.

(2) Incidence complexes satisfying the diamond condition (I4P) were originally called *incidence polytopes* (see [13, 34]). During the writing of [23] the new name *abstract polytopes* was adopted in place of *incidence polytopes*, and the name *(string) C-groups* ('C' standing for 'Coxeter') was coined for the type of groups that are automorphism groups of abstract regular polytopes. Also, some

of the original terminology of [13] was changed; for example, the term 'rank' was used in place of 'dimension' (the term 'dimension' was reserved for geometric realizations of abstract regular polytopes [23, Chap. 5] and [24, 25]).

(3) Incidence complexes which are lattices were originally called *non-degenerate* complexes, indicating a main focus on lattices consistent with ordinary polytope theory. As abstract polytope theory developed, this distinction played less of a role, in part also because the lattice property did not translate into an elegant property for the automorphism group (see, for example, condition (d) in Theorem 6.1).

(4) Danzer's original definition of an incidence complex used the original connectivity condition (I3). The equivalent condition (I3′) for strong flag-connectedness was first introduced in [33].

(5) Our current condition (I4) is weaker than Danzer's original defining condition, which required that there be numbers $k_0, \ldots, k_{n-1}$ such that, for any $i = 0, \ldots, n-1$ and for any $(i-1)$-face F and $(i+1)$-face G with $F < G$, there are exactly k_i i-faces H with $F < H < G$. For regular (and many other kinds of highly symmetric) incidence complexes the two conditions are equivalent (see Eq. (2)).

(6) In [34], the faces of the regular n-complex $\mathcal{K}(\Gamma)$ of Sect. 4 for a given group Γ were denoted by formal symbols of the form $[\varphi, F_i]$, where $\varphi \in \Gamma$ and F_i is from a fixed $(n+2)$-set (in a way, the base flag). This description of the faces is equivalent to the coset description of faces adopted above, with $[\varphi, F_i]$ corresponding to $\Gamma_i\varphi$ (see the footnote on p. 40 of [34]). In [34], there were no explicit analogues of the conditions of Lemma 3.6(c) and (14) both of which describe the partial order in terms of intersections of cosets (and make these structures into coset geometries as defined by Tits [1, 4, 5, 29, 44–46]). The analysis in [33, 34] was carried out in terms of the conditions of Lemma 3.6(b) and (13).

Acknowledgements I am grateful to the referees for their careful reading of the original manuscript and their helpful suggestions that have improved this article.

References

1. M. Aschbacher, Flag structures on Tits geometries. Geom. Dedicata **14**, 21–32 (1983)
2. M. Aschbacher, *Finite Group Theory*, 2nd edn. (Cambridge University Press, Cambridge, 2002)
3. F. Buekenhout, Diagrams for geometries and groups. J. Comb. Theory Ser. A **27**, 121–151 (1979)
4. F. Buekenhout, A.M. Cohen, *Diagram Geometry* (Springer, New York, 2013)
5. F. Buekenhout, A. Pasini, Finite diagram geometries extending buildings, in *Handbook of Incidence Geometry*, ed. by F. Buekenhout (Elsevier Publishers, Amsterdam, 1995), pp. 1143–1254
6. M. Conder, G.A. Jones, J. Širáň, T.W. Tucker, *Regular Maps*, in preparation
7. H.S.M. Coxeter, *Regular Polytopes*, 3rd edn. (Dover, New York, 1973)

8. H.S.M. Coxeter, *Regular Complex Polytopes*, 2nd edn. (Cambridge University Press, Cambridge, 1991)
9. H.S.M. Coxeter, Regular honeycombs in hyperbolic space, in *Twelve Geometric Essays* (Southern Illinois University Press, Carbondale, 1968), pp. 199–214
10. H.S.M. Coxeter, W.O.J. Moser, *Generators and Relations for Discrete Groups*, 4th edn. (Springer, Berlin, 1980)
11. G. Cunningham, D. Pellicer, Chiral extensions of chiral polytopes. Discrete Math. **330**, 51–60 (2014)
12. L. Danzer, Regular incidence-complexes and dimensionally unbounded sequences of such, I, in *Convexity and Graph Theory (Jerusalem 1981)*, North-Holland Mathematics Studies, vol. 87 (North-Holland, Amsterdam, 1984), pp. 115–127
13. L. Danzer, E. Schulte, Reguläre Inzidenzkomplexe, I. Geom. Dedicata **13**, 295–308 (1982)
14. A.C. Duke, E. Schulte, Cube-like incidence complexes and their groups. Proc. Steklov Inst. Math. **288**, 226–242 (2015)
15. M.E. Fernandes, D. Leemans, A. Ivic Weiss, Hexagonal extensions of toroidal hypermaps, preprint
16. B. Grünbaum, *Convex Polytopes*, Graduate Texts in Mathematics (Springer, Berlin, 2003)
17. B. Grünbaum, Regularity of graphs, complexes and designs, in *Problèmes combinatoires et théorie des graphes*, Coll. Int. C.N.R.S. vol. 260 (Orsey, 1977), pp. 191–197
18. B. Grünbaum, Regular polyhedra—old and new. Aequationes Math. **16**, 1–20 (1977)
19. M.I. Hartley, All polytopes are quotients, and isomorphic polytopes are quotients by conjugate subgroups. Discrete Comput. Geom. **21**, 289–298 (1999)
20. J.E. Humphreys, *Reflection Groups and Coxeter Groups* (Cambridge University Press, Cambridge, 1990)
21. D. Leemans, *Polytopes, Incidence Geometries, and Computation* (title tentative), in preparation
22. R.C. Lyndon, P.E. Schupp, *Combinatorial Group Theory* (Springer, Berlin, 1977)
23. P. McMullen, E. Schulte, *Abstract Regular Polytopes, Encyclopedia of Mathematics and its Applications*, vol. 92 (Cambridge University Press, Cambridge, UK, 2002)
24. P. McMullen, Realizations of regular polytopes. Aequationes Math. **37**, 38–56 (1989)
25. P. McMullen, B.R. Monson, Realizations of regular polytopes, II. Aequationes Math. **65**, 102–112 (2003)
26. B. Monson, E. Schulte, Semiregular polytopes and amalgamated C-groups. Adv. Math. **229**, 2767–2791 (2012)
27. B. Monson, E. Schulte, Finite polytopes have finite regular covers. J. Algebraic Comb. **40**, 75–82 (2014)
28. B. Monson, D. Pellicer, G. Williams, Mixing and monodromy of abstract polytopes. Trans. Am. Math. Soc. **366**, 2651–2681 (2014)
29. A. Pasini, *Diagram Geometries* (Oxford University Press, Oxford, 1994)
30. D. Pellicer, Extensions of regular polytopes with preassigned Schläfli symbol. J. Comb. Theory Ser. A **116**, 303–313 (2009)
31. D. Pellicer, E. Schulte, Regular polygonal complexes in space, I. Trans. Am. Math. Soc. **362**, 6679–6714 (2010)
32. D. Pellicer, E. Schulte, Regular polygonal complexes in space, II. Trans. Am. Math. Soc. **365**, 2031–2061 (2013)
33. E. Schulte, *Reguläre Inzidenzkomplexe*, Doctoral Dissertation, University of Dortmund, Germany, 1980
34. E. Schulte, Reguläre Inzidenzkomplexe, II. Geom. Dedicata **14**, 33–56 (1983)
35. E. Schulte, Reguläre Inzidenzkomplexe, III. Geom. Dedicata **14**, 57–79 (1983)
36. E. Schulte, On arranging regular incidence-complexes as faces of higher-dimensional ones. Eur. J. Comb. **4**, 375–384 (1983)
37. E. Schulte, Extensions of regular complexes, in *Finite Geometries* ed. by C.A. Baker, L.M. Batten, Lecture Notes Pure Applied Mathematics, vol. 103 (Marcel Dekker, New York, 1985), pp. 289–305

38. E. Schulte, Regular incidence-polytopes with Euclidean or toroidal faces and vertex-figures. J. Comb. Theory A **40**, 305–330 (1985)
39. E. Schulte, Amalgamation of regular incidence-polytopes. Proc. Lond. Math. Soc. **S3–56**, 303–328 (1988)
40. E. Schulte, A.I. Weiss, Chiral polytopes, in *Applied Geometry and Discrete Mathematics (The Victor Klee Festschrift)*, ed. by P. Gritzmann, B. Sturmfels, DIMACS Series in Discrete Mathematics and Theoretical Computer Science, vol. 4 (American Mathematical Society and Association for Computing Machinery, 1991), pp. 493–516
41. E. Schulte, A.I. Weiss, Free extensions of chiral polytopes. Can. J. Math. **47**, 641–654 (1995)
42. G.C. Shephard, Regular complex polytopes. Proc. Lond. Math. Soc. **s3–2**, 82–97 (1952)
43. R.P. Stanley, *Enumerative Combinatorics* vol. 1 (Wadsworth & Brooks/Cole, Monterey, CA, 1986)
44. J. Tits, *Buildings of Spherical Type and Finite BN-Pairs*, Lecture Notes in Mathematics, vol. 386 (Springer, New York, 1974)
45. J. Tits, A local approach to buildings. In *The Geometric Vein: The Coxeter Festschrift*, ed. by C. Davis, B. Grünbaum, F.A. Sherk (Springer, New York, 1981), pp. 519–547
46. R.M. Weiss, *The Structure of Spherical Buildings* (Princeton University Press, Princeton, NJ, 2004)

The manufacturer's authorised representative in the EU is Springer Nature Customer Service Centre GmbH, Europaplatz 3, 69115 Heidelberg, Germany. If you have any concerns regarding our products, please contact ProductSafety@springernature.com

Printed and bound by CPI Group (UK) Ltd, Croydon, CR0 4YY
01/10/2026
02225975-0014